Recommendations of the Committee for Waterfront Structures Harbours and Waterways EAU 2004

8th Edition
Translation of the 10th German Edition

Issued by the Committee for Waterfront Structures of the Society for Harbour Engineering and the German Society for Soil Mechanics and Foundation Engineering

D1695484

Ernst & Sohn
A Wiley Company

The original German edition was published under the title
Empfehlungen des Arbeitsausschusses „Ufereinfassungen"
Häfen und Wasserstraßen EAU 2004
by Ernst & Sohn, Berlin

Cover picture: Container terminal Bremerhaven (Photo: bremenports)

Bibliographic information published by Die Deutsche Bibliothek
Die Deutsche Bibliothek lists this publication in the Deutsche Nationalbibliografie;
detailed bibliographic data is available in the internet at http://dnb.ddb.de

ISBN-13: 978-3-433-01666-4
ISBN-10: 3-433-01666-6

Typesetting: Manuela Treindl, Laaber
Printing: betz-druck GmbH, Darmstadt
Binding: Litges & Dopf Buchbinderei GmbH, Heppenheim

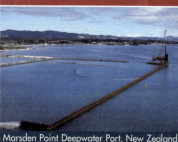

Recommendations of the Committee for Waterfront Structures Harbours and Waterways EAU 2004

Ernst & Sohn
A Wiley Company

Members of the Committee for Waterfront Structures

At present the working committee "Waterfront Structures" has the following members:

Professor Dr.-Ing. *Werner Richwien,* Essen, Chairman
Baudirektor Dipl.-Ing. *Michael Behrendt,* Bonn
Project Manager Ir. *Jacob Gerrit de Gijt,* Rotterdam
Prof. Dr.-Ing. *Jürgen Grabe,* Hamburg
Baudirektor Dr.-Ing. *Michael Heibaum,* Karlsruhe
Professor Dr.-Ing. *Stefan Heimann,* Berlin
Managing Director Ir. *Aad van der Horst,* Gouda
Dipl.-Ing. *Hans-Uwe Kalle,* Hagen
Professor Dr.-Ing. *Roland Krengel,* Dortmund
Dipl.-Ing. *Karl-Heinz Lambertz,* Duisburg
Dr.-Ing. *Christoph Miller,* Hamburg
Dr.-Ing. *Karl Morgen,* Hamburg
Managing Director Dr.-Ing. *Friedrich W. Oeser,* Hamburg
Managing Director Dipl.-Ing. *Emile Reuter,* Luxemburg
Managing Director Dr.-Ing. *Peter Ruland,* Hamburg
Dr.-Ing. *Wolfgang Schwarz,* Schrobenhausen
Leitender Baudirektor Dr.-Ing. *Hans Werner Vollstedt,* Bremerhaven

Preface to the 8th Revised Edition

This, the 8th English edition of the Recommendations of the Committee for Waterfront Structures, in the translation of the 10th German edition of the recommendations, which was published at the end of 2004. Now the full revision of the collected published recommendations which began with EAU 1996 is concluded. The concept of partial safety factors stipulated in EC 7 and DIN 1054 has been incorporated in the EAU's methods of calculation. At the same time, the revised recommendations also take account of all the new standards and draft standards that have also been converted to the concept of partial safety factors and had been published by mid-2004. Like with EAU 1996, further details concerning the implementation of the partial safety factor concept can be found in section 0. The incorporation of the partial safety factor concept of DIN 1054 called for a fundamental reappraisal of the methods of calculation and design for sheet piling structures contained in sections 8.2 to 8.4 and the methods of calculation for sheet piles contained in section 13. Extensive comparative calculations had to be carried out to ensure that the established safety standard of the EAU was upheld when using methods of analysis according to the concept of partial safety factors. This has been achieved by adapting the partial safety factors and by specifying redistribution diagrams for active earth pressure. The use of the new analysis concept for the design of sheet piling structures therefore results in component dimensions similar to those found by designs to EAU 1990.

Now that the inclusion of the European standardisation concept has been concluded, the 10th German edition of the EAU (and hence also the 8th English edition) satisfies the requirements for notification by the EU Commission. It is therefore registered with the EU Commission under Notification No. 2004/305/D.

A component of the notification is the principle of "mutual recognition", which must form the basis of contracts in which the EAU or individual provisions thereof form part of the contract. This principle is expressed as follows:

"Products lawfully manufactured and/or marketed in another EC Member State or in Turkey or in an EFTA State that is a contracting party to the Agreement on the European Economic Area that do not comply with these technical specifications shall be treated as equivalent – including the examinations and supervisory measures carried out in the country of manufacture – if they permanently achieve the required level of protection regarding safety, health and fitness for use."

The following members of the working committee have been involved with the German edition EAU 2004 since the summer of 2000.

Prof. Dr.-Ing. Dr.-Ing. E. h. Victor Rizkallah, Hannover (Chairman)
Dipl.-Ing. Michael Behrendt, Bonn (since 2001)
Ir. Jakob Gerrit de Gijt, Rotterdam
Dr.-Ing. Hans Peter Dücker, Hamburg
Dr.-Ing. Michael Heibaum, Karlsruhe
Dr.-Ing. Stefan Heimann, Bremen/Berlin (since 2002)
Dipl.-Ing. Wolfgang Hering, Rostock
Dipl.-Ing. Hans-Uwe Kalle, Hagen (since 2002)
Prof. Dr.-Ing. Roland Krengel, Dortmund (since 2004)
Dipl.-Ing. Karl-Heinz Lambertz, Duisburg (since 2002)
Prof. Dr.-Ing. habil. Dr. h. c. mult. Boleslaw Mazurkiewicz, Gdańsk
Dr.-Ing. Christoph Miller, Hamburg (since 2002)
Dr.-Ing. Karl Morgen, Hamburg
Dr.-Ing. Friedrich W. Oeser, Hamburg
Dr.-Ing. Heiner Otten, Dortmund (until 2002)
Dipl.-Ing. Martin Rahtge, Bremen (since 2004)
Dipl.-Ing. Emile Reuter, Luxembourg (since 2002)
Dipl.-Ing. Ulrich Reinke, Bremen (until 2002)
Prof. Dr.-Ing. Werner Richwien, Essen (Deputy Chairman)
Dr.-Ing. Peter Ruland, Hamburg (since 2002)
Dr.-Ing. Helmut Salzmann, Hamburg
Dr.-Ing. Roger Schlim, Luxembourg (until 2002)
Prof. Dr.-Ing. Hartmut Schulz, Munich
Dr.-Ing. Manfred Stocker, Schrobenhausen
Dipl.-Ing. Hans-Peter Tzschucke, Bonn (until 2002)
Ir. Aad van der Horst, Gouda
Dr.-Ing. Hans-Werner Vollstedt, Bremerhaven

The fundamental revisions contained in EAU 2004 also made detailed discussions with colleagues and specialists outside the committee necessary, even to the extent of setting up temporary study groups for specific topics. The committee thanks all those colleagues who in this way made significant contributions to EAU 2004.

In addition, numerous contributions presented by the professional world and recommendations from other committees and international technical–scientific associations have been incorporated in these recommendations.

These contributions and the results of the revision work mean that EAU 2004 now conforms with the current international standard. It provides the construction industry with an adapted, updated set of recommendations brought into line with European standards that will continue to act as a valuable aid for design, tendering, placing orders, technical processing, economic and ecological construction, quality control and settlement of contracts, and will thus enable harbour and waterway construction projects to be carried out according to the state of the art and according to uniform conditions.

The committee thanks all those whose contributions and suggestions have helped to bring the recommendations up to their present state, and wishes the EAU 2004 the same success as its earlier editions.

Vote of thanks goes to Prof. Dr.-Ing. Dr.-Ing. E. h. Victor Rizkallah, who was chairman of the committee until the end of 2004 and thus the 10th German edition of the recommendations have been prepared and published under his responsibility. In the translation works very valuable advices and help came from Prof. Dr.-Ing. Martin Hager, who was chairman of the committee up to the end of 1996. Finally a very special vote of thanks goes to my co-worker, Dipl.-Ing. Carsten Pohl, who assisted me in the extensive preparation of this edition and in the review of the text with great dedication and diligence.

Further special thanks are owed to the publisher Ernst & Sohn for the good cooperation and the meticulous care with which all drawings, tables and equations were prepared, providing once again an excellent printing quality and layout of the 8th revised English edition of EAU 2004.

Hannover, November 2005 *Prof. Dr.-Ing. Werner Richwien*

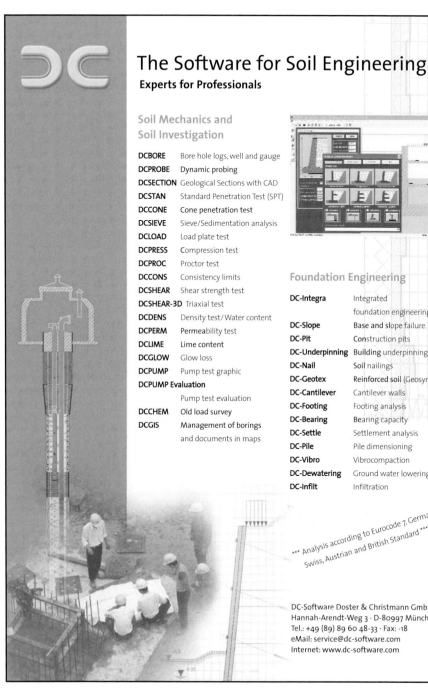

Contents

XII

XVI

XVII

1 0 0 Jahre BetonKalender
1906 – 2006

Concrete constructions: fundamentals, examples, standards

Hochqualifizierte Beiträge, detaillierte Informationen, stets aktuelle Hinweise und Erläuterungen zu Regelwerken sowie alle Neuerungen rund um den Baustoff Beton. Herausgeber: Konrad Bergmeister, Johann-Dietrich Wörner.

Jährliche Schwerpunkte seit 2003:

Turmbauwerke und Industriebauten

Beton-Kalender 2006

2005. Ca. 1100 Seiten.
1050 Abb. 260 Tab. Geb.
Ca. € 159,-* / sFr 251,-
Fortsetzungspreis:
Ca. € 139,-* / sFr 220,-
ISBN 3-433-01672-0
Erscheint November 2005

Fertigteile und Tunnelbauwerke

Beton-Kalender 2005

2004. XL, 1348 Seiten,
1057 Abb. 258 Tab. Geb.
€ 159,-* / sFr 251,-
Fortsetzungspreis:
€ 139,-* / sFr 220,-
ISBN 3-433-01670-4

Brücken und Parkhäuser

Beton-Kalender 2004

2003. XXXIII, 1156 Seiten,
836 Abb. 239 Tab. Geb.
€ 159,-* / sFr 251,-
Fortsetzungspreis:
€ 139,-* / sFr 220,-
ISBN 3-433-01668-2

Hochhäuser und Geschossbauten

Beton-Kalender 2003

2002. 1100 Seiten. Geb.
€ 159,-* / sFr 251,-
Fortsetzungspreis:
€ 139,-* / sFr 220,-
ISBN 3-433-01645-3

Ernst & Sohn
Verlag für Architektur und
technische Wissenschaften GmbH & Co. KG

1 0 0 Jahre BetonKalender
1906 – 2006

Ernst & Sohn
A Wiley Company
www.ernst-und-sohn.de

Für Bestellungen und Kundenservice:
Verlag Wiley-VCH
Boschstraße 12
69469 Weinheim
Telefon: +49(0) 6201 / 606-400
Telefax: +49(0) 6201 / 606-184
E-Mail: service@wiley-vch.de

List of Recommendations in the 8th Edition

XX

XXIII

Das Standardwerk zum Leitungstunnelbau

Dietrich Stein

Grabenloser Leitungsbau

Dietrich Stein
Grabenloser Leitungsbau
2003. XXII, 1144 Seiten,
968 Abb. 329 Tab.
Gebunden
€ 199,-* / sFr 314,-
ISBN 3-433-01778-6

Ernst & Sohn
Verlag für Architektur und
technische Wissenschaften GmbH & Co. KG

Für Bestellungen und Kundenservice:
Verlag Wiley-VCH
Boschstraße 12
69469 Weinheim
Telefon: (06201) 606-400
Telefax: (06201) 606-184
Email: service@wiley-vch.de

Ernst & Sohn
A Wiley Company

www.ernst-und-sohn.de

Der grabenlose Leitungsbau macht es möglich, Kabel- und Rohrleitungen für die sichere Versorgung mit Wasser, Gas, Fernwärme, Elektrizität und Telekommunikation sowie eine umweltfreundliche Abwasserentsorgung unabhängig vom Leitungsdurchmesser sowie den geologischen und hydrogeologischen Randbedingungen zu verlegen. In diesem Buch werden erstmals die zur Verfügung stehenden Verfahren umfassend beschrieben. Es wurde darauf Wert gelegt, neben der Beschreibung von Arbeitsweise und -ablauf sowie Ausrüstung insbesondere auch die jeweiligen Einsatzbereiche und Anwendungsgrenzen nach neuesten Erkenntnissen darzustellen. Die vielfältigen Fachinformationen dieses Standardwerks helfen bei der Planung und Ausführung von Leitungsbaumaßnahmen und erlauben eine wirtschaftlich und technisch optimale Auswahl der Verfahrenstechnik für den jeweiligen Anwendungsfall in Abhängigkeit der zahlreichen örtlichen und systembedingten Gegebenheiten. Das Arbeiten mit diesem Buch wird zu einem unentbehrlichen Hilfsmittel im beruflichen Alltag.

Recommendations

0   Structural calculations

0.1   General

Up until the 8th German edition (EAU 1990), the basis for the 6th English edition, the earth static calculations in the recommendations of the "Committee for Waterfront Structures", which had been developed in over 50 years of work by the Committee, were based on reduced values of soil properties, known as "calculation values", with the prefix "cal". Results of calculations using these calculation values then had to fulfil the global safety criteria in accordance with R 96, section 1.13.2a, of EAU 1990. This safety concept distinguished between three different load cases (R 18, section 5.4), and proved its worth over the years.

The introduction of EAU 1996 resulted in a changeover to the concept of partial safety factors. Since then, development of the Eurocodes has led to extensive changes to the DIN pre-standards, which have been taken into account in the 2004 edition of the EAU.

As part of the realisation of the Single European Market, the "Eurocodes" (EC) are currently being drawn up and will form harmonised directives for fundamental safety requirements for buildings and structures. These Eurocodes are as follows:

DIN EN 1990, EC 0: Basis of structural design
DIN EN 1991, EC 1: Actions on structures
DIN EN 1992, EC 2: Design of concrete structures
DIN EN 1993, EC 3: Design of steel structures
DIN EN 1994, EC 4: Design of composite steel and concrete structures
DIN EN 1995, EC 5: Design of timber structures
DIN EN 1996, EC 6: Design of masonry structures
DIN EN 1997, EC 7: Geotechnical design
DIN EN 1998, EC 8: Design provisions for earthquake resistance
                        of structures
DIN EN 1999, EC 9: Design of aluminium structures

Verification of structural safety must always be carried out according to these standards. EC 0 to EC 9, but in particular EC 7, are significant for proof of stability to EAU 2004. However, until the aforementioned Eurocodes are adopted, the following national DIN standards should be used in Germany. Some of these DIN standards have already been converted to the concept of partial safety factors on the basis of DIN 1055-100. These include the following standards:

DIN 1054:   Subsoil – Verification of the safety of earthworks and foundations
DIN 4017:   Soil – Calculation of design bearing capacity of soil beneath shallow foundations
DIN 4019:   Subsoil; settlement calculations for perpendicular central loading
DIN 4084:   Subsoil – Calculation of embankment failure and overall stability of retaining structures
DIN 4085:   Subsoil – Calculation of earth pressure

The previous standards covering design and construction will be superseded by new Eurocodes under the heading of "Implementation of special geotechnical works":

Previous standard	New standard
DIN 4014: Bored piles	DIN EN 1536: Bored piles
DIN 4128: Grouted piles	
DIN 4125: Ground anchors	DIN EN 1537: Ground anchors
DIN 4126: Stability analysis of diaphragm walls	DIN EN 1538: Diaphragm walls
	DIN EN 12 063: Sheet-pile walls
DIN 4026: Driven piles	DIN EN 12 699: Displacement piles
DIN 4093: Ground treatment by grouting	DIN EN 12 715: Grouting
	DIN EN 12 716: Jet grouting

From now on, EAU 2004 will make reference to the quantitative statements regarding methods of calculation with partial safety factors contained in these Eurocodes and the national standards, DIN 1054 in particular.

Wherever standards are cited in the recommendations, no mention is made of their respective status. However, Annex I.3 contains a list of the standards and their status as of October 2004.

0.2 Safety concept

0.2.1 General

The failure of a structure can occur as a result of exceeding the limit state of bearing capacity (LS 1, failure in the soil or structure) or the limit state of serviceability (LS 2, excessive deformation).

Analyses of limit states (verification of safety) are carried out according to the provisions of DIN EN 1991-1 for actions or action effects and those of DIN EN 1992–1999 for actions or action effects and resistances. DIN EN 1990 has been incorporated into DIN 1055-100, DIN EN 1997-1 into DIN 1054. DIN EN 1997-1 permits three options for assessing the verification of safety. These are designated "methods of verification 1 to 3".

DIN 1054 distinguishes between three cases for analysing limit state 1 (ultimate limit state of bearing capacity):

LS 1A: limit state of loss of support safety
LS 1B: limit state of failure of structures and components
LS 1C: limit state of loss of overall stability
        (after EC 7 method of verification 3)

In DIN 1054, method of verification 2 has been incorporated for earth static analyses of LS 1B, and method of verification 3 for analyses of LS 1C. Only one method is given for limit state LS 1A in DIN EN 1997-1; EAU 2004 makes use of this method of analysis.

The partial safety factors associated with these three cases are reproduced in tables R 0-1 and R 0-2 for loading cases 1 to 3 of DIN 1054. In the analysis of limit states 1B and 1C to DIN 1054, adequate ductility of the complete system of subsoil plus structure is assumed (DIN 1054, 4.3.4).

0.2.2 Design situations for geotechnical structures

0.2.2.1 Combinations of actions

Combinations of actions (CA) are permutations of the actions to which a structure may be subjected simultaneously at the limit states according to cause, magnitude, direction and frequency. We distinguish between:

a) Standard combination CA 1:
    Permanent actions and those variable actions occurring regularly during the functional lifetime of the structure.

b) Rare combination CA 2:
    Seldom or one-off planned actions apart from the actions of the standard combination.

c) Exceptional combination CA 3:
An exceptional action that may occur at the same time and in addition to the actions of the standard combination, especially in the event of catastrophes or accidents.

0.2.2.2 Safety classes for resistances

Safety classes (SC) take into account the different safety requirements for resistances depending on the duration and frequency of the critical actions. We distinguish between:

a) Conditions of safety class SC 1:
Conditions related to the functional lifetime of the structure.

b) Conditions of safety class SC 2:
Temporary conditions during the construction or repair of the structure and temporary conditions during building measures adjacent to the structure.

c) Conditions of safety class SC 3:
Conditions occurring just once or probably never during the functional life of the structure.

0.2.2.3 Loading cases

The loading cases (LC) for limit state LS 1 are derived from the combinations of actions in conjunction with the safety classes for the resistances. We distinguish between:

• Loading case LC 1:
Standard combination CA 1 in conjunction with the conditions of safety class SC 1. Loading case LC 1 corresponds to the "permanent design situation" according to DIN 1055-100.

• Loading case LC 2:
Rare combination CA 2 in conjunction with the conditions of safety class SC 1, or standard combination CA 1 in conjunction with the conditions of safety class SC 2. Loading case LC 2 corresponds to the "temporary design situation" according to DIN 1055-100.

• Loading case LC 3:
Exceptional combination CA 3 in conjunction with the conditions of safety class SC 2, or rare combination CA 2 in conjunction with the conditions of safety class SC 3. Loading case LC 3 corresponds to the "exceptional design situation" according to DIN 1055-100. The loading case categories for waterfront structures are dealt with in R 18, section 5.4.

0.2.2.4 Partial safety factors

0.2.2.4.1 Partial safety factors for actions and action effects

Table 0-1. Partial safety factors for actions for the ultimate limit states of bearing capacity and serviceability for permanent and temporary situations

Action or action effect	Symbol	Loading case		
		LC 1	LC 2	LC 3
LS 1A: limit state of loss of support safety				
Favourable permanent actions (dead load)	$\gamma_{G,stb}$	0.90	0.90	0.95
Unfavourable permanent actions (uplift/buoyancy)	$\gamma_{G,dst}$	1.00	1.00	1.00
Flow force in favourable subsoil	γ_H	1.35	1.30	1.20
Flow force in unfavourable subsoil	γ_H	1.80	1.60	1.35
Unfavourable variable actions	$\gamma_{Q,dst}$	1.00	1.00	1.00
LS 1B: limit state of failure of structures and components				
General permanent actions	γ_G	1.35	1.20	1.00
Hydrostatic pressure in certain boundary conditions[a]	$\gamma_{G,red}$	1.20	1.10	1.00
Permanent actions due to steady-state earth pressure	γ_{E0g}	1.20	1.10	1.00
Unfavourable variable actions	γ_Q	1.50	1.30	1.00
LS 1C: limit state of loss of overall stability				
Permanent actions	γ_G	1.00	1.00	1.00
Unfavourable variable actions	γ_Q	1.30	1.20	1.00
LS 2: limit state of serviceability				

$\gamma_G = 1.00$ for permanent actions or action effects
$\gamma_Q = 1.00$ for variable actions or action effects

[a] According to DIN 1054, section 6.4.1(7), the partial safety factors γ_G for hydrostatic pressure may be reduced as specified for waterfront structures in which larger displacements can be accommodated without damage if the conditions according to section 8.2.0.3 are complied with.

0.2.2.4.2 Partial safety factors for resistances

Table 0-2. Partial safety factors for resistances for the ultimate limit state of bearing capacity for permanent and temporary situations

Resistance	Symbol	Loading case		
		LC 1	LC 2	LC 3
LS 1B: limit state of the failure of structures and components				
Soil resistances				
Earth resistance	γ_{Ep}	1.40	1.30	1.20
Earth resistance for determining bending moment[a]	$\gamma_{Ep,red}$	1.20	1.15	1.10
Ground failure resistance	γ_{Gr}	1.40	1.30	1.20
Sliding resistance	γ_{Gl}	1.10	1.10	1.10
Pile resistances				
Pile compression resistance for test load	γ_{Pc}	1.20	1.20	1.20
Pile tension resistance for test load	γ_{Pt}	1.30	1.30	1.30
Pile resistance in tension and compression based on empirical values	γ_P	1.40	1.40	1.40
Grouted anchor resistances				
Resistance of steel tension member	γ_M	1.15	1.15	1.15
Pull-out resistance of grout	γ_A	1.10	1.10	1.10
Resistances of flexible reinforcing elements				
Material resistance of reinforcement	γ_B	1.40	1.30	1.20
LS 1C: limit state of loss of overall stability				
Shear strength				
Friction angle tan φ' of drained soil Cohesion c' of drained soil	γ_φ	1.25	1.15	1.10
Shear strength cu of undrained soil	γ_{c},γ_{cu}	1.25	1.15	1.10
Pull-out resistances				
Ground or rock anchors, tension piles	γ_{N},γ_{Z}	1.40	1.30	1.20
Grout of grouted anchors	γ_A	1.10	1.10	1.10
Flexible reinforcing elements	γ_B	1.40	1.30	1.20

[a] Reduction for calculating the bending moment only. According to DIN 1054, section 6.4.2(6), the partial safety factors γ_{Ep} for earth resistance may be reduced as specified for waterfront structures in which larger displacements can be accommodated without damage if the conditions according to section 8.2.0.2 are complied with.

0.2.3   Ultimate limit state LS 1: bearing capacity

The calculated verification of adequate stability is always carried out for the ultimate limit state of bearing capacity 1 (LS 1) with the help of design values (index d) for actions or action effects and resistances. The design values are derived from the characteristic values (index k) of the actions or action effects and resistances as follows:

- the characteristic actions or action effects are multiplied by partial safety factors,
  e.g. $E_{a,d} = E_{a,k} \cdot \gamma_G$ (earth pressure component from permanent loads).

- the characteristic resistances are divided by the partial safety factors,
  e.g. $E_{p,d} = E_{p,k} / \gamma_{Ep}$ (earth resistance) in LS 1B or $c'_d = c'_k / \gamma_c$ (effective cohesion) in LS 1C.

The characteristic value of a parameter is the value of a, normally scattered, physical variable, e.g. the friction angle φ' or the cohesion c' or c_u of a component resistance (e.g. the pull-out force of an anchor pile of the earth resistance force in front of the base of sheet piling), that is to be used in calculations. It is the cautious anticipated mean value lying on the safe side and is defined in DIN 1054.

The verification of safety is assessed according to the following fundamental equation:

$$E_d \leq R_d$$

E_d is the design value of the actions or action effects derived from the characteristic values of the actions or action effects, multiplied by the respective safety factors (e.g. foundation load).

R_d is the design value of the resistances derived as a function of the characteristic soil resistances, or from structural elements, divided by the associated partial safety factors according to the corresponding method of calculation (e.g. ground failure to DIN 4017).

The partial safety factors to be used are to be taken from tables R 0-1 and R 0-2 and from the corresponding building materials and components standards, provided no other partial safety factors are specified in the corresponding recommendations of EAU 2004.

0.2.3.1   Limit state LS 1A

Proceed as follows for analyses of stability for limit state LS 1A:

a)   Firstly, calculate the design values of the actions from the characteristic actions. In doing so, distinguish between favourable and unfavourable actions. Resistances do not play a role determining the support safety (LS 1A).

b) Secondly, compare the design values of the favourable and unfavourable actions and verify that the respective limit state condition is complied with. For further information see DIN 1054.

0.2.3.2 Limit state LS 1B
The following procedure is useful for analysing stability for limit state LS 1B:

a) Firstly, apply the characteristic actions to the chosen structural system and hence determine the characteristic action effects (e.g. internal forces).
b) Secondly, convert the characteristic action effects with the partial safety factors for actions into design values for action effects, the characteristic resistances to design values for resistances.
c) Finally, compare the design values of the action effects with the design resistances and show that the limit state equation is complied with for the failure mechanism under investigation.

This method assumes that a linear–elastic calculation is generally possible. Refer to DIN 1054, section 4.3.2 (3) when calculating the stability of non-linear problems for LS 1B. According to this standard the action effects calculated from the most unfavourable combination of permanent and variable actions may be broken down into a component comprising permanent actions and a component comprising variable actions in each case owing to a sufficiently accurate criterion.

0.2.3.3 Limit state LS 1C
Proceed as follows for analyses of stability for limit state LS 1C:

a) Firstly, calculate the design values of the actions from the characteristic actions for the failure mechanism under investigation.
b) Secondly, convert the characteristic shear strengths and, if necessary, component resistances with the partial safety factors for resistances into design resistances.
c) Finally, show that the limit state equation is complied with using the design values for actions and resistances for the failure mechanism under investigation.

0.2.4 Limit state LS 2: serviceability
Deformation verification is to be provided for all structures whose function can be impaired or rendered useless through deformations. The deformations are calculated with the characteristic values of actions and soil reactions and must be less than the deformations permissible for perfect functioning of a component or the whole structure. Where applicable, the calculations should include the upper and lower limit values of the characteristic values.

In the case of deformation verification in particular, the progression of influences over time must be taken into account in order to allow for critical deformation states during various operating and construction stages.

0.2.5 Geotechnical categories

The minimum requirements in terms of scope and quality of geotechnical investigations, calculations and supervisory measures are described by three geotechnical categories in accordance with EC 7, these designating a low (category 1), normal (category 2) and high (category 3) geotechnical difficulty. These are reproduced in DIN 1054, section 4.2. Waterfront structures are to be allocated to category 2, or category 3 in the case of difficult subsoil conditions. A specialist geotechnical engineer should always be consulted.

0.2.6 Probabilistic analysis

Even though its origins lie in the idea of a probabilistic concept of verification, the concept of partial safety factors to EC 0 or DIN 1054 is, by its very nature, deterministic. A failure mechanism's compliance with the limit state equation merely shows that there is a sufficient probability that the mechanism investigated will not occur. If, on the other hand, a stability analysis is to include a statement on the likelihood of the occurrence of a limit state, it is necessary to carry out an analysis based on probabilities. If the scatters of the effects and the independent parameters of the resistances are known, which is frequently the case in harbour engineering, analyses of stability based on probabilities can then lead to more economic structures than is possible when using a deterministic analysis concept.

A probability-based analysis assumes that the independent variables describing the actions or action effects and resistances are inserted into the limit state equation as variables of their distribution densities $f(R)$ and $f(E)$ for each of the limit states to be considered. The solution to the limit state equation $f(Z)$ itself then represents a function of a scattered variable:

$$f(Z) = f(R) - f(E)$$

The probability of failure P_f, or rather the reliability $1 - P_f$ of a design, is calculated from the integral of the function $f(Z)$ for negative arguments of Z.

In the case of a large number of mechanisms to be investigated, it is useful to determine the critical mechanism for the failure of a design by means of a tree analysis [213, 214]. Here, the mechanisms not relevant to failure are systematically eliminated, but correlations between the mechanisms are taken into account [215].

0.3 Calculations for waterfront structures

Waterfront structures should always be designed as simply as possible in structural terms with clearly defined paths for transferring the loads and forces. The less uniform the soil, the greater the need for statically determinate designs in order to avoid, as far as possible, any additional stresses from unequal deformations, which cannot be properly taken into account. Accordingly, wherever possible, analyses of stability should be broken down into clear, straightforward steps.

Verifying the stability of a waterfront structure must include the following elements in particular:

- details of the use of the structure,
- drawings of the structure with all essential planned dimensions,
- a brief description of the structure including, in particular, all details that are not readily identifiable from the drawings,
- design value of the bottom depth,
- characteristic values of all actions,
- soil strata and associated characteristic soil parameters,
- critical water levels, referenced to marine chart datum or mean sea level or a local gauge zero, together with corresponding groundwater levels (out of reach of high water, out of reach of flooding),
- combinations of actions or load cases,
- partial safety factors necessary/used
- intended building materials and their strengths or resistance values,
- all data about building schedules and building execution with principal temporary states,
- description and justification of the intended verification procedures,
- information about literature used and other calculation aids.

The actual analyses of stability and serviceability must take into account that in foundation and hydraulic engineering, appropriate soil investigations, shearing parameters, load assumptions, the ascertaining of hydrodynamic influences and unconsolidated states, a favourable bearing system and a realistic computing model are more important than exaggeratedly accurate numerical calculations.

Furthermore, reference is made to the Additional Technical Contract Conditions ZTV-ING [216], ZTV-W (LB 215) for hydraulic structures of plain and reinforced concrete [118], and ZTV-W (LB 202) for technical processing [164].

1   Subsoil

1.1   Mean characteristic soil properties (R 9)

1.1.1   General

The soil properties given in table R 9-1 are cautious mean empirical values for a larger area of soil. They may be used as characteristic values in the meaning of DIN 1054, which is why they are given the index k. Without verification, the values in the table for low strength should be assumed for natural sands. Medium strength can only be expected after compaction, apart from geologically older sediments. Without verification, the values for soft consistency apply for cohesive soils.

Designs should always be based on soil properties measured locally (see DIN 4020 and R 88, section 1.4). These can lie either above or below the values given in table R 9-1.

The effective shear parameters φ' and c' of a cohesive soil (shear parameters for drained soil, see DIN 18 137 part 2) are ascertained from undisturbed soil samples in a direct shear test (DIN 18 137 part 3) and where applicable also in a triaxial test (DIN 18 137 part 2).

According to [142] it can be presumed that the friction angle φ' for non-cohesive soils in the plane strain state amounts to 9/8 of the friction angle measured in the triaxial shear test. Therefore, this can be increased by up to 10% for dense soils in calculations for long waterfront structures, with the consent of the test laboratory.

The characteristic values for the shear parameters φ'_k and c'_k for cohesive soils apply to calculations for final stability (consolidated condition = final strength).

The characteristic values of the shear parameters for unconsolidated soil $\varphi_{u,k}$ and $c_{u,k}$ are the shear parameters for the unconsolidated initial state. In water-saturated soils $\varphi_{u,k} = 0$.

1.2   Layout and depth of boreholes and penetrometer tests (R 1)

1.2.1   General

The aim of boreholes and penetrometer tests is to ascertain the nature and structure of the subsoil and obtain soil samples for soil mechanics tests in the laboratory. The extent of boreholes and penetrometer tests must always such that all the characteristics of the subsoil relevant to the planning are established and a sufficient number of suitable soil samples is obtained for the laboratory tests.

When determining the number and type of boreholes and penetrometer tests, geological maps and, if necessary, the findings of earlier boreholes and penetrometer tests should also be taken into account.

Subsoil investigations by means of boreholes are performed in accordance with DIN 4020 "Geotechnical investigations for civil engineering

Table 9-1. Characteristic soil parameters (empirical)

No. 1	Soil type	Soil group as per DIN 18 196[1]	Penetration resistance q_c	Strength resp. consistency in initial state	Weight density γ_k / submerged weight density γ'_k		Compressibility[2] Initial loading[3] $E_s = v_e\,\sigma_{at}(\sigma/\sigma_{at})_e^w$		Shear parameters of the drained soil φ'_k	c'_k	Shear parameters of the undrained soil $c_{u,k}$	Permeability factor k_k
2			MN/m²		γ_k kN/m³	γ'_k kN/m³	v_e	w_e	degrees	kN/m²	kN/m²	m/s
3	Gravel, uniform	GE U[4] < 6	<7,5 / 7,5–15 / >15	low / medium / high	16,0 / 17,0 / 18,0	8,5 / 9,5 / 10,5	400 / 900	0,6 / 0,4	30,0–32,5 / 32,5–37,5 / 35,0–40,0			$2\cdot10^{-1}$ to $1\cdot10^{-2}$
4	Gravel, non-uniform or intermittent	GW, GI $6 \leq$ U[4] ≤ 15	<7,5 / 7,5–15 / >15	low / medium / high	16,5 / 18,0 / 19,5	9,0 / 10,5 / 12,0	400 / 1100	0,7 / 0,5	30,0–32,5 / 32,5–37,5 / 35,0–40,0			$1\cdot10^{-2}$ to $1\cdot10^{-6}$
5	Gravel, non-uniform or intermittent	GW, GI U[4] > 15	<7,5 / 7,5–15 / >15	low / medium / high	17,0 / 19,0 / 21,0	9,5 / 11,5 / 13,5	400 / 1200	0,7 / 0,5	30,0–32,5 / 32,5–37,5 / 35,0–40,0			$1\cdot10^{-2}$ to $1\cdot10^{-6}$
6	Sandy gravel, $d < 0,006$ mm < 15 %	GU, GT	<7,5 / 7,5–15 / >15	low / medium / high	17,0 / 19,0 / 21,0	9,5 / 11,5 / 13,5	400 / 800 / 1200	0,7 / 0,6 / 0,5	30,0–32,5 / 32,5–37,5 / 35,0–40,0			$1\cdot10^{-5}$ to $1\cdot10^{-6}$
7	Gravel-sand-fine grain $d < 0,06$ mm > 15 %	G$\bar{\text{U}}$, G$\bar{\text{T}}$	<7,5 / 7,5–15 / >15	low / medium / high	16,5 / 18,0 / 19,5	9,0 / 10,5 / 12,0	150 / 275 / 400	0,9 / 0,8 / 0,7	30,0–32,5 / 32,5–37,5 / 35,0–40,0			$1\cdot10^{-7}$ to $1\cdot10^{-11}$
8	Sand, uniform coarse sand	SE U[4] < 6	<7,5 / 7,5–15 / >15	low / medium / high	16,0 / 17,0 / 18,0	8,5 / 9,5 / 10,5	250 / 475 / 700	0,75 / 0,60 / 0,55	30,0–32,5 / 32,5–37,5 / 35,0–40,0			$5\cdot10^{-3}$ to $1\cdot10^{-4}$
9	Sand, uniform, fine sand	SE U[4] < 6	<7,5 / 7,5–15 / >15	low / medium / high	16,0 / 17,0 / 18,0	8,5 / 9,5 / 10,5	150 / 225 / 300	0,75 / 0,65 / 0,60	30,0–32,5 / 32,5–37,5 / 35,0–40,0			$1\cdot10^{-4}$ to $2\cdot10^{-5}$
10	Sand, non uniform or intermittent	SW, SI $6 \leq$ U[4] ≤ 15	<7,5 / 7,5–15 / >15	low / medium / high	16,5 / 18,0 / 19,5	9,0 / 10,5 / 12,0	200 / 400 / 600	0,70 / 0,60 / 0,55	30,0–32,5 / 32,5–37,5 / 35,0–40,0			$5\cdot10^{-4}$ to $2\cdot10^{-5}$

No.	1	2	3	4	5	6	7	8	9	10
11	Sand, non uniform or intermittent	SW, SI $U^{(+)} > 15$	< 7,5 7,5–15 > 15	low medium high	17,0 19,0 21,0	9,5 11,5 13,5	200 400 600	0,70 0,60 0,55	30,0–32,5 32,5–37,5 35,0–40,0			$1 \cdot 10^{-4}$ to $1 \cdot 10^{-5}$
12	Sand, $d < 0,06$ mm $< 15 \%$	SU, ST	< 7,5 7,5–15 > 15	low medium high	16,0 17,0 18,0	8,5 9,5 10,5	150 350 500	0,80 0,70 0,65	30,0–32,5 32,5–37,5 35,0–40,0			$2 \cdot 10^{-5}$ to $5 \cdot 10^{-7}$
13	Sand, $d < 0,06$ mm $> 15 \%$	SŪ, SṮ	< 7,5 7,5–15 > 15	low medium high	16,5 18,0 19,5	9,0 10,5 12,0	50 250	0,9 0,75	30,0–32,5 32,5–37,5 35,0–40,0			$2 \cdot 10^{-6}$ to $1 \cdot 10^{-9}$
14	Silt, anorganic cohesive soils with low plasticity ($w_L < 35 \%$)	UL		safe stift semi-fine	17,5 18,5 19,5	9,0 10,0 11,0	40 110	0,80 0,60	27,5–32,5	0 2–5 5–10	5–60 20–150 50–300	$1 \cdot 10^{-5}$ to $1 \cdot 10^{-7}$
15	Silt, anorganic cohesive soils with medium plasticity ($50 \% > w_L > 35 \%$)	UM		safe stift semi-fine	16,5 18,0 19,5	8,5 9,5 10,5	30 70	0,90 0,70	25,0–30,0	0 5–10 10–15	5–60 20–150 50–300	$2 \cdot 10^{-6}$ to $1 \cdot 10^{-9}$
16	Clay, anorganic cohesive soils with low plasticity ($w_L < 35 \%$)	TL		safe stift semi-fine	19,0 20,0 21,0	9,0 10,0 11,0	20 50	1,0 0,90	25,0–30,0	0 5–10 10–15	5–60 20–150 50–300	$1 \cdot 10^{-7}$ to $2 \cdot 10^{-9}$
17	Clay, anorganic cohesive soils with medium plasticity ($50 \% > w_L > 35 \%$)	TM		safe stift semi-fine	18,5 19,5 20,5	8,5 9,5 10,5	10 30	1,0 0,95	22,5–27,5	5–10 10–15 15–20	5–60 20–150 50–300	$5 \cdot 10^{-8}$ to $1 \cdot 10^{-10}$
18	Clay, anorganic cohesive soils with high plasticity ($w_L > 50 \%$)	TA		safe stift semi-fine	17,5 18,5 19,5	7,5 8,5 9,5	6 20	1,0 1,0	20,0–25,0	5–15 10–20 15–25	5–60 20–150 50–300	$1 \cdot 10^{-9}$ to $1 \cdot 10^{-11}$
19	Silt or clay, organic	OU und OT		very soft soft stift	14,0 15,5 17,0	4,0 5,5 7,0	5 20	1,00 0,85	17,5–22,5	0 2–5 5–10	2 – < 15 5–60 20–150	$1 \cdot 10^{-9}$ to $1 \cdot 10^{-11}$

No.	1	2	3	4	5		6	7	8	9	10
20	Peat [5]	HN, HZ		very soft soft stiff semi firm	10,5 11,0 12,0 13,0	0,5 1,0 2,0 3,0	[5]	[5]	[5]	[5]	[5]	$1 \cdot 10^{-5}$ to $1 \cdot 10^{-8}$
21	Mud [6], Faulschlamm	F		very soft soft	12,5 16,0	2,5 6,0	4 15	[6]	0	1,0 0,9	<6 6-60	$1 \cdot 10^{-7}$ $1 \cdot 10^{-9}$

Explanations:

[1] Code letters for the main and secondary components:

G gravel   U silt   O organic components   F mud
S sand   T clay   H peat (humus)

Codes for characteristic physical soil properties:

Grain size distribution: Plastic properties:
W well-graded grain   L low plasticity
   size distribution   M medium plasticity
E uniform grain size   A high plasticity
   distribution
I intermittent graded
   grain size distribution

Degree of decomposition of peat:
N not deomposed or scarcely
Z decomposed peat

[2] v_c: stiffness factor, empirical parameter
     w_c: empirical parameter
     σ: load in kN/m^2

[3] σ_{at}: atmospheric pressure ($= 100 \; kN/m^2$)
     v_c-values for repeated load until 10-times higher,
     w_c goes towards 1.

[4] U uniformity coefficient

[5] The values of compressibility and the shear para-
     meters of peat scatter in such a range that mean
     empirical values can not be given.

[6] The effective friction angle of consolidated mud can
     be very high. Relevant is the value with respect of the
     degree of consolidation, which can only be determined
     by laboratory tests.

purposes". DIN EN 1536, 1538 and 12 063 must also be taken into account for pile structures and sheet piling.

When the subsoil conditions are generally known, subsoil investigations frequently begin with an exploratory investigation by way of static or dynamic penetrometer tests, if necessary in conjunction with non-destructive geophysical surface measurements. They enable an initial, rough evaluation of the types of soil. The principal borehole programme, together with a further penetrometer test programme if necessary, can be stipulated on the basis of the results of such penetrometer tests. Boreholes and penetrometer tests can also provide information about any possible hindrances in the subsoil, although it should not be assumed that there are no hindrances in the subsoil if the boreholes and penetro-meter tests do not reveal any as such.

In many cases static penetrometer tests can frequently be carried out economically and quickly, which means it is often useful to supplement the boreholes with penetrometer tests and to perform these at an early stage of the planning. The penetrometer resistance thus allows for allocation as per table R 9-1, section 1.1, for early draft designs. Please refer to DIN 4094-1 and [1], [2] and [106] for the determination of shear strength and coefficients of compressibility from the results of static penetrometer tests.

Geophysical surface measurements will give additional two dimensional information on the geological profile, the ground-water level and in some cases on hindrances in the subsoil. Geophysical measurements can be performed up to large depth, an interpretation is however not possible without direct information from boreholes.

1.2.2 Principal boreholes

Principal boreholes should preferably lie on the later axis of the structure (shoreline). For non-anchored walls they should be sunk to about twice the height of the difference in ground level or as far as a known geological stratum. Recommended borehole spacing = 50 m. For expediency, in heavily stratified soils, above all in varved soils, hose core bores to DIN 4020 are sunk, also for taking undisturbed soil samples of quality grade 2 (DIN 4021), or quality grade 1 in favourable conditions. Piezometers and pore water pressure gauges can be installed subsequently in some principal borings to ascertain the water level.

1.2.3 Intermediate boreholes

Depending on the findings of the principal boreholes or the earlier penetrometer tests, intermediate boreholes are also sunk to the depth of the principal boreholes, or to a depth at which a known, uniform soil stratum is encountered (as revealed by the principal boreholes/penetro-meter tests). Typical borehole spacing is again about 50 m.

1.2.4 Penetrometer tests

Penetrometer tests are generally executed according to the layout shown in fig. R 1-1. They are generally made to the the same depth as the principal boreholes but continue at least into a known, loadbearing geological stratum whose properties can be reliably assessed. Please refer to DIN 4094 parts 1 and 3 for details of the equipment for and execution of penetrometer tests.

Static penetrometer tests should also enable the measurement of the local skin friction and pore water pressure.

In the case of soft, cohesive soils, vane shear tests in accordance with DIN 4094-4 are recommended for determining the undrained shear strength (see R 88, section 1.4).

For the calibratoin of penetrometer tests some of them should be performed directly aside of a borehole. These penetrometer tests have to be performed in advance to the boreholes to ensure that the results of the penetrometer tests are not distorted.

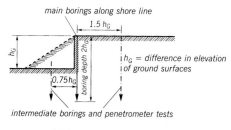

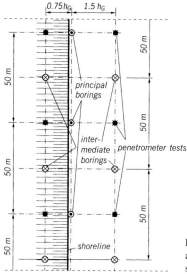

Fig. R 1-1. Example of layout of boreholes and penetrometer tests for waterfront structures

16

1.3 Preparation of subsoil investigation reports, expert opinions and foundation recommendations for waterfront structures (R 150)

1.3.1 General

In line with the provisions of DIN 1054, waterfront structures are always assigned to geotechnical category 2, or category 3 in the case of difficult subsoil conditions. A soil mechanics engineer should therefore be consulted for every project, in fact as soon as the necessary field tests have been established according to R 1, section 1.2.

The duties of the soil mechanics engineer include arranging the subsoil investigations, supervising these and preparing reports and expert opinions to communicate the findings of the tests and recommend types of foundation.

R 1 contains recommendations regarding the nature and extent of subsoil investigations.

DIN EN 1997-1, DIN 1054, DIN 4020 and [7] contain general rules for the production of reports and expert opinions on subsoil investigations and foundations.

1.3.2 Subsoil investigation reports, expert opinions and foundation recommendations for waterfront structures

The subsoil investigation report and the foundation recommendations can be provided separately or together in one document.

Apart from a precise description of the objective of the investigation and a list of the documents on the planned structure(s) that were made available, the subsoil investigation report must include, above all:

- details of the construction project, e.g. location, foundation depth, loads, structural system,
- details of the general geological and, if applicable, also the hydro-geological conditions,
- the findings of the boreholes and penetrometer tests,
- the findings of the laboratory tests and, if applicable, simulations carried out,
- a clear compilation of the test results,
- a comprehensive assessment of the subsoil.

The subsoil expert opinion and foundation recommendations must include, first and foremost:

- an assessment of the subsoil with regard to the specific construction measures,
- the establishment of characteristic soil parameters and, if applicable, methods of calculation too,
- details of the groundwater levels or design water levels,

- if required, also details of driving hindrances or the method of installing piles and sheet piles,
- if applicable, details of earthquake risks,
- general foundation proposals with the results of associated rough earth static calculations or settlement calculations,
- if necessary, details of the strength of the in situ soil after loosening (soil classification for dredging works, DIN 18 311).

1.3.3 Reproducing the findings of the field and laboratory tests in the subsoil investigation report

1.3.3.1 Field tests

The exact positions of the soil explorations and field tests must be shown on a scale drawing that also indicates the outline of the planned structure(s). Reference dimensions to permanent reference points or lines should also be included. The dates and times of the tests and boreholes and any particular observations must also be noted on the drawing.

The investigation-methods used should be explained in the subsoil investigation report, but in the case of standardised procedures it is sufficient to quote the relevant standard. If the work carried out deviates from a standardised method, the reasons for this must be stated and the procedures described.

If the subsoil investigation report does not include the stratigraphical legend for boreholes according to DIN 4022, there should at least be a note indicating where this can be found. This also applies to the soil samples taken.

Colour photographs should accompany any soil samples taken. However, such colour photographs cannot replace the characterization and assessment of soil samples.

The results of penetrometer tests must be recorded according to DIN 4094 parts 1 to 5. You are recommended to enter the results of penetrometer tests next to neighbouring soil profiles and using a generally applicable reference system for levels (e.g. mean sea level or lowest astronomical tide).

1.3.3.2 Laboratory tests

The findings of laboratory tests should be recorded and described in full according to soil properties (e.g. grading curves, results of compression and shear tests) so that the reader can interpret the findings properly. The test apparatus used in each case should also be described. In the case of standardised tests it is sufficient to quote the relevant standard. The results of compression tests must always be shown as pressure–settlement curves and time–settlement curves, with details of the loading stages and the duration of consolidation. As the test apparatus for the compression test is currently not covered by any standard (DIN 18 135 is in preparation), details of the apparatus dimensions and method of installing the soil in the apparatus must be included with the results. The results of shear tests are to be shown in accordance with the relevant standards.

Finally, show the results of laboratory tests in tabular form arranged according to borehole, sampling depth and sample number.

1.3.3.3 Summary of results of investigations

The findings of the field and laboratory tests are summarised in the subsoil investigation report in order to reach a conclusion about the subsoil. Grading curves are grouped into the grading ranges of the principal soil types. The ranges and characteristic values for the soil parameters are specified for the principal soil types.

1.3.4 Content and statements in the subsoil expert opinion and foundation recommendations

The findings summarised in the subsoil investigation report form the basis for the subsoil expert opinion and foundation recommendations prepared by the soil mechanics engineer. These always contain an assessment of the subsoil both in structural–constructional and in earthworks engineering terms together with a comprehensive description of the geological structure, the properties of the soil strata established and their soil mechanics parameters. Also included are, above all, details about the particle distributions, the in situ density of non-cohesive soils, the state of cohesive soils and an assessment of the shear parameters and coefficients of compressibility given in the subsoil investigation report. The subsoil expert opinion and foundation recommendations also lay down the critical characteristic soil parameters, e.g. weight densities, coefficients of compressibility and, in particular, shear parameters, relevant for the earth static calculations. In doing so, the interaction between structure and subsoil must be taken into account, and also the method of calculation to be selected. If necessary, the soil mechanics engineer agrees these values in advance with the developer, the designer, the contractor and the building authority responsible or the engineer checking the structural design.

The subsoil expert opinion must include the water and/or groundwater levels established, plus the design water levels derived from these, in the context of the structural–constructional description.

The subsoil expert opinion and foundation recommendations should also contain details of the soil groups to DIN 18 196 and soil classification to DIN 18 300.

In seismic regions the soil mechanics engineer – if necessary in collaboration with an earth quake expert for the region concerned – should also provide details of the anticipated seismic effects.

19

One important task of the subsoil expert opinion and foundation recommendations is also the assessment of the in situ soil with respect to the methods used to install piles and sheet piles. If the scope of the investigations does not permit an evaluation in the meaning of R 154, section 1.8, this should be expressly stated. In such cases appropriate supplementary investigations must be carried out at a later date.

The subsoil expert opinion and foundation recommendations must also include an assessment of driving hindrances, above all along the driving axes for waterfront structures. In doing so, besides the findings of boreholes and penetrometer tests, it is mainly the geological findings and older maps indicating earlier building measures that are useful here.

1.4 Determination of undrained shear strength c_u in field tests (R 88)

1.4.1 Vane shear test

The vane shear test in accordance with DIN 4094-4 is suitable for measuring the shear strength of undrained, stone-free, soft cohesive soils. The vane assembly is pressed directly into the soil either at the surface or at the bottom of a borehole. Experience has shown that the vane shear test gives reliable results for soft normal consolidated soils as well as for slightly overconsolidated soils.

The shear strength c_{fu} of the undrained soil can be determined from the maximum shear stress c_{fv} of the soil upon initial shearing and taking into account correction factors μ:

$$c_{fu} = \mu \cdot c_{fv}$$

The correction factors μ for soft normal consolidated soils are given in table R 88-1. DIN 4094 part 4 contains other correction factors for overconsolidated soils.

Table R 88-1. Correction factors μ for soft normal consolidated soils

I_p	0	30	60	90	120
μ	1.0	0.8	0.65	0.58	0.50

1.4.2 Cone penetration test

The following relationships exist for example between the vane shearing strength c_{fu} and the toe resistance q_c:

in clay: $$c_{fu} \approx \frac{1}{14} \cdot q_c$$

in overconsolidated clay: $$c_{fu} \approx \frac{1}{20} \cdot q_c$$

in soft clay: $$c_{fu} \approx \frac{1}{12} \cdot q_c$$

1.4.3 Plate bearing tests

The c_u values for soil strata close to the surface can be determined in the field with the help of plate bearing tests to DIN 18 134 using plates of at least 30 cm diameter.

The respective c_u value for saturated soil is derived from the following equation:

$$c_u = \frac{1}{6} p_{\text{failure}}$$

p_{failure} = mean normal bottom stress at ground failure

If a pronounced failure point does not appear on the pressure–settlement curve of the test, settlement = $^1/_{10}$ of the plate diameter will be assumed as having caused failure.

However, plate bearing tests supply correct values only for a depth equal to about two times the diameter of the plate. Therefore, to determine the shear strength of the undrained soil in lower strata the tests must be carried out at the bottom of trial pits or excavations. For this, a trial pit must be square with a side length equal to at least three times the diameter of the plate.

1.4.4 Pressuremeter tests

The c_u values can also be determined in the field by means of pressure-meter tests. Refer to DIN 4094-5 for details of such tests.

1.5 Investigation of the degree of density of non-cohesive backfill for waterfront structures (R 71)

1.5.1 Need for investigation

Generally, special demands are placed on backfill for waterfront structures, either because crane tracks, bollards or other structures are sited on the backfill on shallow foundations, or because the design of the waterfront structure assumes soil parameters that attribute a particular strength to the backfill. These values must be verified by means of appropriate investigations.

The same applies when anchors can be subjected to additional strains due to settlement and subsidence of the backfill.

The respective requirements should always be taken into account when installing and compacting the backfill (see R 175, section 1.6). The scope of random investigations and tests should be such that it is possible to make an unambiguous assessment. Areas in which experience has shown compaction to be difficult, e.g. in the immediate proximity of structures and foundations, must always be given special attention.

1.5.2 Methods of investigation

It is possible to check the compaction achieved in backfill to waterfront structures using non-cohesive soil by determining its degree of density D in accordance with DIN 18 126; the degree of density is directly related to the limit values of "loosest density" and "densest density" specific to the soil.

DIN 18 126 defines degree of density thus:

$$D = \frac{\max n - n}{\max n - \min n}$$

where

$\max n = $ porosity of dry soil in loosest state
$\min n = $ porosity of soil in densest state
$n \quad = $ porosity of soil being tested

Contrasting with this, the degree of compaction D_{pr} (DIN 18 127) is a relationship that specifies the extent of the compaction in situ in comparison with the best compaction achieved in the laboratory with a certain compaction intensity. It is initially not possible to generate a relationship to the aforementioned "loosest density" and "densest density" limit values, but is possible after calculating the degree of density D from the degree of compaction D_{pr}.

The following relationship exists between degree of density D and degree of compaction D_{pr}:

$$D = A + B \cdot D_{pr}$$

where

$$A = \frac{\max n - 1}{\max n - \min n}$$

$$B = \frac{1 - n_{pr}}{\max n - \min n}$$

$$D_{pr} = \frac{\rho_d}{\rho_{pr}} = \frac{1 - n}{1 - n_{pr}}$$

$\rho_d = $ dry density
$\rho_{pr} = $ dry density with optimum water content in Proctor test
$n_{pr} = $ porosity with optimum water content in Proctor test

In the above equations the porosity n is the dimensionless ratio of pore volume to total volume of the soil (DIN 18 125). However, in the international literature it is usually the voids ratio e that is used as the reference variable. The void ratio e is the dimensionless ratio of pore volume to volume of solids. Porosity n and voids ratio e are related as follows:

$$e = \frac{n}{1-n}$$

The parameter formed with the void ratio is called the specific degree of density I_D:

$$I_D = \frac{\max e - e}{\max e - \min e}$$

where

$\max e$ = void ratio for loosest density in dry state
$\min e$ = void ratio for densest density
e = void ratio of soil to be checked

There is no direct relationship between D and I_D, but as an approximation for values of e or n relevant for practical building purposes, $I_D \cong 1.1 \cdot D$. The degree of density D can also be determined with the help of static and dynamic penetrometer tests (depth of test > 1 m). DIN 4094 parts 1 and 3 contain correlations between the penetrometer resistance and the degree of density D for a number of non-cohesive soils. These can be determined for specific conditions by way of comparative investigations.

1.6   Degree of density of hydraulically filled, non-cohesive soils (R 175)

1.6.1   General
The loadbearing capacity of port areas is essentially determined by the degree of density and strength of the uppermost 1.5–2.0 m of the in situ soil. The degree of density of hydraulically filled ground depends primarily on the following factors:

- Granulometric composition, especially silt content of fill material. To achieve the best possible degree of density, it is important to limit the proportion of fine particles < 0.06 mm to max. 10%. This can be guaranteed through, for example, correct scow loading (R 81, section 7.3), correct shaping and installation of the backfill site, correct type of hydraulic filling process and the channelling of the flush water.
- Type of extraction and further processing of the fill material.
- Shaping and installation of the backfill site.

- Location and type of the hydraulic flush water discharge.
- During hydraulic filling above water, a greater degree of density is generally achieved without additional measures than is the case below water.
- The influence of tides and waves often compacts the hydraulically filled sand within a short time, which therefore achieves a very high degree of density.

1.6.2   Influence of the hydraulic fill material

Experience has shown that hydraulic filling below water achieves the following approximate degrees of density D:

Fine sand with different uniformity coefficients and a mean grain size $d_{50} < 0.15$ mm:

$$D = 0.35 \text{ to } 0.55$$

Medium sand with different uniformity coefficients and a mean grain size $d_{50} = 0.25$ to 0.50 mm:

$$D = 0.15 \text{ to } 0.35$$

As the granulometric composition and silt content of the material do not remain constant during execution of the work, the aforementioned empirical values represent only a rough guide.

The influence of tides and waves can considerably increase the degree of density within a short time, but the relevant relationships have not yet been fully investigated.

The choice of a hydraulic fill material is generally limited by economic and technical restraints.

1.6.3   Required degree of density

Hydraulically filled, non-cohesive soils in port areas should exhibit the following degrees of density D depending on the respective use of the area:

Table R 175-1. Use-related degrees of density required for port areas

Type of utilisation	D	
	Fine sand $d_{50} < 0.15$ mm	Medium sand $d_{50} = 0.25$ to 0.50 mm
Storage areas	0.35–0.45	0.20–0.35
Traffic areas	0.45–0.55	0.25–0.45
Structure areas	0.55–0.75	0.45–0.65

Therefore, for the same loads, fine sand always requires a higher degree of density than medium sand.

1.6.4 Checking the degree of density

The degree of density in the upper part of hydraulic backfill can be determined with the customary tests for density determination to DIN 4021, as a rule using equivalent methods as well as plate bearing tests to DIN 18 134 or radiometric penetration sounding devices. At greater depths, the degree of density can be determined through static or dynamic penetrometer tests to DIN 4094 parts 1 and 3, or with a radiometric depth sounder.

The cone penetration test (CPT) is ideal for checking the degree of density of hydraulically filled sands. However, the heavy dynamic penetration test (DPH) is also suitable when, for instance, surfaces are inaccessible for the cone penetration test. The light dynamic penetration test (DPL) is used for exploration depths of just a few metres. The values given in table R 175-2 are empirical values for the correlation between the respective test findings in fine and medium sands and the degree of density, and only apply from about 1.0 m below the application point of the test.

Table R 175-2. Correlation between degree of density D, toe resistance q_c of cone penetration test and dynamic penetration resistances for number of blows N_{10} in hydraulically filled sands (empirical values for non-uniform fine sand and uniform medium sand).

Type of utilisation		Storage areas	Traffic areas	Structure areas
Degree of density D	fine sand	0.35–0.45	0.45–0.55	0.55–0.75
	medium sand	0.20–0.35	0.25–0.45	0.45–0.65
Cone penetration test q_c MN/m^2	fine sand	2–5	5–10	10–15
	medium sand	3–6	6–10	> 15
Heavy dynamic penetration test DPH 15, N_{10}	fine sand	2–5	5–10	10–15
	medium sand	3–6	6–15	> 15
Light dynamic penetration test DPL 10, N_{10}	fine sand	6–15	15–30	30–45
	medium sand	9–18	18–45	> 45
Light dynamic penetration test DPL 5, N_{10}	fine sand	4–10	10–20	20–30
	medium sand	6–12	12–30	> 30

1.7　Degree of density of dumped, non-cohesive soils (R 178)

1.7.1　General

This recommendation is essentially supplementary to recommendations R 81, section 7.3, R 73, section 7.4, and R 175, section 1.6.

The dumping of non-cohesive soils generally leads to a more or less pronounced segregation of the material. Dumped, non-cohesive soils can be loosened by failures in the foundation, slope or embankment if the embankments formed during dumping are too steep. Embankments with a slope of 1 : 5 or less are stable.

Dumped, non-cohesive soils can be further compacted by tide and wave effects too.

1.7.2　Influences on the achievable degree of density

The degree of density of dumped, non-cohesive soils depends primarily on the following factors:

a) Granulometric composition and silt content of the dumped material. Generally, a non-uniform granular structure produces a higher degree of density than a uniform one. The silt content should not exceed 10%.

b) Water depth
   The greater the depth of water, the greater the segregation, especially in non-cohesive soils with a uniformity coefficient $U > 5$, which leads to an alteration in the grain distribution.

c) Flow velocity in the dumping area
   The greater the flow velocity in the dumping area, the greater the segregation and the more irregular the settlement of the soil. Refer to R 109, section 7.9, for information on dumping in silty flowing water.

d) Method of dumping
   A somewhat higher degree of density is achieved with vessels with slotted flaps than with split-hopper barges.

Owing to the fact that these influences cancel each other out to a certain extent, the degree of density of dumped, non-cohesive soils can vary considerably. The geostatic overburden has only a minimal influence on the degree of density of dumped, non-cohesive soils. Even with increasing overburden heights, there is generally scarcely any change to the degree of density. For small overburden heights, the density will generally be only low, provided the soil is not compacted by some other means.

1.8 Assessment of the subsoil for the installation of sheet piles and piles and methods of installation (R 154)

1.8.1 General

Construction materials, shape, size, length and installation batter of piles and sheet piles all play a decisive role when choosing a method for installation sheet piles and piles. Essential references may be found in:

- R 16, section 9.5,   Design and installation of driven steel piles
- R 21, section 8.1.2,   Design and driving of reinforced concrete sheet piling
- R 22, section 8.1.1,   Design and driving of timber sheeting
- R 34, section 8.1.3,   Steel sheet piling
- R 104, section 8.1.12,   Driving of combined steel sheet piling
- R 105, section 8.1.13,   Observations during the installation of steel sheet piles, tolerances
- R 118, section 8.1.11,   Driving corrugated steel sheet piles

In view of the great significance, these recommendations draw particular attention to the need for the selection of the element (material, cross-section, etc.) to take account not only of structural requirements and economic aspects but, above all, also the stresses and strains involved in installing in the respective subsoil. Therefore, the subsoil investigation report must always include an assessment of the in situ subsoil with respect to the installation of piles and sheet piles (see R 150, section 1.3).

1.8.2 Report and expert opinion on the subsoil

The nature and scope of the soil exploration must be specified in such a way that the findings enable the subsoil to be assessed in terms of installing piles and sheet piles (see R 1, section 1.2). In order to ensure an assessment of the ground with respect to driving and vibrating, the boreholes and penetrometer tests must be positioned at those places where driving and vibrating will take place later. Where this is not possible, then a direct statement regarding installation and, above all, any hindrances that may be encountered, is not possible.

The field and laboratory tests provide information on the following aspects:

- stratification of the subsoil,
- grain shape,
- existing inclusions, e.g. rocks > 63 mm, blocks, old backfill, tree trunks or other obstacles and their depth,
- shear parameters,
- porosity and void ratio,
- weight density above water and submerged weight density,

- degree of density,
- compactability of the soil upon installing the elements,
- cementation of non-cohesive soils, incrustation,
- preloading and swelling characteristics of cohesive soils,
- groundwater level upon installation,
- artesian groundwater in certain strata,
- water permeability of the soil,
- degree of saturation in cohesive soils, especially in silts,

The corresponding findings are to be recorded in the subsoil investigation report and assessed, especially in terms of the installation of piles and sheet piles.

The shear parameters have only limited relevance on the behaviour of the subsoil during the installation of piles and sheet piles. For example, a rocky lime marl can posses comparatively low shear parameters due to its open seams, but must be viewed as difficult driving soil.

Increasingly more difficult installation must be expected when the number of blows N_{10} exceeds 30 for 10 cm penetration with the heavy dynamic penetrometer (DPH, DIN 4094-3), or N_{30} exceeds 50 for 30 cm with borehole dynamic probing (BDP, DIN 4094-2). In general it can be presumed that driven elements can be embedded under difficult driving when the number of blows N_{10} is 80-100 per 10 cm penetration. Deeper driving is possible in individual cases. For more detailed data see [5] and [6]. In non-homogeneous soils the penetrometer results can exhibit a wide scatter and lead to erroneous conclusions.

Static penetration tests in uniform fine and medium sands at a depth of at least 5 m furnish good information on the strength and degree of density in non-cohesive soils (for further data see DIN 4094-1). Conclusions regarding the difficulty of driving can be derived from this too.

1.8.3 Evaluation of types of soil as regards method of installation

1.8.3.1 Driving

Easy driving may be expected in soft or very soft soils, such as bog, peat, silt, clay containing sea-silt, etc. Furthermore, easy driving may generally also be anticipated in loosely deposited medium and coarse sands, as well as in gravels without rock inclusions, unless cemented layers are interspersed.

Medium-difficult driving is experienced in medium densely deposited medium and coarse sands, fine gravels, stiff clay and loam.

Difficult to most difficult driving may be expected in most cases of densely deposited medium and coarse gravels, densely deposited fine-sandy and silty soils, interposed cemented strata, semi-firm and firm clays, rubble and moraine layers, boulder clay, eroded and soft to medium-hard rock. Damp or dry soils exhibit higher penetration resistance during driving than those subject to uplift. The same applies to partly saturated cohesive soils, primarily silts.

1.8.3.2   Vibration

During vibration the skin friction and the base resistance of the section being installed are substantially reduced. Therefore, in comparison with driving, vibration enables the elements to be quickly installed to the required depth. For further details see R 202, section 8.1.22.

Gravels and sands with a rounded grain shape, as well as very soft soil types with low plasticity, are ideal for vibration. Gravels and sands with an angular grain shape or strongly cohesive soils are much less suitable. Dry fine sands and stiff loams and clays are especially critical because they absorb the energy of the vibrator to a large extent.

If the soil is compacted during vibration, the penetration resistance increases so sharply that the sections may not be able to be installed to the required depth. The risk of this is particularly high with closely spaced sheet piles or piles and when vibrating in non-cohesive soils. In such cases, the vibration work should be stopped (see R 202) unless auxiliary means as described in section 1.8.3.4 are employed.

In non-cohesive soils in particular, the installation of elements by vibration can cause local settlements, the magnitude and lateral influence of which depends on the output of the vibrator, the elements, the duration of vibration and the soil. When working in the immediate vicinity of existing structures, an analysis must be carried out to establish whether such settlements could cause damage. If necessary, the method of vibration must be adapted or another method of embedment should be used.

1.8.3.3   Jacking

For jacking to be successful there must be no obstacles in the soil, or these must be removed before jacking starts.

Jacking (generally hydraulic) is suitable for installing slim sections in cohesive soils free from obstacles, or in loose, non-cohesive soils. In dense, non-cohesive soils, sections can only be jacked in when the soil is loosened beforehand.

1.8.3.4   Installation aids

In dense sands and gravels in particular, also hard and stiff clays, jetting can be employed to decrease the amount of energy required, or indeed enable installation in otherwise impossible conditions.

Further helpful measures can consist of loosening boreholes or local soil replacement with advancing large-diameter boreholes and the like. In rocky soils, targeted blasting can make the soil so "driveable" that the intended depth can be achieved with an appropriate choice of profile. For further details see R 183, section 8.1.10.

1.8.4   Installation equipment, installation elements, installation methods

Installation equipment, installation elements and installation methods should be chosen to suit the respective subsoils, see R 104, section 8.1.12, R 118, section 8.1.11, R 202, section 8.1.22, and R 210, section 7.14. Slow-stroke free-drop, diesel or hydraulic hammers are suitable for cohesive and non-cohesive soils. Rapid-stroke and vibration hammers transfer more moderate stresses to the driven elements, but are generally only particularly effective in non-cohesive soils with rounded grains. When driving in rocky soil, even after previous loosening blasting, rapid-stroke hammers or heavy pile hammers with small drop heights are to be preferred.

Interruptions during the installation, e.g. between pre-driving and re-driving, can facilitate or aggravate subsequent driving, depending on type of soil, water saturation and duration of interruption. As a rule, preliminary tests can make the respective tendency apparent.

The evaluation of the subsoil for the embedment of piles and sheet piles presupposes special knowledge and relevant experience in this field. Information on construction sites with comparable conditions, especially regarding the subsoil, can prove very useful.

1.8.5   Testing the installation method and the bearing capacity under difficult conditions

In large-scale projects, if there are any doubts about achieving the structurally required depth for sheet piles without damaging the piles, or doubts about the adequacy of the rated pile length for accommodating the working loads, then the intended installation methods should be tested and test loadings carried out beforehand. At least two tests should be carried out per method in order to furnish meaningful information.

Such testing can also be necessary to forecast settlement of the soil and the spread and extent of vibrations caused by the method of installation.

2 Active and passive earth pressures

2.0 General

Verification of the stability of waterfront structures according to these recommendations is usually provided according to DIN 1054. Total active earth pressures are calculated according to DIN 4085. Verification of slope failure safety is carried out according to DIN 4084, ground failure to DIN 4017, unless simplifications are stated in these recommendations. The verification of stability must indicate that failure does not occur for limit state 1B (LS 1B) nor limit state 1C (LS 1C) to DIN 1054.

In the case of retaining structures whose stability depends on the passive earth pressure in front of the structure, a lowering of the ground surface must be taken into account amounting to 10% of the wall height on the passive side or 10% of the height underneath the lowest support in the case of supported walls, but 0.5 m at the most. This does not apply when the design depth of the harbour bottom is stipulated according to R 36, section 6.7.

The procedures and principles for determining active and passive earth pressure stated in the following sections apply generally, i.e. for determining the earth pressures with both the characteristic and design values of the shear parameters. The shear parameters are therefore given in diagrams and equations without any index, unless the characteristic value or design value (index k and index d respectively) is referred to explicitly. Section 8.2 covers the design of sheet piling.

2.1 Assumed apparent cohesion (capillary cohesion) in sand (R 2)

Cohesion in cohesive soils may be considered in active and passive earth pressure calculations when the following conditions are fulfilled:

- The soil must be undisturbed; in backfill with cohesive material the soil must be compacted without any voids.
- The soil should be permanently protected against drying out and freezing.
- The soil should not become pulpy when kneaded.

If these requirements cannot be met or are fulfilled in part only, cohesion should be considered only if justified on the basis of special investigations.

2.2 Assumed apparent cohesion (capillary cohesion) in sand (R 3)

Apparent cohesion c_c (capillary cohesion to DIN 18 137 part 1) in sand is caused by the surface tension of the pore water. It is lost when the soil is completely wet or dry. As a rule, it is not included in the active or passive earth pressure calculations, but remains as a reserve factor for stability. The apparent cohesion may be taken into account for temporary conditions if it can be guaranteed that it remains effective throughout the period concerned. Table R 3 contains characteristic values for the apparent cohesion.

Table R 3. Characteristic values for apparent cohesion

Type of soil	Designation to DIN 4022-1	Apparent cohesion $c_{c,k}$ [kN/m^2]
Gravelly sand	G, s	0–2
Coarse sand	g S	2–4
Medium sand	m S	4–6
Fine sand	f S	6–8

2.3 Assumed angle of earth pressure and adhesion (R 4)

See section 8.2.4

2.4 Determination of the active earth pressure using the CULMANN method (R 171)

(See DIN 4085, section 6.3.1.7, figs. 10–12)

2.4.1 Solution for uniform soil without cohesion (fig. R 171-1)
In the CULMANN method the COULOMB vector polygon (fig. R 171-1) is turned through an angle of $90° - \varphi'$ to the perpendicular, whereby the dead load G is applied in the line of the slope. If a line parallel to the "active earth pressure line" is drawn at the start of the dead load G, it intersects the relevant slip plane at a point on the CULMANN active earth pressure line (fig. R 171-1).

The distance of this point of intersection from the slope line measured in the direction of the active earth pressure line is the respective active earth pressure for the sliding wedge investigated at the angle of earth pressure δ_a selected. This is now repeated for various failure planes. The maximum of CULMANN's active earth pressure line represents the relevant earth pressure required.

In uniform soil the CULMANN method can be used for any ground surface form and the surcharges prevalent there. Any groundwater table present in any case is also allowed for through the corresponding application of the sliding wedge loads with γ or γ'. The same also applies to any other changes in weight density, provided φ' and δ_a remain the same.

The earth pressure loads on a wall are then determined section by section, starting at the top, and are preferably plotted as loads per unit area over

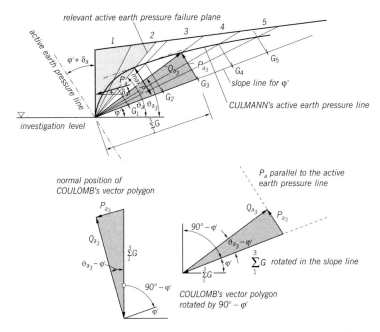

Fig. R 171-1. System sketch for determining the active earth pressure according to Culmann in uniform soil without cohesion

the height of the section. Section 8.2 contains further information on earth pressure distribution.

2.4.2 **Solution for uniform soil with cohesion (fig. R 171-2)**

In the case with cohesion, besides the soil reaction force Q, the cohesion force $C' = c'_k \cdot l$ also acts in the failure plane with length l. In the Coulomb polygon of forces C' is applied before the dead load G. In the Culmann method C' rotated through an angle of $90° - \varphi'$ is also placed on the slope line of the dead load G. The line parallel to the active earth pressure line is drawn through the initial point of C' and continued to intersect with the associated slip plane, through which the new associated point of Culmann's earth pressure line is found. After investigating several failure planes, the relevant active earth pressure is the maximum distance of Culmann's earth pressure line from the line joining the initial points of C', measured in the direction of the active earth pressure line (fig. R 171-2).

The applicability of straight failure lines must be verified for greater cohesion, particularly with sloping ground. In such cases, curved or

33

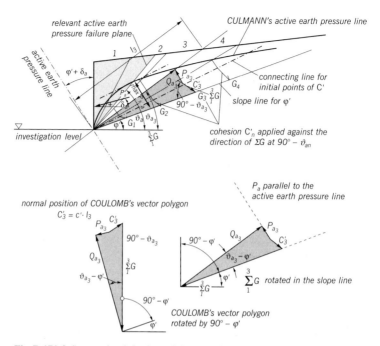

Fig. R 171-2. System sketch for determining the active earth pressure according to Culmann in uniform soil with cohesion

cranked failure lines frequently cause greater earth pressure loads. See R 198, section 2.5.

The active earth pressures calculated are again converted into loads per unit area.

2.4.3   Solution with stratified soil (fig. R 171-3)

In stratified soil the active earth pressure is generally determined for a continuous straight failure line. Exceptions are the top strata with high cohesion, where curved or cranked failure lines are critical (see R 198). For irregular surfaces, groundwater tables falling away towards the quay wall or when additional loads are to be included in calculation of the active earth pressures, the earth pressure may be calculated according to one of the methods stated in DIN 4085.

Fig. R 171-3a shows an example for determining the resultant of the active earth pressure on a wall with three layers and a straight failure line. In this example the internal earth pressure forces are applied horizontally at the lamella boundaries (see fig. R 171-3b).

34

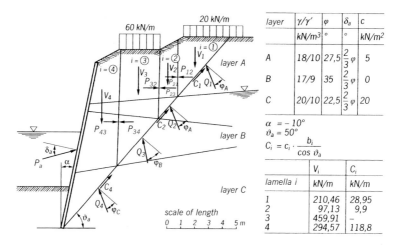

layer	γ/γ'	φ	δ_a	c
	kN/m³	°	°	kN/m²
A	18/10	27,5	$\frac{2}{3}\varphi$	5
B	17/9	35	$\frac{2}{3}\varphi$	0
C	20/10	22,5	$\frac{2}{3}\varphi$	20

$\alpha = -10°$
$\vartheta_a = 50°$
$C_i = c_i \cdot \dfrac{b_i}{\cos \vartheta_a}$

lamella i	V_i	C_i
	kN/m	kN/m
1	210,46	28,95
2	97,13	9,9
3	459,91	–
4	294,57	118,8

Fig. R 171-3a. Example for determining active earth pressures in stratified soil using the lamella method. Geometry and projection of the forces

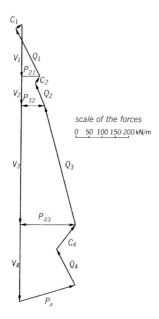

Fig. R 171-3b. Vector polygon of the graphical determination of the earth pressure force divided into lamella

Fig. R 171-3. Determining active earth pressure for stratified soil

35

The relevant failure line combination is that for which the earth pressure load P_a is greatest (not examined in fig. R 171-3). A weighted earth pressure angle or weighted adhesion is to be used for the gradient of the resulting earth pressures at the retaining wall. It is easiest to obtain the weighting from a layer-by-layer calculation of the resulting earth pressures. See section 8.2 for details of earth pressure distribution. The analytical solution for straight failure lines to fig. R 171-3 is:

$$P_a = \left[\sum_{i=1}^{n} \left(V_i \frac{\sin(\vartheta_a - \varphi_i)}{\cos \varphi_i} - \frac{c_i \cdot b_i}{\cos \vartheta_a} \right) \right] \cdot \frac{\cos \overline{\varphi}}{\cos(\vartheta_a - \overline{\varphi} - \overline{\delta}_a + \alpha)}$$

where

i = consecutive number of lamella
n = number of lamellas
V_i = weight under uplift including surcharges of the lamellas
ϑ_a = inclination of the failure line to the horizontal
φ_i = friction angle in the line of slope in the lamellas i
c_i = cohesion in the lamellas i
b_i = width of the lamellas i
α = inclination of the waterfront wall, defined according to DIN 4085

$\overline{\varphi}$ = mean value of friction angle along failure line:

$$\overline{\varphi} = \arctan \frac{\displaystyle\sum_{i=1}^{n} V_i \cos \vartheta_a \cdot \tan \varphi_i}{\displaystyle\sum_{i=1}^{n} V_i \cos \vartheta_a}$$

$\overline{\delta}_a$ = mean value of earth pressure angle over wall height. $\overline{\delta}_a$ may be approximated with $\overline{\delta}_a = 2/3\,\overline{\varphi}$ for horizontal strata and comparatively low surcharges. For more precise examinations, the mean must be formed using the earth pressure calculated layer by layer

2.5 Determination of active earth pressure in a steep, paved embankment of a partially sloping bank construction (R 198)

A steep embankment case exists if the inclination of the slope β is greater than the effective angle of friction φ' of the in situ soil. The stability of the embankment is only guaranteed if the soil has a permanently effective cohesion c' and is permanently protected from surface erosion, e.g. by means of dense turf or a revetment. Verification of safety against

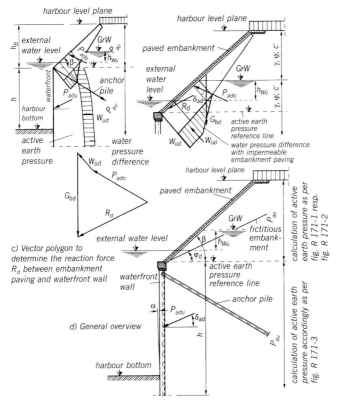

a) System sketch

b) Effects of force on a paved, steep embankment fortification

c) Vector polygon to determine the reaction force R_d between embankment paving and waterfront wall

d) General overview

Fig. R 198-1. Partially sloping bank with a steep, paved embankment

embankment failure can then be made in accordance with DIN 4084, for example.

If the cohesion is inadequate to guarantee the relevant limit states of stability (1B or 1C) for the embankment, the embankment requires strengthening, e.g. in the form of paving, which must be able to support downslope forces and be structurally connected to the waterfront wall. The embankment strengthening must be designed in such a way that the resultant force of the applied actions always lies within the core of the strengthening cross-sections. The active earth pressure for the embankment area down to the top of the supporting beam for the embankment (active earth pressure reference line in fig. R 198-1) can be calculated for non-preponderant cohesion thus:

37

$$\frac{c'}{\gamma \cdot h} < 0.1$$

according to R 171, section 2.4, whereby the dead load of the embankment strengthening is not taken into consideration.

In doing so, any water pressure difference must be allowed for in addition to the active earth pressure P_a. Fig. R 198-1a illustrates this for impermeable paving. The water pressure difference is somewhat lower in the case of permeable paving. The loading assumptions for embankment strengthening are shown in fig. R 198-1b. The reaction force R_d between the embankment strengthening and the waterfront wall is given by the vector polygon in fig. R 198-1c.

The resultant force R_d must be taken into account fully in the calculations for the waterfront wall and its anchoring. The active earth pressure P_{au} can be determined similarly in accordance with fig. R 171-3 from the active pressure reference line (imagined stratum boundary) downwards in the case:

$$\frac{c'}{\gamma \cdot h} < 0.1$$

In doing so, it should be noted that the design value of the active earth pressure force P_{ao} and the dead load of the embankment strengthening are already included in the reaction force R and are carried directly by the waterfront wall and its anchoring. Further calculation proceeds according to R 171, section 2.4. By way of approximation, the active earth pressure P_{au} below the active earth pressure reference line of fig. R 198-1 can also be determined with a wall projecting above the active earth pressure reference line by the fictitious height:

$$\Delta h = \frac{1}{2} \cdot h_B \cdot \left(1 - \frac{\tan \varphi'}{\tan \beta} \right)$$

with an embankment simultaneously inclined at the fictitious angle φ' (fig. R 198-2).

In the case of preponderant cohesion

$$\frac{c'}{\gamma \cdot h} \geq 0.1$$

calculation with straight failure planes in accordance with fig. R 171-2 or R 171-3 results in an inadequate active earth pressure P_a. In this case you are recommended to determine the active earth pressure both for the sections above and below the active earth pressure reference line with curved or cranked failure lines.

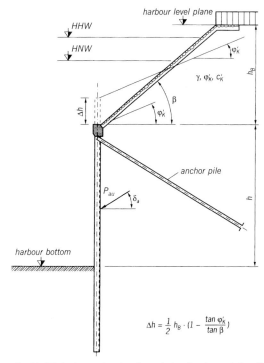

$$\Delta h = \frac{1}{2} h_B \cdot (1 - \frac{\tan \varphi'_k}{\tan \beta})$$

Fig. R 198-2. Approximation formulation for determining P_{au}

2.6 Determination of active earth pressure in saturated, non- or partially consolidated, soft cohesive soils (R 130)

2.6.1 Calculation with total stresses

This method is for use with saturated soils only. In this case an additional surcharge is initially carried by the pore water only, the effective stress σ' remains unchanged. Only when the subsequent consolidation starts is the surcharge also transferred to the grain structure of the soil.

The shear strength of the soil at the time of loading is the shear strength of the undrained soil c_u (DIN 18 137 part 1).

For a vertical wall ($\alpha = 0$) and horizontal ground ($\beta = 0$) with an earth pressure angle $\delta_a = 0$ (because $\varphi_u = 0$), the horizontal component of the active earth pressure stress is:

$$e_{ah} = \sigma \cdot K_{agh} - c_u \cdot K_{ach}$$

Here, σ is the total stress, i.e. the sum of geostatic surcharge plus additional load.

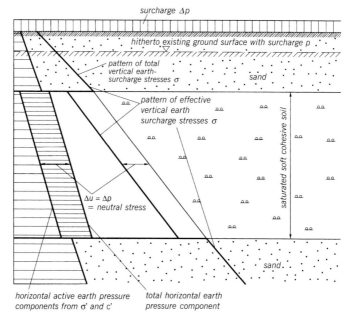

surcharge Δp

hitherto existing ground surface with surcharge p

pattern of total vertical earth-surcharge stresses σ

sand

pattern of effective vertical earth surcharge stresses σ

saturated soft cohesive soil

$\Delta u = \Delta p$ = neutral stress

sand

horizontal active earth pressure components from σ' and c'

total horizontal earth pressure component

Fig. R 130-1. Example for determining the horizontal component of the active earth pressure distribution for the initial state with shear parameters of the drained soil

As $\varphi_u = 0$ for saturated soil, the coefficient of active earth pressure $K_{agh} = 1$ and thus:

$$e_{ah} = \sigma - 2c_u$$

This simple approach is, however, for saturated soils only. The shear strength c_u of the undrained soil must be ascertained from soil samples of quality grade 1 to DIN 4021 (special samples) or in field tests according to R 88, section 1.4.

Redistribution of earth pressure according to 8.2.2.3 is only allowed for loads, for which the soil is fully consolidated.

2.6.2   Calculations with effective shear parameters

If the shear strength c_u of the undrained soil is not known or cannot be ascertained, the active earth pressure due to an additional load must be calculated with the help of the effective shear parameters.

The effective stress σ', the effective shear parameters φ' and c', and the size of the additional surcharge Δp are all critical factors influencing the magnitude of the active earth pressure. It is assumed in the initial state that the full additional surcharge is superimposed on the active earth pressure due to the effective stress:

$$e_{ah} = \sigma' \cdot K_{agh} - c' \cdot K_{ach} + \Delta p$$

(with K_{agh} and K_{ach} to DIN 4085)

The active earth pressure distribution is depicted in the example shown in fig. R 130-2 for the case of a uniformly distributed surcharge Δp extending indefinitely towards the land side, i.e. a plane state of stress and deformation exists. In the soft cohesive layer, every active earth pressure coordinate is produced as a result of the respective effective stress σ' and the cohesion c', increased by the full surcharge Δp.

After complete consolidation, the full surcharge Δp is added to the effective stress. Intermediate states can be taken into account by determining the degree of consolidation of that component of Δp where the soil has already been consolidated. This is then used in the above equation for the effective stress, the component not yet consolidated is superimposed in full on the active earth pressure.

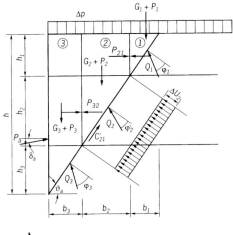

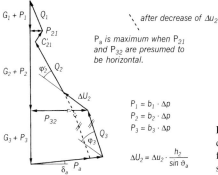

after decrease of Δu_2

P_a is maximum when P_{21} and P_{32} are presumed to be horizontal.

$P_1 = b_1 \cdot \Delta p$
$P_2 = b_2 \cdot \Delta p$
$P_3 = b_3 \cdot \Delta p$

$\Delta U_2 = \Delta u_2 \cdot \dfrac{h_2}{\sin \vartheta_a}$

Fig. R 130-2. Example for graphical determination of active earth pressures for a non-consolidated layer under surcharge

41

If the structure lies partially in the groundwater, the pattern of the effective vertical stresses is determined taking account of the submerged weight density. Differing levels between outer water and groundwater create a pressure difference which must also be taken into account.

In difficult cases, e.g. sloping strata, irregular ground surface, non-uniform surcharges, etc., it is possible to use a graphical method with variations of the slip plane inclination ϑ_a. In doing so, it should be noted that the surcharges for which the soil in the failure plane section has not been consolidated create a pore water pressure normal to the failure plane (fig. R 130-2, e.g. layer 2).

If the soil is not saturated, an additional surcharge leads to an increase in the effective stress from the start, but this can only be quantified on the basis of measurements. The unconsolidated active earth pressure represents an upper limit value, the consolidated (for the full surcharge) active earth pressure a lower limit value for the actual magnitude of the active earth pressure. Redistribution of earth pressure according to 8.2.2.3 is only allowed for loads, for which the soil is fully consolidated.

2.7 Effect of artesian water pressure under harbour bottom or river bed on active and passive earth pressure (R 52)

Artesian water pressure occurs where the harbour bottom or river bed consists of a cohesive layer of low permeability lying on a groundwater-bearing, non-cohesive layer, and at the same time the free low water level lies below the concurrent hydraulic head of the groundwater. The effects of this artesian water pressure on the active and passive earth pressure must be taken into account in the design.

The artesian water pressure acts on the surface layer from below and thereby reduces its effective dead load. As a result, the passive earth pressure decreases in the surface layer and in the non-cohesive soil below. Given a level difference Δh between the hydraulic head of the ground-water and the free water level, and the weight density of water γ_w, this results in an artesian water pressure of $\Delta h \cdot \gamma_w$.

2.7.1 Influence on the passive earth pressure

If artesian pressure acts under a surface layer of thickness d_s and submerged weight density γ', the passive earth pressure is calculated as follows:

2.7.1.1 Case with preponderant dead load on the surface layer
$(\gamma' \cdot d_s > \gamma_w \cdot \Delta h)$ (**fig. R 52-1**)

(1) Assuming a linear decrease in the artesian water pressure through the thickness of the surface layer, the passive earth pressure in the surface layer due to soil friction is calculated with the reduced weight density thus: $\gamma_v = \gamma' - \gamma_w \cdot \Delta h / d_s$

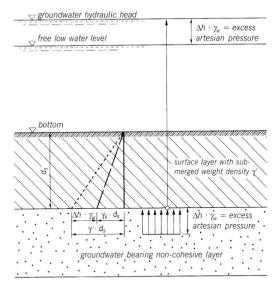

Fig. R 52-1. Artesian pressure in groundwater with preponderant weight load of surface layer

(2) The passive earth pressure in the surface layer as a result of cohesion is not reduced by the artesian water pressure.

(3) The passive earth pressure in the soil below the surface layer is calculated for the surcharge $\gamma_v \cdot d_s$.

2.7.1.2 Case with preponderant artesian water pressure $(\gamma' \cdot d_s < \gamma_w \cdot \Delta h)$
This case can occur in tidal areas, for example. At low water the surface layer becomes detached from the non-cohesive subsoil due to the effect of the artesian water pressure acting from below, and slowly begins to enter a state of suspension as the groundwater influx takes effect. During the subsequent high water it is again pressed down onto its bed. This process is generally not dangerous with thick surface layers. However, if the surface layer is weakened by dredging or the like, heave-like eruptions of the surface layer can occur, leading to local disturbances in the vicinity of the eruption, but also easing of the artesian pressure.
Similar conditions can also occur in excavations involving sheet piling. The following design principles are then applicable:

(1) Passive earth pressure due to soil friction may not be taken into account in the surface layer.

(2) Passive earth pressure in the surface layer due to cohesion c' of the drained soil may only be used if bottom eruption cannot occur (e.g. as a result of structural countermeasures, such as surcharges).

43

(3) Passive earth pressure of the soil below the surface layer is to be calculated for an unloaded free surface at the underside of the surface layer. The decrease in the passive earth pressure as a result of the flow pressure should be considered when vertical flow is possible.

In this context, the reader is referred to R 114, section 2.9.

The above approaches for calculating the passive earth pressure also apply to sheet pile calculations and investigations into ground and foundation failures.

2.7.2 Influence on active earth pressure

Artesian water pressure also reduces the active earth pressure. But the influence is generally so marginal that it can be neglected, particularly in view of the fact that applying the active earth pressure without taking into account this influence it certainly on the safe side.

2.7.3 Influence on water pressure difference

The water pressure difference can be taken as zero in permeable subsoil under the surface layer. In the cohesive surface layer there is a linear transition from the artesian to the groundwater pressure level. Waterfront walls that end in the surface layer require special investigations to determine the water pressure difference at the foot of the wall appropriately. This can include determining a flow and potential line network according to R 113, section 4.7.

2.8 Use of active earth pressure and water pressure difference, and construction advice for waterfront structures with soil replacement and fouled or disturbed dredge pit bottom (R 110)

2.8.1 General

When waterfront structures are constructed with soil replacement as per R 109, section 7.9, the effects of a fouled dredge pit bottom and the inshore dredge pit slope in the existing soft soil on the active earth pressure and water pressure difference must be carefully analysed and taken into account in the design, calculation and dimensioning of the waterfront structure. In the interests of economic efficiency, uncontrolled intermediate states should be prevented from becoming effective in the design.

2.8.2 Calculation for determining active earth pressure

Besides the customary calculation of the structure for the improved soil conditions and the ground failure investigations as per DIN 4084, the boundary and disturbing influences arising from a failure plane created by the dredging as per fig. R 110-1 must also be taken into account.

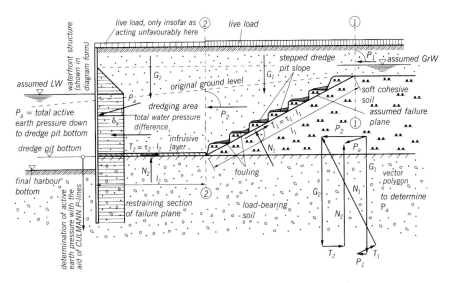

Fig. R 110-1. Determination of active earth pressure P_a on the waterfront structure

In this context, the waterfront structure is to be designed using the partial safety factors for limit state 1B to DIN 1054. Verification of ground failure is to be based on limit state 1C with the partial safety factors for influence and resistances contained in DIN 1054.

For the active earth pressure P_a acting on the structure down to the dredge pit bottom, the following are the most important factors:

(1) Length, and insofar as any exists, inclination of the restraining section l_2 of that part of the dredge pit bottom that is a possible failure plane.

(2) Thickness, shear strength τ_2 and effective soil surcharge load of the intrusive layer on l_2.

(3) The possibility of dowelling section l_2 by means of piles and the like.

(4) Thickness of the adjacent soft cohesive soil along the inshore edge of the dredge pit, its shear strength, together with nature and inclination of the dredge pit slope.

(5) Sand surcharge and live load, especially on the dredge pit slope.

(6) Characteristics of the fill soil.

The distribution of the active earth pressure P_a down to the dredge pit bottom depends on the deformations and the type of waterfront structure. The earth pressure below the dredge pit bottom can be determined, for example, with the aid of CULMANN lines. In doing so, the shear forces in section l_2, including any dowelling, are to be taken into account.

45

At any stage of construction, including the original dredging of the harbour bottom, or any subsequent deepening of the harbour, the shear stress τ_2 existing at the time in section l_2 of the intrusive layer can be determined for the material in the layer using the equation:

$$\tau = (\sigma - u) \cdot \tan \varphi' \approx \sigma' \cdot \tan \varphi'$$

where σ' is the effective vertical surcharge stress effective at the place and time of the investigation, and φ' is the effective angle of friction of the material of the intrusive layer. The final shear strength after consolidation is then

$$\tau = \sigma'_a \cdot \tan \varphi'$$

where σ'_a represents the effective surcharge stress of the investigation area of section l_2 at full consolidation ($u = 0$).

Separate calculations are necessary to consider the effects of dowelling section l_2 to the loadbearing soil by means of piles [11].

If dredging has been properly done in larger steps, the failure plane passes through the rear corners of the steps and thus lies in undisturbed soil (fig. R 110-1). Due to the long consolidation period in soft cohesive soils, the shear strength used for the failure plane is equal to the initial shear strength of the soil before excavation. If the soft cohesive soil proves to have layers with different initial shear strengths, these must be taken into account.

Should the dredge pit slope in soft soil be severely disturbed, excavated in small steps or unusually fouled, the shear strength of the disturbed slip layer must be used in calculations instead of the initial shear strength of the natural soil. This is usually less than the initial shear strength and should therefore be determined in supplementary laboratory tests.

On account of the very slow consolidation of the soft cohesive soil underneath the dredge pit slope, an increase in shear strength in this area can be taken into account in the analyses only if the soil is drained with closely spaced drains.

2.8.3 **Calculation for determining the water pressure difference**

The total difference in level between the groundwater table at reference line 1-1 (fig. R 110-1) and the lowest attendant outer water level is to be taken into account. Permanently effective backwater drainage behind waterfront structures can lower the groundwater level in the catchment area, and thus reduce the difference in levels.

The total water pressure difference can be applied in the usual approximated trapezoidal form (fig. R 110-1). However, it can be determined more accurately using a flow net because the actual pore pressures derived from the flow net are available for use in the investigation planes (R 113, section 4.7, and R 114, section 2.9).

2.8.4 Advice for the design of waterfront structures

Investigations of dredge pit bottoms have shown that intrusive layers of up to around 20 cm thick deposited on section l_2 (fig. R 110-1) of the failure plane, even if only partly drained, will have become sufficiently consolidated to support a surcharge, during the construction period, including dredging the harbour bottom. If the intrusive layer is thicker, full consolidation cannot generally be expected in this period. In such cases the shear strength of the intrusive layer must be calculated from an estimation of the degree of consolidation. However, in order that partially consolidated intermediate states do not become effective for the design, it may be necessary to plan certain construction measures, e.g. initial dredging or later deepening of the harbour, so that consolidation is completed before the loads associated with these measures become effective.

Anchoring forces are best transmitted fully into the loadbearing soil through the dredge pit bottom and into the subsoil below by means of piles or other loadbearing members. Anchoring forces introduced above the dredge pit bottom place additional loads on the sliding wedge.

Apart from its structural functions, section l_2 (fig. R 110-1) is to be designed with a length to accommodate all structural piles so that the bending stresses in the piles due to settlement resulting from the backfill are kept to a minimum.

When silt deposits are so heavy that thick intrusive layers of soft cohesive material and/or very loose layers of sand cannot be avoided despite careful soil replacement as per R 109, section 7.9, thus possibly causing severe deflection of the piles accompanied by stresses up to the yield point, brittle fracture must be avoided by using only piles of double killed steel (R 67, section 8.1.6.1 and R 99, section 8.1.18.2).

If foundation piles have been driven so as to produce dowelling of the failure plane [11] in section l_2 (fig. R 110-1) and verify the stability of the overall system as per DIN 4084, the maximum working stress in the stress verification for these piles due to axial loads, shear and bending must in no case exceed 85% of the yield point. The dowelling calculation only allows for such pile deflections that are compatible with the other movements of the structure, i.e. only a few centimetres. Effective dowelling is therefore impossible in yielding, soft cohesive soil (fig. R 110-1). Piles with anticipated possible stresses up to the yield point due to settlements of the subsoil or fill soil may not be used for dowelling.

In order to prevent intrusive layers in section l_2 (fig. R 110-1), the failure plane and the dredge pit slope resulting in an unnecessarily large structure, it is essential for the dredge pit bottom to be as clean as possible, section l_2 of the failure plane must be adequately long and/or the angle of the dredge pit slope must be correspondingly shallow (see effects in the vector polygon in fig. R 110-1). If only a relatively thin intrusive layer

47

is expected, the shearing strength of section l_2 can be substantially improved if it is cleaned and then covered with a layer of rubble. If sufficient time is available, closely spaced drains in the soft cohesive soil up to the top of the dredge pit slope can also relieve the load on the structure. A temporary reduction in the live load on the backfill over the dredge pit slope, and/or a temporary lowering of the groundwater table to a point behind reference plane 1-1 can also be used to overcome unfavourable initial conditions.

If the restraining section l_2 (fig. R 110-1) is to be omitted in areas with silty soils without imposing additional loads on the structure, the dredge pit slope may only be about 1 : 4, provided that the soil replacement has been efficiently and carefully carried out in all respects. The additional loads should be verified by calculation independently of this recommendation.

2.9   Effect of percolating groundwater on water pressure difference, active and passive earth pressures (R 114)

2.9.1   General

If water percolates around a structure, the flowing groundwater exerts a flow pressure on the soil masses of the sliding wedges for the active and passive earth pressures, thus changing the magnitude of these forces. The total effects of the groundwater flow on P_a and P_p can be determined with a flow net as described in R 113, section 4.7.7 (fig. R 113-2). All other water pressures acting on the boundary surfaces of the sliding sections of earth are determined, and are taken into account in the COLOUMB vector polygon for the active earth pressure (fig. R 114-1a) and the passive earth pressure (fig. R 114-1b). These figures show the forces that are to be included for the general case of a straight failure line. G_a and G_p are the dead loads of the sliding wedges for the saturated soil. W_1 is the corresponding free water surcharge on the sliding wedges, W_2 the resultant water pressure acting directly on the structure in the sliding wedge area, W_3 the resultant water pressure acting in the failure plane, determined according to the flow net (R 113, section 4.7.7, fig. R 113-2). Q_a and Q_p are the soil reactions at an angle φ' to the slip plane normal, and P_a or P_p the total active and passive earth pressures acting at an angle δ_a or δ_p respectively to the wall normal, taking into account all the flow effects. In this approach the resultant water pressure difference to be taken into account is the difference between the water pressures acting directly on the structure from inside and from outside. The result is all the more applicable, the better the flow net coincides with the natural conditions.

As the solution according to fig. R 114-1 supplies only the total values of P_a and P_p but not their distribution, separate consideration of the

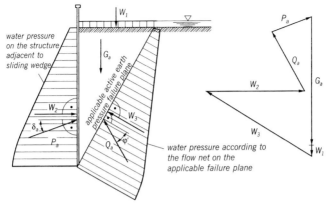

a) Determination of active earth pressure P_a

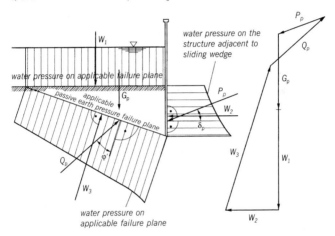

b) Determination of passive earth pressure P_p

Fig. R 114-1. Determination of active and passive earth pressures P_a and P_p, taking into account the effects of flowing groundwater

horizontal and vertical flow pressure effects is recommended in practice. In this case the horizontal effects are added to the water pressure difference which is related to the respective failure plane for the active or the passive earth pressure (fig. R 114-2). The vertical flow pressure effects are added to the vertical soil pressures from the dead load of the soil, reduced by the uplift, or are approximated by assuming a modified effective soil density. These methods of calculation are dealt with in the following.

49

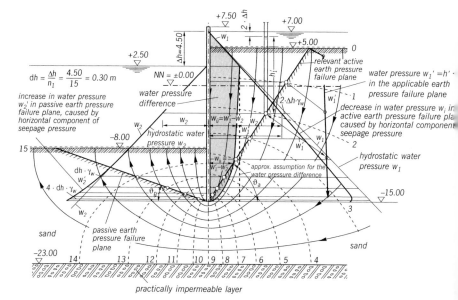

Fig. R 114-2. Determination of water pressure difference acting on a sheet pile structure, with the flow net as per R 113, section 4.7.7

2.9.2 Determination of the water pressure difference

In order to explain the calculation, the flow net as per R 113, section 4.7.7, fig. R 113-2 is used, and the calculation sequence is shown in fig. R 114-2. Accordingly, the first requirement is to determine the water pressure distribution in the failure planes for the active and passive earth pressures. This is shown in fig. R 114-2 for the applicable active earth pressure failure plane only. It is determined in each case for the points of intersection of the equipotential lines with the pertinent failure plane. The water pressure corresponds in each case to the product of the density of the water and the height of the water column in the standpipe (fig. R 114-2, right side) installed at the investigation point. If the water pressures measured at the points of intersection considered are plotted horizontally from a vertical reference line, the result is the horizontal projection of the water pressures acting in the failure plane investigated. The horizontal water pressure difference acting on the structure, in which the seepage pressure influences are already included, is the result of superimposing the horizontal water pressure acting from outside and inside.

A good approximate solution can also be found in section 2.9.3.2. This takes account of the fact that percolation around the wall results in decreased potential on the active side and increased potential on the passive side. Both effects can be equated through a decrease or increase of $\Delta\gamma_w$ in the weight density γ_w of the water.

These $\Delta\gamma_w$ values are calculated with inverted signs using the same equations as the $\Delta\gamma'$ values according to section 2.9.3.2. They lead to a decrease in the hydrostatic water pressure distribution on the land side and a corresponding increase on the water side. The difference in the water pressure modified in this way then supplies a good approximation of the water pressure difference acting on the sheet piling.

In most cases, however, a differentiated calculation of water pressure difference is not necessary if calculated according to R 19, section 4.2, or – if the inflow is mainly horizontal – according to R 65, section 4.3. However, the influence can be considerable for larger water pressure differences.

2.9.3   Determination of the effects on active and passive earth pressures when the flow is mainly vertical

2.9.3.1   Calculation using the flow net

The flow net described in R 113, section 4.7.7, fig. 113-2 is used to explain the calculations.

The calculation process is shown in detail in fig. R 114-3. The hydrostatic level difference per net field is equivalent to a vertical seepage pressure in each case. The seepage pressure increases on the active earth pressure side from top to bottom, and decreases on the passive earth pressure side from bottom to top. If dh is the hydrostatic level difference between equipotential lines in the flow net and n is the number of fields beginning with the applicable boundary equipotential line, the result is a additional vertical stress of $n \cdot \gamma_w \cdot dh$ on the active earth pressure side due to the flow pressure, and consequently an increase in the horizontal component of the earth pressure stress amounting to

$$\Delta p_{ahn} = +n \cdot \gamma_w \cdot dh \cdot K_{ag} \cdot \cos\delta_a$$

On the passive earth pressure side the corresponding decrease in the horizontal component of the passive earth pressure stress amounts to

$$\Delta p_{phn} = -n \cdot \gamma_w \cdot dh \cdot K_{pg} \cdot \cos\delta_p$$

In contrast to the reduced water pressure on the active earth pressure side, there is usually an increase in the earth pressure amounting to about one-third of the water pressure decrease. As a result of the substantially larger K_p value, the decrease in the earth pressure is considerably greater on the passive earth pressure side.

The effect of the horizontal component of the seepage pressure on the active and/or passive earth pressure is taken into account by determining

$$dh = \frac{\Delta h}{n_1} = \frac{4.50}{15} = 0.30 \, m$$

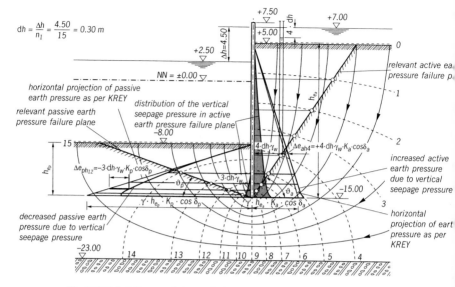

Fig. R 114-3. Influence of the vertical seepage pressure on the active and passive earth pressure with chiefly vertical flow, determined with the flow net for the sheet piling structure, as per R 113, section 4.7.7

the water pressure difference according to section 2.9.2, fig. R 114-2, with the water pressure applied to the critical active and/or passive earth pressure failure plane.

2.9.3.2 Approximate solution based on the assumption of changed effective soil weight densities on the active and passive earth pressure sides

When water percolates primarily vertically around a sheet pile wall, the increase in the active earth pressure – or rather the decrease in the passive earth pressure – due to vertical components of the seepage pressures can be approximated by assuming suitable changes in the weight density of the soil.

The increase $\Delta\gamma'$ in the weight density on the active earth pressure side and its decrease on the passive earth pressure side can be determined approximately for exclusively vertical flow and homogeneous subsoil after [12] using the following equations:

• on the active earth pressure side:

$$\Delta\gamma' = \frac{0.7 \cdot \Delta h}{h_{so} + \sqrt{h_{so} \cdot h_{su}}} \cdot \gamma_w$$

• on the passive earth pressure side:

$$\Delta\gamma' = -\frac{0.7 \cdot \Delta h}{h_{su} + \sqrt{h_{su} \cdot h_{so}}} \cdot \gamma_w$$

The symbols in the above equations and in fig. R 114-4 have the following meanings:

Δh = hydrostatic difference in level (potential difference)

h_{so} = vertical depth of soil flowed through on land side of sheet piling down to the pile toe in which a decrease in potential occurs

h_{su} = driving depth or soil stratum on the water side of the sheet pile wall in which a decrease in potential occurs

γ' = submerged weight density of soil for uplift condition

γ_w = weight density of water

The above equations apply when there is also soil underneath the base of the sheet pile wall that, in a seepage situation, contributes to decreasing the potential to the same extent as the soil in front of and behind the sheet pile wall (see section 4.9.3).

Otherwise, the general remarks made in section 2.9.3.1 apply similarly. In the case of horizontal flow towards the wall, the residual potential at the sheet pile wall increases considerably, and therefore the above approximation should not be used in such cases.

R 165, section 4.9.4, contains advice concerning the calculation of the active and passive earth pressures for stratified soils.

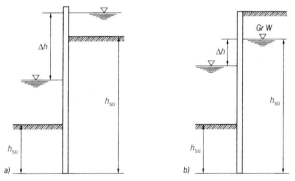

Fig. R 114-4. Dimensions for the approximate determination of the effective weight density of the soil in front of and behind a sheet piling structure, as changed by seepage pressure

2.10 Determining the amount of displacement required for the mobilisation of passive earth pressure in non-cohesive soils (R 174)

2.10.1 General

As a rule, a considerable amount of displacement is required for the mobilisation of full passive earth pressure. The amount is chiefly contingent on the embedded length of the pressing wall area and on the strength of the soil, as well as on the ratio of wall height h to wall width b. On the basis of large-scale model tests [13-15], it is possible to estimate the amount of displacement required for given wall loads or the magnitude of the mobilised partial passive earth pressure for a given displacement. The general relationships were derived for the passive earth pressure in front of beams, but can also be transferred to the passive earth pressure in front of walls within the scope of these recommendations.

2.10.2 Calculations

According to [13] the correlation

$$P_{p(s)} = w_e \cdot P_p$$

is applicable for a parallel displacement of walls (fig. R 174-1);

where:

P_p = passive earth pressure to DIN 4085
$P_{p(s)}$ = mobilised partial passive earth pressure contingent on displacement s
w_e = displacement coefficient; $w_e = f(s/s_B)$
s = displacement
s_B = required displacement for mobilisation of P_p (failure displacement)

The failure displacement s_B according to [13] and [14] for walls with $h/b < 3.33$ (fig. R 174-2) is as follows:

$$s_B = 100 \cdot (1 - 0.6\,D) \cdot \sqrt{h^3}$$

According to [15] the following equation applies for narrow walls subject to pressure ($h/b \geq 3.33$):

$$s_B = 40 \cdot \frac{1}{1 + 0.5\,D} \cdot \frac{h^2}{\sqrt{b}}$$

where:

D = degree of density to DIN 18 125
h = wall height or embedded depth of wall
b = wall width

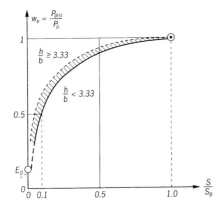

Fig. R 174-1. Passive earth pressure $P_{p(s)}$ contingent on displacement s

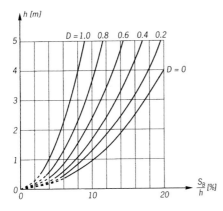

Fig. R 174-2. Failure displacement s_B contingent on the wall height or embedded length h and the degree of density D for $h/b < 3.33$

2.10.3 Method of calculation

The displacement s related to the wall loading, or rather the mobilised passive earth pressure $P_{p(s)}$ related to the allowable displacement, may be determined directly with the help of figs. R 174-1 and R 174-2 for $h/b < 3.33$, the case occurring most frequently.

The value s_B/h and thus the failure deformation s_B result from fig. R 174-2, with the wall height or embedded length h contingent on D. The value s/s_B can then be determined from fig. R 174-1 for a given wall compressive force $P_{p(s)}$ and thus the displacement s to be expected or the mobilised passive earth pressure

$$P_{p(s)} = w_e \cdot P_p$$

can be ascertained for a given allowable displacement.

2.11 Measures for increasing the passive earth pressure in front of waterfront structures (R 164)

2.11.1 General

An increase in the passive earth pressure in front of waterfront structures generally requires building measures below water. The following measures, for example, may be considered for this:

(1) Replacement of in situ soft cohesive soil with non-cohesive material.
(2) Compaction of loosely deposited, non-cohesive in situ soil or fill, if need be with added surcharge.
(3) Drainage of soft cohesive soils.
(4) Placing of a fill.
(5) Consolidation of the natural soil.
(6) A combination of measures (1) to (5).

Measures that do not prevent or hinder later deepening, e.g. driving in front of a wall, are to be preferred. Various measures are explained below.

2.11.2 Soil replacement

When replacing soft cohesive subsoil with non-cohesive material, see R 109, section 7.9, insofar as the works themselves are concerned. In determining the passive earth pressure, any depositing of intrusive layers in the boundary plane is to be taken into account. The pertinent remarks in R 110, section 2.8, apply accordingly.

The extent of soil replacement in front of the waterfront structure is, as a rule, determined according to earth statical aspects. In order to exploit the maximum passive earth pressure that can be achieved with the replacement soil deposited, these measures have to include the entire area of the passive earth pressure sliding wedge.

2.11.3 Soil compaction

Non-cohesive soil can be compacted by means of vibroflotation. The reciprocal spacing of the vibration points (grid width) depends on the natural subsoil and the desired mean degree of density. The greater the desired improvement in the degree of density and the finer the grain structure of the soil to be compacted, the narrower the grid width must be. A mean grid width of 1.80 m is recommended. The compaction must extend down as far as the dredge pit bottom and so that effective dowelling is possible. The deep compaction should cover the entire area of the passive earth pressure sliding wedge in front of the structure and thereby adequately penetrate the applicable passive earth pressure failure plane starting from the theoretical base of the sheet piling. If doubt arises, curved or cranked slip planes should be used to verify that the area of compaction is sufficiently large.

Compaction by means of vibroflotation causes these soil in the immediate vicinity of the vibratory plant to be temporarily liquefied, but it is then compacted due to the effect of the soil surcharge. Therefore, the effect of compaction depends on the soil surcharge, and the layers near the surface (about 2–3 m thick) can only be compacted when a temporary surcharge (in the form of fill) is introduced during compaction.

Vibroflotation can also be used for subsequent strengthening of waterfront structures. However, when using this method it must be guaranteed that the temporary local liquefaction of the soil does not lead to critical conditions regarding the stability of the structure. Experience has shown that extensive and prolonged liquid states can occur, in particular, in loosely deposited fine-grained, non-cohesive soils and in fine sand.

Soil compaction can reduce the risk of liquefaction quite effectively in earthquake zones. However, this should be carried out before erecting the structure.

2.11.4 Soil surcharge

Under certain conditions, e.g. refurbishment of an existing waterfront structure, it may be practical to improve the support of the structure by placing a fill with high weight density and high angle of internal friction in the region of passive earth pressure. Suitable materials here are steel mill slag or natural stone. The submerged weight density of the materials is critical; steel mill slag can attain values of $\gamma' \geq 18$ kN/m^3. The characteristic angle of internal friction may be assumed to be $\varphi'_k = 42.5°$.

In the case of existing soft subsoil, care must be taken to ensure that the fill material does not sink by suitably grading the grains of the fill material, installing a filter layer between fill and existing soil, or limiting the thickness of the fill.

The material to be placed must be checked constantly for compliance with the conditions and specifications. This applies particularly to weight density.

The remarks in sections 2.11.2 and 2.11.3 apply accordingly regarding the necessary extent of the works.

Additional vertical drains can be installed to accelerate the consolidation of soft strata below the fill material.

2.11.5 Soil stabilisation

If readily permeable, non-cohesive soils (e.g. gravel, gravelly sand or coarse sand) are available in the passive earth pressure area, the soil can be stabilised by injecting cement as well. In less permeable, non-cohesive soils, stabilisation by means of high-pressure injection is another popular solution. Stabilisation with chemicals such as water glass is also possible, provided the chosen stabilising medium can cure properly – allowing for the chemical properties of the pore water. Generally, however, chemical stabilisation is too expensive, which restricts it mainly to special boundary conditions and tight construction schedules.

Prerequisite for all types of injection is an adequate surcharge, which must be placed in advance when an adequate surface layer is lacking, and then possibly removed again.

The required dimensions of the stabilisation area can be stipulated in accordance with sections 2.11.2 and 2.11.3. Subsequent driving works, later dredging for harbour deepening and the like must be taken into consideration. Cored soil samples and/or penetrometer tests are always required to verify the success of soil stabilisation measures.

2.12 Passive earth pressure in front of sheet piles in soft cohesive soils, with rapid loading on the land side (R 190)

2.12.1 General

The same principles apply when determining passive earth pressure in front of sheet piling with rapidly applied additional loading on the land side as for determining active earth pressure for this load case (R 130, section 2.6.2). As given in 2.6.2 for the active earth pressure, the passive earth pressure, too, can be calculated or determined graphically with total stresses when using the shear parameters (c_u, φ_u) of the undrained soil or with effective stresses if use is made of the shear parameters (c', φ') of the drained soil, depending on the situation. As the passive earth pressure is influenced by the excess pore water pressure, it is better to use the shear parameters c_u and φ_u from the undrained triaxial test (DIN 18 137 part 2).

However, the shear parameters that suit the loading situation should always be chosen for the passive earth pressure on the one hand and the active earth pressure on the other.

2.12.2 Method of calculation

With a rapidly applied additional load on the active earth pressure side, a wall bearing pressure increased by the amount ΔP_p is generated on the passive earth pressure sliding wedge (fig. R 190-1b).

When calculating with c_u and φ_u ($\varphi_u = 0$ for saturated soil), the passive earth pressure can be calculated according to fig. R 190-1a.

If only shear parameters c' and φ' are known, the total available passive earth pressure $P_{p\,(t=\infty)}$ at time t = 0 (time of application of additional load) is reduced by the uniformly distributed assumed excess pore pressure $\Delta u_2 = -\Delta e_p$. The effective passive earth pressure only becomes critical once again when the excess pore pressure decreases (fig. R 190-1c).

The total available passive earth pressure $P_{p\,(t=\infty)}$ is expediently determined at time $t = 0$ in such a way that a uniform distribution of Δu_2 (fig. R 190-1b) is assumed and the sheet piling is considered as a statically

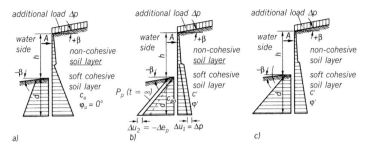

Fig. R 190-1. Active earth pressures on sheet piling in unconsolidated, soft cohesive soil as a result of rapidly applied ground surcharge of unlimited extent
a) Method with soil properties for undrained soil samples
b) Method with soil properties for drained soil samples
c) Method after consolidation

determinate system (beam on two supports). The magnitude of Δu_2 is determined with the aid of $\Sigma M = 0$ around A. The embedment depth of the wall must initially be assumed here, and subsequently checked iteratively.

The available passive earth pressure $P_{p\,(t=0)}$ at the time of applying the load ($t = 0$) is equal to the passive earth pressure $P_{p\,(t=\infty)}$ for the fully consolidated state, minus the pressure area $\Delta u_2 \cdot d$, calculated from the pore water pressure:

$$P_{p\,(t=0)} = P_{p\,(t=\infty)} - \Delta u_2 \cdot d$$

For time $t = \infty$, $\Delta u_2 = 0$, and the passive earth pressure achieves the value for the fully consolidated state.

2.12.3 Graphic procedure

If only shear parameters φ' and c' are known, the passive earth pressure ΔP_p available for accommodating a rapid additional load on the land side can be determined graphically by specifying a failure plane which is inclined at less than $\vartheta_p = 45° - \varphi'/2$, where $\delta_p = 0$ and $\alpha = \beta = 0$ (fig. R 190-2).

This solution is based on the premise that the passive earth pressure P_{p0} has already been mobilised by prior loading, but, in total, the maximum passive earth pressure $P_{p\,(t=\infty)}$ from effective parameters φ' and c' is available following full consolidation. The pore water pressure U_2 (fig. R 190-2) is generated at the time of applying the additional loading. In doing so, sufficient reserve P_p must always exist between P_{p0} and P_p. The pressure P_{p0} is the supporting force at the base required to balance the earth and water pressures prior to loading, and must be determined from the equilibrium condition ΣH or $\Sigma M = 0$ around the anchor point.

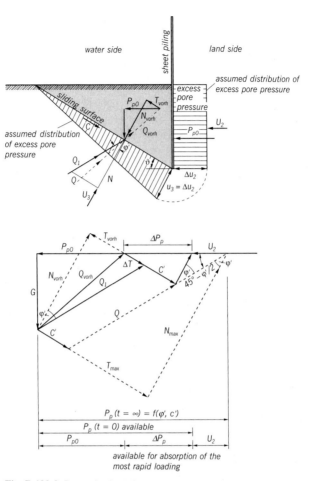

Fig. R 190-2. Determination of available passive earth pressure in unconsolidated, soft, cohesive soil as a result of rapidly applied additional loading on the land side of the sheet piling

Initially, the resultant slip plane force Q_{avail} is determined for the assumption $\delta_p = 0$ from the weight of the sliding body G and the passive earth pressure P_{p0} already exploited at the failure plane (fig. R 190-2). The mobilisable friction force ΔT is generated with the direction (φ' to the normal) of the mobilisable slip plane force Q_1. The additional loading can now be accommodated only by the cohesion force C and ΔT, whereby due to equilibrium the pore water pressure U_3 reaches such a magnitude that the vector of U_3 in the vector polygon ends with the influence line of the passive earth pressure. This can result in a non-linear distribution of the pore water pressure difference in the passive earth resistance failure line. In the vector polygon this results in the available reserve force ΔP_p for accommodating the rapid loading.

As shown in fig. R 190-2, virtually the same result is achieved with this graphical procedure as with the mathematical procedure: the passive earth pressure P_p calculated from the effective shear parameters φ' and c' is reduced by the amount U_2. For $\varphi' = 30°$, $U_3 = U_2$ exactly. There are slight differences when φ' deviates from this.

2.13 Effects of earthquakes on the design and dimensioning of waterfront structures (R 124)

2.13.1 General

In practically all countries where earthquakes can be expected, there are various standards, directives and recommendations, especially for buildings, that prescribe requirements for design and calculation in more or less detail. Consult DIN 4149 with regard to the Federal Republic of Germany, and [16], for example, for harbour installations in particular. The intensity of the earthquakes to be expected in the various regions is generally expressed in such publications by means of the horizontal seismic acceleration a_h which occurs during a seismic event. Any simultaneous vertical acceleration a_v is generally negligible compared to the acceleration due to gravity g.

The acceleration a_h causes not only immediate loads on the structure but also has an influence on the active and passive earth pressures, the water pressure and, in some cases, the shear strength of the earth masses at foundation level as well. In unfavourable circumstances, this shear strength may disappear completely temporarily.

Higher demands are placed on the calculations when earthquake damage can endanger human lives or cause destruction of vital supply facilities etc. The additional actions occurring during an earthquake are normally taken into account in the actual structure in such a manner that the additional horizontal forces

$$\Delta H = \pm k_h \cdot V$$

that act at the centre of gravity of each accelerated mass are applied simultaneously with the other loads.

In the above equation:

k_h = a_h/g = seismic coefficient = ratio of horizontal earthquake acceleration to gravitational acceleration

V = dead load of the structural member or sliding wedge considered, including pore water

61

The magnitude of k_h depends on the intensity of the earthquake, the distance from the epicentre and the subsoil. The first two factors are taken into account in most countries by dividing the endangered areas into earthquake zones with appropriate k_h values (DIN 4149). When doubt arises, agreement on the magnitude of k_h to be used is to be reached by consulting an experienced seismic expert.

In the case of tall, slender structures at risk of resonance, i.e. when the natural oscillation period and the earthquake period are nearly identical, the inertia forces must also be taken into account in the calculations. This, however, is not generally required for waterfront structures. The principal requirements which must be met in the design of an earthquake-proof waterfront structure is that the additional horizontal seismic forces will be safely absorbed, even with the attendant reduction in passive earth pressure.

2.13.2 Effects of earthquakes on the subsoil

Waterfront structures in earthquake regions must also take special account of the conditions in deeper subsoil. Thus, for example, it should be understood that earthquake vibrations are most severe where loose, relatively thin deposits overly solid rock (see [7]).

The most sustained effects of an earthquake make themselves felt when the subsoil, especially the foundation soil, is liquefied by the earthquake, i.e. loses its shear strength temporarily. This situation arises when seismic effects transform loosely deposited, fine-grained, non- or weakly cohesive, saturated, marginally permeable soil (e.g. loose fine sand or coarse silt) into a denser deposit (settlement flow, liquefaction). The lower the overburden pressure at the depth in question and the greater the intensity and duration of the seismic shocks, the sooner liquefaction will occur.

When the risk of liquefaction cannot be definitely ruled out, it is advisable to investigate the true liquefaction potential.

Looser soil strata tending to liquefy can be compacted before starting construction work. Cohesive soils do not tend to liquefy.

2.13.3 Determination of the structural effects of earthquakes on active and passive earth pressures

The influence of earthquakes on active and passive earth pressure is also generally determined according to COULOMB; however, the additional forces ΔH (section 2.13.1) created by earthquakes must receive consideration. Therefore, the dead loads of the earth wedges should no longer be applied vertically but rather at a defined angle deviating from the vertical. This is best taken into consideration by referring the inclination of the active and passive earth pressure reference plane and the plane of the ground surface to the new force direction [7]. This results in imaginary

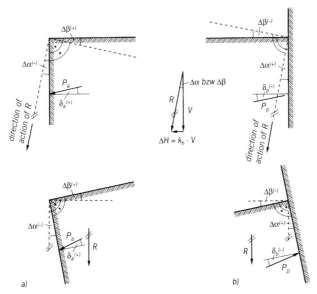

Fig. R 124-1. Determination of the imaginary angles $\Delta\alpha$ and $\Delta\beta$ and depiction of the systems rotated by angle $\Delta\alpha$ or $\Delta\beta$ (signs as per KREY)
a) for calculating the active earth pressure, b) for calculating the passive earth pressure

changes in the inclination of the reference plane $(\pm\,\Delta\alpha)$ and the ground surface $(\pm\,\Delta\beta)$.

$$k_{\mathrm{h}} = \tan\Delta\alpha \text{ and } \tan\Delta\beta \text{ respectively (fig. R 124-1)}$$

The active and passive earth pressures are then calculated on the basis of the imaginary system rotated through an angle of $\Delta\alpha$ or $\Delta\beta$ respectively (reference plane and ground surface).

An equivalent procedure is to calculate on the assumption that the inclination of the wall is $\alpha\pm\Delta\alpha$ and that of the ground surface $\beta\pm\Delta\beta$ (fig. R 124-1).

When determining the active earth pressure below the water table, it must be recognised that the mass of the soil and the mass of the water enclosed in the soil pores are accelerated, but that the reduction in the submerged weight density of the soil remains unchanged. In order to take this into account, a larger seismic coefficient – the so-called apparent seismic coefficient k_{h}' – is used for practical purposes for calculations in the area below the groundwater table.

In the section considered in fig. R 124-2

$$\sum p_{\mathrm{v}} = p + \gamma_1 \cdot h_1 + \gamma_2' \cdot h_2 \text{ and}$$
$$\sum p_{\mathrm{h}} = k_{\mathrm{h}} \cdot [p + \gamma_1 \cdot h_1 + (\gamma_2' + \gamma_{\mathrm{w}}) \cdot h_2]$$

63

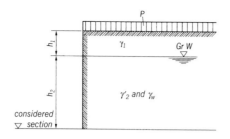

Fig. R 124-2. Sketch showing the arrangement for calculating the value of k'_h

The apparent seismic coefficient k'_h for determining active earth pressure below the water table is thus:

$$k'_h = \frac{\sum P_h}{\sum P_v} = \frac{p + \gamma_1 \cdot h_1 + (\gamma'_2 + \gamma_w) \cdot h_2}{p + \gamma_1 \cdot h_1 + \gamma'_2 \cdot h_2} \cdot k_h$$

A similar procedure can be employed for passive earth pressure.

For the special case where the groundwater level is at the surface and where there is no ground surcharge, the result for the active earth pressure side with $\gamma_w = 10$ kN/m³ is:

$$k'_h = \frac{\gamma' + 10}{\gamma'} \cdot k_h = \frac{\gamma_r}{\gamma_r - 10} \cdot k_h \cong 2 k_h$$

where:

γ' = submerged weight density of soil
γ_r = weight density of the saturated soil

For the sake of simplicity, the unfavourable value of k'_h determined in this way for the active earth pressure side is customarily also used as the basis for calculation in other cases, even when the groundwater table is lower and also for live loads.

In calculating the active earth pressure coefficient K_{ah}, which is determined by k_h and k'_h, a sudden change takes place theoretically in the active earth pressure at the level of the groundwater table, as shown in fig. R 124-3. If a more accurate calculation of the value of k'_h, which changes with depth, and the change in K_{ah} is deemed unnecessary, the active earth pressure can be applied in simplified form as shown in fig. R 124-3.

In difficult cases for which tabular figures are not available for calculating the active and passive earth pressures, it is possible to determine the influences of both the horizontal and – if present – vertical earthquake accelerations on active and passive earth pressures by an extension of the CULMANN method. The forces resulting from earthquake accelerations

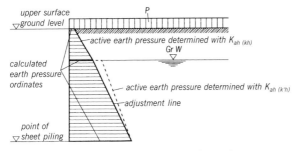

Fig. R 124-3. Simplified disposition of the active earth pressure

and correspondingly acting on the wedges investigated must then also be taken into account in the vector polygons. Such an approach is recommended for larger horizontal accelerations, above all if the soil lies partly below the groundwater table.

2.13.4 Water pressure difference

The application of water pressure difference for waterfront structures in earthquake regions may be approximated as in the normal case, i.e. according to R 19, section 4.2, and R 65, section 4.3, because the effects of the earthquake on the pore water have already been taken into account when determining the active earth pressure with the apparent vibration coefficient k_h' as per section 2.13.3. It must be observed in the earthquake case, however, that the applicable active earth pressure failure plane is inclined at a shallower angle to the horizontal than in the normal case. For this reason, increased water pressure difference may act on the failure plane.

2.13.5 Live loads

Because the simultaneous occurrence of earthquake, full live load and full wind load is improbable, it suffices to combine the loads caused by the earthquake with only half the live load and half the wind load (see DIN 4149, explanatory notes). The crane wheel loads caused by wind and the component of the line pull due to wind should therefore be correspondingly reduced. Loads due to the travel and slewing movements of cranes need not be superimposed on earthquake effects.

However, those loads which in all probability remain constant for a longer period of time, e.g. loads from filled tanks or silos and from bulk cargo storage, may not be reduced.

2.13.6 Safety factors

Earthquake forces may be taken into consideration according to DIN EN 1991 EC1 as an exceptional design situation with the corresponding partial safety factors for actions and resistances, taking account of cases B and C (LS 1B and 1C) according to DIN EN 1997-1.

According to DIN 1054, load case 3 applies for seismic loads.

2.13.7 Advice for the consideration of seismic influences in waterfront structures

Taking into account the foregoing remarks and other recommendations by the EAU, waterfront structures can also be calculated and designed systematically and with adequate stability for earthquake regions. Supplementary references to definite types of construction such as sheet pile structures (R 125, section 8.2.18), waterfront structures in blockwork construction (R 126, section 10.9) and walls on piled foundations (R 127, section 11.8) are given in the those recommendations.

Experience gained from the earthquake in 1995 in Japan is dealt with in [198].

For further information please refer to [257].

3 Overall stability, foundation failure and sliding

3.1 Relevant standards

The following standards are valid for the verification of safety against sliding and foundation failure, and also for determining the active earth pressure:

Sliding: DIN 1054
Foundation failure: DIN 4017
Slope failure: DIN 4084
Earth pressure: DIN 4085

3.2 Safety against failure by hydraulic heave (R 115)

Failure by hydraulic heave occurs when a body of earth in front of a structure is loaded by the upward flow force of the groundwater. In this state the passive earth pressure is reduced. A failure state occurs when the vertical component S of the flow force is equal to or greater than the dead load G of the body of earth under uplift, which lies between the structure and an assumed, theoretical failure surface. The body of earth is then raised and the passive earth pressure is lost completely.

All heave failure surfaces to be considered in the analysis extend – in homogeneous soil – progressively outward from the base of the structure. The surface with the smallest safety factor as determined by test calculations forms the basis for assessing safety. Advice on the approach in the case of a stratified subsoil is given in R 113, section 4.7.7.2.

According to DIN 1054, section 11.5, it should be checked that the following condition applies for limit state LS 1A for the upward flow within the failure body in front of the base of the wall:

$$S'_k \cdot \gamma_H \leq G'_k \cdot \gamma_{G,stb}$$

where:

S'_k = characteristic value of flow force in body of earth in flow

γ_H = partial safety factor for flow force in limit state LS 1A to DIN 1054 table 2

G'_k = characteristic value of weight under uplift for body of earth in flow

$\gamma_{G,stb}$ = partial safety factor for favourable permanent actions in limit state LS 1A to DIN 1054 table 2

The flow force S'_k can be calculated using a flow net as per R 113, section 4.7.7, fig. R 113-2, or R 113, section 4.7.5. S'_k is the product of the volume of the heave failure body multiplied by the weight density of

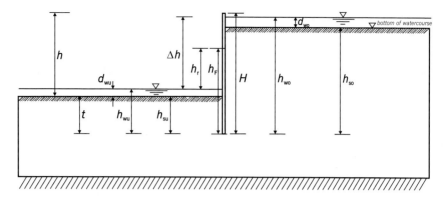

Fig. R 115-1. Safety against failure by hydraulic heave – relevant dimensions

the water γ_w and by the mean flow gradient measured in this body in the vertical.

If water flows upwards through the soil in front of the base of the wall, the flow force should be considered in a body of soil whose width may generally be assumed to be equal to half the embedment depth of the wall (DIN 1054, section 11.5 (4)). In more accurate calculations, other boundaries to the body of soil should be examined as well.

Fig. R 115-2 shows the method according to TERZAGHI-PECK [17, p. 241], fig. R 115-3 the method according to BAUMGART-DAVIDENKOFF [168, p. 61].

In the rectangular failure body with a width equal to half the embedment depth t of the structure, the characteristic value of the vertical flow force S'_k can be found approximately from the following equation:

$$S'_k = \frac{\gamma_w\,(h_1 + h_r)}{2} \cdot \frac{t}{2}$$

where:

h_r = effective hydraulic head at base of wall (difference between standpipe water level at base of sheet piling and underwater table level)

h_1 = effective hydraulic head at boundary of failure body opposite base of wall

According to BAUMGART-DAVIDENKOFF [168, p. 66] verification of safety against hydraulic heave failure can also be calculated in a simplified way as follows. In this instance only one flow channel of the upward flow directly in front of the vertical part of the building element is considered:

$$(\gamma_w \cdot i) \cdot \gamma_H \leq \gamma' \cdot \gamma_{G,stb}$$

where:

γ' = soil weight density, submerged
γ_w = gradient density of water
i = mean hydraulic gradient in path considered ($i = h_r / t$)

In this approach as well, the effective hydraulic head h_r at the base of the sheet piling can be calculated using a flow net according to R 113, section 4.7.4 or 4.7.5.

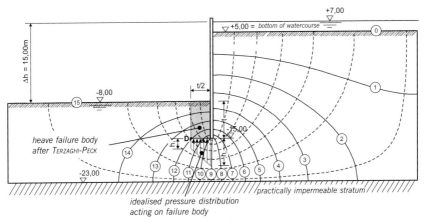

Fig. R 115-2. Safety against a failure by hydraulic heave in a dredge pit bottom according to the TERZAGHI-PECK method, determined with the flow net as per R 113, section 4.7.7

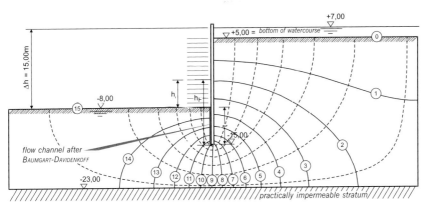

Fig. R 115-3. Safety against a failure by hydraulic heave in a dredge pit bottom according to the BAUMGART-DAVIDENKOFF method

For a sheet piling structure situated in a vertical flow of water, a simplified hydraulic head level above the base of the sheet piling h_F may be assumed (after [12]):

$$h_F = \frac{h_{wu} \cdot \sqrt{h_{so}} + h_{wo} \cdot \sqrt{t}}{\sqrt{h_{so}} + \sqrt{t}}$$

from which we get:

$$h_r = h_F - h_{wu}$$

where:

h_r = difference between standpipe water level at base of sheet piling and lower water level
h_F = standpipe water level at base of sheet piling
h_{so} = depth of soil in flow on upper water side of sheet piling
h_{wo} = water level above base of sheet piling on upper water side
h_{wu} = water level above base of sheet piling on lower water side
t = embedment depth of sheet piling

In the case of a horizontal inflow, the residual hydraulic head at the base of the sheet piling increases considerably and in such cases the approximate approach may not be used!

The decrease in hydraulic head is not linear over the height of the sheet piling and therefore it is not permitted to calculate h_r from the development of the flow path along the sheet piling! Advice on determining the hydraulic head at the base of the sheet piling and the factor of safety against heave failure in stratified subsoils can be found in R 165, section 4.9.4.

The danger of an impending hydraulic heave failure in an excavation may be indicated by wetting of the ground in front of the sheet piling; the ground then seems soft and springy. If this occurs, the excavation should immediately be partly flooded at least, or a surcharge introduced in front of the sheet piling. Only afterwards may corrective measures be introduced, e.g. as suggested in R 116, section 3.3, 5th and subsequent paragraphs, or, if considered preferable, in the form of a local earth surcharge in the excavation, or by relieving the flow pressure from below by lowering the groundwater level.

Generally, safety against hydraulic heave failure is calculated for the final condition, i.e. after the steady flow around the wall has become established. In less permeable soils, the lowering of the pore water pressure in the soil often, however, takes place at a much slower rate than the external pressure changes acting on the soil due to changes in the water level (e.g. tides, waves, lowering of the water level) and

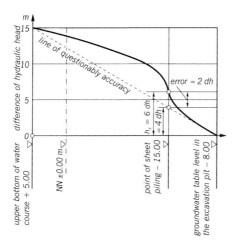

Fig. 115-4. Lowering of the hydraulic head along the sheet piling according to the flow net as per fig. R 115-2

excavation work. The lower the permeability k and the saturation S of the soil and the quicker the changes in pressure, the longer is the delay [210]. A process is "quick" when the rate of application v_{zA} of the external pressure reduction (e.g. lowering of the water level in the excavation) is greater than the critical permeability k of the soil ($v_{zA} > k$) [211]. The soil is considered as unsaturated below the groundwater table as well; in particular, degrees of saturation of the soil amounting to $0,8 < S < 0,99$ are frequently encountered directly below the groundwater table and in the case of a low water depth ($h_w < 4$–10 m). Natural pore water contains microscopic gas bubbles which have a considerable effect on the physical behaviour during pressure changes; such water cannot be considered as an ideal (incompressible) fluid. The pressure compensation in the pore water is therefore always connected with a mass transport, i.e. the (non-steady) flow of pore water in the direction of a lower hydraulic head. Quick changes in the water level therefore always require verification for the non-steady excess pore water pressure $\Delta u(z)$ for the initial condition in addition to verification of adequate safety against heave failure in the final condition with a steady flow. The object of the verification is to avoid the limit states for stability and serviceability. Serviceability can be impaired by unacceptable loosening (heave) of the base of the excavation, or undesirable base and (lateral) wall deformations [211]. The distribution of the excess pore water pressure $\Delta u(z)$ over the depth of soil z can be calculated using the following simplified equation (after Köhler [210]):

71

$$\Delta u(z) = \gamma_{\mathrm{w}} \cdot \Delta h \cdot (1 - e^{-b \cdot z})$$

where:

Δh = lowering of water level (or depth of excavation)
γ_{w} = weight density of water
b = pore water pressure parameter from fig. R 115-5
z = depth of soil

The pore water pressure parameter b (unit of measurement: 1/m) can be found from fig. R 115-5 based on the relevant time period t_{A} and the permeability k of the soil.

For a depth z the dead load G_{Br} of the body of soil with unit width 1, unit length 1 and density γ' is:

$$G_{\mathrm{br}} = \gamma' \cdot z$$

Due to the excess pore water pressure, a vertical, upward, seepage force develops in the soil and acts on the body of soil:

$$W_{\mathrm{instat}} = (1 - e^{-b \cdot z}) \cdot \gamma_{\mathrm{w}} \cdot \Delta h$$

The ratio of weight to seepage force is most unfavourable at the section in which the excess pore water pressure reaches its maximum. In this case this is at depth $z = z_{\mathrm{crit}}$:

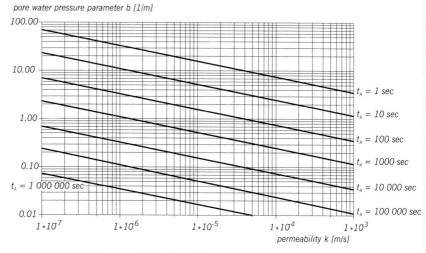

Fig. R 115-5. Parameter b for determining the flow force for unsteady flow depending on time t_{A}

$$z_{crit} = \frac{1}{b} \cdot \ln\left(\frac{\gamma_w \cdot \Delta h \cdot b}{\gamma'}\right)$$

When $z_{crit} < 0$ no further verification is necessary.

If z_{crit} lies below the base of the sheet piling, the verification should be carried out for the base.

If the condition

$$\gamma_w \cdot \Delta h \cdot b < \gamma'$$

is fulfilled, then adequate safety is guaranteed from the outset.

The verification of adequate safety against hydraulic heave failure for non-steady flow processes requires verification of equilibrium at least:

$$G_{Br} \geq W_{unst}$$

3.3   Piping (foundation failure due to erosion) (R 116)

The risk of piping exists when soil at the bottom of a river or excavation can be washed out by a flow of water. The process is initiated when the exit gradient of the water flowing around the waterfront structure is capable of moving soil particles up and out of the soil. This continues in the soil in the opposite direction to the flow of water and is therefore called retrogressive erosion. A channel roughly in the shape of a pipe forms in the ground and develops along the flow path with the steepest gradient in the direction of the upper water level. The maximum gradient always occurs at the interface between structure and wall. If the channel reaches a free upper watercourse, erosion causes widening of the channel within a very short time and leads to a failure similar to heave failure. In doing so, a mixture of water and soil flows into the excavation with a high velocity until equilibrium is achieved between outer water and excavation. A deep crater forms behind the wall.

The presence of loose soils or weak spots in the base support area (e.g. inadequately sealed boreholes) and loose zones in the immediate vicinity of the wall/soil interface behind the wall are the conditions that tend to promote piping, but also a sufficient amount of water (free upper watercourse) and a relatively high hydraulic gradient.

The occurrence of a failure condition in homogeneous non-cohesive soil is shown schematically in fig. R 116-1.

A possible piping failure is first indicated on the lower water table or on the base of the excavation by swelling of the ground and the ejection of soil particles. At this early stage the impending failure can still be prevented by depositing adequately thick, graded or mixed gravel filters to prevent further soil from being washed out.

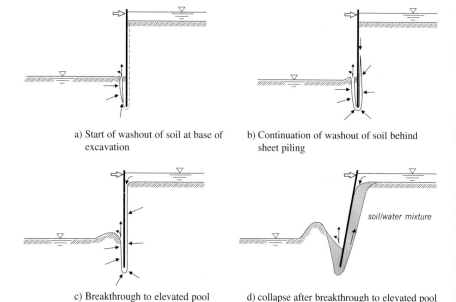

a) Start of washout of soil at base of excavation

b) Continuation of washout of soil behind sheet piling

c) Breakthrough to elevated pool

d) collapse after breakthrough to elevated pool

soil/water mixture

Fig. R 116-1. The development of piping around waterfront structures

However, if the condition has already reached an advanced stage (recognisable through the intensity of the water flow in the erosion channels already formed), the breakthrough to the upper watercourse can no longer be predicted. In such a case immediate equalisation of the water tables must be brought about by raising weir gates, flooding the excavation or similar measures. Only after this has been accomplished is it possible to undertake permanent remedial measures, such as placing a sturdy filter on the low side, grouting the eroded "pipes" from the lower end, vibroflotation or high-pressure injection to compact or stabilise the soil in the endangered area, lowering the groundwater table, or placing a dense covering on the bottom of the upper watercourse which extends well beyond the endangered area.

The risk of piping cannot generally be determined by calculation, and must be considered for every individual case owing to the diversity of the designs and boundary conditions. Other conditions being equal, the risk of piping increases in proportion to the increase in the hydraulic head between the upper and lower water levels, as well as with an increase in the presence of loose, fine-grained, non-cohesive material in the subsoil, particularly when there are embedded sand lenses or veins in soils not otherwise at risk of erosion. In cohesive soil there is generally no risk of piping.

Even if there is no free upper watercourse, erosion channels can still begin from the lower water side. In general, however, these peter out in the subsoil because there is insufficient free water available for a critical erosion effect. If by chance an extraordinarily capacious water-bearing stratum is encountered, water flowing out of this stratum can initiate the erosion process anew.

If conditions prevail that appear to make piping possible, precautions to prevent this should be planned from the very beginning of work on site in order to take appropriate countermeasures immediately if this should prove necessary. In particular, in such cases it is important to drive the sheet piling to an adequate depth in order to reduce the gradient of the flows around the sheet piling. The minimum embedment depth to prevent erosion failures can be determined according to R 113, section 4.7.7.

Defects in the walls (e.g. interlock declutching in sheet piling) shorten the seepage path of the flow around the wall and hence steepen the gradient dramatically. Therefore, such defects are to be considered with respect to a hydraulic heave failure provided there is a real risk of these in the actual circumstances.

3.4 Verification of overall stability of structures on elevated piled structures (R 170)

3.4.1 General
Verification of overall stability of structures on elevated piled structures can be carried out using DIN 4084.

3.4.2 Data
The following must be available for verification:

(1) Data on the design and dimensions of the piled structure, applicable loads and internal forces, most unfavourable water levels and live loads.

(2) Soil mechanics characteristics of the subsoil, especially the weight densities (γ, γ') and the shear parameters (φ', c') of the in situ soil types, which in cohesive soils are also to be determined for the initial state (φ_u, c_u). If required, the time-dependent influences of surcharges or excavations on the shear strengths are also to be taken into consideration for cohesive soils.

3.4.3 Application of the actions (loads)
The following loads in most their unfavourable combinations are to be taken into account:

(1) Loads in or on the sliding body, especially individual actions due to live or other external loads.

(2) Dead load of the sliding body and the soil surcharge thereon, taking into account the groundwater table (weight densities γ, and γ').

(3) Water pressure loads on the slip plane of the failure mechanisms being examined due to pore water pressure, determined as per R 113, section 4.7.

3.4.4 Application of the resistances

(1) The resistances (axial and dowelling forces) of the rows of piles within the pile bent plane are distributed along the equivalent length according to fig. R 170-1. They are calculated with the least favourable values as per DIN 4084, section 6.

(2) The passive earth pressure may be calculated with the kinematically compatible passive earth pressure angle δ_p. In special cases the

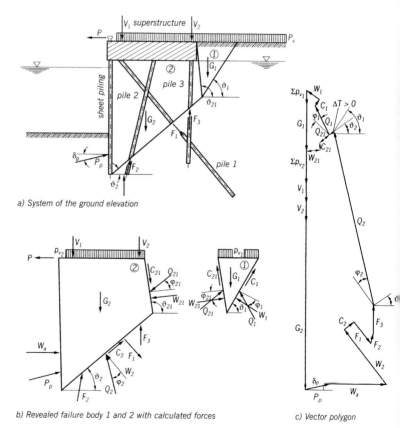

a) System of the ground elevation

b) Revealed failure body 1 and 2 with calculated forces

c) Vector polygon

Fig. R 170-1. Sketch for determining overall stability of an elevated pile-founded structure

question arises as to whether the displacement required for mobilising the passive earth pressure does not already impair the utilisation of the structure. But this is generally not necessary when using design values for the shear parameters. In addition, the supporting section of soil must be permanently available. Any possible future harbour dredging must therefore be taken into account in advance.

(3) The favourable effect of deeper embedded stabilising walls or aprons may be taken into account.

(4) Effects of earthquake actions are to be covered in accordance with R 124, section 2.13. For verification with combined failure mechanisms having straight failure lines as per DIN 4084, the seismic force may be calculated as a horizontal mass force.

4 Water levels, water pressure, drainage

The loads determined in this section are allocated to the load cases as per R 18, section 5.4. Section 5.4.4 is to be observed regarding the partial safety factors.

4.1 Mean groundwater level (R 58)

The allocation of the relevant hydrostatic load situation resulting from changing outer and groundwater levels requires an analysis of the geological and hydrological conditions of the area concerned. Where available, observations over several years are to be evaluated to obtain knowledge about the water tables. The groundwater table behind a waterfront structure is governed by the soil strata and the design of the waterfront structure. In tidal areas, the groundwater table for permeable soils follows the tides in a more or less attenuated manner. Assuming a theoretical groundwater head of 0.3 m above half tide ($T \frac{1}{2} w$) in tidal areas or mean water level in non-tidal areas can only be used as an approximation in preliminary designs. The actual groundwater and flow conditions must be investigated as part of the detailed design work. In areas with stronger groundwater flow from the land side, the mean groundwater level is higher. If this flow is also severely impeded by an extended waterfront structure, the groundwater table may rise considerably. Soils with low permeability can cause water-bearing layers at various levels.

4.2 Water pressure difference in the water-side direction (R 19)

The magnitude of the water pressure difference is influenced by fluctuations in the outer water level, the location of the structure, the groundwater inflow, the permeability of the foundation soil, the permeability of the structure and the efficiency of any backfill drainage available.
The water pressure difference $w_{\ddot{u}}$ amounts to

$$w_{\ddot{u}} = \gamma_w \cdot \Delta h$$

for a difference in height Δh between the critical outer water level and the corresponding groundwater level for the specific weight density of water γ_w.
For the outer water level, use the level of the trough of a wave for wave heights > 0.5 m.
The water pressure difference may be assumed to be as shown in fig. R 19-1 and R 19-2 in the case of weepholes or permeable soil and unhindered circulation around the toe – if the waterfront structure is extended and plane flow can be assumed and without appreciable wave action. It is then allocated to the load cases 1–3.

Situation	Figure	Load cases as per R 18		
		1	2	3
1 Minor water level fluctuations $(h < 0.50 \text{ m})$ with weepholes or permeable soil and structure	MW, MLW, GW, Δh, weepholes, $\Delta h \cdot \gamma_w$	$\Delta h = 0.50$ m	$\Delta h = 0.50$ m	–
2a Major water level fluctuations $(h > 0.50 \text{ m})$ with weepholes or well permeable soil and structure	weepholes, GW, MNW, Δh, $\Delta h \cdot \gamma_w$	$\Delta h = 0.50$ m in frequent elevation	$\Delta h = 1.00$ m in unfavourable elevation	$\Delta h \geq 1.00$ m max. drop in outer water level over 24 h and least favourable elevation
2b Major water level fluctuations without weepholes	MHW, MLW, GW, a, 0.30, Δh, $\Delta h \cdot \gamma_w$	$\Delta h = a + 0.30$ m $a = \dfrac{\text{MHW} - \text{MLW}}{2}$	$\Delta h = a + 0.30$ m	–

Fig. R 19-1. Water pressure difference at waterfront structures for permeable soils in non-tidal area

The water levels given in figs. R 19-1 and R 19-2 are nominal values (design values).

Hydraulic backfilling to waterfront structures can lead to extreme water levels which must be taken into account for the respective condition of the building works.

Special investigations are required where flooding of the banks, stratified soils, highly permeable sheet pile interlocks or artesian pressure occur (R 52, section 2.7.3), likewise if a natural groundwater flow is constricted or cut off by a waterfront structure.

Non-tidal area

Situation	Figure	Load cases as per R 18		
		1	2	3
3a Major water level fluctuations without drainage – normal case		$\Delta h = a + 0.30$ m $a = \dfrac{\text{MHW} - \text{MLW}}{2}$ $d = \text{MLW} - \text{MLWS}$	–	–
3b Major water level fluctuations without drainage – limit case extreme low water level		–	–	$\Delta h = a + 2b + d$ $a = \dfrac{\text{MHW} - \text{MLW}}{2}$ $b = \dfrac{\text{MHWS} - \text{LLW}}{2}$ $d = \text{MLW} - \text{MLWS}$
3c Major water level fluctuations without drainage limit case falling high water		–	–	$\Delta h = 0.30$ m $+ 2a$
3d Major water level fluctuations with drainage		$\Delta h = 1.00$ m $+ e$ for outer water level in MLWS	$\Delta h =$ 0.30 m $+ b + d$ $+ e$	–

Fig. R 19-2. Water pressure difference at waterfront structures for permeable soils in tidal area

80

For details of the water pressure on sheet pile walls to canals, please refer to section 6.4.2 fig. R 106-1.

The relieving action of drainage installations according to R 32, section 4.5, R 51, section 4.4, and R 53, section 4.6, may only be considered if its effectiveness can be constantly monitored and the drainage installation can be reinstated at any time.

4.3 Water pressure difference on sheet piling in front of embankments below elevated decks in tidal areas (R 65)

4.3.1 General

In the case of embankments below elevated decks (fig. R 65-1), a partial water pressure equalisation is possible where the flow is predominantly horizontal. No water pressure difference occurs at the surface of the embankment. In the ground further back, a water pressure difference exists with respect to the outer water level, which depends on the position of the point under consideration, the soil conditions, the magnitude and frequency of water level fluctuations, and the inflow from the land.

The water pressure difference should always be related to the relevant earth pressure failure plane. For this, knowledge of the variation of the groundwater level with respect to the outer water is necessary.

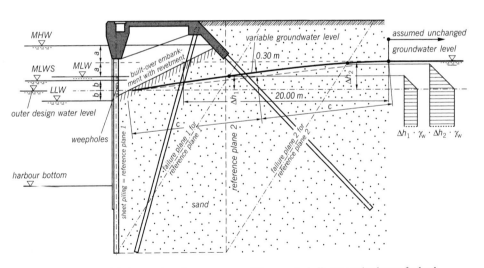

Fig. R 65-1. Assumption of water pressure difference at a built-over embankment for load case 2

81

4.3.2 Approximation for water pressure difference

The approximation for the water pressure difference shown in fig. R 65-1, which incorporates recommendations R 19, section 4.2, and R 58, section 4.1, can be used for conditions generally experienced in the North German tidal area and for a fairly uniform sandy subsoil. Fig. R 65-1 shows load case 2, but can be used similarly for the other load cases. The revetment must be sufficiently permeable and should not cause a rise of the groundwater level.

4.4 Design of filter weepholes for sheet piling structures (R 51)

Filter weepholes should only be used in silt-free water and groundwater with a sufficiently low (i.e. harmless) iron content. Otherwise they would quickly become clogged. Weepholes should also not be used where there is a danger of heavy barnacle growth. When very high outer water levels occur, filter weepholes should not be used because there is a risk of penetrating water causing damage to systems embedded in the ground on the inner side, or other damage due to subsidence of non-cohesive soil.

The filter weepholes must be located below mean water level so that they do not become encrusted. They are constructed with graded gravel filters as per R 32, section 4.5.

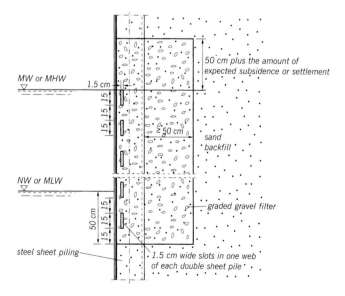

Fig. R 51-1. Filter weepholes for wave-shaped steel sheet piling

Slots 1.5 cm wide and approx. 15 cm high are burned in the sheet pile webs for drainage (fig. R 51-1). The slot ends are to be rounded off. In contrast to round holes, these slots cannot be blocked by pebbles. Please also refer to R 19, section 4.2, last paragraph.

Filter weepholes are considerably less expensive than flap valves (R 32). However, experience shows that they result in only a slight reduction in the water pressure difference in tidal areas because too much water flows behind the sheet piling through the weepholes at high tide. For this reason, filter weepholes are not permitted for quay sheet piling that also serves as flood protection.

Filter weepholes are particularly effective in locations where there is no tide and where a sudden drop in the outer water level, strong ground-water or bank water flows, or flooding of the structure may occur.

Sheet piling must normally be designed to allow for complete failure of the filter weepholes. In this case the sheet piling may be designed for load case 3 according to R 18, section 5.4.3. This verification is not necessary when redundant drainage systems are provided, e.g. a pump system with two independent pumps.

4.5 Design of drainage systems with flap valves for waterfront structures in tidal areas (R 32)

4.5.1 General

Effective drainage is possible only in non-cohesive soils. If drainage is to remain effective in silt-laden harbour water and is to limit the water pressure difference where there is a considerable tidal range, it must be equipped with collector pipes and reliable flap valves which permit the outflow of water from the collector into the harbour water, but hinder the backflow of muddy water.

Simple flap valves can be considered, but preferably in conjunction with drainage chambers.

4.5.2 Flap valves

Flap valves must be positioned to be readily accessible at mean low water (MLW) for checking during periodic inspections of the structure and for easy maintenance and repair. They should be checked at least twice a year and also before each dredging operation, plus, for example, when the transit of heavy goods may cause unusually severe surcharge, and also after heavy wave action.

The flap valves must be designed for maximum reliability. Inspection chambers are to be provided at appropriate intervals (max. 50 m) for maintenance purposes.

The usual spacing of simple flap valves is 7 to 8 m for sheet piling drainage systems.

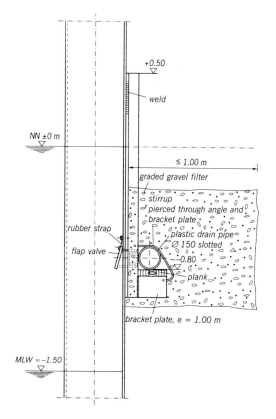

+0.50

weld

NN ±0 m

≤ 1.00 m

graded gravel filter

stirrup
pierced through angle and
bracket plate
plastic drain pipe
Ø 150 slotted
~0.80

rubber strap

flap valve

plank

bracket plate, e = 1.00 m

MLW ≈ -1.50

Fig. R 32-1. Example of a drainage system with flap valves behind a sheet pile wall with plastic drain pipe as collector

In silt-laden water, use only double secured backwater drainage systems. Efficient drains behind the wall can improve the drainage considerably. Plastic drain pipes are usually used as collectors.

For further details see figs. R 32-1 and R 32-2.

Fig. R 32-2 shows a groundwater drainage system for a quay structure in a tidal area. Two drain pipes DN 350 made of PE-HD (DIN 19 666) run the whole length of the quay structure to relieve the water pressure. The pipes are embedded in a graded gravel filter as per section 4.5.3; the filter is separated from the surrounding soil by a non-woven geotextile. The depth is selected to keep the drain pipes permanently in the groundwater; this rules out practically any risk of clogging because air is excluded from the filter. The outlet depth of -4.20 MSL results from the position of the filter and the fall of the outlet pipe to the outer water.

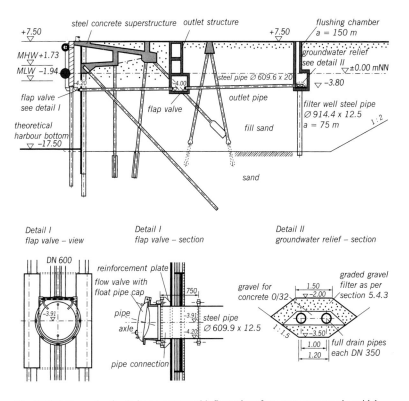

Fig. R 32-2. Example of a drainage system with flap valves for a quay structure in a tidal area

Filter wells are positioned at intervals of 75 m to tap the groundwater from deeper strata. The groundwater collected is drained to the outer water via outlet pipes (steel pipe, $\varnothing\,609.6 \times 20$) at intervals of about 350 m. Two flap valves with adjustable float hollow covers prevent any inflow of silt-laden river water. One flap is positioned immediately at the outlet into the outer water and the second flap is located in an outlet structure, protected from any damage.

4.5.3 Filter

Each collector must be separated from the soil to be drained by means of a carefully installed gravel filter or geotextile filter. Graded gravel filters should only be used when segregation cannot occur during installation.

85

4.6 Relieving artesian pressure under harbour bottoms (R 53)

4.6.1 General

Relieving the artesian pressure under harbour bottoms is best accomplished by means of efficient relief wells of ample capacity. Their effectiveness is independent of the use of mechanical equipment or their power supply. The outlets of the wells should always be placed below LLW. Since there must be a pressure difference when the relief wells are discharging water, it follows that residual artesian pressure will remain under the confining stratum, even under favourable conditions with fully effective, closely spaced wells. The residual pressure difference shall be taken as 10 kN/m^2 in calculations according to R 52, section 2.7.

4.6.2 Calculation

The design of the relief wells must always be verified by calculating the groundwater lowering, e.g. [217]. This must work on the premise that a residual pressure difference of 10 kN/m^2 is maintained at the back edge of the passive earth pressure wedge. Within the wedge, the residual pressure is thus correspondingly lower, therefore providing a desirable safety margin.

In exceptional circumstances, e.g. during alterations to a structure, when the outlet lies above LLW, a residual pressure head of 1.0 m above the outlet must be expected.

4.6.3 Layout

The relief wells are best placed in steel box and steel pipe piles, forming part of the outer face of the sheet piling. This not only simplifies construction but also positions the wells where they are protected and most effective for relief.

In tidal areas, the harbour water level at high tide generally lies above the artesian pressure head of the groundwater. In simple relief wells, the harbour water then flows into the wells and into the subsoil. This leads to rapid silting up of the relief well in silt-laden water because the flushing force at the bottom of the well at low tide is not sufficient to remove a silt deposit. In such cases relief wells must be equipped with effective anti-flood valves. Ball valves have proven ideal for this purpose. The valves must be easily removed for inspection of the wells, and re-installed again without damage to the watertight seal.

In addition, the filter zone of the wells is fitted with an insert which forces the groundwater flowing in the well through a narrow space between the pipe insert and the well bottom. This has the effect of forcefully flushing out any sediment which might accumulate (fig. R 53-1). Dredge cuts in the bottom surface layer are not sufficient for permanent relief of artesian pressure in silt-laden water. They silt up again, as do wells without flap valves.

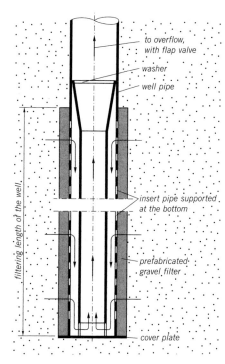

Fig. R 53-1. Flow into a relief well

Labels in figure:
- to overflow, with flap valve
- washer
- well pipe
- insert pipe supported at the bottom
- prefabricated gravel filter
- cover plate
- filtering length of the well

4.6.4 Filter

In order to achieve maximum capacity, the filters of the relief wells should be placed in the most permeable layer. The best filters must be used, completely immune to corrosion and clogging. The wells must be installed by an experienced contractor.

4.6.5 Checking

The effectiveness of the installation must be checked frequently through observation wells which lie inboard of the quay and which extend below the confining stratum.

If the required relief is no longer achieved, the wells must be cleaned and, if necessary, additional wells installed. Therefore, a sufficient number of steel box piles must be installed to be accessible from the deck of the quay.

Please also refer to R 19, section 4.2, last paragraph.

4.6.6 Radius of action

The area affected by a relieving installation is generally so small that harmful effects due to the lowering of the water table are not a problem, at least not in tidal areas. In special cases, however, this radius of action should also be investigated. Relieving wells should not be used if harmful effects are possible.

4.7   Assessment of groundwater flow (R 113)

4.7.1   General

Where a quay wall and other marine structures and their parts are to be built in flowing groundwater, proper planning and design are only possible when the designer is thoroughly familiar with the essential characteristics of the flowing groundwater. Only then is it possible to recognise and avoid risks and arrive at the best solution in terms of engineering and economics.

A laminar flow can be assumed for calculating the groundwater flow, provided a flow gradient $i = 1$ is not exceeded significantly and the flow velocity is less than approx. $6 \cdot 10^{-4}$ m/s (≈ 2 m/h). Other relationships between gradient, flow velocity and permeability can be found in, for example, [217].

In a laminar flow the flow velocity v is the product of the permeability k and the hydraulic gradient i (DARCY's law).

$$v = k \cdot i$$

4.7.2   Prerequisites for the determination of flow nets

A laminar groundwater flow follows the potential theory. Its solution is depicted by two sets of curves which intersect at right-angles and whose net widths have a constant ratio (fig. R 113-2).

In this flow net one set of curves represents the flow lines and the other the equipotential lines.

The flow lines are the paths of the water particles, whereas the equipotential lines are those corresponding to equal standpipe water heads (fig. R 113-2).

4.7.3   Determining the boundary conditions for a flow net

The boundary of a flow net can be a flow line or an equipotential line or, if the groundwater discharges into the air, a seepage line.

The following can be boundary flow lines: the surface of an impermeable layer of soil, the surface of an impermeable structure, the water table if it has a sloping, convex surface (seepage line), etc. (Fig. R 113-1).

The following can be boundary equipotential lines: a horizontal groundwater table, a river bottom, a submerged sloping bank, etc. For clarification, fig. R 113-1 shows the boundary conditions for several characteristic situations.

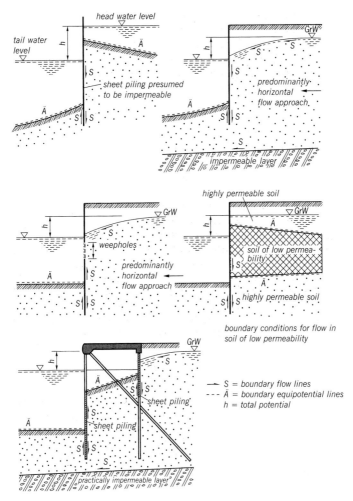

Fig. R 113-1. Boundary conditions for flow nets, characteristic examples with flow around the bottom of sheet piling

4.7.4   **Graphic procedure for determining the flow net**

One method for determining a flow net is the so-called graphic procedure, which still provides a rapid overview of critical flow zones for simple cases and steady flow conditions despite all the groundwater models which can be used easily today. However, it is generally presumed that the water flows through homogeneous soil.

Once all boundary conditions have been determined, the flow net is depicted according to the following rules:

- Flow lines are perpendicular to equipotential lines.
- The total potential difference Δh between the highest and lowest hydraulic potential is divided into equal (equidistant) potential steps dh (in fig. R 113-2 in 15 steps, i.e. every step is equal to a potential difference of 4.50 m/15 = 0.3 m).
- All flow lines pass through the available flow cross-section, i.e. they are closer together in narrower passages and are further apart in wider sections.
- The number of equipotential steps and flow lines is chosen such that neighbouring equipotential and flow lines form squares bounded by curved lines in order to guarantee the geometrical similarity across the whole net. The accuracy can be checked by drawing inscribed circles in the squares.

The procedure continues by trial and error until both the boundary conditions and the requirement for squares bounded by curved lines are fulfilled with adequate accuracy throughout the whole net. Incomplete flow channels or equipotential steps can be accepted along the edges. They are included in the calculations with their cross-section (see section 4.7.7.1).

4.7.5 Use of groundwater models

4.7.5.1 Physical and analogue models
Physical models use natural media (water, sand gravel, clay) on a model scale to be selected. They are mainly used today for research purposes in 3D problems and are no longer relevant for practical applications.
Analogue models use media whose movement is similar to the groundwater flow. Examples here are the movement of viscous materials between two closely spaced plates (gap models), the passage of electric current through conductive paper or a network of electrical resistances (electrical models). Another example is the deformation of thin skins by point load (seepage line with wells). In any case, kinematic similarity factors are required besides geometrical factors in order to convert a potential (e.g. electric voltage) into the hydraulic potential (water table head). However, these procedures, too, have largely disappeared from modern design offices and have been replaced by groundwater models running on powerful PCs.

4.7.5.2 Numerical groundwater models
Groundwater models are methods of calculation in which the entire equipotential field is divided into individual elements and represented by the equipotential heads of a sufficient number of support points. These points are the corner points (finite element method) or the gravity points

(finite differences method) of individual small but finite surfaces. Boundaries and inconstancies (wells, springs, drains, etc.) in the flow field must be represented by node points or element lines.

The user has to select the correct boundary conditions and the geohydraulic parameters, particularly when using programs which take account of unsteady flow conditions.

4.7.6 Calculation of individual hydraulic variables

Whereas the groundwater model calculates the entire hydraulic potential field and from this the distribution of gradients, velocities, discharge, etc., there are procedures that determine only individual variables.

Here are some examples:
- Resistance coefficient method after CHUGAEV for determining gradients and discharges of underseepage [168].
- Fragment procedure after PAVLOVSKY for calculating underseepage [167].
- Diagrams by DAVIDENHOFF & FRANKE for calculating hydraulic heads for excavations with sheet piling [166].

4.7.7 Evaluation of examples

4.7.7.1 Sheet piling with underseepage in homogeneous subsoil

In the flow net in fig. R 113-2 the potential difference Δh of 4.50 m is divided into 15 potential steps $dh = 0.30$ m between MSL +7 m and MSL +2.50 m. At a depth of MSL = –23 m there is an impermeable stratum whose upper surface represents the boundary of the model (boundary flow line).

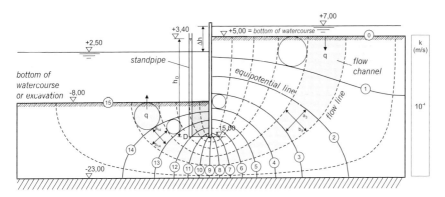

Fig. R 113-2. Example of a groundwater flow net in homogeneous soil with a vertical flow – case 1

91

The following characteristics for the potential field are apparent:

- Hydraulic head at point D (corner point of failure body after TERZAGHI – see R 115, section 3.2):

$$h_D = 7.00 - 12/15 \cdot 4.50 \text{ m} = 3.40 \text{ m } (= 2.50 + 3/15 \cdot 4.50 \text{ m})$$

- Hydraulic head h_F at toe of sheet piling:

$$h_F = 7.00 \text{ m} - 9/15 \cdot 4.50 \text{ m} = 4.30 \text{ m } (= 2.50 \text{ m} + 6/15 \cdot 4.50 \text{ m})$$

- Hydraulic gradients:

$$i_3 = \text{d}h/a_3 = 0.30/6.00 = 0.05; \quad i_{14} = \text{d}h/a_{14} = 0.30/4.3 = 0.07$$

The lengths a_3 and a_{14} were taken from the drawing.

- Discharge:
  The discharge is the same in every flow channel because all rectangles are mathematically similar and as large as the distance between two equipotential lines in a partial section. The discharge q is the product of the flow velocity v and the cross-sectional area A of the flow.

The following equation applies to an individual flow channel:

$$q = v \cdot A = k \cdot i \cdot A, \text{ for the planar case: } q_i = k \cdot i \cdot b_i$$

$$q_i = k \cdot \text{d}h/a_{14} \cdot b_{14} = k \cdot \text{d}h/a_3 \cdot b_3 = k \cdot \text{d}h \cdot b/a$$

$b/a = 1$ for a square net

The total discharge depends on the number of flow channels; incomplete flow channels are taken into account in accordance with their cross-section.
In fig. R 113-2 the cross-sectional area of the boundary flow line is only approx. 10% of that of a complete flow channel. Therefore, the discharge is as follows:

$$q = 6.1 \cdot 10^{-4} \cdot 0.3 \cdot 1 = 1.83 \cdot 10^{-4} \text{ m}^3/(\text{s} \cdot \text{m})$$

The calculation of the hydraulic heave failure at the base of the sheet pile wall is carried out according to R 115, section 3.2.

$$h_1 = h_D - h_{wu} \qquad h_1 = 3.4 - 2.5 = 1.1 \text{ m}$$

$$h_r = h_F - h_{wu} \qquad h_r = 4.3 - 2.5 = 1.8 \text{ m}$$

According to TERZAGHI the vertical flow force is as follows:

$$S_k' = \frac{\gamma_w \cdot (h_1 + h_r)}{2} \cdot \frac{t}{2} = \frac{10 \cdot (1.1 + 1.8)}{2} \cdot \frac{7}{2} = 50.75 \text{ kN/m}$$

The body of soil has a submerged weight amounting to:

$$G_k' = \frac{\gamma_B \cdot t^2}{2} = 10 \cdot 24.5 = 245 \text{ kN/m}$$

Adequate safety against hydraulic ground heave, even with unfavourable subsoil, is guaranteed when using the partial safety factors to DIN 1054 (section 0) for loading case 1:

$$S_k' \cdot \gamma_H \leq G_k' \cdot \gamma_{G,stb}$$

$$50.75 \cdot 1.8 < 245 \cdot 0.9$$

$$91.35 < 220.5$$

4.7.7.2   Sheet piling in flowing groundwater in stratified subsoil

The boundary conditions of fig. R 113-2 are retained, but there is a 2 m thick horizontal layer with a considerably lower permeability at differing depths. Flow and equipotential lines are calculated with a groundwater model.

In fig. R 113-3 we can see the concentration of equipotential lines in the less permeable strata, which considerably reduces the safety against boiling failure in case 2a, but increases it for case 2b. The critical water pressure W_{St} for calculating the heave failure is in each case at the underside of the impermeable stratum.

The condition for this potential distribution is that the stratum with low permeability continues for a sufficient distance in front of and behind the wall. Otherwise, the potential distribution is determined by the flow around and not through this stratum. Impermeable strata in particular require a careful check on the inflow side to ensure that the stratum extends for a sufficient distance and does not terminate at a distance that would enable a flow around the stratum. Such a situation would increase the water pressure below the impermeable stratum on the outflow side considerably. In cases of doubt, ignore the stratum on the inflow side.

To calculate the hydraulic heave failure according to R 115, section 3.2, in stratified soil it is necessary to find the critical section, or rather failure body. If there is a less permeable stratum above a permeable one, then generally the underside of the less permeable stratum forms the critical section, or rather the underside of the critical failure body. The critical hydraulic heads or gradients can be determined from the potential net:

93

The following applies in fig. R 113-3 for a high-level stratum with low permeability (case 2a):

- Hydraulic head at point D (underside of stratum with low permeability):

$$h_D = 7.00 - 1/15 \cdot 4.50 \text{ m} = 3.70 \text{ m} \ (= 2.50 + 4/15 \cdot 4.50 \text{ m})$$

- Mean hydraulic gradient at base of sheet pile wall:

$$i = \frac{\Delta h}{\Delta l} = \frac{3.70 - 2.50}{2.00} = 0.60$$

Case 2a

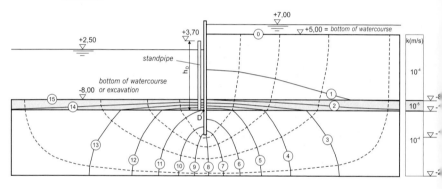

Case 2b

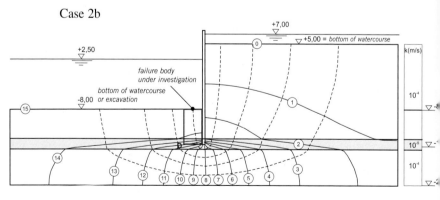

Fig. R 113-3. Potential nets in stratified soil with a vertical flow, stratum with low permeability at high level (case 2a) and at low level (case 2b)

The following applies in fig. R 113-3 for a low-level stratum with low permeability (case 2b):

- Hydraulic head at point D (corner point of failure body under investigation):

$$h_D = 7.00 - 12/15 \cdot 4.50 \text{ m} = 3.40 \text{ m} (= 2.50 + 3/15 \cdot 4.50 \text{ m})$$

- Hydraulic head at base of wall:

$$h_F = 7.00 - 9/15 \cdot 4.50 \text{ m} = 4.30 \text{ m} (= 2.50 + 6/15 \cdot 4.50 \text{ m})$$

- Characteristic value of flow force in body of soil in groundwater flow with a width of 3.0 m:

$$S'_k = [(3.4 - 2.5) + (4.3 - 2.5)]/2 \cdot 10 \cdot 3.0 = 40.50 \text{ kN/m}$$

4.7.7.3 Sheet piling in flowing groundwater with horizontal inflow

Fig. R 113-4 shows the effect of the assumptions regarding the boundary potential: for a horizontal groundwater inflow (case 3, ground surface and groundwater on right side of model at a level of MSL +7 m instead of surface water). The right-hand vertical edge of the model is a boundary potential line in this case.

The water pressure distribution at the wall and the risk of a hydraulic heave failure changes depending on the sequence of strata. Furthermore, the result is also heavily influenced by the distance between the wall in the flow and the boundary potential line with the maximum water level.

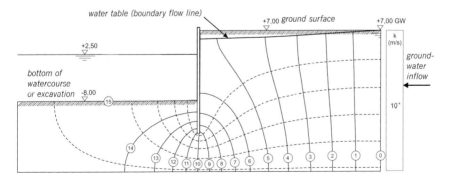

Fig. R 113-4. Potential net for a horizontal inflow (case 3)

4.8   Temporary stabilisation of waterfront structures by groundwater lowering (R 166)

4.8.1   General

The stability of a waterfront structure can be increased for a limited period of time by lowering the groundwater on the land side of the waterfront structure.

First of all, studies are necessary to ensure that the structure itself or other structures in the area influenced by groundwater lowering will not be endangered. In particular, there may be an increase in the negative skin friction at pile foundations.

The increase in stability brought about by groundwater lowering is attributable to:

- the decrease in water pressure difference, whereby even a supporting effect can be attained from the water side, and
- the increase in the effective mass forces of the passive earth pressure wedge through reducing the seepage pressure from below or vice versa through a water surcharge and seepage pressure from above.

These positive influences are counteracted by the following negative factors (although these are considerably less significant):

- the increase in active earth pressure on the structure as a result of increased weight density of the soil through the loss of uplift in the area where the groundwater has been lowered, and
- from case to case, an increase in active earth pressure through seepage pressure acting from top to bottom.

4.8.2   Case with soft, cohesive soil near the ground surface

If there is less permeable soft soil extending from the surface to a greater depth, which is underlain by well-permeable, non-cohesive soil (fig. R 166-1), the soil is initially not consolidated under the additional weight due to the lack of uplift in the range of the lowering depth Δh. Since in this state the active earth pressure coefficient is $K_{ag} = 1$ for the additional load, and since with cohesive soil $\gamma - \gamma' = \gamma_w$, the additional active earth pressure at the level of the lowered groundwater table at the start of consolidation $\Delta e_{ah} = \gamma_w \cdot \Delta h \cdot 1$.

The reduced water pressure difference is therefore fully compensated in the initial state by the increased active earth pressure in soft soil. With increasing consolidation, however, the additional active earth pressure drops to the value

$$\Delta e_a = \gamma_w \cdot \Delta h \cdot K_{ag} \cdot \cos \delta_a$$

On the passive earth pressure side, an increase in the water surcharge caused by groundwater lowering, augmented by flow pressure, has a

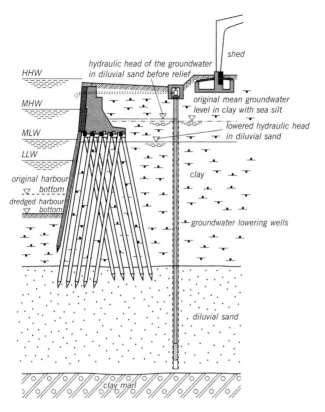

Fig. R 166-1. Example of quay wall stabilisation by means of groundwater lowering

favourable effect, mainly with non-cohesive sublayers (fig. R 166-1). In overlying cohesive soil, the state of consolidation must be taken into consideration accordingly.

4.8.3   Case according to section 4.8.2 but with stronger upper water inflow

If, deviating from fig. R 166-1, a strongly water-bearing non-cohesive stratum exists above the soft cohesive soil behind the waterfront structure, a predominantly vertical potential flow to the lower permeable stratum takes place in the uniformly cohesive sublayer during groundwater lowering. In this case the hydraulic head of the water in the overlying, highly permeable layer is applicable for the water pressure at the upper surface of the cohesive layer, and the hydraulic head of the groundwater in the lower, non-cohesive layer corresponding to the groundwater lowering is applicable for the water pressure at the lower surface of the cohesive layer. The changes in active and passive earth pressures depend on the

97

respective flow conditions and/or water surcharge, in which case the state of consolidation according to section 4.8.2 must be taken into account.

4.8.4 Conclusions for structure stabilisation

The effectiveness of stabilisation works through groundwater lowering is always ensured in the final steady state, but in the initial state is heavily dependent on the soil conditions. When this measure is used, the initial state and intermediate conditions must also be carefully considered. The method may then be used successfully for overloaded waterfront structures and, above all, for compensating any dredging of the harbour bottom in front of a waterfront structure. As a result, the final strengthening of the structure ultimately required, which is generally substantially more expensive, may be postponed to a time possibly more favourable in economic terms.

4.9 Flood protection walls in seaports (R 165)

4.9.1 General

Flood protection walls generally have the function of shielding harbour terrain from flood waters. They are also utilised, however, when flood protection cannot be attained with dykes.

The special additional demands for such types of walls are explained in the following sections.

4.9.2 Design water levels

4.9.2.1 Design water levels for high tide

(1) *Outer water level and theoretical level*

The theoretical height of a flood protection wall is determined according to the design still water level (design water level according to HHW) plus supplementary freeboard as local influences (waves) and, if necessary, wind-induced accumulation.

Due to the higher wave run-up on walls, the tops of flood protection walls are positioned higher than those of dykes, unless brief overflowing of the walls is acceptable.

Walls are generally less vulnerable to overflow than dykes. However, it must be ensured that an overflow of water cannot cause any scouring behind the wall and that the water can be drained away without causing any damage (see section 4.9.6.1).

(2) *Corresponding inner water level*

The corresponding inner water level is to be taken generally as the ground surface, unless other possible water levels, e.g. embankments, are less favourable (fig. R 165-1).

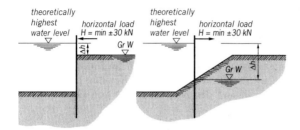

Fig. R 165-1. Design water levels at high water

For water levels as per section 4.9.2.1 (1) and (2), verification of stability for the wall is provided according to load case 3, as long as special forces according to section 4.9.5 are taken into consideration.

4.9.2.2 Design water levels for low tide

(1) Outer water levels

The mean tidal low water (MLW) is to be taken as normal low water in load case 1.

Extraordinarily low outer water levels, which occur only once a year, are to be classified as load case 2.

The lowest tidal low water (LLW) ever measured or a lower outer water level to be expected in future is to be considered as load case 3.

(2) Corresponding inner water levels

As a rule, the inner water level is to be taken as the ground surface, unless a lower water level can be allowed due to more accurate seepage flow investigations, or can be permanently ensured through structural

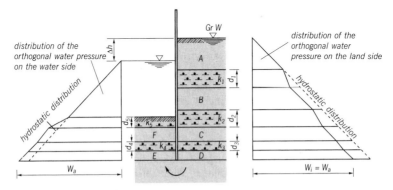

Fig. R 165-2. Example for the reduction of Δh (A, B, C, D, E, F are highly permeable layers)

measures such as drains. However, in case of drain failure, a safety factor ≥ 1.0 (hazard case) must still be available. In individual cases, the critical inner water level – if local circumstances are precisely known – can also be determined on the basis of observations of groundwater levels.

(3) Outgoing high tide
As the tide recedes, water level differences can occur that are similar to the situation at low tide (higher pressure on the land side) but lead to a higher load on the wall, e.g. like for a water level above ground level on the inside of the wall (fig. R 165-2).

4.9.3 Water pressure difference and use of weight density for the design of flood protection walls

4.9.3.1 Applications with almost homogeneous soils
Progression of the water pressure difference coordinates can be determined with the aid of an equipotential flow net according to R 113, section 4.7, or R 114, section 2.9.2. Changes in weight density due to the groundwater flow can be taken into account as per R 114, section 2.9.3.2.

4.9.3.2 Applications with stratified soils
Usually, horizontal or only slightly sloping boundary layers occur, so the following remarks will remain confined to these.

A difference in permeability between two strata is to be taken into account when the ratio of the permeabilities is greater than about 5.

The decrease in water pressure due to primarily vertical flow occurs almost entirely in the layers with lower permeability for permeability ratios $k_1/k_2 > 5$, provided that substantial disturbances do not prevail and that safety against hydraulic failure is ensured.

The flow resistance in the relatively permeable layers can be neglected. In approximately horizontal layers with very different permeabilities, a purely vertical flow may be applied to determine the water pressure coordinates if it can be guaranteed that the inflow takes place only on the surface or in the uppermost permeable stratum. An inflow of groundwater in lower strata leads to a completely different pressure distribution, possibly even to levels with confined groundwater. When applying a water pressure distribution according to a fig. R 165-2 it must also be ensured that the strata with low permeability extend a sufficient distance horizontally to prevent flow around these strata, and that the water inflow takes place exclusively from the uppermost permeable strata (A).

The decrease in Δh in the less permeable strata ΔW_i is proportional to the respective stratum thickness d_i and inversely proportional to the permeability k_i [m/s]. A decrease in potential when there is a flow around the base of the sheet pile wall is not taken into account owing to the higher permeability of the stratum at that position.

$$\Delta W_i = \gamma_w \cdot \Delta h \cdot \frac{d_i}{k_i} \cdot \frac{1}{\sum \dfrac{d_i}{k_i}}$$

Under these conditions, the weight density of the soil with water flowing through is changed practically only in the layers of low permeability. This change is given by the following equation:

$$\left| \Delta \gamma_i \right| = \gamma_w \cdot \Delta h \cdot \frac{1}{k_i} \cdot \frac{1}{\sum \dfrac{d_i}{k_i}}$$

A flow from top to bottom increases $\gamma (+\Delta \gamma_i)$, a flow from bottom to top decreases $\gamma (-\Delta \gamma_i)$.

If cleavage occurs between the wall and a less permeable layer as a result of wall flexure, this layer is to be regarded as fully permeable.
Relief of γ deserves special attention if on the water side the uppermost layer of low permeability is very close to or forms the ground surface. Here, $\Delta \gamma_i$ can be equal to γ_i, which leads to buoyancy of the top stratum. For support of the flood protection wall, passive earth pressure from cohesion only may then be considered for the upper stratum, but should preferably be completely neglected to increase safety.

4.9.4   **Minimum embedment depth of the flood protection wall**
The minimum embedment depth of the flood protection wall results from the structural calculations and the required verification of safety against ground failure. In doing so, take into account the reduction in weight density according to section 4.9.3.2 in the passive earth pressure area. Attention must also be paid to the following:

- The subsoil risks and the workmanship risks with regard to possible leaks (interlock damage) are to be considered, where
  - even just one defect in the flood protection wall can lead to failure of the entire structure, and
  - a suitability test for the design load case high tide (or flood water) is not possible.
- A longer driving depth is to be determined, taking account of a possibly unfavourable slope as per R 56, section 8.2.9.

Therefore, in the flood load case, the flow path in the ground should not fall below the following values:

- In homogeneous soils with relatively permeable soil structure and in the case of cleavage following wall flexure: the 4-fold difference between the design water level and the land-side ground surface (irrespective of the actual inner water level).
- In stratified soils with differences in permeability of at least 2 powers of 10: the 3-fold difference between the design water level and the ground surface (irrespective of the actual inner water level). Horizontal seepage paths may not be included when voids can occur.

4.9.5   Special loads acting on a flood protection wall

Apart from the usual live loads, loads of at least 30 kN deriving from the impact of drifting objects at high tide and the impact of land vehicles are to be considered (see fig. R 165-1). At risky locations with unfavourable current and wind conditions, or good accessibility, the impact loads must be taken to be substantially higher. Distribution of the loads through suitable structural measures is allowed if the functional capability of the flood protection wall is not impaired as a result.

In case of special loads, stability verification is allowed as per load case 3.

4.9.6   Structural measures – requirements

4.9.6.1   Surface stabilisation on the land side of the flood protection wall

Surface stabilisation must be provided to prevent land-side scouring from splashing water in the case of flooding. The width of the stabilisation measures should match at least the land-side height of the wall.

4.9.6.2   Defence road

The installation of a flood protection wall defence road with asphalt roadway, close to the wall, is recommended. It should be at least 2.50 m wide and can also serve as surface protection as per section 4.9.6.1.

4.9.6.3   Stress relief filter

A stress relief filter should be installed on the land side directly at the flood protection wall. The filter should be approx. 0.3–0.5 m wide so that no significant bottom water pressure can develop below the defence road.

In the case of sheet piling structures, it is sufficient for the land-side troughs to be filled with corresponding filter material (e.g. steel mill slag 35/55).

4.9.6.4   Watertightness of the sheet piling

An artificial interlock seal according to R 117, section 8.1.20, should be generally provided for the section of the sheet piling projecting above ground level.

4.9.7 Advice for flood protection walls in embankments

Generally, the low water levels of the outer water are applicable for the design of flood protection walls in or in the vicinity of embankments.

The passive earth pressure outside decreases simultaneously with the increase of loads from inner water pressure difference and increased specific density, also from seepage pressure. The changed water levels also frequently lead to a decrease in slope failure safety.

In the case of stratified soils (cohesive intermediate layers) which are not dowelled by adequately long sheet piles, usual verification of slope failure must be supplemented by verification of the stability of the soil wedge in front of the flood protection wall.

The outer slope is to be protected against scouring by hand-set stone pitching or equivalent measures. Safety against ground or slope failure is to be verified as per DIN 4084, at least for load case 2. Regular checks of these embankments are to be arranged.

4.9.8 Services in the area of flood protection walls

4.9.8.1 General

Services laid in the vicinity of flood protection walls can constitute weak points for several reasons, in particular:

- Leaking liquid service lines reduce internal erosion through the existing seepage path.
- Excavations to replace damaged services reduce the supportive effect of the passive earth pressure and also shorten the seepage path.
- Decommissioned services can leave uncontrolled cavities.

4.9.8.2 Services parallel to a flood protection wall

Services running parallel to a flood protection wall should not be installed within an adequately wide protective strip on both sides of the wall (> 15 m). Existing services should be repositioned or decommissioned. Any resulting or remaining voids must be properly filled.

Special attention must be given to any services that remain in protective strips:

- A trench reaching to the bottom of the pipeline for conceivable work on the buried service is to be considered when determining the passive earth pressure and in particular the seepage paths.
- Trench supports are to be designed for the passive earth pressure used in the wall calculation.
- It must be possible to seal off services carrying liquids on entering and leaving the protective strips by means of suitable shut-off devices.
- Work on services should not be carried out in the season at risk from storm tides, if this can be avoided.

4.9.8.3 Services crossing a flood protection wall

The passage of services through a flood protection wall always constitute potential weak points and should be avoided wherever possible. Therefore:

- The services should be routed over the wall, particularly high-pressure pipes or high-voltage cables.
- Individual services in the ground outside the protective strip should be grouped together and routed through the flood protection wall as a collective service line or bundle.
- Service crossings should be approximately at right-angles to the wall.

Structural measures are to be provided to take account of the differing settlement behaviour of services and flood protection wall (flexible sleeves, articulated pipes). Rigid crossings are not allowed.

Each service crossing is considered individually depending on the type of medium concerned.

- Cable crossings
  Communications cables and power cables may not be routed through the flood protection wall directly but must be protected in a sleeve. The space between cable and sleeve is to be sealed in a suitable manner.

- Pressurised services crossings
  Pressurised services (gas, water, etc.) are to be protected by a sleeve in the whole protective strip in such a way that, upon failure of the pressurised line, they can be replaced without excavations in the protective strip. The sleeve should be able to withstand the operating pressure with the same safety factors as the pressurised line. The same applies to the sealing between service and sleeve required on both sides of the flood protection wall. Use only durable materials for the sealing.

- Canal or culvert crossings
  If there is danger that water will be forced into the polder through canals or culverts during flooding, suitable double shut-off devices must be provided. Here, either a shaft with gate or sluice valve is to be installed on both sides of the flood protection wall, or both installations and a part of the protection wall should be combined in a structure with double gate or sluice valve safeguard. When the risk is less, one of the two valves can also be in the form of a flap valve.

- Within the protective strip, supply and disposal services are to be designed for loads as per load case 3. Stability verification is to be provided for load case 2. Resistance to abrasion, corrosion and other chemical attacks is to be given special attention.

- Dyke openings

  Dyke openings in connection with the flood protection wall are rated appropriately according to the proven design principles of sea dyke openings. The load cases given above also apply in this case.

- Decommissioned services

  Decommissioned services are to be removed from the protective strip. If this is not possible, the service voids are to be properly filled.

5 Ship dimensions and loads on waterfront structures

5.1 Ship dimensions (R 39)

5.1.1 Seagoing vessels

The following typical average ship dimensions given in the tables below may be used for the calculation and design of waterfront structures, fenders and dolphins. It should be remembered that these are average values which may vary up or down by up to 10%. The figures have been essentially determined statistically based on Lloyd's Register of Ships, April 2001 edition, and other unpublished information from Japan and Bremen, Germany, which together represents a very comprehensive database.

Definitions of the customary details concerning vessel sizes:
- Vessel dimensions are based on the Gross Register Tonnage (GRT), a dimensionless variable which is derived from the total volume of the ship.
- The deadweight tonnage (dwt) is specified in metric tonnes and indicates the maximum carrying capacity of a fully equipped, fully operational vessel. There is no mathematical relationship between the carrying capacity and the size of the ship.
- The displacement specifies the actual weight of the ship in metric tonnes, including the maximum loading.
- Container ships are often rated in terms of the number of containers they can accommodate, which is specified in Twenty feet Equivalent Units (TEU). A TEU is the smallest container length in use and is equal to 20 feet (ft) (= 6.10 m).

5.1.1.1 Passenger vessels (table R 39-1.1)

Tonnage	Carrying capacity	Dis-place-ment G	Overall length	Length between perps	Beam	Max. draught
GRT	dwt	t	m	m	m	m
70 000	–	37 600	260	220	33.1	7.6
50 000	–	27 900	231	197	30.5	7.6
30 000	–	17 700	194	166	26.8	7.6
20 000	–	12 300	169	146	24.2	7.6
15 000	–	9 500	153	132	22.5	5.6
10 000	–	6 600	133	116	20.4	4.8
7 000	–	4 830	117	103	18.6	4.1
5 000	–	3 580	104	92	17.1	3.6
3 000	–	2 270	87	78	15.1	3.0
2 000	–	1 580	76	68	13.6	2.5
1 000	–	850	60	54	11.4	1.9

5.1.1.2 Bulk carriers (table R 39-1.2)

Tonnage	Carrying capacity	Dis-place-ment G	Overall length	Length between perps	Beam	Max. draught
	dwt	t	m	m	m	m
–	250 000	273 000	322	314	50.4	19.4
–	200 000	221 000	303	294	47.1	18.2
–	150 000	168 000	279	270	43.0	16.7
–	100 000	115 000	248	239	37.9	14.8
–	70 000	81 900	224	215	32.3	13.3
–	50 000	59 600	204	194	32.3	12.0
–	30 000	36 700	176	167	26.1	10.3
–	20 000	25 000	157	148	23.0	9.2
–	15 000	19 100	145	135	21.0	8.4
–	10 000	13 000	129	120	18.5	7.5

5.1.1.3   General cargo freighters (table R 39-1.3)

Tonnage	Carrying capacity	Displacement G	Overall length	Length between perps	Beam	Max. draught
	dwt	t	m	m	m	m
–	40 000	51 100	197	186	28.6	12.0
–	30 000	39 000	181	170	26.4	10.9
–	20 000	26 600	159	149	23.6	9.6
–	15 000	20 300	146	136	21.8	8.7
–	10 000	13 900	128	120	19.5	7.6
–	7 000	9 900	115	107	17.6	6.8
–	5 000	7 210	104	96	16.0	6.1
–	3 000	4 460	88	82	13.9	5.1
–	2 000	3 040	78	72	12.4	4.5
–	1 000	1 580	63	58	10.3	3.6

There appears to be no trend towards construction of larger general cargo freighters. If necessary, the figures given in section 5.1.1.2 may be used accordingly.

5.1.1.4   Container ships (table R 39-1.4)

Carrying capacity	Displacement G	Overall length	Length between perps	Beam	Max. draught	No. of containers	Generation
dwt	t	m	m	m	m	TEU	
100 000	133 000	326	310	42.8	14.5	7 100	6th
90 000	120 000	313	298	42.8	14.5	6 400	6th
80 000	107 000	300	284	40.3	14.5	5 700	5th
70 000	93 600	285	270	40.3	14.0	4 900	5th
60 000	80 400	268	254	32.3	13.4	4 200	4th
50 000	67 200	250	237	32.3	12.6	3 500	3rd
40 000	53 900	230	217	32.3	11.8	2 800	3rd
30 000	40 700	206	194	30.2	10.8	2 100	2nd
25 000	34 100	192	181	28.8	10.2	1 700	2nd
20 000	27 500	177	165	25.4	9.5	1 300	2nd
15 000	20 900	158	148	23.3	8.7	1 000	1st
10 000	14 200	135	126	20.8	7.6	600	1st
7 000	10 300	118	109	20.1	6.8	400	1st

The width of a container ship depends on the maximum number of rows of containers that can be stacked on deck adjacent to one another. It should be remembered that the above table is based on data from the year 2001. At that time only 20 container ships with a deadweight of 90 000 or more were in operation worldwide.

However, the size of container ships is undergoing a very dynamic development. It is to be expected that ships with a TEU of 12 000 will be built in the foreseeable future. Such ships will have a carrying capacity of about 160 000 dwt, a displacement of about 220 000 t and will measure about 400 × 53 × 16 m (length × beam × design draught).

5.1.1.5 Ferries (table R 39-1.5)

Carrying capacity	Displacement G	Overall length	Length between perps	Beam	Max. draught
dwt	t	m	m	m	m
40 000	30 300	223	209	31.9	8.0
30 000	22 800	201	188	29.7	7.4
20 000	15 300	174	162	26.8	6.5
15 000	11 600	157	145	25.0	6.0
10 000	7 800	135	125	22.6	5.3
7 000	5 500	119	110	20.6	4.8
5 000	3 900	106	97	19.0	4.3
3 000	2 390	88	80	16.7	3.7
2 000	1 600	76	69	15.1	3.3
1 000	810	59	54	12.7	2.7

The dimensions of ferries depend heavily on the routes they sail and their particular purpose. The above dimensions should therefore be used for preliminary investigations only.

5.1.1.6 Ro-Ro ships (table R 39-1.6)

Carrying capacity	Displace-ment G	Overall length	Length between perps	Beam	Max. draught
dwt	t	m	m	m	m
30 000	45 600	229	211	30.3	11.3
20 000	31 300	198	182	27.4	9.7
15 000	24 000	178	163	25.6	8.7
10 000	16 500	153	141	23.1	7.5
7 000	11 900	135	123	21.2	6.6
5 000	8 710	119	109	19.5	5.8
3 000	5 430	99	90	17.2	4.8
2 000	3 730	85	78	15.6	4.1
1 000	1 970	66	60	13.2	3.2

5.1.1.7 Oil tankers (table R 39-1.7)

Carrying capacity	Displace-ment G	Overall length	Length between perps	Beam	Max. draught
dwt	t	m	m	m	m
300 000	337 000	354	342	57.0	20.1
200 000	229 000	311	300	50.3	17.9
150 000	174 000	284	273	46.0	16.4
100 000	118 000	250	240	40.6	14.6
50 000	60 800	201	192	32.3	11.9
20 000	25 300	151	143	24.6	9.1
10 000	13 100	121	114	19.9	7.5
5 000	6 740	97	91	16.0	6.1
2 000	2 810	73	68	12.1	4.7

5.1.1.8 LNG tankers (table R 39-1.8)

Carrying capacity	Capacity	Displacement G	Overall length	Length between perps	Beam	Max. draught
dwt	m³	t	m	m	m	m
100 000	155 000	125 000	305	294	50.0	12.5
70 000	110 000	100 000	280	269	45.0	11.5
50 000	77 000	75 000	255	245	38.0	10.5
20 000	30 500	34 000	195	185	30.0	8.5
10 000	15 000	19 000	148	135	26.0	7.0

5.1.1.9 LPG tankers (table R 39-1.9)

Carrying capacity	Capacity	Displacement G	Overall length	Length between perps	Beam	Max. draught
dwt	m³	t	m	m	m	m
70 000	105 000	90 000	260	250	38.0	14.0
50 000	65 000	65 000	230	220	35.0	13.0
20 000	20 000	27 000	170	160	25.0	10.5
10 000	10 000	15 000	130	120	21.0	9.0
5 000	5 000	8 000	110	100	18.0	6.8
2 000	2 000	3 500	90	75	13.0	5.5

5.1.2 River-sea ships (table R 39-2)

Tonnage	Carrying capacity	Displacement G	Overall length	Beam	Max. draught
GRT	dwt	t	m	m	m
999	3 200	3 700	94.0	12.8	4.2
499	1 795	2 600	81.0	11.3	3.6
299	1 100	1 500	69.0	9.5	3.0

5.1.3 Inland vessels (table R 39-3 and table R 39-3.1)
(see pages 112–114)

Table R 39-3.1. Classification of the European inland waterways

According to ECE resolution no. 30 dated 12.11.1992 – TRANS/SC 3R.153, the following classification applies to European waterways:

Type of inland waterway	Class of inland waterway	Motor vessels and barges in tow Type of vessel: general features					Push tow Type of pushed lighter: general features					Vertical clearance under a bridge [m] [2]	Graphical symbol on the map
		Designation	Max. length L [m]	Max. beam B [m]	Draft d [m] [7]	Tonnage T [t]	Formation	Length L [m]	Beam B [m]	Draft d [m] [7]	Tonnage T [t]		
1	2	3	4	5	6	7	8	9	10	11	12	13	14
Elbe river west of the — of regional significance	I	Peniche	38.5	5.05	1.8–2.2	250–400						4.0	
	II	Kempenaar	50–55	6.6	2.5	400–650						4.0–5.0	
Elbe river east of the — of regional significance	I	Large Finow	41	4.7	1.4	180						3.0	
	II	BM-500	57	7.5–9.0	1.6	500–630						3.0	
	III	6)	67–70	8.2–9.0	1.6–2.0	470–700		118–132[1]	8.2–9.0[1]	1.6–2.0	1000–1200	4.0	
of international significance	IV	Johann Welker	80–85	9.5	2.5	1000–1500		85	9.5[5]	2.50–2.80	1250–1450	5.25 or 7.00[4]	
	Va	large Rhine ship	95–110	11.40	2.50–2.80	1500–3000		96–110[1]	11.40	2.50–4.50	1600–3000	5.25 or 7.00[4]	
	Vb							172–185[1]	11.40	2.50–4.50	3200–6000	7.00 or 9.10[4]	
	VIa							95–110[1]	22.80	2.50–4.50	3200–6000	7.00 or 9.10[4]	
	VIb	3)	140	15.00	3.90			185–195[1]	22.80	2.50–4.50	6400–12000	7.00 or 9.10[4]	
	VIc							270–280[1] 195–200[1]	22.80 33.00–34.20[1]	2.50–4.50	9600–18000 9600–18000	9.10[4]	
	VII							285	33.00– ^)	2.50–4.50	14500–	9.10[4]	

Footnotes for the classification table:

1) The first number considers the current situation, whereas the second shows both future developments and, in some cases, the existing situation.
2) Considers a safety clearance of approx. 30 cm between the highest fixed point of the ship or its cargo and a bridge.
3) Considers the dimensions of vessels under own power expected in Ro-/Ro- and container traffic. The stated dimensions are approximate values.
4) Rated for transporting containers:
   – 5.25 m for ships with two layers of containers,
   – 7.00 m for ships with three layers of containers,
   – 9.10 m for ships with four layers of containers.
   – 50 % of the containers can be empty, otherwise ballast is required.
5) Some existing waterways can be allocated to class IV on account of the greatest permissible length of ships and barges, although the greatest beam is 11.40 m and the largest draft 4.00 m.
6) Vessels used in the region of the Oder and on the waterways between Oder and Elbe.
7) The draft for a specific federal waterway is to be ascertained according to the local conditions.
8) On certain sections of waterways in class VII, push tows can be used consisting of a larger number of lighters. Here the horizontal dimensions can exceed the values stated in the table.

113

Table R 39-3. Inland vessels

Designation	Carrying capacity	Displacement G	Length	Beam	Draught
	t	t	m	m	m
Motor freighters:					
Large Rhine ship	4 500	5 200	135.0	11.4	4.5
2600-ton class	2 600	2 950	110.0	11.4	2.7
Rhine ship	2 000	2 385	95.0	11.4	2.7
"Europe" ship	1 350	1 650	80.0	9.5	2.5
Dortmund-Ems-Canal ship	1 000	1 235	67.0	8.2	2.5
Large-Canal-Class ship	950	1 150	82.0	9.5	2.0
Large-"Plauer"-Class ship	700	840	67.0	8.2	2.0
BM-500 ship	650	780	55.0	8.0	1.8
Kempenaar	600	765	50.0	6.6	2.5
Barge	415	505	32.5	8.2	2.0
Peniche	300	405	38.5	5.0	2.2
Large-Saale-Class ship	300	400	52.0	6.6	2.0
Large-Finow-Class ship	250	300	41.5	5.1	1.8
Push lighters:					
Europe IIa	2 940	3 275	76.5	11.4	4.0
	1 520	1 885			2.5
Europe II	2 520	2 835	76.5	11.4	3.5
	1 660	1 990			2.5
Europe I	1 880	2 110	70.0	9.5	3.5
	1 240	1 480			2.5
Carrier ship lighters:					
Seabee	860	1 020	29.7	10.7	3.2
Lash	376	488	18.8	9.5	2.7
Push tows:					
with one lighter Europe IIa	2 940	3 520[1]	110.0	11.4	4.0
	1 520	2 130[1]			2.5
with 2 lighters Europe IIa	5 880	6 795[1]	185.0	11.4	4.0
			110.0	22.8	4.0
	3 040	4 015[1]			2.5
with 4 lighters Europe IIa	11 760	13 640[2]	185.0	22.8	4.0
	6 080	8 080[2]			2.5

[1] Push vessel 1 480 kW; approx. 245 t displacement
[2] Push vessel 2963–3333 kW; approx. 540 t displacement

5.1.4   Displacement

The displacement G [t] is the product of the length between perpendiculars, the beam, the draught, the block coefficient c_B and the density ρ_w [t/m^3] of water. The block coefficient varies between 0.50 and 0.90 for seagoing vessels, 0.80 and 0.90 for inland vessels, and 0.90 and 0.93 for push lighters.

5.2   Assumed berthing pressure of vessels at quays (R 38)

In preparing the design, accidental impacts need not be taken into consideration but only the usual berthing loads. The magnitude of these berthing loads depends on the ship dimensions, the berthing velocity, the fenders and the deformation of the ship's hull and the structure.

If the failure of the waterfront structure due to an accident (e.g. ship impact) could lead to high risks, e.g. for other buildings and structures immediately inland of the waterfront, this should be given due consideration in the planning work and appropriate measures should be agreed between the designers, developers and authorities concerned.

In order to give the quay sufficient stability against normal berthing loads, but on the other hand to avoid unnecessarily large dimensions, it is recommended that the front wall be designed for a concentrated impact load equal to the magnitude of the relevant line pull force at any position, without the total stresses exceeding the permissible limits. Berthing impact for quay walls in seaports according to R 12, section 5.12.2, with the values in table R 12-1, and for quay walls in inland harbours 100 kN according to R 102, section 5.13.2.

This concentrated force may be distributed according to the fendering available; if there are no fenders, then distribution over a square with a side length of 0.50 m is recommended. In sheet pile walls without solid superstructures, only the walings and bolts need be designed for this force.

The berthing loads on dolphins are dealt with in R 128, section 13.3.

5.3   Berthing velocities of vessels transverse to berth (R 40)

When vessels make their approach transverse to a berth, it is recommended that the following berthing velocities, which correspond to the Spanish ROM [197], be taken into consideration when designing the corresponding fendering:

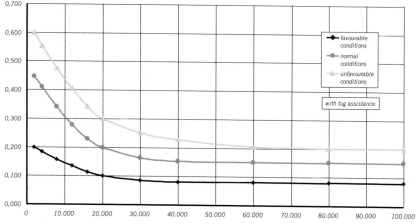

Fig. R 40-1. Berthing velocities with tug assistance

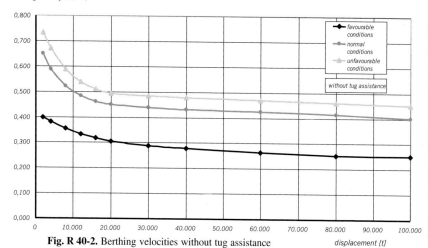

Fig. R 40-2. Berthing velocities without tug assistance

5.4 Load cases (R 18)

DIN 1054, section 6.3.3, defines load cases for verifying stability and allocating the appropriate partial safety factors. These arise from the combinations of actions in conjunction with the safety classes for the resistances. The following classifications apply for waterfront structures:

5.4.1 Load case 1

Loads due to active earth pressure (in unconsolidated, cohesive soils, separately for initial and final states) and to water pressure differences where unfavourable outer and inner water levels occur frequently (see R 19, section 4.2), earth pressures resulting from the normal live loads, from normal crane loads and pile loads, direct surcharges from dead loads and normal live loads.

5.4.2 Load case 2

As for load case 1, but with restricted scour caused by currents or ship's screw action, and together with the following, insofar as they can occur simultaneously: water pressure difference for rare, unfavourable outer and inner water levels (see R 19, section 4.2), water pressure difference caused by anticipated regular flooding of the waterfront structure, the suction effect of passing ships, loads and active earth pressure from exceptional local surcharges; combination of earth and water pressures with wave loads from frequent waves (see R 136, section 5.6.4); combinations of earth and water pressure differences with short-term horizontal tension, compression and impact loads such as line pull on bollards, pressure on fenders or lateral crane loads, loads from temporary construction conditions.

5.4.3 Load case 3

As for load case 2, but with exceptional design situations such as surcharges on larger areas not previously allowed for, unusual slumping of an underwater slope in front of the sheet piling, unusual scouring due to currents or ship's screws, water pressure differences following extreme water level situations (see R 19, section 4.2, and R 165, section 4.9), water pressure difference following exceptional flooding of the banks, combination of earth and water pressures with wave loads from rare waves (see R 136, section 5.6.4); combination of earth and water pressures with impact of drifting materials according to section 4.9.5, all loading combinations in conjunction with ice formation or ice pressure.

5.4.4 Exceptional cases

In exceptional cases it may be appropriate to use partial safety factors $\gamma_F = \gamma_R = 1.00$ for actions and resistances. Examples of this are the simultaneous occurrence of extreme water levels and extreme wave loads due to plunging breakers according to R 135, section 5.7.3, extreme water levels with simultaneous complete failure of the drainage system (see R 165, section 4.9.2), combinations of three simultaneous short-term events, e.g. high tide (HHW, see R 165, section 4.9.2), rare waves (see R 136, section 5.6.4) and impact of drifting materials (see R 165, section 4.9.5).

5.5   Vertical live loads (R 5)

All quantitative loads (actions) stated in this section are characteristic values.

5.5.1   General

Vertical live loads (variable loads in the meaning of DIN ENV 1991-1) are essentially the loads resulting from stored materials and the loads from vehicular traffic. The load actions of rail-mounted or vehicular mobile cranes must be considered separately, insofar as they exert any effect on the waterfront structures. At waterfront structures in inland ports, the latter is generally only the case for waterfront structures that are expressly intended for handling heavy loads with mobile cranes. In seaports, in addition to the rail-mounted quayside cranes, mobile cranes are being used increasingly for general cargo handling, i.e. not only for heavy loads.

We distinguish between three different basic types (table R 5-1) for the dynamic influences of the loads:

In basic type 1 the loadbearing members of the structures are driven over directly by the vehicles and/or loaded directly by the stacked materials, e.g. piers (table R 5-1a).

In basic type 2 the load from vehicles and the stacked materials acts on a more or less deep bedding course, which distributes and transmits the loads to the structural members. This type of design is used, for example, below elevated structures over slopes with a load-distributing bedding layer on the pier slab (table R 5-1 b).

In basic type 3 the load from vehicles and the stacked goods acts only on the solid mass of earth fill behind the waterfront structure, which consequently is subject only indirectly to additional stresses from the live loads as the result of increased active earth pressure. Simple sheet piling bulkheads or partially sloped banks are typical examples (table R 5-1c).

In addition, there are also hybrid types, e.g. piled structures on piles with a short pile cap.

If complete and reliable calculations are available, the live loads should normally be taken at the anticipated magnitude. Any increases in the live loads subsequently necessary can be better accommodated within the permissible limits, the greater the dead load share and the better the distribution of loads in the structure. Loadbearing systems according to basic type 2, and in particular basic type 3, offer particular advantages in this respect.

Please refer to R 18, section 5.4, for allocation of the corresponding loads to load cases 1, 2 and 3.

Table R 5-1. Vertical live loads

Basic type	Traffic life loads[1]					Storage area outside water front cargo-handling area
	Railroad	Roads				
		Vehicles	Road bound cranes	Light weight traffic		
a) BT 1	Loads as per RIL 804 resp. DIN Special Report 101	Loads as per DIN 1055 or DIN-report 101	Fork lift loads as per DIN 1055 Clow loads for mobile cranes as per section 5.5.5 and 5.14.3	$5 kN/m^2$		Loads according to the use actually anticipated in accordance with Section 5.5.6
	Impact factor: The parts exceeding 1.0 can be decreased by half					
b) BT 2	As BT 1, but further reduction on the impact factor to 1.0 at bedding layer, thickness $h = 1.00$ m. For bedding layer thickness $h \geq 1.50$ m uniformly distributed surface load of $20 kN/m^2$					
c) BT 3	Loads as in BT 2 with a bedding layer thickness of more than 1.50 m					

[1] Crane loads are to be taken as stipulated in R 84, Section 5.14.

5.5.2 Basic type 1

Railway live loads correspond to load diagram 71 of DIN Special Report 101. The load assumptions according to DIN 1055 or DIN Special Report 101 are to be applied for road traffic. Generally, load model 1 is to be assumed here. In the dynamic coefficients given for railway bridges, with which the live loads are to be multiplied, the parts exceeding 1.0 can as a rule be halved because of the slow speed. Load model 1 already assumes a slow speed (congestion situation) for road bridges and so a further reduction is not permissible. For piers in seaports, loads from fork-lifts are to be taken according to DIN 1055 and the outriggers of mobile cranes (see Section 5.5.5 and table R 84-1, section 5.14.3).

5.5.3 Basic type 2

Essentially the same as basic type 1. The dynamic coefficients for railway bridges, however, may be further reduced linearly according to bedding layer thickness, and completely ignored when the bed is at least 1.00 m thick (measured from the top of the rails when the rails are embedded in the pavement). The load section by section is, however, still to be taken into account.

If the bedding layer thickness is at least 1.50 m, the total live load from railroads can be replaced by a uniformly distributed load of 20 kN/m^2.

5.5.4 Basic type 3

Loads as for basic type 2, with a bedding layer thickness of more than 1.50 m.

5.5.5 Load assumptions directly behind the head of the waterfront structure

When working with heavy vehicular cranes or similar heavy-duty vehicles and heavy construction plant, such as crawler excavators and similar, which manoeuvre directly behind the front edge of the waterfront structure, the following are to be applied for the design of the uppermost parts of the structure, inclusive of any upper anchoring:

a) live load = 60 kN/m^2 from rear edge of wall, inboard for a width of 2.0 m, or

b) live load = 40 kN/m^2 from rear edge of wall, inboard for a width of 3.50 m.

Both a) and b) include the effects from an outrigger load up to $P = 550$ kN insofar as the distance between the axis of the waterfront structure and the axis of the outrigger is at least 2.0 m. For higher outriggers loads see section 5.14.3.

5.5.6 Loads outside the waterfront cargo handling area

Outside the waterfront cargo handling area, the following live loads are taken as the basis in accordance with [140], working on the basis of 300 kN gross load for 40 ft containers and 200 kN for 20 ft containers.

- Light traffic (cars)   5 kN/m^2
- General traffic (HGVs)   10 kN/m^2
- General cargo   20 kN/m^2
- Containers:
  - empty, stacked 4 high   15 kN/m^2
  - full, stacked 2 high   35 kN/m^2
  - full, stacked 4 high   55 kN/m^2
- Ro-Ro loads   30–50 kN/m^2
- Multipurpose facilities   50 kN/m^2
- Offshore supply bases   55–150 kN/m^2
- Paper   depending on the
- Timber products   bulk/stacking height

- Steel   recommended values
- Coal   for specific density
- Ore   to DIN 1055

Further details regarding the material properties of bulk and stacked goods are to be found in the tables of ROM 02.-90 [197].

When calculating the active earth pressure of retaining structures, as a rule the differing loads in the cargo handling and container areas can be grouped together to produce an average distributed load of 30 to 50 kN/m^2.

5.6   Determining the "design wave" for maritime and port structures (R 136)

5.6.1   General

The wave loads on maritime and port structures are essentially a result of the sea conditions as created by the wind. The significance of this for the design in accordance with the local boundary conditions must be investigated. At the shore it is generally not just the local sea conditions that are relevant to the design because low water depths and the lengths over which the wind is effective help to limit the amount of energy in the waves. It is far more relevant to consider the local sea conditions (wind-generated waves) in conjunction with the sea conditions on the open sea beyond the area of the project together with the sea conditions (swell) approaching the shore.

The following descriptions are limited to fundamental processes and simplified approaches for determining hydraulic boundary conditions and the loads on the structure. Detailed information on this can be found in, for example, EAK 2002 [46].

It is recommended that an institute or consulting engineers experienced in coastal engineering should be consulted when it comes to investigating the wave conditions in the planning area and the specific loads on the structure. The need for detailed physical or numerical simulations should be examined carefully prior to beginning the detailed design work.

5.6.2   Description of the sea conditions

The natural sea conditions can basically be described as an irregular chronological sequence of waves of different height (or amplitude), period (or frequency) and direction representing a chronological and spatial superimposition of various short- and long-period sea condition components. The direct influence of the wind causes irregular sea conditions with a short period also known as wind-generated waves. Long-period irregular sea conditions are caused by the superimposition of wave components with a uniform direction in which a sorting of the waves caused by various interactions takes place and the sea conditions are no longer influenced by the wind directly.

The natural irregular sea conditions prevailing in the area of a project are comprised of the local short-period sea conditions (wind-generated waves) plus the long-period sea conditions (swell), which originally evolved as wind-generated waves outside the area of the project.

Given the fact that it is necessary to consider the actual sea conditions that exist and are relevant to the design as a loading variable in an existing design method, it is first necessary to parameterise the irregular sea conditions because, as a rule, only individual characteristic sea condition parameters (see section 5.6.3) can be included in the calculations. This parameterisation of the irregular sea conditions [46] can take place both

(1) in a time domain (direct short-term statistical evaluation of the time series) by determining and depicting characteristic wave parameters (wave heights and periods) as arithmetical mean values, and also

(2) in a frequency domain (Fourier analysis) by determining and depicting a wave spectrum, where the energy content of the sea conditions is included as a function of the wave frequency.

The parameterisation of the sea conditions relevant to the design inevitably results in the loss of some information regarding the wave time series, its statistics and the wave spectrum. It is necessary to consider the outcome of examining the sea conditions in terms of time and frequency domain analyses in conjunction with the location of the project when planning and designing maritime and port structures. In the majority of cases it is normally sufficient to parameterise the sea conditions relevant to the design and characterise them by way of the individual parameters wave height, period and direction. However, complex wind and wave conditions, and shallow water regions with breaking waves especially, may render it necessary to examine further local sea condition characteristics in order to define the loading variables reliably for the design method [46].

5.6.3 Determining the sea condition parameters

5.6.3.1 General

The sea condition parameters define and quantify certain properties of the irregular sea conditions that change with place and time. Depending on the method of evaluation chosen (see section 5.6.2) these are

(1) in the time domain mean values of individual parameters such as wave heights or periods or combinations of these, and

(2) in the frequency domain striking frequencies or integral variables from the spectral density of the sea conditions spectrum.

The wave conditions in the project area must be analysed with respect to the theoretical probability of their occurrence based on measurements or observations gathered over a sufficiently long period of time. To do this, the significant wave parameters such as wave heights, periods and approach directions resulting from the short-term statistical analysis are to be determined (depending on the task in hand) according to their seasonal frequencies or long-term maximum values in order to derive meaningful information for the design. If such measurements are not available, empirical–theoretical or numerical methods for determining the wave parameters from the wind data (hindcasting) must be employed and verified, if possible, by using any wave measurement figures available.

The parameterisation of the natural sea conditions is carried out on the premise that there are statistical relationships between the heights of individual natural waves recorded in the measurements. According to LONGUET-HIGGINS [26] these relationships can be described by the Rayleigh distribution, assuming a narrow-band wave spectrum and a multitude of different waves (see section 3.7.4 in EAK 2002 [46], and CEM [225]).

In deep water ($d \geq L/2$) there is very good agreement between the wave height distributions based on measurements and the Rayleigh distribution, even in the case of broadband wave spectra.

In shallow water ($d \leq L/20$) there are larger deviations between the measured wave height distribution and the theoretical Rayleigh distribution because the waves are affected by the shallow water effect (see section 5.6.5). The wave spectrum in shallow water no longer exhibits a narrow band structure and the associated wave height distribution can differ substantially from the Rayleigh distribution owing to the breaking of the waves.

The deviation from the Rayleigh distribution increases with the height of the waves and decreases as the spectral bandwidth narrows. The Rayleigh distribution tends to overestimate the height of waves in all water depth ranges.

The determination of the sea condition parameters in the time and frequency domains and their designations is explained in detail below.

5.6.3.2 Sea condition parameters in the time domain

In the time domain, evaluation of the wave heights and periods recorded during a period of observation are described by way of stochastic variables of the frequency distribution. The Rayleigh function for describing the probability $P(H)$ of the occurrence of a wave with height H (individual probability) or the probability $P(H)$ of the occurrence of a number of waves up to a height H (cumulative probability) can be used as a good approximation for the wave heights:

$$P(H) = 1 - e^{-\frac{\pi}{4}\left(\frac{H}{H_m}\right)_2}$$

Probability of not attaining this value:

$$P(H < H_s) = 1 - \exp\left[-2 \cdot \left(\frac{H}{H_s}\right)^2\right]$$

Fig. R 136-1 is modified after [229], where

n	=	frequency of the wave heights H during the period of observation, expressed as a percentage
H_m	=	mean value of all wave heights recorded during a period of observation
H_d	=	most frequent wave height
$H_{1/3}$	=	mean value of 33% of the highest waves
$H_{1/10}$	=	mean value of 10% of the highest waves
$H_{1/100}$	=	mean value of 1% of the highest waves
$\max H$	=	maximum wave height

The frequency distribution of the wave heights results in the following approximate correlations after [24] and [26], assuming the theoretical wave height distribution of the sea conditions corresponding to the Rayleigh distribution. These theoretical correlations match well with those determined from actual measurements despite the possible presence of a wave spectrum bandwidth larger than that assumed by the Rayleigh distribution:

$$H_m = 0.63 \cdot H_{1/3}$$
$$H_{1/10} = 1.27 \cdot H_{1/3}$$
$$H_{1/100} = 1.67 \cdot H_{1/3}$$

The maximum wave height H_{max} principally depends on the number of waves recorded within the period of measurement available. According to LONGUET-HIGGINS [26] the use of the approach

$$H_{max} = 0.707 \cdot \sqrt{\ln(n)} \cdot H_{1/3}$$

results in a maximum wave height $H_{max} = 1.86 \, H_{1/3}$ for $n = 1000$ waves. For practical engineering purposes, the maximum wave height can be adequately estimated with

$$H_{max} = 2 \cdot H_{1/3}$$

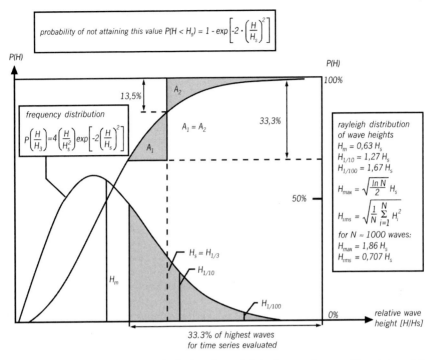

$$P(H < H_s) = 1 - exp\left[-2 \cdot \left(\frac{H}{H_s}\right)^2\right]$$

probability of not attaining this value

P(H)

13,5%

A_2

frequency distribution

$$P\left(\frac{H}{H_3}\right) = 4\left(\frac{H}{H_s^2}\right)exp\left[-2\left(\frac{H}{H_s}\right)^2\right]$$

A_1

$A_1 = A_2$

33,3%

H_m

$H_s = H_{1/3}$

$H_{1/10}$

$H_{1/100}$

P(H)

100%

rayleigh distribution of wave heights
$H_m = 0,63\ H_s$
$H_{1/10} = 1,27\ H_s$
$H_{1/100} = 1,67\ H_s$

$$H_{max} = \sqrt{\frac{ln\ N}{2}}\ H_s$$

$$H_{rms} = \sqrt{\frac{1}{N}\sum_{i=1}^{N}H_i^2}$$

for $N \approx 1000$ waves:
$H_{max} = 1,86\ H_s$
$H_{rms} = 0,707\ H_s$

50%

0%

relative wave height [H/Hs]

33.3% of highest waves
for time series evaluated

Fig. R 136-1. Rayleigh distribution of wave heights for natural sea conditions (schematic)

Another sea condition parameter frequently encountered in practice is the wave height H_{rms} (rms = root mean square). In the sea conditions with a Rayleigh distribution this results in the correlation $H_{rms} = 0.7\ H_{1/3}$. The wave periods in the time domain can be estimated on the basis of natural measurements in a similar way to the wave height correlations [46]. Among the factors that affect the true wave height and weight period correlations are the actual wave height distribution, the actual form of the sea conditions spectrum and the period of measurements, and in shallow water in particular the correlations can deviate from the aforementioned theoretical values owing to the actual distribution of the waves and their asymmetry. Short measuring periods of, for example, 5 or 10 minutes can lead to considerable errors when determining the correlations. Therefore, EAK 2002 recommends a measuring period of at least 30 minutes for sea conditions in order to take account of statistical regularities.

125

5.6.3.3 Sea condition parameters in the frequency domain

The parameterisation of the irregular sea conditions in the frequency domain involves converting the time series of sea condition measurements into an energy density spectrum by superimposing the individual wave components, and the associated wave phases into a corresponding phase spectrum (Fourier transformation). The common depiction of the sea conditions spectrum for all wave directions is known as a "one-dimensional spectrum" and the separate depiction for various wave directions as a "directional spectrum".

The sea conditions spectrum enables the following (and other) characteristic sea condition parameters to be specified as a function of the frequency f [Hz] taking into account spectral moments of the nth order

$$m_n = \int S(f) \cdot f^n \, df \quad \text{where } n = 0, 1, 2 \dots$$

and also as a function of the wave approach direction:

H_{m0} = characteristic wave height = $4\,m_0^{1/2}$ where m_0 is the area below the wave spectrum
T_{01} = mean period = m_0/m_1
T_{02} = mean period = m_0/m_2
T_p = peak period = wave period at maximum energy density

There is a fixed relationship between the wave periods T_{01} and T_{02}, which depends on the form of the wave spectrum and describes the bandwidth of the wave spectrum.

The wave spectrum helps to identify, in particular, long-period wave components such as swell waves, wave components transformed because of the structure, or changes to the spectrum due to shallow water. Such aspects can be significant when defining the hydraulic boundary conditions for the design of maritime and port structures.

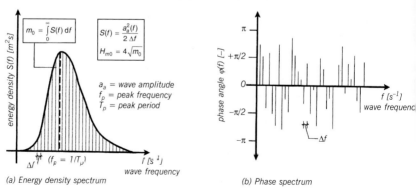

(a) Energy density spectrum   (b) Phase spectrum

Fig. R 136-2. Parameters of a wave spectrum – sketches of definitions [229]

5.6.3.4 Correlations between sea condition parameters in the time and frequency domains

The "significant wave height" H_s was introduced to characterise the irregular sea conditions for practical engineering applications [228]. Assuming a Rayleigh distribution of the sea conditions, it is assumed that the "significant wave height" H_s can be determined by the wave height $H_{1/3}$ of the time domain or the wave height H_{m0} of the frequency domain for practical engineering purposes.

$$H_s = H_{1/3} = H_{m0}$$

Furthermore, the wave periods T_m (time domain) and T_{02} (frequency domain) are theoretically equal. For further correlations between sea condition parameters in the time and frequency domains, which can always vary depending on the respective wave spectrum, please refer to EAK 2002.

5.6.4 Design concepts and definition of design parameters

The design sea conditions are taken to be the sea condition event that leads to the relevant loading of a structure or structural component, or describes its characteristic effect and is the result of a relevant combination of various influencing variables.

In terms of the design of structures, we distinguish between
- the structural design as verification of the stability for an extreme event, and
- the functional design, which deals with the effect and the influence of the structure on its environment (see section 3.7 in EAK 2002 [46]).

The irregularity of the sea conditions and the description of these as input variables for appropriate methods of design is critical for determining the true magnitudes of the loads that occur. Depending on the structure to be designed and the method of design to be employed, the design sea conditions

- can be used as a characteristic single wave in order to determine a defined low (deterministic method; possible application: load on a flood protection wall), or
- can be considered as a characteristic wave time series, where the result is a time series of the loads acting on the structure which are evaluated statistically and with respect to the maximum and total loads (stochastic method; possible application: structures resolved into individual members, e.g. offshore platforms), or
- can be input as a completely statistical distribution from which a probability of failure of the structure can be determined taking into account various limit states of the structure (probabilistic method).

127

However, deterministic methods of design prevail in practical engineering design and these will be looked at in detail below. EAK 2002 provides advice on how to take into account the irregularity of the sea conditions based on regular waves in investigations, calculations and design analyses.

As the measurements of both wave and wind conditions seldom match the intended period of use of the structure or the return periods associated with extreme sea conditions, it is necessary to extrapolate the available wave data over a longer period of time (frequently 50 or 100 years) by employing a suitable theoretical distribution (e.g. Weibull). Extrapolation over a period of time equal to three times the period of measurement should not be carried out. The theoretical return period and hence the parameters of the design wave H_d should be specified taking into account the potential damage or the permissible risk of flooding or severe damage to the structure (type of failure), but also the database and other aspects (structural planning).

In terms of functional planning, considerably shorter return periods must sometimes be used in order to estimate the average restrictions on usage and risk situations to be expected.

The ratio of design wave height H_d to significant wave height $H_{1/3}$ should be taken as 2.0 in the case of high safety requirements. However, a more accurate analysis of the true loads and stability properties of the structure by way of hydraulic model tests is recommended if reliable and economic solutions are to be achieved.

If the frequency distribution used is based on long periods of observation or corresponding extrapolation (approx. 50 years), or corresponding theoretical or computational work, the design wave determined from this can be defined as a rare wave in load case 3 according to section 5.4.3. In the case of shorter periods of observation or periods of analysis, the design wave determined should be defined as a frequent wave in load case 2 according to section 5.4.2.

Table R 136-1. Recommendations for specifying the design wave height

Structure	$H_d / H_{1/3}$
Breakwater	1.0 to 1.5
Sloping moles	1.5 to 1.8
Vertical moles	1.8 to 2.0
Flood protection walls	1.8 to 2.0
Quay walls with wave chambers	1.8 to 2.0
Excavation enclosures	1.5 to 2.0

5.6.5 Transformation of the sea conditions

Only in exceptional circumstances do we know the wave conditions in the immediate vicinity of the structure to be designed. Generally, therefore, the sea conditions from deep water have to be transformed to the planning segment on the shore. As the waves reach shallow water, or strike obstacles, various effects become relevant:

(1) Shoaling effect
When a wave touches bottom, its velocity and wavelength are thereby decreased. After a local, insignificant diminution, the height therefore increases steadily as the wave approaches the shore, caused by an energy equilibrium (up to the point of breaking). This is known as the shoaling effect [28].

(2) Bottom friction and percolation
The wave height is decreased by frictional losses and by an exchange process at the bottom. These effects are usually negligible for design purposes [23].

(3) Refraction
Due to the varying contact with the bottom in the case of waves approaching at an angle to the shore (or rather to the depth contours), the waves turn towards the shore owing to the varying local shoaling effect in such a way that, depending on the shape of the shoreline, the effective wave energy may be reduced – but also increased – if the wave energy is concentrated on, for example, a spit of land.

(4) Breaking waves
Generally, waves can break either when the limiting steepness is exceeded (parameter H/L) or the wave height reaches a certain dimension compared with the depth of water (parameter H/d). The height of deep water waves running into shallow water is limited by the breaking action when the associated limiting water depth is exceeded. Generally, the ratio of breaker height H_b to limiting water depth d_b is

$$0.8 < H_b/d_b < 1.0 \text{ (breaker criterion)}.$$

However, higher values have been observed in special cases [33]. The outcome of the different wave heights in a sea conditions spectrum is that the breaking of the waves mostly takes place in a so-called surf zone, the position and extent of which is determined by seabed topography, tides and other factors.

Strictly speaking, the ratio of breaker height H_b to water depth d_b is a function of the shore slope α and the steepness of the deep water wave H_0/L_0. These parameters are taken into account in the refraction coefficient ξ, which specifies, approximately, the type of breaker for regular waves (i.e. surging/collapsing, plunging or spilling breakers). For more detailed information please refer to [34, 35] and [46].

129

Table 136-2. Definitions of breaker types (the values are based on investigations with slopes with gradients from 1 : 5 to 1 : 20)

Breaker form	ξ_0	ξ_b
Surging/collapsing breaker	> 3.3	> 2.0
Plunging breaker	0.5 bis 3.3	0.4 bis 2.0
Spilling breaker	< 0.5	< 0.4

The refraction coefficient ξ can be used for both the deep water wave height H_o (ξ_0) and also the wave height at the breaking point H_b (ξ_b) (see table R 136-2). The following relationships apply:

α = angle of slope of bottom [°]
H/L_0 = wave steepness
H = local wave height
L_0 = length of incoming deep water wave
ξ = $\dfrac{\tan \alpha}{\sqrt{H / L_0}}$ = refraction coefficient

Structures with a high reflective effect and also influences due to the coastal geometry can have a considerable influence on the breaking process. Appropriate breaker criteria are then required (see, for example, section 5.7.3).

(5) Diffraction
Diffraction occurs when waves encounter obstacles (structures, but also, for example, offshore islands). Following the change of direction before reaching the obstacle, the waves spill into the wake of the obstacle in such a way that energy transport takes place along the crest of the wave, which generally reduces the height of the wave. At certain points beyond the wake, however, increased wave heights (caused by superimposition of diffraction waves in the vicinity of, for example, closely positioned obstacles) are also possible [21].

(6) Reflections caused by the structure
Waves approaching the shore or the structure are reflected to a certain extent, which essentially depends on the properties of the reflecting contours (including inclination, roughness, porosity) and the depth of the water in front of the structure. Waves that approach vertically but do not break are repelled almost completely when striking a vertical structure, forming, theoretically, a stationary wave with twice the height of the incoming wave. The reflection coefficient for a sloping structure is also heavily dependent on the steepness of the wave and hence is variable for the waves contained in the wave spectrum.

130

As the aforementioned influences depend on many factors related to the structure and the local conditions, it is not possible to make a general statement. For further information see [46, 218] and [229].

5.7 Wave pressure on vertical waterfront structures in coastal areas (R 135)

5.7.1 General

The wave pressure and/or wave movements on the front of a waterfront structure are to be taken into account for:

- blockwork walls in uplift and in joint water pressure,
- embankments below elevated decks with non-backfilled front wall, taking into account the effective water pressure difference on both sides of the wall,
- non-backfilled sheet piling walls,
- flood protection walls,
- stresses during construction,
- backfilled structures in general, also because of the lowered outer water level in the wave trough.

Furthermore, waterfront structures are loaded by waves through line pulls, vessel impacts and fender pressures from ship movements.

When taking into account wave pressure on vertical waterfront structures, we distinguish between three loading modes as follows:

(1) The structure is subject to load from non-breaking waves.
(2) The structure is subject to load from waves breaking on it.
(3) The structure is subject to load from waves that have already broken at some distance from the wall.

Which of these three loading modes is relevant depends on the water depth, the sea conditions, and the morphological and topographical conditions in the area of the planned structure.

Loading assumptions for the above loading modes are explained in the following sections. The method of determining the loads due to stationary, breaking or broken waves in natural sea conditions after GODA [26] and [46) can also be used. The empirically determined dynamic pressure increase coefficient after TAKAHASHI [224] can be used to calculate dynamic pressure loads. The disadvantage of this method is that it covers only load components facing the shore.

5.7.2 Loads due to non-breaking waves

A structure with a vertical or approximately vertical front wall, in water of such a depth that the highest incoming waves do not break, is loaded as a result of reflection by the increased water pressure difference on the water side at the crest of the reflected wave, and on the landward side at the wave trough.

Standing waves are formed when incoming waves are superimposed on the backwash. In reality, true standing waves never occur: the irregularity of the waves creates certain wave impact loads, but these are generally negligible compared to the following load assumptions and so are considered to be practically static. The wave height doubles as a result of reflection when the waves meet a vertical or approximately vertical wall and if no losses are incurred (reflection coefficient $\kappa_R = 1.0$). A reduction in the wave height due to partial reflection ($\kappa_R < 0.9$) at vertical walls should only be taken into consideration when verified through large-scale model tests. Otherwise, please refer to the reflection coefficients in the EAK.

SAINFLOU's method [20], as shown in fig. R 135-1, is recommended for calculating wave action when the wave approach is at right-angles to the wall. However, the loads calculated by this method are marginally too large if the waves are steep, whereas the loads from very-long-period, shallow waves are underestimated. For further details and other methods of design, e.g. MICHE-RUNDGREN, please refer to CEM [225] and EAK 2002 [46].

Definition of symbols used in fig. R 135-1:

H = height of incoming wave

L = length of incoming wave

h = difference in level between still water level and mean water level in the reflected wave on the face of the wall,

$$= \frac{\pi \cdot H^2}{L} \cdot \coth \frac{2 \cdot \pi \cdot d}{L}$$

Δh = difference between the still water level in front of the wall and the groundwater or rearward harbour water table

d_s = depth of groundwater or rearward harbour water table

γ = specific density of water

p_1 = pressure increase (wave crest) or decrease (wave trough) at the base of the structure due to the wave effect

$$= \gamma \cdot H / \cosh \frac{2 \cdot \pi \cdot d}{L}$$

p_0 = maximum water pressure difference ordinate at the level of the land-side water table according to fig. R 135-1c)

$$= (p_1 + \gamma \cdot d) \cdot \frac{H + h - \Delta h}{H + h + d}$$

p_x = water pressure difference ordinate at the level of the wave trough according to fig. R 135-1d)

$$= \gamma \cdot (H - h + \Delta h)$$

132

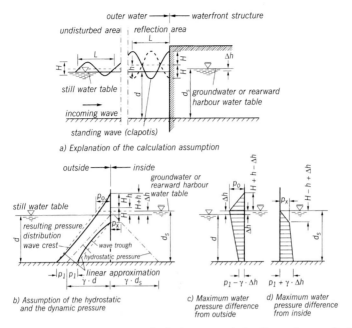

a) Explanation of the calculation assumption

b) Assumption of the hydrostatic and the dynamic pressure

c) Maximum water pressure difference from outside

d) Maximum water pressure difference from inside

Fig. R 135-1. Dynamic pressure distribution on a vertical wall at total wave reflection, in accordance with SAINFLOU [20], as well as water pressure difference at wave crest and wave trough

Application of this procedure to the case of oblique wave approach is dealt with in [30]. According to that, the assumptions for right-angled wave approach should also be used for an acute-angled wave approach, especially for long structures.

5.7.3   Loads due to waves breaking on the structure

Waves breaking on a structure can exert extreme impact pressures of $10\,000$ kN/m^2 and more. These pressure peaks, however, are very localised and of very brief duration (1/100 s to 1/1000 s).

Owing to the huge pressure impulses and dynamic loads that can occur, the structure should be suitably arranged and built to ensure that – as far as possible – high waves do not break immediately at the structure. If this is not possible, model studies at the largest possible scale are recommended for the final design. For further information regarding design for pressure impacts, please refer to EAK 2002, section 4.3.23 [46], and CEM [225].

The following method of calculation may be used for simple geometries.

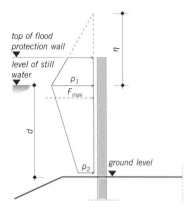

top of flood
protection wall

level of still
water

η

p_1

F_{max}

d

p_2 ground level

Fig. R 135-2. Loads due to plunging breakers

Tests on a large-scale hydraulic model of a caisson structure founded on rubble resulted in the following approximation for the impact pressure load on vertical walls [218, 220].

The maximum static horizontal force F_{max} on the quay wall based on fig. R 135-2 is calculated using the following equation:

$$F_{max} = \varphi \cdot 8.0 \cdot \rho \cdot g \cdot H_b^2 \; [kN/m]$$

The point of application of this force lies just below the still water level. An approximation for reducing the load as a result of overwash is explained in [218].

- **Breaking wave height H_b**

    The wave steepness-related breaker criterium developed for relatively steep embankments [218, 219] leads to the following equation:

    $$H_b = L_b \cdot [0.1025 + 0.0217 \, (1 - \chi_R) \, / \, (1 - \chi_R)] \tanh (2 \, \pi \, d_b \, / \, L_b)$$

    where:

    χ_R = reflection coefficient of quay wall
    d_b = depth of water at breaking point
    L_b = wavelength of breaking wave = $L_0 \tanh (2 \, \pi \, d_b \, / \, L_b)]$

    Taking a reflection coefficient of 0.9 and assuming that the depth of water d_b and the wavelength L_b are approximately equal to the corresponding values onshore (depth of water d at the wall and wavelength L_d), we get:

    $$H_b \cong 0.1 \cdot L_0 \cdot [\tanh (2 \, \pi \, d \, / \, L_d)]^2$$

    where:

    L_0 = wavelength in deep water = $1.56 \cdot T_p^2$
    T_p = peak period in wave spectrum
    L_d = wavelength in depth of water $d \cong L_0 \cdot [\tan h \, (2 \, \pi \, d \, / \, L_0)^{3/4}]^{2/3}$

134

- **Impact factor φ**

  The impact factors given below were derived from calculations for the dynamic interaction of impulse-type, wave pressure impact loads varying over time with the action effect and deformation conditions of the structure and the subsoil [218].

  The impact factor $\varphi = M_{dyn}/M_{stat}$ depends on the level of the section being analysed (wall moment for impact-like load/wall moment for quasi-static load) [218, 220, 221].

  Walls with yielding support in the subsoil (e.g. cantilever walls, see example in fig. R 135-2, or walls supported at a level deeper than 1.50 m below ground level) [222]:

  $\varphi = 1.2$ for all analyses above 1.50 m below ground level
  $\varphi = 0.8$ for all analyses below 1.50 m below ground level

  Walls with rigid support (e.g. concrete walls on quay structures) or walls supported at a level higher than 1.50 m below ground level:

  $\varphi = 1.4$ for all analyses above 1.50 m below ground level
  $\varphi = 1.0$ for all analyses below 1.50 m below ground level

- **Pressure ordinate p_1 at still water level**

  $$p_1 = F_{max} / [0.625 \cdot d_b + 0.65 \cdot H_b]$$

  η = level of pressure figure (difference in height between top of wave pressure load and still water level) = $1.3 \cdot H_b$

- **Pressure ordinate p_2 at ground level**

  $$p_2 = 0.25 \cdot p_1$$

5.7.4 Load from broken waves

An approximate calculation of the forces from a broken wave is possible according to [21]. In this case it is assumed that the broken wave continues with the same height and velocity it had at breaking; however, this overestimates the actual loads. For a more accurate analysis of the actual loads, EAK 2002 proposes a correction to the characteristic values based on the method by CAMFIELD [226], which is not discussed further here.

5.7.5 Additional loads caused by waves

If the structure standing on a permeable bedding course does not have a watertight face, e.g. is in the form of an impervious diaphragm, additional bottom water pressure from the effects of the waves must also be taken into account along with the water pressure on the wall surfaces. This also applies to the water pressure in joints between blocks.

5.8 Loads arising from surging and receding waves due to inflow or outflow of water (R 185)

5.8.1 General

Surging and receding waves arise in bodies of water as a result of temporary or temporarily increased inflow/outflow of water. However, surging and receding waves manifest themselves significantly only with wetted cross-sections of bodies of water that are small in comparison with the inflow/outflow volume per second. Therefore, great importance is generally attached to allowing for surging and receding waves and their effects on waterfront structures only in navigation channels. In such cases, the effects of changes in water levels on embankments, the linings to bodies of water, revetments and other facilities must be taken into account.

5.8.2 Determination of wave values

Surging and receding waves are shallow-water waves in the range:

$$\frac{d}{L} < 0.05$$

The wavelength depends on the duration of the water inflow or outflow. The wave propagation velocity can be roughly calculated with:

$$c = \sqrt{g \cdot (d \pm 1.5\,H)}\ \text{[m/s]} \quad \begin{cases} + \text{ for surging} \\ - \text{ for receding} \end{cases}$$

where:

g = acceleration due to gravity
d = depth of water
H = rise in the case of surging or fall in the case of receding as compared with still water level

If the H/d ratio is small,

$$c = \sqrt{g \cdot d}$$

can be used.

The rise or fall in the water level is roughly

$$H = \pm \frac{Q}{c \cdot B}$$

where:

Q = volume of water inflow/outflow per second
B = mean breadth of water level

The wave height can increase or decrease as a result of reflections or subsequent surging or receding waves. The wave attenuation is small in the case of uniform channel cross-sections and smooth channel linings in particular, which means that the waves can run backwards and forwards several times, especially with short reaches.

In navigation channels the most frequent cause of surging and receding phenomena is the inflow or outflow of lockage water. In order to prevent extreme surging and receding phenomena, the volume of lockage water is generally restricted to 70 to max. 90 m^3/s.

Locking of vessels at intervals equal to the reflection time or a multiple of this can lead to superimposition of the waves and hence to an increase in the surging and receding values, especially in canals.

5.8.3 Load assumptions

Load assumptions for waterfront structures must take account of the hydrostatic load arising from the height of the surging or receding wave and its possible superimposition on reflected or subsequent waves, as well as fluctuations in the water table possible simultaneously, e.g. from wind-induced water build-up, ship's waves, etc., in the least favourable combination. Owing to the long-period nature of surging and receding waves, the effect of the resulting flow gradient of the groundwater must be examined in the case of permeable revetments as well.

The dynamic effects of surging and receding waves can be ignored because of the mostly low flow rates caused by such waves.

The resulting loads are characteristic values which are to be multiplied by the partial safety factors for load case 2 (see R 18, section 5.4.2) according to DIN 1054.

5.9 Effects of waves from ship movements (R 186)

5.9.1 General

Waves of different types are generated by the bow and stern of every moving vessel. Depending on the local circumstances, these cause different loads on waterfront structures or their structural members. The moving ship first causes a tensioning of the water surface in front of the bow, the size of which in the direction of movement can amount to several times the length of the ship. A further, local build-up occurs directly in front of the ship's bow. At the ship itself the previously undisturbed cross-section of the watercourse is reduced by the cross-section of the ship. The flow around the ship must now take place within a reduced flow cross-section. In line with the laws of hydraulics, this causes an acceleration of the flow. The increase in the flow velocity past the ship compared to the ship's velocity through the water is called backflow. This is in turn linked with a lowering of the water level adjacent to the

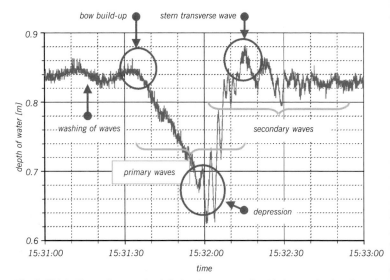

Fig. R 186-1. Change in water level during the passage of a ship in a navigation channel of limited width

ship. This lowering of the water level further reduces the flow cross-section and leads to a further lowering. The flow conditions are equalised again at the stern of the ship, characterised by a rise in the water level – the stern transverse wave. The entire depression alongside the ship has the character of a long wave and is designated as a primary wave. Its wavelength is equal to the length of the ship. In canals the depression stretches across the full width of the canal. In wider navigation channels the depression (effective backflow cross-section) decreases with increasing distance from the ship. The width of the depression in a first approximation is equal to a multiple of the length of the ship (see [129]). At the same time, regular, short-period waves, designated secondary waves (fig. R186-1), are generated at the bow of the ship. These are, on the one hand, divergence waves, which propagate at an angle to the axis of the ship, and, on the other hand, transverse waves, which are approximately perpendicular to the axis of the ship (figs. R 186-2 and R 186-3). The superimposition of these two systems generates an interference line, which exhibits a characteristic angle to the axis of the ship which depends on the speed of the ship. At customary subcritical ship speeds, this angle is approx. 19°; as we approach the headwater wave velocity it reaches a maximum of 90°. The headwater wave velocity is the wave velocity of shallow water waves. Inland vessels do not usually reach this speed, especially when the depth and width of the watercourse

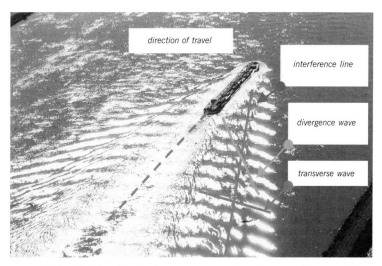

direction of travel

interference line

divergence wave

transverse wave

Fig. R 186-2. Aerial view showing the wave pattern on the surface around a moving inland vessel; the secondary waves and wash of the ship's screw can be clearly seen

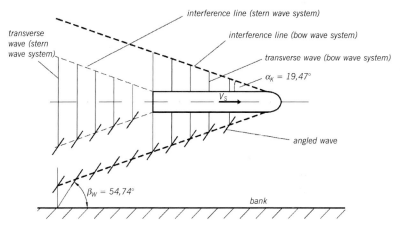

interference line (stern wave system)

transverse wave (stern wave system)

interference line (bow wave system)

transverse wave (bow wave system)

$\alpha_K = 19{,}47°$

v_S

angled wave

$\beta_W = 54{,}74°$

bank

Fig. R 186-3. Wave pattern (schematic)

is restricted. Their speed is limited by the critical speed. At this speed the flow around the ship changes from normal flow to supercritical flow. The signs of this are a stern low in the water and the occurrence of a breaking stern wave. The fact that the speed of the ship is limited means that the aforementioned angle of approx. 19° for the propagation of the interference waves usually applies as we approach the critical ship speed as well.

5.9.2 Wave heights

It is necessary to take into account the effects of ship wave systems on embankments and revetments, for restricted navigation channels in particular. The relevant design loads are given by the increase and decrease in pressure in the region of the depression of the primary wave system and the breaking of the waves from the primary (stern wave) and secondary wave system (angled waves and stern transverse waves) at the transition to the shallow water area at the embankment, and depending on the direction of the incoming waves. The rock size required for the embankment is generally governed by the load due to breaking stern transverse waves. Due to their hydrostatic pressure fluctuations, the headwater wave and lowering of the water level influence the pore water pressures in the subsoil via a temporary pressure build-up and lead to a destabilisation of the bank strengthening measures. Consequently, these normally determine the necessary thickness of the revetment [129]. If reflections are possible, e.g. in short branches with a perpendicular junction (e.g. lock basins), the height of the build-up or depth of the lowering can increase by a factor of two. More accurate figures can be obtained from model tests. The chronological progression of the build-up or the lowering may need to be considered in terms of its effect on groundwater movements in the case of permeable waterfront structures. The reader's attention is drawn to the possible effects on automatic gates, e.g. dyke culverts (sudden opening and closing of the gates as a result of sudden changes in pressure) and lock gates.

The build-up in front of the ship can be treated as a "single wave", i.e. as a wave with just one crest above still water level. The height of the build-up is generally low and seldom exceeds 0.2 m above still water level.

The height of the waves on the interference line can be estimated as follows (after [129]):

$$H_{sec} = A_W \frac{v_S^{8/3}}{g^{4/3}(u')^{1/3}} f_{cr}$$

where:

A_W = wave height coefficient, depending on form and dimensions of ship, loaded draught and depth of water
A_W = 0.25 for conventional inland vessels and tugs
A_W = 0.35 for unladen, single-row push tows
A_W = 0.80 for fully laden, multi-row push tows
f_{cr} = speed coefficient
$f_{cr} = 1$ is valid for $v_S/v_{crit} < 0.8$
g = gravitational acceleration
H_{sec} = secondary wave height
u' = distance between ship's side and bank

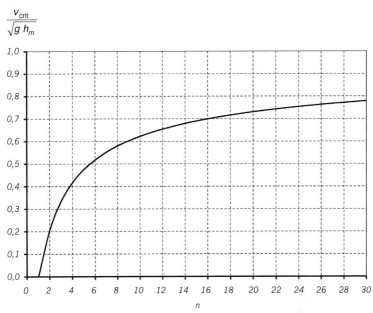

Fig. R 186-4. Relationship between critical ship speed v_{crit}, average depth of water h_m, gravitational acceleration g and cross-section ratio n for canals

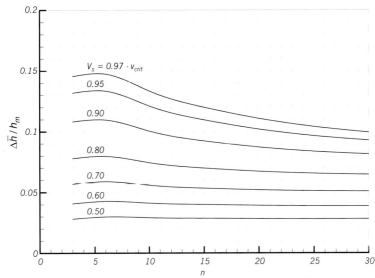

Fig. R 186-5. Relationship between average lowering of water depth $\Delta\bar{h}$, average depth of water h_m, and relative, i.e. v_{crit}-related, ship speed v_S for various cross-section ratios

141

The lowering of the water level corresponds with the backflow below and alongside the immersed hull of the ship and in terms of form and size depends on the shape, drive and speed of the ship as well as the conditions in the navigation channel (ratio n of watercourse cross-section effective for backflow to immersed main frame cross-section of the ship, proximity and form of bank). The maximum lowering seldom exceeds approx. 15% of the depth of the water, even upon reaching the critical ship speed. Figs. R 186-4 and R 186-5 enable the designer to make a cautious estimate of the magnitude of the lowering depending on n and the critical ship speed. Lowering and wave height change with the distance from the ship and, primarily, in the proximity of the banks (see, for example, [129]). The maximum values obtained can be used as the design values.

5.10   Wave pressure on pile structures (R 159)

5.10.1   General

In the design of pile structures, the loads originating from the wave motion are to be taken into account with respect to the loads on each individual pile as well as on the entire pile structure, insofar as this is required by local conditions. The superstructures should be located above the crest of the design wave wherever possible. Otherwise, large horizontal and vertical loads from direct wave action can affect the superstructures, the determination of which is not the object of this recommendation because reliable values for such cases can only be obtained from model studies. The level of the crest of the design wave is to be determined taking into account the highest still water level occurring simultaneously, where applicable also taking into account the wind-raised water level, the influence of the tides and the raising and steepening of the waves in shallow water.

The superimposition method according to Morison, O'Brien, Johnson & Schaaf [38] is suitable for slender structural members, whereas the method for wider structures is based on the diffraction theory [39].

The object of this recommendation is the superimposition method according to Morison [21], which is applicable for non-breaking waves. Due to lack of accurate calculation methods for breaking waves, a makeshift method is proposed in section 5.10.5.

The method according to Morison furnishes useful values, if

$$\frac{D}{L} \le 0.05$$

for the individual pile

where:

D = pile diameter or – for non-circular piles – characteristic width of structural member (width transverse to direction of incoming wave)

L = length of "design wave" in accordance with R 136, section 5.6, in conjunction with table R 159-1, No. 3

This criterion is mostly fulfilled.

For determination of the wave loads, please refer to [42] and [21], which contain tables and diagrams for performing the calculation. The diagrams in [21] were developed on the basis of the stream-function theory and may be applied to waves of differing steepness up to the breaking limit, whereas the diagrams in [42] are applicable only when assuming a linear wave theory.

Other design methods are valid for offshore structures, e.g. according to API (American Petroleum Institute).

5.10.2 Calculation method according to MORISON [38]

The wave load on an individual pile consists of the components

- force due to the water particle velocity (drag force), and
- force due to the water particle acceleration (inertia force),

which must be determined separately and superimposed according to phases.

According to [41, 43] and [21], the horizontal total load per unit length for a vertical pile is:

$$p = p_D + p_M = C_D \cdot \frac{1}{2} \cdot \frac{\gamma_w}{g} \cdot D \cdot u \cdot |u| + C_M \cdot \frac{\gamma_w}{g} \cdot F \cdot \frac{\partial u}{\partial t}$$

Accordingly, for a pile with circular cross-section, this is

$$p = C_D \cdot \frac{1}{2} \cdot \frac{\gamma_w}{g} \cdot D \cdot u \cdot |u| + C_M \cdot \frac{\gamma_w}{g} \cdot \frac{D^2 \cdot \pi}{4} \cdot \frac{\partial u}{\partial t}$$

where:

p_D = pressure due to the water particle velocity caused by the flow resistance per unit length of pile

p_M = inertia force due to unsteady wave motion per unit length of pile

p = total load per unit length of pile

C_D = drag coefficient taking into account resistance of pile against flow pressure

C_M = inertia coefficient taking into account resistance of pile against acceleration of water particles

Table R 159-1. Linear wave theory. Physical relationships [28]

	Shallow water $\dfrac{d}{L} < \dfrac{1}{20}$	Transition area $\dfrac{1}{20} < \dfrac{d}{L} < \dfrac{1}{2}$	Deep water $\dfrac{d}{L} > \dfrac{1}{2}$
1. Profile of the free surface	General equation $\eta = \dfrac{H}{2} \cdot \cos\vartheta$		
2. Wave velocity	$c = \dfrac{L}{T} = \dfrac{g}{\omega}\,kd = \sqrt{gd}$	$c = \dfrac{L}{T} = \dfrac{g}{\omega}\tanh(kd) = \sqrt{\dfrac{g}{k}\tanh(kd)}$	$c = \dfrac{L}{T} = \dfrac{g}{\omega} = \sqrt{\dfrac{g}{k}}$
3. Wave length	$L = c\cdot T = \dfrac{g}{\omega}\,kdT = \sqrt{gd}\cdot T$	$L = c\cdot T = \dfrac{g}{\omega}\tanh(kd)\cdot T = \sqrt{\dfrac{g}{k}\tanh(kd)}\cdot T$	$L = c\cdot T = \dfrac{g}{\omega}\cdot T = \sqrt{\dfrac{g}{k}}\cdot T$
4. Velocity of the water particles			
a) Horizontal	$u = \dfrac{H}{2}\sqrt{\dfrac{g}{d}}\cdot\cos\vartheta$	$u = \dfrac{H}{2}\cdot\omega\cdot\dfrac{\cosh[k(z+d)]}{\sinh(kd)}\cdot\cos\vartheta$	$u = \dfrac{H}{2}\cdot\omega\cdot e^{kz}\cdot\cos\vartheta$
b) Vertical	$w = \dfrac{H}{2}\cdot\omega\cdot\left(1+\dfrac{z}{d}\right)\sin\vartheta$	$w = \dfrac{H}{2}\cdot\omega\cdot\dfrac{\sinh[k(z+d)]}{\sinh(kd)}\cdot\sin\vartheta$	$w = \dfrac{H}{2}\cdot\omega\cdot e^{kz}\cdot\sin\vartheta$
5. Acceleration of the water particles			
a) Horizontal	$\dfrac{\partial u}{\partial t} = \dfrac{H}{2}\cdot\omega\cdot\sqrt{\dfrac{g}{d}}\cdot\sin\vartheta$	$\dfrac{\partial u}{\partial t} = \dfrac{H}{2}\cdot\omega^2\cdot\dfrac{\cosh[k(z+d)]}{\sinh(kd)}\cdot\sin\vartheta$	$\dfrac{\partial u}{\partial t} = \dfrac{H}{2}\cdot\omega^2\cdot e^{kz}\cdot\sin\vartheta$
b) Vertical	$\dfrac{\partial w}{\partial t} = -\dfrac{H}{2}\cdot\omega^2\cdot\left(1+\dfrac{z}{d}\right)\cos\vartheta$	$\dfrac{\partial w}{\partial t} = \dfrac{H}{2}\cdot\omega^2\cdot\dfrac{\sinh[k(z+d)]}{\sinh(kd)}\cdot\cos\vartheta$	$\dfrac{\partial w}{\partial t} = \dfrac{H}{2}\cdot\omega^2\cdot e^{kz}\cdot\cos\vartheta$

g	=	gravitational acceleration
γ_w	=	weight density of water
u	=	horizontal component of velocity of water particles at pile location examined
$\dfrac{\partial u}{\partial t} \approx \dfrac{du}{dt}$	=	horizontal component of acceleration of water particles at pile location examined
D	=	pile diameter or – for non-circular piles – characteristic width of structural member
F	=	cross-sectional area of the pile in the flow at area examined in direction of flow

The velocity and acceleration of the water particles are calculated from the wave equations. Different wave theories may be used. For the linear wave theory, the relationships required are given in table R 159-1. Please refer to [21, 44] and [46] for the use of theories of a higher order.

In table R 159-1:

$$\vartheta \;=\; \frac{2\pi \cdot x}{L} - \frac{2\pi \cdot t}{T} = k\,x - \omega\,t \qquad \text{(phase angle)}$$

$$k \;=\; \frac{2\pi}{L}, \;\; \omega = \frac{2\pi}{T}, \;\; c = \frac{\omega}{k}$$

t	=	time duration
T	=	wave period
c	=	wave velocity
k	=	wave number
ω	=	wave angular frequency

5.10.3 Determination of the wave loads on a vertical individual pile

Since the velocities and, accordingly, accelerations of the water particles are also a function of the distance of the location examined from the still water level, the wave load diagram according to fig. R 159-1 results from the calculation of the wave pressure load for various values of z.

The origin of the system of coordinates lies at the still water level, but can be fixed on the x-axis.

z	=	ordinate of the point investigated ($z = 0$ = still water level)
x	=	x-axis of the point investigated
η	=	water level varying over time, related to still water level (water surface displacement)
d	=	water depth below still water level
D	=	pile diameter
H	=	wave height
L	=	wavelength

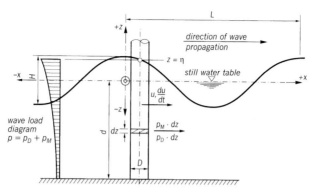

Fig. R 159-1. Wave action on a vertical pile

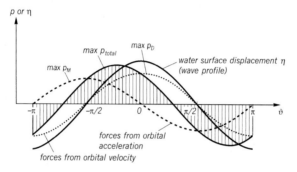

Fig. R 159-2. Variation of forces from velocity and acceleration over one wave period

Please note that the components of the wave load max p_D and max p_M occur out of phase. The calculation therefore has to be carried out for differing phase angles and the maximum load determined by an in-phase superimposition of the components from flow velocity and flow acceleration. Thus, for instance, when the linear wave theory is used, the acceleration force is out of phase by 90° ($\pi/2$) as compared to the flow velocity compressive force, which matches the phase of the wave profile (fig. R 159-2).

5.10.4 Coefficients C_D and C_M

5.10.4.1 Drag coefficient for flow pressure C_D
The drag coefficient C_D taking into account the resistance against flow pressure is determined from measurements. C_D depends on the shape of the body in the flow, the REYNOLDS number R_e, the surface roughness of the pile and the initial degree of turbulence of the current [42, 43, 47].

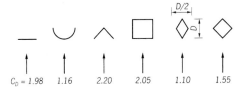

Fig. R 159-3. C_D values of pile cross-sections with stable separation points [41]

The location of the separation point of the boundary layer is decisive for flow pressure. The C_D value is practically constant for piles where the separation point is prescribed by corners or break-away edges (fig. R 159-3).

In the case of piles without stable separation points, e.g. circular cylindrical piles, we distinguish between a subcritical range of the REYNOLDS number with laminar boundary layer and a supercritical range with turbulent boundary layer.

However, in general high REYNOLDS numbers prevail under natural conditions, which means that in the case of smooth surfaces a constant value of $C_D = 0.7$ is recommended [21, 42]. For further details, please consult [162].

Larger C_D values can be expected on rough surfaces, see [48], for example.

5.10.4.2 Inertia or mass coefficient C_M for flow acceleration

The potential flow theory provides the value $C_M = 2.0$ for the circular cylindrical pile, whereas C_M values up to 2.5 have also been ascertained on the basis of tests on a circular cross-section [49].

Normally, the theoretical value $C_M = 2.0$ can be used. Otherwise, please refer to [21, 48, 162].

5.10.5 **Forces from breaking waves**

At present there is no viable calculation method available for determining the forces from breaking waves reliably. The MORISON equation is therefore used again for these waves, but under the assumption that the wave acts on the pile as a water mass with high velocity and without acceleration. In doing so, the inertia coefficient C_M is set to 0, whereas the drag coefficient C_D is increased to 1.75 [21].

5.10.6 **Wave load for pile groups**

When determining the wave loads on pile groups, the phase angle ϑ relevant for the respective pile location is to be taken into account.

Using the designations of R 159-4, the horizontal total load on a pile structure of N piles is:

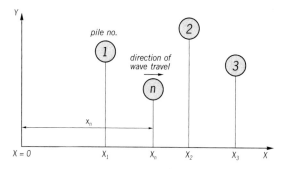

Fig. R 159-4. Definitions for a pile group (layout) (according to [21])

$$\text{total } P = \sum_{n=1}^{N} P_n \, (\vartheta_n)$$

where:

N = number of piles

$P_n \, (\vartheta_n)$ = wave load on an individual pile n, taking into account the phase angle $\vartheta = k \cdot x_n - \omega \cdot t$

x_n = distance of the pile n from the y-z plane

It should be noted that for piles at a spacing less than about four pile diameters, there is an increase in the load on adjacent piles transverse to the wave direction, and a decrease on piles parallel with the wave direction.

The correction factors given in table R 159-2 are proposed for this condition [49]:

Table R 159-2. Multiplier for small pile distances

Pile centre-to-centre distance e / Pile diameter D	2	3	4
for piles in rows parallel to the wave crest	1.5	1.25	1.0
for piles in rows perpendicular to the wave crest	0.7[1]	0.8[1]	1.0

1) Reduction does not apply to the front pile directly exposed to wave action.

5.10.7 Inclined piles

In the case of inclined piles, it should also be noted that the phase angle ϑ for the local coordinates x_0, y_0, z_0 differs for the individual pile sections d_s. The pressure on the pile at the section considered is to be determined with the coordinates x_0, y_0 and z_0 according to fig. R 159-5.

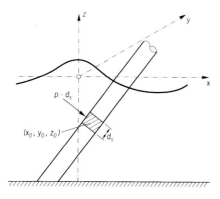

Fig. R 159-5. For calculating the wave forces on an inclined pile [21]

The local force $p \cdot d_s$ due to the velocity and acceleration of the water particles on the pile element d_s ($p = f[x_0, y_0, z_0]$) can be equated with the horizontal force on an assumed vertical pile at (x_0, y_0, z_0) according to [21]. However, when the pile slopes at a steeper angle, a check should be carried out to ascertain whether the determination of the load furnishes more unfavourable values, taking into account the components of the resulting velocity acting perpendicular to the pile axis

$$v = \sqrt{u^2 + w^2}$$

and the resultant acceleration

$$\frac{\partial v}{\partial t} = \left(\frac{\partial u}{\partial t}\right)^2 + \left(\frac{\partial w}{\partial t}\right)^2$$

5.10.8   Safety factors
The design of pile structures to withstand wave action is heavyly dependent on the selection of the "design wave" (R 136, section 5.6, in conjunction with table R 159-1, No. 3). Also of influence are the wave theory used and the associated coefficients C_D and C_M. This applies particularly to pile structures in shallow water. In order to allow for such uncertainties, the designer is recommended to multiply the loads calculated by increased partial safety factors according to [21].
Consequently, when the "design wave" occurs only rarely, i.e. in normal cases with deep water conditions, the resulting wave load on piles is to be increased by a partial safety factor $\gamma_d = 1.5$ When the "design wave" occurs frequently, which is usually the case in shallow water conditions, a partial safety factor $\gamma_d = 2.0$ is recommended.

149

Please refer to [162] and [46] regarding the possibility of using the coefficients C_D and C_M depending on the REYNOLDS and KEULEGAN-CARPENTER numbers and a corresponding reduction in the partial safety factor.

Critical vibrations may occur occasionally in pile structures, especially when separation eddies act transverse to the flow direction, or the inherent frequency of the structure is close to the wave period, giving rise to resonance phenomena. In this case regular waves smaller than the "design wave" can be less favourable. Special investigations are required in such cases.

5.11 Wind loads on moored ships and their influence on the design of mooring and fendering facilities (R 153)

5.11.1 General

This recommendation is applicable as a supplement to the proposals and advice concerning the planning, design and dimensioning of fender and mooring facilities, especially R 12, section 5.12, R 111, section 13.2, and R 128, section 13.3.

The loads for mooring installations, such as bollards or quick-release hooks with their associated anchorages, foundations, retaining structures, etc., which result from this recommendation, replace the load values in R 12, section 5.12, only when the influence of swell, waves and currents on ship berths can be neglected. Otherwise, these latter factors must be specially checked and taken into consideration as well.

R 38, section 5.2, is not affected by this recommendation. In determining the "normal berthing loads" dealt with therein, reference to R 12, section 5.12.2, therefore remains applicable without restriction.

5.11.2 Relevant wind velocity

Owing to the mass inertia of ships, it is not the short-term peak gusts (order of magnitude = seconds) that are relevant for determining the line pull forces, but rather the average wind over a period of time T. The value of T should be taken as 0.5 min for ships up to 50 000 dwt, and 1.0 min for larger ships. The wind intensity of the maximum wind averaged over one minute is generally about 75% of that of the value over one second. The designer is recommended to use wind measurements when determining the relevant wind speed. If wind data for the immediate area of the project is not available, wind data from more remote measuring stations can be used by means of interpolation or numerical methods of calculation and taking into account the orography. The period of the wind measurements should be used to produce statistics of extreme values. A 50-year return wind is recommended for design purposes.

If there is no other specific data available on the wind conditions for the area of the ship berth, the values given in DIN 1055 part 4 can be used as relevant wind velocities v for all wind directions.

This basic value can be differentiated according to wind directions if more detailed data is available.

5.11.3 Wind loads on the moored vessel

The loads quoted are characteristic values.

Wind load components:

$$W_t = (1 + 3.1 \sin \alpha) \cdot k_t \cdot H \cdot L_{\ddot{u}} \cdot v^2$$
$$W_l = (1 + 3.1 \sin \alpha) \cdot k_l \cdot H \cdot L_{\ddot{u}} \cdot v^2$$

Equivalent loads for $W_t = W_{tb} + W_{th}$:

$$W_{tb} = W_t \cdot (0.50 + k_e)$$
$$W_{th} = W_t \cdot (0.50 - k_e)$$

Force diagram (schematic):

where:

H	=	greatest freeboard height of ship (ballasted or unladen) and additional height of load above freebord
$L_{\ddot{u}}$	=	overall length
v	=	relevant wind velocity
W_i	=	wind load components
k_t and k_l	=	wind load coefficients
k_e	=	coefficient of eccentricity

As international experience has shown, the load and eccentricity coefficients may be applied in accordance with tables R 153-1 and R 153-2. Please refer to the tables in [232] for more detailed data on various types of ship.

Forces diagram:

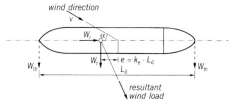

Fig. 153-1. Application of wind loads on the moored vessel

5.11.4 Loads on mooring and fender facilities

In order to determine the mooring and fender forces, a static system has to be introduced formed by the ship, the hawsers and the mooring and fender structures. The elasticity of the hawsers, which is contingent on material, cross-section and length, has to be taken into account to the same extent as the angles of the hawsers in the horizontal and vertical directions, subject to variable loads and water level conditions.

The elasticity of the mooring and fender structures is to be ascertained at all support and bearing points of the static system. Anchored sheet pile walls and structures with raking pile foundations may therefore be considered as rigid elements. Please note that the static system can alter if individual lines fall slack or fenders remain unloaded under specific load situations. All characteristic mooring and fender loads determined on the basis of the wind loads given in section 5.11.3 are to be multiplied by a partial safety factor of $\gamma_d = 1.25$ to cover dynamic and other influences not readily ascertainable.

The wind-shielding effect of structures and facilities may be taken into account to a reasonable extent.

Table R 153-1. Load and eccentricity coefficients for ships up to 50 000 dwt

α^0	Ships up to 50 000 dwt		
	k_t [kN · s²/m⁴]	k_e [I]	k_1 [kN · s²/m⁴]
0	0	0	$9.1 \cdot 10^{-5}$
30	$12.1 \cdot 10^{-5}$	0.14	$3.0 \cdot 10^{-5}$
60	$16.1 \cdot 10^{-5}$	0.08	$2.0 \cdot 10^{-5}$
90	$18.1 \cdot 10^{-5}$	0	0
120	$15.1 \cdot 10^{-5}$	−0.07	$-2.0 \cdot 10^{-5}$
150	$12.1 \cdot 10^{-5}$	−0.15	$-4.1 \cdot 10^{-5}$
180	0	0	$-8.1 \cdot 10^{-5}$

Table R 153-2. Load and eccentricity coefficients for ships over 50 000 dwt

α^0	Ships over 50 000 dwt		
	k_t [kN · s²/m⁴]	k_e [I]	k_1 [kN · s²/m⁴]
0	0	0	$9.1 \cdot 10^{-5}$
30	$11.1 \cdot 10^{-5}$	0.13	$3.0 \cdot 10^{-5}$
60	$14.1 \cdot 10^{-5}$	0.07	$2.0 \cdot 10^{-5}$
90	$16.1 \cdot 10^{-5}$	0	0
120	$14.1 \cdot 10^{-5}$	−0.08	$-2.0 \cdot 10^{-5}$
150	$11.1 \cdot 10^{-5}$	−0.16	$-4.0 \cdot 10^{-5}$
180	0	0	$-8.1 \cdot 10^{-5}$

5.12 Layout and loading of bollards for seagoing vessels (R 12)

5.12.1 Layout

For simple and straightforward structural treatment, a bollard spacing of about 30 m (equal to the normal section length) is recommended for quay walls and piled walls of plain or reinforced concrete. Generally, the bollard should be placed in the centre of a section. If there are two bollards in each section, they should be placed symmetrically about the centre of the section at the outer quarter points. Proceed accordingly for shorter section lengths. The distance of the bollard from the face of the wall should be as recommended in R 6, section 6.1.2.

The bollards can be designed as single or double bollards. They can accommodate several hawsers simultaneously. They should be designed to allow for easy repair or replacement.

5.12.2 Loads

The hawsers placed around the bollards are generally not all fully stressed at the same time, and the hawser forces partially cancel each other out in action, meaning that the line pull forces according to table R 12-1 may be used for both single and double bollards:

The loads above are characteristic values. For designing the bollard, use the partial safety factors for load and material strength, $\gamma_Q = 1.3$ and $\gamma_M = 1.1$, irrespective of the loading case. Design the bollard anchorage for 1.5 times the load. At berths for large ships with a strong current, the bollard line pull forces as given in table R 12-1 should be increased by 25% for all ships with a displacement of 50 000 t or more.

Table R 12-1. Line pull forces

Displacement t	Line pull force kN
up to 2 000	100
up to 10 000	300
up to 20 000	600
up to 50 000	800
up to 100 000	1 000
up to 200 000	1 500
> 200 000	2 000

5.12.3 Direction of the line pull force

The line pull force may occur at any angle towards the water-side. An inboard line pull force is not assumed, unless the bollard also serves a waterfront structure lying in that direction, or has a special purpose as a corner bollard. When designing the waterfront structure, the line pull force is usually assumed to act horizontally.

When designing the bollard itself and its connection to the waterfront structure, upward inclines of up to 45° with corresponding line pull forces are to be taken into consideration.

5.13 Layout, design and loading of bollards in inland harbours (R 102)

This recommendation has been brought into line with DIN 19 703 "Locks for waterways for inland navigation – principles for dimensioning and equipment", insofar as the principles of this standard can be applied to waterfront structures.

The term bollard has been used to cover all mooring facilities, and includes edge bollards, recess bollards, dolphin bollards, mooring hooks, mooring lugs, mooring rings and similar facilities.

5.13.1 Layout and Design

In inland harbours, ships should be moored to the shore by three hawsers: a bow line, a breast line and a stern line. An appropriate number of bollards should be provided on the bank.

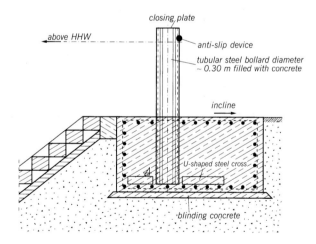

Fig. R 102-1. Foundation for a bollard on the port ground surface (design to statical requirements)

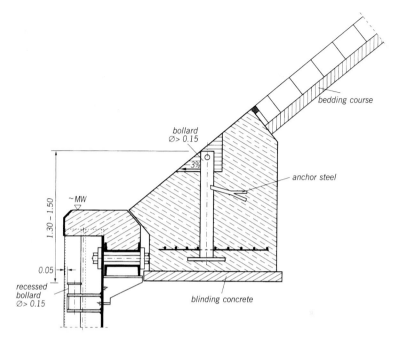

Fig. R 102-2. Bollard foundation for a partially sloped bank
(shown as an example, design to statical requirements)

Bollards must be arranged on and above the port ground level, with their upper surface above HHW (fig. R 102-1). The bollard diameter must be more than 15 cm. If the bollard does not project sufficiently above HNW, slippage of the line must be prevented with a crosspiece. In addition to bollards along the top of the bank, in river ports other bollards must be installed at various levels to suit local water level fluctuations. Only then can the ship's crew moor the ship without any difficulty at any water level and any freeboard height.

In the case of vertical waterfront structures, the bollards are positioned at differing levels but in line vertically. The position of the rows depends on the position of the ladders. To avoid undue tension on the ladders, a row of bollards is located to the left and right of every ladder at a distance of approx. 0.85 m to 1.00 m from the centre of the ladder for solid walls and twice the spacing of the bollards for sheet pile walls. The spacing of the ladders and rows of bollards should be about 30 m. In the case of steel pile walls, the precise centre-to-centre dimension is determined by the system dimension of the piles, and by the section lengths in the case of solid walls.

155

The lowest bollard is located at approx. 1.50 m above LLW, in tidal areas above MLWS (max. 1.0 m above the lowest low water level for locks on inland waterways). The vertical distance between this and the upper edge of the waterfront structure is divided up by further bollards at a distance of 1.30 to 1.50 m (up to 2.00 m in borderline cases).

In the case of waterfront structures of reinforced concrete, the bollards are positioned in recesses, the housings of which are anchored and concreted in position. In the case of steel pile walls, the bollards can be bolted or welded in position. The front edge of the bollard cone should be 5 cm behind the front edge of the waterfront structure. Appropriate clearance is to be left at the side behind and over the bollard cone so that the ships' hawsers can be looped over and removed again easily. The edges to the waterfront structure are to be rounded off to prevent any damage to the hawsers and the waterfront structure.

In the case of sloping and partially sloping banks, the bollards are positioned on both sides next to the steps (fig. R 102-2). The steps are located as extensions of the ladders.

In this arrangement the bollard foundation is best located under the steps for both bollards.

5.13.2 Loads

The line pull forces that occur essentially depend on the ship's size, the distance from and speed of passing ships, the velocity of water currents at the berth and the quotient of the water cross-section to the submerged ship cross-section.

The characteristic load is to be taken as 200 kN per bollard and 300 kN for their anchorages. For designing the bollard, use the partial safety factors for load and material strength, $\gamma_Q = 1.5$ and $\gamma_M = 1.1$, irrespective of the loading case.

Ships in motion are not permitted to brake at bollards, so the corresponding effect is not taken into consideration in the load assumptions (actions).

5.13.3 Direction of line pull forces

Line pull forces can be applied only from the water side, mostly at an acute angle and only rarely at right-angles to the bank. The calculations must, however, take account of every reasonable horizontal and vertical angle to the bank.

5.13.4 Calculation

Stability verification is to be provided for a single line pull force acting at the most unfavourable angle. Stability verification can also be provided by means of test loadings.

5.14 Quay loads from cranes and other transhipment equipment (R 84)

The following loads are characteristic values which are to be multiplied by the partial safety factors for the relevant load cases (see R 18, section 5.4) according to DIN 1054.

5.14.1 Customary general cargo harbour cranes

5.14.1.1 General

The customary general cargo harbour cranes constructed in Germany are chiefly full portal level-luffing cranes spanning one, two or three railway tracks. Occasionally, however, half portal cranes are manufactured. The lifting capacity varies between 7 and 50 t at a working radius of 20 to 45 m.

The axis of rotation of the crane superstructure should lie as close as possible to the outboard crane rail in order to achieve a good working radius from the centre of rotation. However, to avoid any collision between a crane and a heeling ship, care must be taken to ensure that neither the crane operator's cabin nor the rear counterweight project out over an inboard inclined plane sloping upwards by about 5° measured from the quay edge.

The distance between the outboard crane rail and the face of the quay wall shall be in accordance with R 6, section 6.1. The corner spacing of small cranes is approx. 6 m. Minimum corner spacing should not be less than 5.5 m as otherwise excessive corner loads occur and the crane must be equipped with excessive central ballast. The length over buffers is approx. 7 to 22 m, depending on the size of the crane. If this results in an excessive wheel load, lower wheel loads can be achieved by increasing the number of wheels. Today, however, there are general cargo handling facilities whose craneways have been built for especially high wheel loads.

General cargo harbour cranes are as a rule classified as lifting class H 2 and load group B 4 or B 5 in accordance with DIN 15 018 part 1. Please also refer to the F.E.M. 1001 [154]. The vertical wheel loads from dead load, live load, inertia forces and wind forces are to be applied when designing the craneway (DIN 15 018 part 1). Vertical inertia forces from the travel or from the lifting or setting down of the live load are to be allowed for through use of impact factor, which is about 1.2 for lifting class H 2. The foundation of the craneway may be dimensioned without taking an impact factor into account. All crane booms can slew through 360°. The associated corner load changes accordingly. With higher wind loads and crane not in operation, load case 3 may be used, if necessary, for designing the quays and craneways.

157

5.14.1.2 Full portal cranes

The portal of light harbour cranes with low lifting capacities has either four or three legs, each one of which has between one and four wheels. The number of wheels in each case depends on the permissible wheel load. General cargo heavy-lift cranes have at least six wheels per leg. On straight quay stretches, the centre-to-centre spacing of the crane rails is at least 5.5 m, but generally 6, 10 or 14.5 m, depending on whether the portal spans one, two or three tracks. The dimensions 10 m and 14.5 m result from the theoretical minimum dimension of 5.5 m for one track, to which one or two times the track spacing of 4.5 m is added.

5.14.1.3 Half portal cranes

The portal of these cranes has only two legs, which run on the outboard crane rail. On the inboard side it is supported by a short leg on an elevated craneway, thus providing unrestricted access to any section of the quay area. The remarks in section 5.14.1.2 apply to the number of wheels under inboard and outboard legs.

5.14.2 Container cranes

True container cranes are constructed as full portal cranes with cantilever beams and crab (loading gantries), whose legs have eight to ten wheels as a rule. The crane rails of existing container terminals are in general 15.24 m (50 ft) or 18.0 m centre-to-centre. A track width of 30.48 m (100 ft) is often selected for new facilities. The clear leg spacing (= clear space between the corners longitudinal) is 17 to 18.5 m at an overall dimension over buffers of about 27.5 m (fig. R 84-1). It should generally be assumed that three container cranes operate buffer to buffer. If the handling of 20 ft containers requires a shorter dimension over buffers, a minimum corner spacing of up to 12 m is possible. The length over buffers is then 22.5 m. The corner spacing in this case is not the same as the portal leg spacing. The lifting capacity of the cranes is selected between 45 and 75 t, in exceptional cases up to 105 t, including spreader. The maximum corner load is affected, in particular, by the design and length of the jib. The working range usual hitherto (38 to 41 m, corresponding to the width of Panmax vessels) is insufficient for the post-Panmax vessels, which cannot pass through the Panama Canal owing to their width. Jib lengths of at least 44.5 m are necessary for this type of vessel. The maximum corner loads for container cranes in operation can be up to 4500 kN for Panmax ships and up to 9000 kN for post-Panmax ships. However, the tendency in the development of container gantries is to jib lengths for handling 22-23 rows of containers on the ships. This results in jib lengths of up to 66 m measured from the water-side rail. The designer is recommended to obtain more accurate planning data from container terminal operators because the large number of possible solutions does not permit the inclusion of more accurate information in this book.

5.14.3 Loads for harbour cranes

The support structure always consists of a portal-type substructure, either with a rotating, luffing jib, or a rigid beam which can be folded upwards when not in use. The portal usually stands on four corner points with several wheels arranged in rockers, depending on the magnitude of the corner load. The corner load is distributed as evenly as possible over all wheels of the corner point. Table R 84-1 summarises the general loads and dimensions to supplement the details given in section 5.14.1 and 5.14.2.

Table R 84-1. Dimensions and characteristic loads of slewing and container cranes

	Rotating cranes	Container cranes and other transhipment gear
Bearing capacity [t]	7–50	10–80
Dead weight [t]	180–350	200–2000
Portal span [m]	6–19	9–45
Clear portal height [m]	5–7	5–13
Max. vertical corner load [kN]	800–3000	1200–8000
Max. vertical wheel surcharge load [kN/m]	250–600	250–700
Horizontal wheel load		
transverse to the direction of the rail	up to approx. 10% of vertical load	
in the direction of the rail	up to approx. 15% of the vertical load of the braked wheels	
Claw load[1] [kN]	mobile cranes up to 2600	

[1] Prerequisite is a zone of 40 m^2 subject to no other loads; the claw load can be taken as distributed over 10 m^2.

5.14.4 Notes

Further details about harbour cranes are to be found in the AHU recommendations and reports [185] E 1, E 9, B 6 and B 8, in ETAB [45] recommendation E 25, and in VDI directive 3576 [184].

160

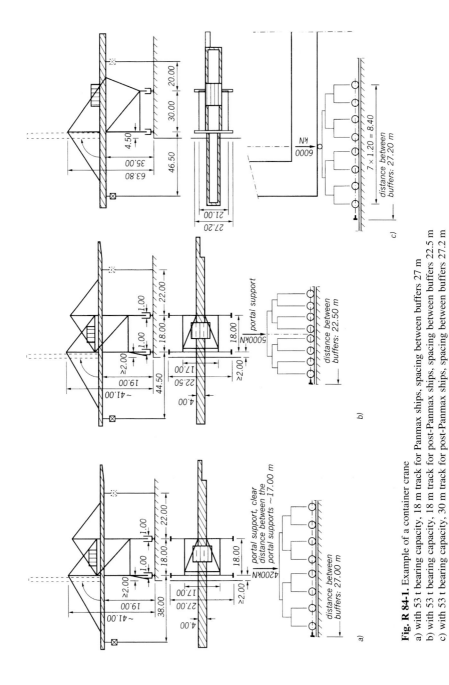

Fig. R 84-1. Example of a container crane

a) with 53 t bearing capacity, 18 m track for Panmax ships, spacing between buffers 27 m

b) with 53 t bearing capacity, 18 m track for post-Panmax ships, spacing between buffers 22.5 m

c) with 53 t bearing capacity, 30 m track for post-Panmax ships, spacing between buffers 27.2 m

5.15 Impact and pressure of ice on waterfront structures, fenders and dolphins in coastal areas (R 177)

5.15.1 General

Loads on hydraulic engineering facilities due to the effects of ice can occur in various ways:

a) as ice impact from collisions with ice floes carried along by the current or by the wind,

b) as ice pressure exercising its effect through ice thrusting against an ice layer adjacent to the structure or through vessel movements,

c) as ice pressure exercising its effect on the structure through an unbroken ice layer as a result of thermal expansion,

d) as live ice loads on the structure when ice forms on the structure or as live or lifting loads when the water levels fluctuate.

The magnitude of the possible load effects depends on the following (and other) factors:

- the shape, size, surface condition and resilience of obstacles against which the ice mass collides,
- the size, shape and propagation speed of the ice mass,
- the nature of the ice and ice formation,
- the salt content of the ice and the ice strength dependent on this,
- the angle of impact,
- the relevant strength of the ice (compression, bending and shear strength),
- the loading rate,
- the ice temperature.

Whenever possible, it is recommended that the relevant load values for waterfront structures, including piled structures, be checked with the assumptions for facilities already built, which have proven their worth, or with local ice pressure measurements.

The ice loads determined below are characteristic values. A partial safety factor of 1.0 can be used because of the usually marginal probability of occurrence.

The reader is directed to the explanations in [46] and [148]. Advice from other international publications (from USA, Canada, Russia, etc.) can be found in [231].

5.15.2 Ice loads on waterfront structures

In the northern German coastal region, an ice thickness of 50 cm and an ice compressive strength of $\sigma_0 = 1.5$ MN/m^2 with temperatures around freezing point can generally be assumed as the starting point for

161

determining the horizontal ice loads on structures with an extensive surface area. This gives rise to the assumptions:

a) 250 kN/m as the mean linear load acting horizontally at the least favourable level of the water levels under consideration, assuming that the maximum load calculated from the ice compressive strength of 750 kN/m on average acts on only 1/3 of the length of the structure (contact coefficient $k = 0.33$).

b) 1.5 MN/m^2 as local surface load.

c) 100 kN/m as the mean linear load acting horizontally at the least favourable level of the water levels under consideration for groynes and waterfront revetments in tidal regions if a broken ice layer is created as a result of water level fluctuations.

The simultaneous action of ice effects with wave loads and/or vessel impact need not be considered.

5.15.3   Ice loads on piles of piled structures or on individual piles

5.15.3.1 Principles for determining the ice load

The ice loads acting on piles depend on the shape, inclination and layout of the piles, and on the ice compression, bending and shear strength applicable for fracturing the ice. The magnitude of the load also depends on the nature of the load – whether primarily a dead load or impact load from colliding ice floes. In the case of North Sea ice, it can generally

Table R 177-1. Measured maximum ice thicknesses as standard values for measurement

North Sea	max h (cm)	Baltic Sea	max h (cm)
Helgoland	30–50	Kiel Canal	60
Wilhelmshaven	40	Flensburg (outer fjord)	32
"Hohe Weg" lighthouse	60	Flensburg (inner fjord)	40
Büsum	45	Schleimünde	35
Meldorf (harbour)	60	Kappeln	50
Tönning	80	Eckernförde	50
Husum	37	Kiel (harbour)	55
Wittdünn harbour	60	Bay of Lübeck	50
		Wismar harbour	50
		Wismar – bay	60
		Rostock – Warnemünde	40
		Stralsund – Palmer Ort	65
		Saßnitz – harbour	40
		Koserow – Usedom	50

be assumed that the mean compressive strength does not exceed $\sigma_0 = 1.5$ MN/m^2, for Baltic Sea ice $\sigma_0 = 1.8$ MN/m^2 and for freshwater ice $\sigma_0 = 2.5$ MN/m^2. These values apply to a specific expansion rate $\dot{\varepsilon} = 0.003$ s^{-1}, at which the compressive strength achieves its maximum value according to tests [108].

If more accurate ice strength investigations are not available, the bending tensile strength σ_B can be assumed to be approx. $1/3\sigma_0$ and the shear strength τ approx. $1/6\sigma_0$. Standard values in accordance with table R 177-1 apply for the ice thicknesses h for the German North Sea and Baltic Sea coasts.

On the German North Sea coast, the ice load is frequently taken at a height of 0.5 m to 1.5 m above MHW for free-standing piles.

The assumptions below apply to slim components up to 2 m wide with flat ice. In the event of the formation of compression ice ridges, the ice loads listed below must be doubled.

5.15.3.2 Ice load on vertical piles

Irrespective of the form of the pile cross-section, the horizontal ice load from the effect of drifting ice is calculated on the basis of the investigations in accordance with [108]:

$$P_p = 0.36 \, \sigma_0 \cdot d^{0.5} \cdot h^{1.1}$$

where:

σ_0 = ice compressive strength in MN/m^2 at specific expansion rate $\dot{\varepsilon} = 0.003$ s^{-1}
d = width of individual pile [cm]
h = thickness of ice [cm]
P_p = ice load [kN]

If the case of incipient ice movement with tightly packed ice is to be considered, the following load assumptions are applicable:
for a round or half-round pile:

$$P_i = 0.33 \, \sigma_0 \cdot d^{0.5} \cdot h^{1.1} \text{ [kN]}$$

for a square pile:

$$P_i = 0.39 \, \sigma_0 \cdot d^{0.68} \cdot h^{1.1} \text{ [kN]}$$

for a wedge-shaped pile:

$$P_i = 0.29 \, \sigma_0 \cdot d^{0.68} \cdot h^{1.1} \text{ [kN]}$$

5.15.3.3 Ice load on inclined piles

With inclined piles, the fracture of the ice floe by shearing off or bending can occur earlier than the crushing of the ice. In accordance with [109], the smaller ice load is relevant in each case. Piles with an inclination steeper than 6 : 1 ($\beta \geq$ approx. 80°) require the ice load to be calculated in accordance with section 5.15.3.2.

In the case of shear fracture, the horizontal ice load is:

$$P_s = c_{fs} \cdot \tau \cdot k \cdot \tan \beta \cdot d \cdot h \text{ [kN]}$$

where:

P_s = horizontal load at shear fracture [kN]
τ = shear strength [MN/m^2]
c_{fs} = shape coefficient as per table R 177-2 [1]
k = contact coefficient, generally approx. 0.75 [1]
β = angle of inclination of pile measured from the horizontal [°]
d = pile width [cm]
h = thickness of ice [cm]

In the case of bending fracture, the horizontal ice load is:

$$P_b = c_{fb} \cdot \sigma_B \cdot \tan \beta \cdot d \cdot h \text{ [kN]}$$

where:

P_b = horizontal ice load at bending fracture [kN]
σ_B = bending tensile strength [MN/m^2]
c_{fb} = shape coefficient as per table R 177-3 [1]

Table R 177-2. Shape coefficient c_{fs} for round pile, square pile or wedge-shaped edge with $2\,\alpha$ = edge angle, measured in the horizontal plane

Edge angle $2\,\alpha$ [°]	Shape coefficient c_{fs}
45	0.29
60	0.22
75	0.18
80	0.17 (= round pile)
90	0.16
105	0.14
120	0.13
180	0.11 (= square pile)

Table R 177-3. Shape coefficient c_{fb} for round pile, square pile or wedge-shaped edge with 2α edge angle, measured in the horizontal plane

Edge angle 2α [°]	Shape coefficient c_{fb}				
	at inclination angle β [°]				
	45	60	65	70	75
45	0.019	0.024	0.029	0.037	0.079
60	0.017	0.020	0.022	0.026	0.038
from 75	0.017	0.019	0.020	0.021	0.027

5.15.3.4 Horizontal load on pile groups

The ice load on pile groups arises from the sum of the ice loads on the individual piles. Generally, assuming the sum of the ice loads that act on the piles facing the ice drift will suffice.

5.15.4 Live ice loads

The live ice load must be assumed in accordance with local conditions. Without more detailed checks, a minimum live ice load of 0.9 kN/m^2 can be regarded as sufficient [110]. In addition to the live ice load, the usual snow load of 0.75 kN/m^2 should also be considered. Conversely, traffic loads that have no effect during thick ice formation generally do not need to be applied at the same time.

5.15.5 Vertical loads with a rising or falling water level

With rising or falling water levels, additional vertical forces from immersed or projecting ice act on frozen structures or piles. Ice adhering to the side of the structure with a strip width $b = 5$ m and ice thickness h and any ice present below the structure can be included with its full volume for rough calculations. Taking the specific density of the ice $\gamma_E = $ approx. 9 kN/m^3 and a falling water level, the ice volume thus determined, V_E, generates the load $P = V_E \cdot \gamma_E$ acting vertically downwards, and with the difference between the specific densities of water and ice $\Delta\gamma_E = 1$ kN/m^3, it generates the load $P = V_E \cdot \Delta\gamma_E$ acting vertically upwards.

5.15.6 Supplementary information

The above recommendations for ice loads on structures are rough assumptions that apply to German conditions, i.e. not to Arctic regions. Considerably reduced values apply to protected areas (bays, harbour basins, etc.) and in seaports with clear tidal action and considerable traffic.

Corresponding reductions in the assumed loads also apply if measures are introduced to reduce the ice load, such as early breaking or blasting of the ice, influencing the current, use of air bubbling facilities, heating or other thermal measures, etc., or when the ice field is only small in size.

Ice formation and ice loads also depend extensively on the wind direction, current and shear zone arrangement in the ice. This must be taken into consideration, particularly in the layout of harbour entrances and the alignment of harbour basins, for example. In narrow harbour basins, considerable ice loads arising from deformation can occur as a result of temperature changes in the ice. In accordance with [111], it can be assumed, in view of the ice temperatures across the North German coastal region, which are generally not very low, that the thermal ice pressure does not exceed 400 kN/m^2.

In individual cases where it is important to determine the ice loads more precisely, experts should be consulted and, if applicable, model tests should be carried out.

If the ice loads in the case of dolphins significantly exceed the loads from vessel impact or bollard tension, a check should be carried out to ascertain whether such dolphins should be designed for the higher ice loads, or whether rare overstresses can be simply accepted for economic reasons.

5.16 Impact and pressure of ice on waterfront structures, piers and dolphins in inland areas (R 205)

5.16.1 General

The data stated in recommendation R 177, section 5.15, is generally applicable to inland areas. This is valid for the general statements and for the loads because they depend on the dimensions of the structure, the ice thickness and the strength properties of the ice.

As the heat balance of inland waterways is nowadays in most cases influenced by the influx of cooling and waste water, it can be assumed that as far as inland waterways and inland harbours are concerned, situations of extreme cold are seldom and thus the probability of ice formation and thick ice are considerably reduced.

The ice loads determined are characteristic values in accordance with R 177, section 5.15.1, and are to be multiplied by a partial safety factor of 1.0.

5.16.2 Ice thickness

The ice thickness can be deduced according to [109] from the sum of daily degrees below zero during an ice period – the so-called "cold sum".

So, for example, according to BYDIN [107], $h = \sqrt{\Sigma |t_L|}$, where h is the ice thickness in cm and $\Sigma |t_L|$ the sum of the absolute values of the negative mean diurnal air temperature in °C for the period of time considered.

If no more detailed statistics or measurements are available, they theoretical ice thickness of $h \leq 30$ cm can usually be presumed, provided the conditions stated under section 5.16.1 apply.

5.16.3 Ice strength

The ice strength depends on the ice temperature t_E, the average ice temperature being half the ice temperature at the surface because 0 °C can always be achieved at the underside of ice.

For moderate ice temperatures, the compressive strength of freshwater ice can be assumed to be $\sigma_0 = 2.5$ MN/m^2 according to R 177, section 5.15.3.1.

If the average temperature of the ice drops below –5 °C, the compressive strength increases according to [148] by approx. 0.45 MN/m^2 per degree below zero.

At ice temperatures above –5 °C, the compressive strength of the ice can also be determined according to [109]:

$$\sigma_0 = 1.1 + 0.35 |t_E| \quad [\text{MN/m}^2]$$

5.16.4 Ice loads on waterfront structures and other structures of larger extent

Generally, according to R 177, section 5.15.2:

$$p_0 = 10 \cdot k \cdot \sigma_0 \cdot h \quad [\text{kN/m}]$$

where:

p_0 = ice load [kN/m]
k = contact coefficient, generally approx. 0.33
σ_0 = compressive strength of the ice [MN/m^2]
h = thickness of the ice [cm]

On sloped surfaces, the horizontal ice load according to [109] can be taken to be:

$$p_h = 10 \cdot k \cdot \sigma_B \cdot h \cdot \tan \beta \, [\text{kN/m}]$$

where:

σ_B = bending tensile strength of the ice [MN/m^2]
$\tan \beta$ = angle of slope [1]

5.16.5 Ice loads on narrow structures (piles, dolphins, bridge and weir piers, ice deflectors)

The loads for vertical or inclined piles according to R 177, sections 5.15.3.2 and 5.15.3.3, are valid in the same way when considering the respective ice strength for the inland area. They are also applicable for pier constructions and ice deflectors, taking into account the form of the cross-section and the surface as well as the inclination.

5.16.6 Ice loads on groups of structures

The indications in R 177, section 5.15.3.4, are valid.

For structures built in water, a distance of at least

$$l = \frac{1.1 \cdot \sigma \cdot d}{v^2}$$

where:

l = spacing of piers [m]

σ = $10 \cdot \dfrac{P}{d \cdot h}$ for an ice load P [kN] according to section 5.16.5

v = drifting speed of the ice [m/s]

d = pier diameter [cm]

is recommended according to [109] in order to avoid obstructions when hauling away the ice.

Otherwise, the possibilities of the formation of pack ice can be estimated according to [148].

Pack ice does not necessarily result in an increase in the ice load when the failure conditions of the packing ice are applicable; changes to load distribution and height of load application must also be considered, together with additional loads from water build-up and changes in current due to cross-section restrictions.

5.16.7 Vertical loads with a rising or falling water level

The indications in R 177, section 5.15.5, are valid.

5.16.8 Supplementary information

The information in R 177, section 5.15.6, are to be observed. Approximate values for the thermal ice pressure depending on the initial temperature and the hourly temperature rise are stated in [148]. The thermal ice pressure remains under 200 kN/m^2 for moderate temperatures.

5.17 Loads on waterfront structures and dolphins from the reaction forces of fenders (R 213)

The energy that can be absorbed by the fender is determined according to the deterministic calculation corresponding to R 60, section 6.14.

The reaction force of a fender can be computed using the corresponding diagrams and/or tables provided by the manufacturer of the fender type selected as well as the calculated energy to be absorbed that acts as a maximum on the waterfront structure or the fender dolphins. This reaction force should be taken as a characteristic value.

In normal cases the reaction force does not lead to any additional load on the quay wall, and it is only necessary to investigate the local effects of the load, unless special constructions, e.g. separate, suspended fender panels etc., are required for the fenders.

6 Configuration of cross-section and equipment for waterfront structures

6.1 Standard dimensions of cross-section of waterfront structures in seaports (R 6)

6.1.1 Walkways (towpaths)

The walkway space (towpath) in front of the outboard crane rail is necessary to provide room for bollards, storing of gangways, as a path and working space for line handlers, for access to the berths and for the accommodation of the outboard portion of the crane gantry. Consequently, it is of special importance in harbour operations. In selecting its width, accident prevention regulations must also be considered.

The greater the width required, the further the crane must be moved back from the edge of the quay. This in turn requires a longer jib. Although a longer jib makes cargo handling more expensive, on the other hand, ample clearance is becoming increasingly desirable because of the growing number of ships that are being built with superstructures hanging over the hull. This overhang is especially hazardous to crane operation when the ship is heeling. For this reason, all positions of the crane operator's cab must clear the vertical plane through the face of the quay

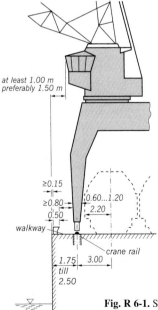

Fig. R 6-1. Standard cross sections of waterfront structures in seaports (supply channels are not illustrated)

by at least 1.00 m and preferably 1.50 m (fig. R 6-1). If need be, this dimension should be measured from the face of the timber fender, fender pile or fender system.

Railings are not required at the edges of quays for mooring and cargo handling services. However, the edges of such quays should be provided with adequate nosing edge protection as per R 94, section 8.4.6. The edges of quays with public access and which are not used for mooring or cargo handling should be provided with a railing.

6.1.2 Edge bollards

Edge bollards that are positioned flush with the face of the quay wall have caused difficulties in handling heavy rope hawsers when the ship is lying close to the quay. Therefore, the face of the bollards must lie at least 0.15 m behind the front face of the quay wall. The width of the head of a bollard may be taken as 0.50 m. The bogies of modern harbour cranes may be assumed to be about 0.60 to 1.20 m wide.

6.1.3 Other equipment

The dimensions given in fig. R 6-1 are recommended for the installation of new harbours and conversion of existing harbours, taking all relevant factors into account. The dimension of 1.75 m between the crane rail and the edge of the wall is recommended as a minimum. It is better to allow 2.50 m for new construction and deepening of berths, particularly when very wide outboard crane bogies affect safety aspects for mooring or for access to and from the gangway.

Railway traffic requires adherence to the safety standards regarding clearances at the crane, even when the outboard bogies are independent; therefore, the centre line of the first track must lie at least 3.00 m inboard of the outer crane rail. However, railway tracks on quays are becoming a rarity.

6.1.4 Arrangement of top of quay wall at container terminals

Owing to the safety requirements as well as higher demands on productivity at container terminals, it is recommended to allow for a greater clearance between the front edge of the quay and the axis of the outboard crane rail. This strip of quayside should be wide enough to store gangways and other equipment parallel to the ship as well as park service and delivery vehicles, and thus keep the container handling and ship service areas separate. It is accepted that the container cranes will require longer outreach. If the container handling system includes automatic transport of the containers between container gantries and storage area, it is essential to separate the ship service traffic and the container handling areas for safety reasons. In such cases the front edge of the quay is located at such a distance in front of the outboard crane rail that all service vehicle lanes can be accommodated in this strip, or one lane here and another adjacent to the rail beneath the portal of the gantry crane. A fence separates the container handling area from the service area.

6.2 Top edge of waterfront structures in seaports (R 122)

6.2.1 General

The elevation of the top of waterfront structures in seaports is determined by the operations level of the port. When fixing the operations level, the following principal factors are to be observed:

(1) Water levels and their fluctuations, especially heights and frequencies of possible storm tides, wind-induced water build-up, tidal waves, the possible effects of flow from upstream water and other actions mentioned in section 6.2.2.2.
(2) Mean level of the groundwater table including frequency and magnitude of fluctuations.
(3) Ship traffic, harbour installations and cargo handling operations, live loads.
(4) Ground conditions, subsoil, fill material and possible mass compensation.
(5) Available options for designing waterfront structures.
(6) Environmental concerns.

According to the requirements of the harbour with respect to operations, economy and construction methods, the relative importance of these criteria must be adjusted in making decisions, in order to arrive at the optimum result.

6.2.2 Levels and frequency of high water

A fundamental distinction must be made here between wet-dock harbours and open harbours with or without tide.

6.2.2.1 Wet-dock harbours

In flood-free wet-dock harbours, the port operations level must be designed as high above the officially determined mean working water level as is required

(1) for the prevention of flooding of the port terrain at the highest possible working water level,
(2) for an adequate height above the highest groundwater table in the port terrain associated with the mean working water level, and
(3) for practical handling of cargo.

The ground surface level should generally be 2.00 to 2.50 m but at least 1.50 m above the mean working water level.

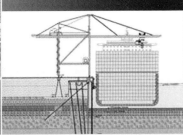

6.2.2.2 Open harbours

The height and frequency of high tides are decisive when it comes to selecting the suitable port operations level.

Wherever possible, planning tasks must use frequency lines for surpassing the mean high tide level. In addition to the main influencing factors stated in section 6.2.1 (1), the following influences must be taken into consideration:

- wind-induced water build-up in the harbour basin,
- oscillating movements of the harbour water due to atmospheric influences (seiching),
- wave run-up along the shore (so-called Mach effect),
- resonance of the water level in the harbour basin,
- secular rise in the water table, and
- long-term coastal lifting or settlement.

If inadequate water level measurement records are available, as many readings as possible must be obtained in situ during the design period and these integrated with existing record frequency lines of high tide levels in adjacent areas.

6.2.3 Effects of elevation and changes to the groundwater table in the port area

The mean elevation of the groundwater table and its local seasonal and other changes, as well as their frequency and magnitude, must be taken into consideration, particularly where the construction of proposed pipelines, cables, roads and railways or anticipated live loads etc. may be affected. In this context, the need for drainage means that the course of the groundwater table to the harbour water must also be given attention.

6.2.4 Port operations level depending on cargo handling procedures

(1) General cargo and container handling

In general, a flood-free operations level is essential. Exceptions should be allowed only in special cases.

(2) Bulk cargo handling

Because of the diversity of cargo handling methods and types of storage, as well as the sensitivity of the goods and susceptibility to damage of the handling gear, it is not possible to give a general guiding principle here. An effort should nevertheless be made to provide a flood-free surface, particularly in view of the environmental problems involved.

(3) Special cargo handling equipment

For ships with side doors for truck-to-truck handling, bow or stern doors for roll-on/roll-off handling or other special types of equipment, the top level of the waterfront structure must be compatible with the type of ship and equipped with either a fixed or movable connecting ramp. In this case the top of the waterfront structure need not be at the same level as the general ground surface. In tidal areas it may be necessary to adjust the levels in the vicinity of the ramps. Floating pontoons or similar may even be necessary. In any case, the requirements of the types of the ship that will use the port facilities must be taken into account.

(4) Cargo handling with ship's gear
In order to achieve adequate working clearance under the load hooks, even for low-lying vessels, the height of the quay must generally be designed lower than one where cargo is handled by quay cranes.

6.3   Standard cross-sections of waterfront structures in inland harbours (R 74)

6.3.1   Port operations level
The port operations level in inland harbours should normally be arranged above the level of the highest water table. In the case of flowing waters with high water table fluctuations, this is frequently only possible at considerable expense. As far as handling sites for solid bulk goods are concerned, occasional flooding can be accepted, but the risk of contamination of the water during such flooding must be taken into consideration. In cargo handling ports on inland canals with lesser fluctuations in the water level, the port operations level should be at least 2.00 m above the normal canal water level.

6.3.2   Waterfront
As far as possible, the waterfronts in inland ports should be straight with a straight, or near straight, face (R 158, section 6.6). Corrugated sheet piling is ideal as a waterfront structure apart from a few exceptions (R 176, section 8.4.15).
It is important that the outermost structural part of the crane does not protrude as far as or beyond the front edge of the waterfront structure. A crane leg width of 0.60–1.00 m is assumed.

6.3.3   Clearance profile
When designing crane tracks and cargo handling cranes, care must be taken to comply with the required side and overhead safety clearances as stipulated in the relevant specifications (EBO, BOA, UVV Deutsche Bahn AG, E 25 in [45]), see fig. R 74-1.
As far as roadways under crane portals are concerned, the recommendations given in fig. R 74-2 apply (EBO, EBA, UVV, E 25 in [45]).

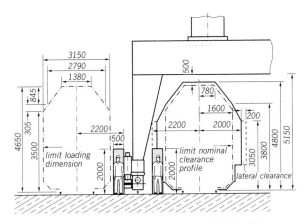

Fig. R 74-1. Lateral and upper safety clearances for the railways

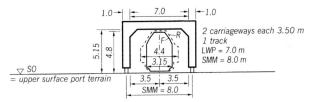

Fig. R 74-2. Recommended track middle dimension (SMM) and width clearance (LWP) for crane portals over roadways and covered tracks

6.3.4 Layout of the outboard crane rail

The aim is to place the outboard crane rail as close as possible to the waterfront edge. This reduces the length of the crane jib to a minimum and saves valuable storage space near the waterfront. The walkway necessary is to be arranged inboard of the crane portal leg (fig. R 74-1). Otherwise, a walkway of 0.80 m is to be provided between crane portal and waterway edge. The vertical waterfront structure can consist of

- reinforced concrete retaining wall
- sheet piling
- combination of sheet piling and reinforced concrete retaining wall on bored piles

Reinforced concrete retaining walls comply with the specifications in section 6.3.2.

In the case of sheet piling, the crane rail should be planned in the axis line of the sheet piling wall (fig. R 74-3).

175

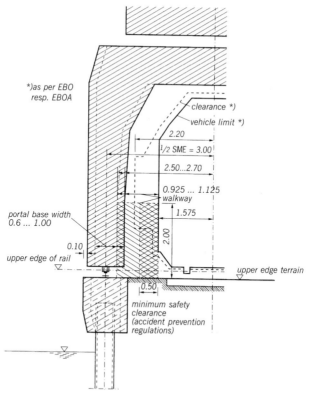

Fig. R 74-3. Standard cross section dimensions with sheet piling structures in inland harbours (crane rail in axis line of sheet piling)

The necessary geometrical requirements (section 6.3.2) can make it necessary to support the crane rail off-centre (fig. R 74-4).

A combined solution (steel sheet piling/reinforced concrete wall with mooring piles) offers the advantage of the separate introduction of the load. In addition, the crane rail can be routed close to the edge without vessel impacts having any influence on the crane. Ladders can be placed in ideal positions for access to the ship, fig. R 74-5 and recommendation E 42 in [45].

6.3.5 Mooring facilities

Sufficient mooring facilities for the ships are to be installed on the outboard side of the waterfront structure (R 102, section 5.13).

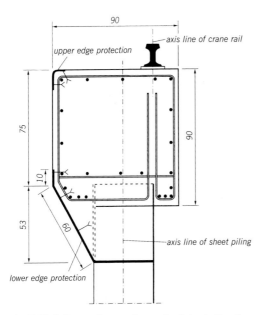

Fig. R 74-4. Crane rail eccentric to axis of sheet piling (example)

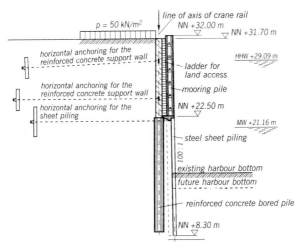

Fig. R 74-5. Anchored sheet piling with mooring piles/reinforced concrete wall, example

6.4   Sheet piling waterfront on canals for inland vessels (R 106)

6.4.1   General

When canals are to be constructed or extended in areas where only limited space is available, waterfront structures of anchored steel sheet piling are frequently the best engineering solution, and considering the reduced land and maintenance costs, also the most economical. This is especially true for stretches that must be made impervious. The sheet piling interlocks can be sealed to improve watertightness, see R 117, section 8.1.20.

Fig. R 106-1 shows a typical example of such a structure.

If operating conditions allow, the upper edge of the sheet piling should remain under the water table for reasons of corrosion protection and landscaping. Please refer to [52] for details of cross-section design.

6.4.2   Stability verification

The waterfront structure and its components are analysed and designed according to the pertinent recommendations. Please refer to R 19, section 4.2, and R 18, section 5.4. In the case of vertical live loads, in contrast to R 5, section 5.5, a uniformly distributed ground live load of 10 kN/m^2 (characteristic value) is to be assumed (fig. R 106-1). Please also refer to R 41, section 8.2.10, and R 55, section 8.2.8.

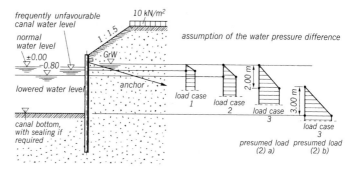

Fig. R 106-1. Cross-section for the sheet piling waterfront of the normal stretch of an inland waterway canal with the most essential load assumptions

6.4.3   Load assumptions

The loads attributed to the load cases are *characteristic* values which are to be multiplied by the partial safety factors in DIN 1054.

The water pressure difference to be used in load case 1 is that which can be expected to occur frequently due to unfavourable canal and ground-water levels. The groundwater table is often assumed to be at the elevation of the top of the sheet piling.

If the groundwater table is inclined towards the sheet piling, the water pressure difference is to be related to the relevant active earth pressure failure plane (R 65, section 4.3, fig. R 65-1, and R 114, section 2.9, fig. R 114-2).

Load case 2 takes account of a lowering of the canal water in front of the sheet piling by 0.80 m due to passing ships.

Load case 3 takes the following loads into consideration:

(1) In canal sections which are purposely emptied at times (e.g. between two gates), the canal water level is to be assumed to coincide with the elevation of the canal bottom, while the groundwater level will depend on local conditions.

(2) In the other areas (normal stretches), it need not be assumed that the canal will be completely dewatered while the surrounding groundwater table remains at its normal level.

If in exceptional circumstances local conditions are such that a rapid and severe drop of the canal water level can be expected in the case of serious damage to the canal, the following two load cases must be investigated:

   a) the canal water level is assumed to be 2.00 m below the groundwater table,

   b) the canal water level is assumed to be at the canal bottom while the groundwater table is 3.00 m above the canal bottom.

(3) In locations where failure of the canal wall would destroy the stability of bridges, loading facilities or other important installations along the canal, the sheet piling must be designed for the load case "emptied canal", or must otherwise be adequately reinforced by means of special structural measures.

In the structural investigations, the planned bottom level of the canal or excavation (e.g. underside of bottom protection) may be assumed to be the actual one for calculation purposes. Overdepth dredging down to 0.30 m below the nominal bottom is generally admissible without special calculations if the EAU recommendations are observed and when the wall is fully fixed in the ground (see R 36, section 6.7). This does not apply to unanchored walls and anchored walls with simple toe support. If in exceptional cases greater deviations are to be expected and severe scour damage due to ship's screw action is probable, the calculations should assume a bottom elevation which is at least 0.50 m below the nominal bottom.

6.4.4   Embedment depth

If impervious soil is encountered at an attainable depth in dam stretches that are to be made watertight, the sheet piling is driven to a uniform depth and firmly embedded in the impervious layer. This renders sealing of the canal bottom unnecessary.

6.5 Partially sloped waterfront construction in inland harbours with extreme water level fluctuations (R 119)

6.5.1 Reasons for partially sloped construction

The mooring, tying up, lying off and departure of unmanned vessels must be possible at every water level without the use of anchors, and port and operations personnel should at all times have safe access to moored vessels. Owing to water level fluctuations, this is only possible when vessels are moored alongside a vertical wall. Fully sloped banks are therefore not suitable as cargo handling sites and can be used as berths only in conjunction with dolphins.

Vertical banks are required at cargo handling areas with a flood-free operations level. As far as the handling of bulk goods at the upper part of the waterfront is concerned, a vertical structure is not necessary and frequently also undesirable.

The partially sloped bank is therefore ideal for inland harbours with extreme water fluctuations. It consists of a vertical quay wall for the lower part and an adjoining upper slope (figs. R 119-1 and R 119-2 show examples).

6.5.2 Design principles

The level of the transition from the vertical to the sloping waterfront is highly significant for cargo handling operations on a partially sloping bank. The transition should be at such a level that it is not below water for more than 60 days (long-term mean).

On the lower Rhine, for example, this corresponds to a transition level of about 1 m above mean water level (MW) (fig. R 119-1). The level of this transition should always remain uniform throughout the harbour basin.

For embankments with a higher harbour ground surface, the elevation of the top of the sheet piling should be selected so that the slope height is limited to a maximum of 6 m in order to avoid difficult operating and unfavourable structural conditions (fig. R 119-2).

Guide piles at intervals of about 40 m are advisable along the vertical bank section at berths and push tow coupling quays without cargo handling operations for unmanned vessels in river ports with extreme water level fluctuations. These serve for marking, safe mooring and as slope protection. They should extend 1.00 m above HHW without projecting over the water side (fig. R 119-1).

The vertical bank section is generally constructed of single-anchored sheet piling with fixed earth support.

The coping could consist of a 0.70 m wide steel or reinforced concrete capping beam (figs. R 119-1 and R 119-2) which also suffices as a berm, which can be walked on safely at the ladder recesses. This width ensures

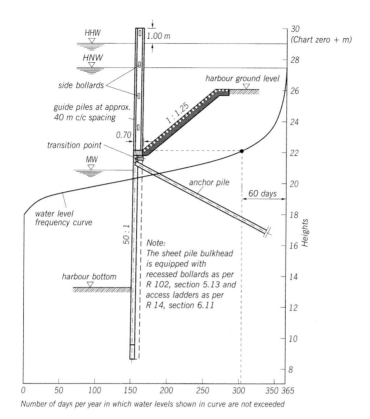

Fig. R 119-1. Partially sloped bank of berths, particularly for push lighters where harbour ground level is subject to flooding

there is still no danger of ships grounding with falling water levels, provided the vessels are properly maintained.

The berm must be constructed continuously behind guide piles.

The water-side edge of the reinforced concrete capping beam is to be protected from damage with a steel plate as described in R 94, section 8.4.6.

The slope must not be steeper than 1 : 1.25 so as to permit the installation of safe, properly designed stairways. Gradients of 1 : 1.25 to 1 : 1.50 are chiefly used.

Bollards on partially sloped banks are constructed according to R 102, section 5.13.

181

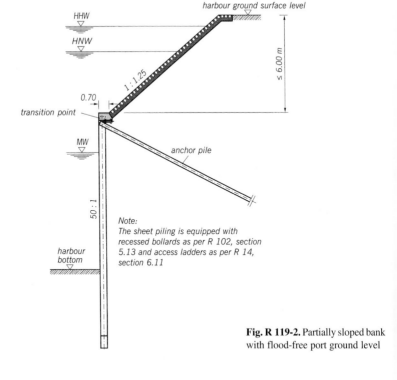

Fig. R 119-2. Partially sloped bank with flood-free port ground level

6.6 **Design of waterfront areas in inland ports according to operational aspects (R 158)**

6.6.1 **Requirements**

The requirements for the layout of the waterfront areas result principally from navigation and cargo handling conditions, but also partly from railway and road operations, as far as the design of the upper part is concerned. In order to ensure proper shipping operations, it must be possible for ships to moor on the waterfront and to cast off safely and easily, and to lie calmly, so that even passing ships or push tows do not create disadvantageous effects, and the mooring cables or ropes can slacken as the water level changes. The waterfront structure shall also act as a guide when berthing the ship. There must be the possibility of direct access for passengers between shore and ship, or for the safe placing of a gangway (E 42 in [45]). Operations with push lighters move relatively large masses; in addition, push lighters are box-shaped with angular boundaries. This results in increased requirements for a waterfront structure face which is as straight as possible.

For cargo handling operations, conditions are to be created which enable the ship to be loaded and unloaded rapidly and safely. During these operations, the ship should move as little as possible. On the other hand, if necessary it should be possible to move or shift the vessel without difficulty. Economic and operational aspects determine the waterfront cross-section. The crane operator must have a clear view.

6.6.2 Planning principles

Waterfront structures with long, straight stretches are required because of the great length of the ships and push tows, but also to improve the guidance of the ship. If changes in direction are unavoidable, they should be designed in the form of cranks (polygons) and not continuously (circular arcs). The cranks are to be spaced so that the intermediate straight stretches match the ship or push tow lengths.

The shape of the ships and their operation demand the straightest possible structures without protruding installations and recesses in which the vessels can catch. The front face can be sloped, partially sloped, inclined or vertical. In the longitudinal direction, however, it should be as straight as possible.

6.6.3 Waterfront cross-sections

(1) Embankments

Embankment surfaces should be designed as straight as possible. Intermediate landings should be avoided if possible. Stairs should be installed at right-angles to the shoreline. Bollards and mooring rings may not protrude beyond the embankment surface. If intermediate berms are unavoidable on high embankments, they must not be located in the area of frequent water level fluctuations but rather situated in the high water zone. The transition from the sloping to the vertical bank is to be positioned in the same manner (see R 119, section 6.5.2). On sloping banks safe guiding of vessels can be guaranteed only in conjunction with closely spaced mooring dolphins.

(2) Vertical waterfronts

Vertical or slightly inclined waterfront structures of solid construction are chiefly suitable for new waterfront facilities constructed in a dry excavation. They present a straight front face. When erected in or over water, diaphragm or bored diaphragm walls may be used, for instance. The shape of their front face, however, does not generally meet operational requirements after dredging. Measures for achieving a straight surface in the ship contact area are then required according to R 176, section 8.4.16.

The sheet piling method of construction represents a proven and economical solution for a waterfront structure. Where special loads from shipping operations are concerned, it may even be necessary to require a smooth surface instead of the usual corrugated one (see R 176, section 8.4.15).

(3) Partially sloping banks
Partially sloping banks have also proved to be suitable for cargo handling. However, in such cases the position of the craneway at the top edge of the embankment means that a longer crane jib is necessary.

6.7 Nominal depth and design depth of harbour bottom (R 36)

6.7.1 Nominal depth in seaports
The nominal depth is the water depth below a defined reference elevation, which should be observed.
When stipulating the nominal depth of the harbour bottom at quay walls, the following factors should be considered:

(1) The draught of the largest, fully loaded ship berthing in the port, taking account of salinity of the harbour water and heeling of the ship.
(2) The safety clearance between ship's keel and nominal depth should generally be a minimum depth of 0.50 m.

The reference elevaton for the water depth is a statistical justified low water level (LW), which often equals the chart zero (SKN). In regions without tidal influence as for example the Baltic Sea the low water level is derived from empirical data.
In tidal regions an appropriate consideration of the tidal heave is needed, to ensure that for berthing ships the needed water depth is guarantied with the necessary frequency. In this case very often SKN is used for the reference elevation head.
Up to the end of 2004 SKN in Germany was derived from mean spring tide low water level (MLWS). From 2005 on SKN is defined as the lowest astronomical tide (LAT). LAT defines the lowest possible water level caused by astronomical influences. For the German North Sea LAT is about 0.50 m below MLWS. This definition of SKN is valid for all states at the North Sea and thus international defined.
Besides of this general definition the reference elevation has to be fixed with respect to local requirements, it can be different form SKN. The reference elevation level has to be fixed in advance of the design by the involved partners conjointly.

6.7.2 Nominal depth in inland harbours
The nominal depth of the harbour and the harbour entrance is to be selected so that the ships can travel on the waterway with the greatest possible load draught. In inland harbours on rivers, the water depth should as a rule be 0.30 m deeper than the nominal bottom of the adjoining waterway, to rule out any dangers for ships at all water levels.

6.7.3　Design depth at quay walls

If dredging is to be carried out in front of quay walls because of silt, sand, gravel or rubble deposits, the dredging must be deeper than the planned nominal depth of the harbour bottom stipulated in sections 6.7.1 and 6.7.2 (fig. R 36-1).

The design depth is made up of the nominal depth of the harbour bottom, the maintenance dredging zone down to the planned dredging depth, plus dredging tolerance and other allowances for special conditions. The dredging depth is influenced by the following factors, with reference to recommendation R 139, section 7.2:

(1) Extent of the silt mass, sand drift, gravel or rubble deposits per dredging period.
(2) Depth to which the soil may be dredged or disturbed under the nominal depth of the harbour bottom.
(3) Costs of every interruption to cargo handling operations caused by dredging works.
(4) Constant or only occasional availability of the required dredging apparatus.
(5) Costs of dredging work with regard to the level of the maintenance dredging zone.
(6) Extra costs of a quay wall with deeper harbour bottom.

Due to the importance of all the aforementioned factors, the scope for dredging at quay walls must be stipulated with care. On the one hand, inadequate scope results in high costs for maintenance dredging and more interruptions to operations, but on the other, excess scope causes higher construction costs and possibly also additional sedimentation.

It is practical to attain the harbour bottom depth first in at least two dredging cuts executed at intervals. A maximum cut thickness of 3 m must be observed.

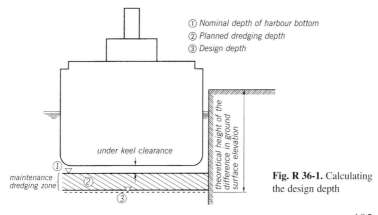

① Nominal depth of harbour bottom
② Planned dredging depth
③ Design depth

under keel clearance

maintenance dredging zone

theoretical height of the difference in ground surface elevation

Fig. R 36-1. Calculating the design depth

185

Table R 36-1. Maintenance dredging depths and minimum tolerances, recommended values in [m]

Depth of water below lowest water level m	Thickness of maintenance dredging zone m	Minimum tolerance* m	Theoretical design depth m
5	0.5	0.2	5.7
10	0.5	0.3	10.8
15	0.5	0.4	15.9
20	0.5	0.5	21.0
25	0.5	0.7	26.2

*) depends on dredging equipment

Table R 36-1 shows the thickness of the maintenance dredging zones under the nominal depth of the harbour bottom for various water depths, with the corresponding minimum tolerances (only intended as a general guide). Note the recommendations of the Dredging Technology Committee and see section 7.2.3.1.

The theoretical design depth in table R 36-1 already takes account of the allowances required in DIN EN 1997-1 (see also section 2.0).

In the case of greater bottom erosion, the design depth is to be increased, or suitable measures are to be taken to prevent erosion.

6.8 Strengthening of waterfront structures to deepen harbour bottoms in seaports (R 200)

6.8.1 General

Developments in vessel dimensions mean that occasionally it is necessary to deepen the harbour bottom in front of existing quay walls. Greater crane and live loads are often an additional factor. The possibility of deepening the harbour bottom in such cases depends on:

a) the design of the quay wall,
b) deformation in the wall since it was built,
c) the structural condition of the quay wall,
d) the extent of deepening required, particularly in respect of the design depth of the harbour bottom,
e) the possibility of reducing the permissible live loads behind the quay wall,
f) the anticipated service life of the quay wall after any strengthening,
g) the availability of the structural calculations performed earlier, with all associated loads, theoretical soil values and water levels, and the design drawings,
h) the costs of strengthening in comparison with the costs of different solutions (e.g. new construction elsewhere).

For a) and b) please refer to R 193, section 15. With respect to g), it may be useful to perform new soil investigations to establish, for example, the consolidation figure for the cohesive soil, and to check and increase/decrease the theoretical soil values.

Once the new loads, water levels, soil values and the increased theoretical depth are known, a structural analysis can then be carried out for the design of a reinforced quay wall.

If old calculations and drawings for the existing wall are no longer available, it is recommended to combine any bottom deepening necessary with a reduction in live loads. The deformation behaviour of such a quay wall plays an important role in this case and special account must therefore be taken of this.

6.8.2 Design of strengthening measures

There are numerous possibilities for reinforcing quay walls for deepening the harbour bottom. In the case of combined walls, care must be taken to ensure that the compensating piles also have a sufficient embedment length. The following sections illustrate a few typical solutions, depending on the factors stated in section 6.8.1.

6.8.2.1 Measures to increase the passive earth pressure

Please refer to R 164, section 2.11, and R 109, section 7.9.

a) Replacement of soft, cohesive soil in front of the quay wall with non-cohesive material of high specific density and shear strength (fig. R 200-1).

The transition must be executed to ensure filter stability. Construction work must be especially carefully supervised, and, if necessary, surveys of the wall should be performed during the construction work in order to detect any outward deflection of the structure during

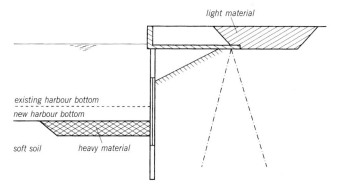

Fig. R 200-1. Soil replacement in front of and/or behind the structure

187

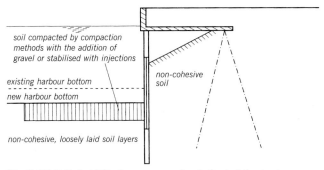

Fig. R 200-2. Soil stabilisation or compaction in front of the structure

the dredging. In such cases the waterfront structure can be relieved by excavating the backfill behind the wall.

Settlement in front of the waterfront structure is to be expected because the soil is generally not sufficiently consolidated for the higher surcharges.

b) Soil compaction for non-cohesive soil (fig. R 200-2).

c) Soil stabilisation by injections for permeable, non-cohesive soil (fig. R 200-2).

6.8.2.2 Measures to reduce active earth pressure

a) Erect a reinforced concrete relieving platform on piles (fig. R 200-3).

b) Replace the backfill with a lighter material; please refer to R 187, section 7.11 in this respect.

c) Use injections to stabilise non-cohesive backfill with good permeability (fig. R 200-2).

6.8.2.3 Measures involving the quay wall

a) Use of supplementary anchors, inclined or horizontal (fig. R 200-4).

b) Drive more deeply and raise the existing waterfront structure (fig. R 200-5).

c) Drive new sheet piling directly in front of the quay wall. The sheet piling can then be anchored in different ways:

• by means of a new superstructure on piles over the existing relieving platform (fig. R 200-6),

• by means of inclined anchor piles or horizontal anchors (fig. R 200-7).

d) Forward extension with reinforced concrete relieving platform on piles, provided sufficient space is available. Please refer to R 157, section 11.5 (fig. R 157-1) in particular. An additional bonus with this method is that the quay surface area is enlarged, to the advantage of cargo handling operations (fig. R 200-8).

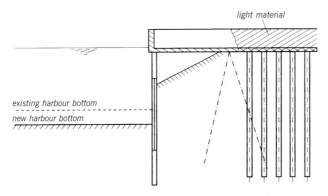

Fig. R 200-3. Stabilisation with a relieving construction on piles

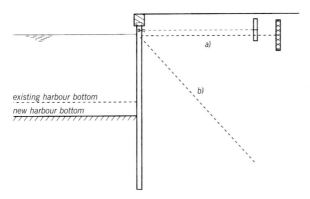

Fig R 200-4. Use of supplementary anchors horizontally (a) or inclined (b)

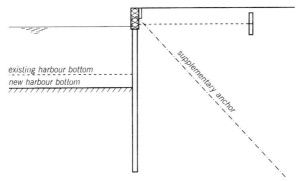

Fig. R 200-5. Drive more deeply and build up existing water front structure plus supplementary anchoring

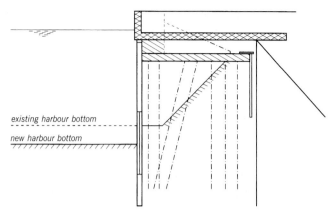

Fig. R 200-6. Forward extension of sheet piling and new superstructure

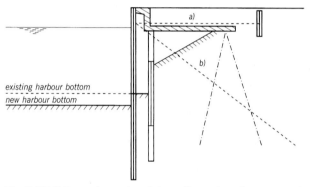

Fig. R 200-7. Forward extension of sheet piling and supplementary anchoring a) or b)

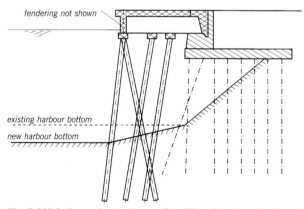

Fig. R 200-8. Forward extension on piles with underwater embankment

6.9 Redesign of waterfront structures in inland harbours (R 201)

6.9.1 General

Initially, everything said in R 200, section 6.8, about quay wall strengthening in seaports generally applies accordingly here. However, the reasons for redesigning waterfront structures in inland harbours are frequently different. It is mostly the erosion of the river bottom which gives rise to the need for deepening the harbour bottom in side basins. In canals and rivers regulated by damming, extension to accommodate a greater unloading depth can also necessitate deepening. In individual cases, increases in the crane and live loads can lead to a redesign.

In the case of sloping banks, a deepening of the river bottom results in a reduction in the harbour basin width and water cross-section. Together with the increasing length of the crane jib, this generally results in a need for a partially sloping or vertical extension. Harbour bottom deepening or increased live loads result in higher loads on individual components, which in some circumstances are then no longer adequately designed.

6.9.2 Possibilities for redesign

It is generally possible to construct a new waterfront structure from, or instead of, the old one. However, it is often sufficient to renew or reinforce certain parts of the waterfront, or to implement other design measures. Thus, for example, sheet piling can be driven deeper and a new superstructure built. Increased anchoring forces can be absorbed by supplementary anchoring. In non-cohesive soils, compaction of the harbour bottom leads to an increase in passive earth pressure. The stability of an embankment can be improved by needling with the subsoil by embedding driven components.

6.9.3 Construction examples

Figs. R 201-1 to R 201-6 show typical examples of redesigns for waterfront structures in inland harbours. The levels are related to MSL.

- Bank construction by replacing the sloped bank with a partially sloped bank (fig. R 201-1).
- Bank construction by driving more deeply and increasing the superstructure on existing bank sheet piling (fig. R 201-2).
- Bank construction by supplementary anchoring of temporarily remaining sheet piling (fig. R 201-3).
- Bank construction by pre-driving new sheet piling (fig. R 201-4).
- Bank construction by compaction of non-cohesive soil to increase the passive earth pressure in front of the sheet piling (fig. R 201-5).
- Bank construction by needling the slope (fig. R 201-6).

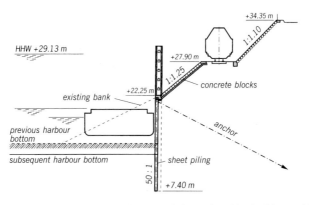

Fig. R 201-1. Bank construction by replacing a sloped bank with a partially sloped bank

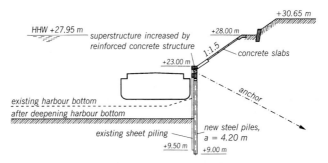

Fig. R 201-2. Bank construction by driving more deeply and building up the superstructure on the existing bank sheet piling

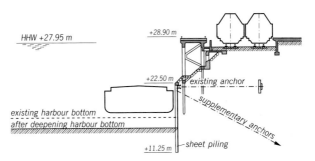

Fig. R 201-3. Bank construction by supplementary anchoring of remaining sheet piling

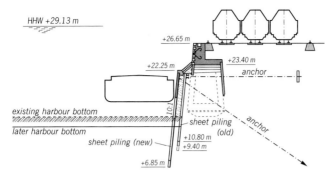

Fig. R 201-4. Bank construction by pre-driving new sheet piling

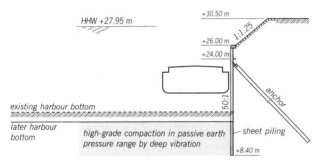

Fig. R 201-5. Bank construction by vibration of non-cohesive soil in passive earth pressure range in front of sheet piling

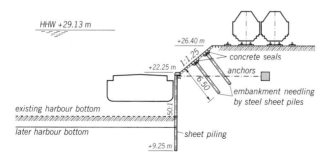

Fig. R 201-6. Bank construction with bank stabilisation by needling

6.10   Provision of quick-release hooks at berths for large vessels (R 70)

Quick-release hooks are provided instead of bollards only in exceptional cases at special berths for large vessels where mooring takes place according to a defined mooring system. The range of movement is essentially defined according to the mooring system. Heavy-duty quick-

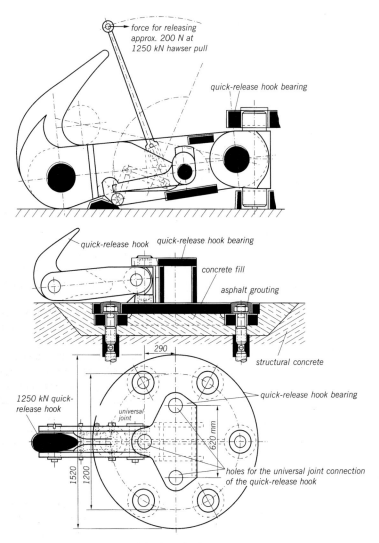

Fig. R 70-1. Example of a quick-release hook

release hooks for loads of 30–3000 kN with manual or hydraulic (oil) release mechanism and remote control guarantee easy mooring and quick release of hawsers even when heavy steel hawsers are used.

Fig. R 70-1 shows an example of a quick-release hook for 1250 kN maximum load capacity with manual release mechanism. It can be used with several hawsers and releases them at full load and also at lesser loads by operating the handle with little effort.

The quick-release hooks are attached to a bearing via a universal joint. The number of quick-release hooks depends on the hawser tension given in R 12, section 5.12, and on the directions from which the principal hawser pulls may occur simultaneously. Several quick-release hooks can be installed on one bearing. The range of movement must ensure that the hook can meet all anticipated operational requirements without jamming. The swivel range is max. 180° horizontally and 45° vertically. It is easier to fasten heavy-duty hawsers when the quick-release hook is combined with a capstan.

6.11 Layout, design and loading of access ladders (R 14)

6.11.1 Layout

Access ladders are primarily intended to provide access to the mooring facilities, and, in emergencies, to enable persons who have fallen into the water to climb onto the quay. Only in exceptional circumstances are such ladders intended for embarkation or disembarkation. Qualified and experienced shipping and cargo handling personnel can also be expected to use the access ladders even when there are greater differences in water levels.

Access ladders in waterfront structures of reinforced concrete should be placed at approx. 30 m intervals. The position of the ladder in a normal section depends on the position of the bollard because the ladders must not be obstructed by mooring lines. If joints between sections have been included, it is advisable to place the ladders near these joints. Proceed in a similar manner for sections lengths shorter than 30 m. In the case of waterfront structures of steel piling, it is recommended to position the access ladders in the pile troughs.

Mooring facilities are to be installed on both sides of each ladder (R 102, section 5.13.1).

6.11.2 Practical dimensions

In order to be accessible from the water at all times, even under LW or LLW, the ladders must extend down to 1.00 m below LW or LLW. For easy installation and replacement of the ladders, the lowest ladder mounting is designed as a plug-in item into which the side pieces can be inserted from the top. The transition between the top of the ladder and

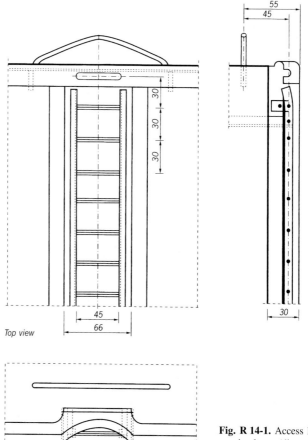

Top view

55
45

30
30
30

45
66

30

Fig. R 14-1. Access ladder in steel capping beam (dimensions in cm)

the deck of the quay must ensure that ascent and descent of the ladder can be accomplished safely. At the same time, the ladder must not be a hazard to traffic on the quay. These two requirements can best be met by dishing in the nosing of the wall at each ladder. In addition, a hand grip should be placed at the head of the ladder, at least for flood-free waterfront structures. This hand grip should be of material 40 mm in diameter with the top 30 cm above the deck of the quay and its longitudinal axis 55 cm from the face of the wall (fig. R 14-1). If the hand grips should present an obstacle during cargo handling, other suitable aids to climb the ladder must be provided. Fig. R 14-2 shows a proven design of this type. The top rung of the ladder in this solution is 15 cm below the deck of the quay.

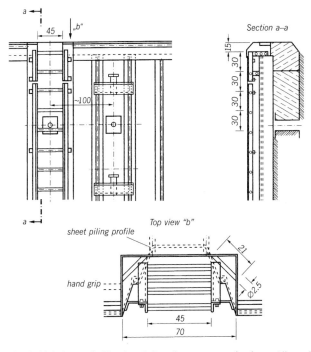

Fig. R 14-2. Access ladders in reinforced concrete capping beam (dimensions in cm)

The ladders should be installed with centre of rungs min. 10 cm behind the face of the wall, consisting of square steel bars 30 × 30 mm installed not flat but turned through 45° so that one edge points upwards. This reduces the danger of slipping due to ice or dirt. The rungs are fastened to the side pieces at a centre-to-centre distance of 30 cm. The clear width between the side pieces of the ladder should be 45 cm.

Please refer to DIN 19 703 for further details.

6.12   Layout and design of stairs in seaports (R 24)

6.12.1   Layout

Stairs are used in seaports where access for the public is required. They must be safe for use by persons not acquainted with the conditions in harbours. The upper exit of the stairs should be placed so that there is little or no interference with foot traffic and cargo handling. The approach to the stairs must be clearly visible and thus permit the smooth flow of foot traffic. The lower end of the stairs is to be positioned so that ships can berth easily and safely with safe passage between ship and stairs.

197

6.12.2 Practical dimensions

Stairs should be 1.50 m wide so that they are clear of the outer crane track on quays for seagoing vessels and do not conflict with the fixings of the crane rails at a distance of 1.75–2.50 m from the edge of the quay. The pitch of the stairs should be determined by the well-known formula $2\,s + a = 59$ to 65 cm (rise s, going a). Concrete steps should have a rough, granolithic concrete finish and the edges of the steps should be fitted with a steel nosing for protection.

6.12.3 Landings

For tidal ranges of greater magnitude, landings should be installed at 0.75 m above MLW, MW and MHW respectively. Additional landings may be required, depending on the height of the structure. Intermediate landings are to be positioned after max. 18 steps; the length of the landing should be 1.50 m or equal to the stair width.

6.12.4 Railings

The stairs should be fitted with a handrail 1.10 m above the front edge of the tread. When harbour operations permit, the stairs should be enclosed by a 1.10 m high railing, which could be removable if necessary.

6.12.5 Mooring equipment

The quay wall next to the lowest landing is equipped with cross poles (R 102, section 5.13). In addition, a recessed bollard or mooring hook is positioned below each landing. Recessed bollards are used for solid quay walls or quay wall members, cross poles generally for steel piling structures.

6.12.6 Stairs in sheet piling structures

Stairs in sheet piling structures are frequently made of steel. The sheet piling wall is driven with a recess large enough to contain the stairs. The stairs must be protected against underrunning by suitable means (e.g. fender piles).

6.13 Equipment for waterfront structures in seaports with supply and disposal facilities (R 173)

6.13.1 General

The supply facilities serve to supply the required operational resources, power, etc. to existing public installations and facilities, but also to businesses located in the port, as well as to the docked ships and the like. The disposal facilities serve to drain any waste water and operational resources.

In the planning of such facilities, consideration must be given to the fact that they must also be located in the immediate vicinity of the waterfront structures and sometimes directly in these structures themselves.

Adequate openings for these services are to be provided in underground structural members, such as craneway beams and the like. Therefore, in order to avoid unnecessary costs, consultation of all participants must take place during the planning of such structural members. Spare openings are to be included to allow for any later expansion.

The supply facilities include:

- water supply facilities
- electric power facilities
- communications and remote control facilities
- other facilities

The discharge facilities include:

- rainwater drainage
- waste water drainage
- fuel and oil interceptors

The respective statutory requirements are to be observed.

6.13.2 Water supply facilities
The water supply facilities provide drinking and process water, and can also be used for extinguishing purposes in case of fire.

6.13.2.1 Drinking and process water supplies
In order to safeguard the drinking and process water supply system in the port, at least two supply points are required, independent of each other, for each port section, whereby the lines are laid out as ring nets to guarantee a permanent flow.

Hydrants are installed at approx. 100–200 m intervals; 60 m (measured parallel to the quayside) is a typical spacing for water hydrants for supplying ships. Underground hydrants are placed on quay walls and in paved crane and rail areas so as not to hinder operations. The hydrants are to be arranged so that there is no danger of being crushed by railborne cranes and vehicles, even when standpipes have been fitted.

When using underground hydrants, special attention is to be paid to the fact that the connection coupling is protected from impurities even in case of any possible flooding of the quay wall. An additional shut-off valve is required to isolate the hydrant from the supply line. The hydrants must accessible at all times. They must be situated in areas where storage of goods is not possible for operational reasons.

The pipes are generally laid with an earth cover of 1.5 to 1.8 m, and at least 1.5 from the front face of the quay wall to protect against frost. In loaded areas with tracks of the port railway, the lines are to be placed in protective ducts.

In quay walls with concrete superstructures, the lines may be placed in the concrete construction. Here, the differing deformation behaviour of the individual construction sections must be taken into account, together with the differing settlement behaviour of deep- or shallow-founded structures. For the drinking water supply, crossings with tracks must be accepted. When ring mains are laid on the land side of the superstructure, drainable connection lines are to be installed between the ring main and the hydrants located on the quay wall. Lines not constantly in operation pose a risk to drinking water hygiene.

In order to avoid major excavation work in operations areas in case of a pipe failure, as far as possible the mains are to be laid not under reinforced concrete surfaces but under strips reserved for such services.

6.13.2.2 Separate fire-fighting water

When there is a high fire risk in a certain section of the port, it is frequently recommended to supplement the drinking and process water supply system with an independent fire-fighting system. The water is taken directly from the harbour basin with pumps. The necessary pump rooms can be located in a chamber of the quay wall below ground so as not to disturb cargo handling.

It is also possible to feed the fire-fighting water supply system through the pumps of the fire-fighting boats at special connection points.

In sheet piling quay walls, the suction pipes may be placed in the sheet piling troughs, whereby the fire-fighting water can be obtained through mobile pumps belonging to the fire department. These suction pipes are adequately protected from the impact of a vessel. The same is also possible in recessed slits in concrete superstructures.

The fire-fighting pipeline system has to satisfy the same requirements as for the drinking and process water supplies.

6.13.3 Electric power supply facilities

The electric power supply facilities provide power to the administration buildings, port installations, crane installations, lighting installations of the rail areas, roads, operations areas, open areas, quays, berths and dolphins, etc.

Only cables are used for the high- and low-voltage supply systems to the port, except during construction. The cables are to be laid in the ground with an earth cover of approx. 0.80–1.00 m, in quay walls and operational areas in plastic ducts with concrete cable drawpits capable of accepting vehicular traffic. The advantage of such a duct system is that the cable installations can be augmented/modified without interrupting port operations.

Where there is a risk of frequent flooding of the quay walls, the power connection points are to be installed in raised, flood-free stands.

Power connection points are generally installed in the quay wall coping at intervals of 100–200 m. They must be capable of accepting vehicular traffic and have a drainage pipe. These connection points are used to provide the power for welding generators carrying out minor repairs on ships and cranes, as well as the power for emergency lighting, and other purposes.

Contact line ducts, cable channels and crane power feeding points must be provided in the quay areas for the power supply to the cranes. The drainage and ventilation of these facilities is particularly important. In quay walls with concrete superstructures, these facilities can be included in the concrete construction.

Special attention is drawn to the fact that the electric power supply networks must be provided with equipotential bonding facilities. This is to prevent unduly high voltages from occurring in crane rails, sheet piling or other conductive components of the quay wall area due to a fault in any electrical facilities (e.g. a crane). Such equipotential bonding systems should be installed about every 60 m.

In the case of craneways integrated in the quay wall superstructure, the equipotential bonding lines are normally concreted in during construction of the superstructure, for reasons of cost. However, they must be laid in protective ducts in areas in which differential settlement may be expected.

6.13.4 Other facilities

These include all supply facilities not mentioned in sections 6.13.2 and 6.13.3, as required, for example, at shipyard quay walls. These include: gas, oxygen, compressed air and acetylene lines, as well as steam and condensate lines in ducts. The layout and installation of such facilities must comply with the pertinent regulations, particularly the safety regulations.

Connections for telephones are usually placed at a spacing of 70–80 m along the front edge of the quay. However, the growing availability and use of mobile phones means that such telephone points are being relied on less and less.

6.13.5 Disposal facilities

6.13.5.1 Rainwater drainage

The rainwater falling in the quay wall area and also on its land side is drained into the harbour directly through the quay wall. For this, the quay and operations areas are equipped with a drainage system consisting of inlets, cross and longitudinal culverts and a collecting main with outlet into the harbour. The catchment drainage areas depend on local characteristics. An attempt should be made to install the minimum possible number of outlets in the waterfront structure. Such outlets can suffer damage from the impact of vessels.

Suitable valves are to be provided in the drainage system in case of dyke drainage to impair flood protection in the case of extremely high water levels.

The outlets are to be provided with slide valves in quay and operational areas with a risk of leakage of dangerous or toxic substances or fire-fighting water into the drainage system so that the valves can be closed to prevent pollution of the harbour water.

6.13.5.2 Waste water disposal

It is not customary to accept the waste water from seagoing vessels because any claims following the transfer of contaminated waste water are very difficult to prove and trace. The waste water occurring in the port is fed through a special waste water disposal system into the municipal sewer system. It may not be drained into the harbour water. A waste water main is therefore only found in waterfront structures in exceptional cases.

6.13.5.3 Fuel and oil interceptors

Fuel and oil interceptors are placed wherever they are required, just like facilities outside the port area.

6.13.5.4 Disposal regulations for ship waste

According to the MARPOL convention, ports should provide facilities for the disposal of ship waste, such as liquids containing oils and chemicals, solid ship waste (galley waste and packaging refuse) and sanitary waste water.

6.14 Fenders at berths for large vessels (R 60)

6.14.1 General

In order to provide vessels with safe berthing at waterfront structures, it is customary these days to include fenders. They absorb the impact of vessels when berthing and avoid damage to ship and structure while the vessel is moored. For large vessels in particular, fenders are indispensable. Although timber fenders and rubber tyres etc. are still common, other, modern forms of fender are becoming more and more established. The main reasons for this are:

- The use of fenders increases the service life of the waterfront structure (see R 35, section 8.1.8.4).
- The cost of vessels is on the increase and so expensive ships demand good fendering.

- Ships are growing in size and hence also the surface area affected by the wind.
- The requirements placed on cargo handling equipment are on the increase.
- The strength of the outer hull is being reduced further and further.

6.14.2 The fendering principle

A fender is, in principle, the intermediate layer between vessel and waterfront structure. This intermediate layer absorbs part of the kinetic energy of a berthing ship; indeed, an energy-absorbing fender will absorb most of this energy. But of course, the waterfront structure itself and the ship's hull also absorb some of the energy through their elasticity/plasticity. The energy absorption E_f of a fender is shown by its characteristic load–deflection curve, which illustrates the relationship between fender deformation d and fender reaction force R (see fig. R 60-1). The area beneath the curve represents the energy absorption E_f.

All fender designs that are supported by a rigid waterfront structure are generally characterised by a steep increase in the reaction force of the structure once the absorption capacity of the fender has been reached. Therefore, an adequate factor of safety against the fender reaction force must be taken into account when designing the structure.

The dimensions and properties (e.g. characteristic load–deflection curve) of the various elastomer fender elements can be found in the publications of the fender manufacturers. However, it should be remembered that these curves apply only when the fender cannot buckle sideways and when the creep under permanent load is not excessive.

If fenders are used on adequately flexible components and other supporting constructions (e.g. dolphins), then it is not only the work capacity of the fender but also the capacity of these components that must be taken into account, if necessary in a total system.

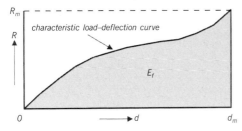

Fig. R 60-1. Energy absorption of a fender

6.14.3 Design principles for fenders

A fender system must be designed with the same level of care and attention as the entire waterfront structure. The right fender system must be considered at an early stage in the planning of the waterfront structure. For comprehensive advice on the design of fenders see [232].

A fender system has to satisfy the following principal requirements:
- Berthing of the vessel without damage
- Mooring of the vessel without damage
- Long, safe and reliable usage
- Prevention of damage to the waterfront structure

Therefore, the following steps must be incorporated in the design process:
- Compilation of the functional requirements
- Compilation of the operational requirements
- Assessment of the local conditions
- Assessment of the boundary conditions for the design
- Calculation of the energy to be absorbed by the fender system
- Selection of a suitable fender system
- Calculation of the reaction force and possible friction forces
- Checking whether the forces that occur in the waterfront structure and in the ship's hull can be accommodated
- Ensuring that all constructional details in the waterfront structure can be accommodated, especially fixings, built-in parts, chains, etc., without any damage being caused to the ship or the waterfront structure due to projecting fixings or other parts of the construction.

The designer can choose from a wide range of fenders and fender systems from diverse manufacturers. The manufacturers frequently offer not only standard products, but also systems tailor-made to suit particular situations.

In order to compare the various fenders on offer, the quality and system data of the manufacturers should comply with the test methods given in [232].

Some fenders require considerable maintenance. Therefore, before selecting and installing fenders, the designer is recommended to check carefully whether and to what extent vessels and/or structures are really at risk and which special requirements a fender system will have to fulfil. When designing a quay wall, pier, etc., and the fender support structures, it is not only the berthing loads that must be taken into account. The horizontal and vertical movements of the ship during berthing and departure, the loading and unloading procedures, swell or fluctuations in the water level, etc. can lead to friction forces in the horizontal and/or vertical direction (provided these movements are not accommodated by the rotation of suitable cylindrical fenders). If lower values cannot be verified, to be on the safe side a friction coefficient $\mu = 0.9$ should be taken for dry elastomer fenders. Polyethylene surfaces result in less friction on the ship's hull; a friction coefficient $\mu = 0.3$ should be assumed in such cases.

The permissible pressure between a fender and ship's hull of a modern large ship is max. 200 kN/m^2. All customary fender types take account of this pressure. This figure should be borne in mind when selecting a type of fender. Special vessels, e.g. navy ships, require softer fenders.

6.14.4   Methods of analysis for fenders

The deterministic method of analysis is usually used for designing fenders. This is based on the energy equation:

$$E = \frac{1}{2} \cdot G \cdot v^2$$

where:

E = kinetic energy of ship [kNm]
G = mass of ship, i.e. displacement [t] according to section 5.1
v = berthing velocity [m/s]

The berthing energy to be absorbed can be determined as follows for a berthing angle of $0°$:

$$E_d = \frac{1}{2} \cdot G \cdot v^2 \cdot C_e \cdot C_m \cdot C_s \cdot C_c$$

where:

E_d = berthing energy to be absorbed [kNm]
G = mass of ship, i.e. displacement [t] according to section 5.1
v = berthing velocity measured perpendicular to quay wall [m/s]
C_e = eccentricity factor [1]
C_m = virtual mass factor [1]
C_s = ship flexibility factor [1]
C_c = waterfront structure attenuation factor [1]

The individual factors are defined as follows:

Mass of ship/Displacement G
The mass of the ship, i.e. its displacement, is required for the energy calculation.

The size of a ship is frequently expressed in terms of its carrying capacity, its "deadweight tonnage" (dwt). The size of passenger vessels, cruise ships and car ferries is based on their Gross Register Tonnage (GRT), whereas gas tankers are often specified in m^3. However, there is no exact mathematical relationship between "deadweight tonnage" (dwt) or Gross Register Tonnage (GRT) and displacement.

Section 5 contains advice on typical ship sizes. If more specific data is not available, the displacement figures given there may be assumed.

Berthing velocity v

The berthing velocity is the significant variable in the calculation of the energy to be absorbed. It is specified as a velocity at right-angles to the fender system or waterfront structure. Measured values for the berthing velocity are not usually available. As a rule, the figures given in section 5.3 can be assumed.

Work carried out in Japan resulted in a berthing angle of, generally, less than 5° for ships with a deadweight of more than 50 000 dwt. To remain on the safe side in calculations, the designer is recommended to assume a berthing angle of 6° for such ships. In the case of smaller vessels, primarily when berthing without tug assistance, an angle of 10–15° should be assumed.

It should be noted that it is the square of the berthing velocity value that is entered in the energy calculation. It is clear from this that the appropriate berthing velocity must be determined with the utmost care. In the general case of the angled approach, the calculation of the energy to be absorbed can be carried out in a similar way to determining the required work capacity of a mooring dolphin according to section 13.3.2.2, but in this case adjusted by the waterfront structure attenuation factor C_c.

Eccentricity factor C_e

The eccentricity factor takes into account the fact that the first contact between ship and fender is not generally at the fender in the centre of the ship. The eccentricity factor is derived from [232] (see R 128, section 13.3.2.2):

$$C_e = \frac{k^2}{k^2 + r^2}$$

where:

k = mass moment of inertia of ship
r = distance of centre of gravity of mass of ship from point of contact on fender

The mass moment of inertia for large ships with a high block coefficient can usually be taken as $0.25 \cdot l$, where l is the length between perpendiculars.

$C_e = 0.5$ can be assumed if more accurate data is not available, and for rough calculations. $C_e = 0.7$ can be assumed when designing dolphin fenders.

$C_e = 1.0$ should be assumed for the bow/stern fenders for Ro/Ro ships that dock with bow or stern.

Virtual mass factor C_m
The virtual mass factor takes into account the fact that a considerable quantity of water is moved together with the ship, and this must be included with the mass of the ship in the energy calculation. Various approaches have been used to determine this factor; see [232].
The assessment and comparison of C_m according to the approaches of diverse authors results in average values between 1.45 and 2.18.

The designer is recommended [232] to use the following values:
- for a large clearance under the keel $(0.5 \cdot d)$ $C_m = 1.5$
- for a small clearance under the keel $(0.1 \cdot d)$ $C_m = 1.8$
where d is the draught of the ship [m].

C_m values for a clearance under the keel between $0.1 \cdot d$ and $0.5 \cdot d$ can be found by linear interpolation.

Ship flexibility factor C_s
The factor for the flexibility of the system takes into account the ratio of the elasticity of the fender system to that of the ship's hull because part of the berthing energy is absorbed by the latter. The following values are normally used:

- for soft fenders and small vessels $C_s = 1.0$
- for hard fenders and large vessels $0.9 < C_s < 1.0$
                                        (e.g. for large tankers: 0.9)

Generally, a value of 1.0 can be assumed, which lies on the safe side.

Waterfront structure attenuation factor C_c
The attenuation factor takes into account the type of waterfront structure. With a closed structure (e.g. vertical sheet pile wall) the water between ship and wall is pressed out and thus absorbs a considerable part of the energy. This aspect depends on various influences:

- layout of waterfront structure
- under keel clearance
- berthing velocity
- berthing angle
- depth of fender system
- cross-section of ship

Experience has shown that the following values can be assumed for C_c:
- for an open waterfront structure: $C_c = 1.0$
- for a closed waterfront structure and parallel berthing: $C_c = 0.9$

Values less than 0.9 should not be used.

The attenuation can reduce considerably at a berthing angle of just 5°, i.e. in such cases assume a value of 1.0.

Calculation programs
The manufacturers can provide software for designing their fenders. However, in order to compare the results obtained from such programs, make sure that the correct factors have been entered.

Additional factors for exceptional berthing manoeuvres
It is left to the discretion of the design engineer to include an additional factor in the calculations to take account of exceptional berthing manoeuvres. Please refer to [232]. In such cases the designer is recommended to choose a factor between about 1.1 and max. 2.0. Exceptional conditions could be, for example, the frequent handling of hazardous goods. Table R 60-1 gives guidance on additional factors.

Table R 60-1. Additional factors for exceptional berthing manoeuvres

Type of vessel	Size of vessel	Additional factor
Tanker, bulk cargo	large	1.25
	small	1.75
Container	large	1.5
	small	2.0
General cargo		1.75
Ro/Ro, ferry		≥ 2.0
Tug, workboat		2.0

Selection of fenders
Once the energy has been calculated, the fenders required can be selected from the relevant manufacturers' publications. However, for detailed planning the designer is advised to consult the manufacturer because many of the construction details cannot be gleaned from the manufacturers' publications. In particular, the construction details regarding the mounting of the fenders.

6.14.5   Types of fender systems
Diverse fender systems are available on the international market. The various types and models can be seen in the catalogues of the manufacturers.

Cylindrical fenders are the most common type and these are available in many different sizes. Floating fenders have proved to be worthwhile for quay walls exposed to considerable water level fluctuations as a result of tides. Berths for ferries are frequently custom solutions with polyethylene-coated rubbing strips on conical fenders or cylindrical fenders loaded along their longitudinal axis. For a comparison of various types of fender, please refer to [232], which also contains further advice and details of advantages/disadvantages. Please note that the designations of the manufacturers can vary for the same type of fender. Test methods for materials and fenders should comply with the data given in [232] in order to be able to assess the equivalence or otherwise of the products of different manufacturers.

The materials for fenders are these days almost exclusively elastomer products or other synthetic products. With the exception of dolphins and rare custom designs, these products guarantee that the energy that occurs during berthing can be transferred to the loadbearing structure in accordance with the calculations and without damage.

For this reason, types of fendering common in the past, e.g. brushwood, vehicle tyres or timber (rubbing strips, timber fenders, fender piles), cannot be designated as fenders because their insufficiently defined material properties prevents them from being included in the energy absorption calculation. Such materials and arrangements are merely suitable for use protecting corners and edges or acting as guides.

6.14.5.1 Elastomer fenders

General

Elastomer elements are used in many ports for fending off impacts of vessels and for absorbing berthing pressures (fig. R 60-1). These elements are generally resistant to seawater, oil and ageing and are not destroyed by occasional overloads, so that they enjoy a long service life.

Elastomer fendering elements of various shapes, dimensions and specific performance characteristics are manufactured, so that it is possible to meet every requirement, from simple fendering for small vessels to fender structures for large tankers and bulk cargo freighters. Special attention must be given to the fendering stresses required for ferry terminals, locks, dry docks and the like.

Elastomers are used either alone as a fender material, against which the ships berth directly, or as suitably designed buffers behind fender piles or fender panels. Occasionally, both types of usage are combined. In such cases it is possible to attain the energy absorption capacity and spring constants best suited to any specific requirement using elements made from commercially available elastomers (R 111, section 13.2).

Cylindrical fenders

Thick-walled elastomer cylinders are frequently used. They can have various diameters from 0.125 m to more than 2 m. They have variable spring characteristics depending on use. Cylinders with smaller diameters are placed with ropes, chains or rods in a horizontal, vertical or, where

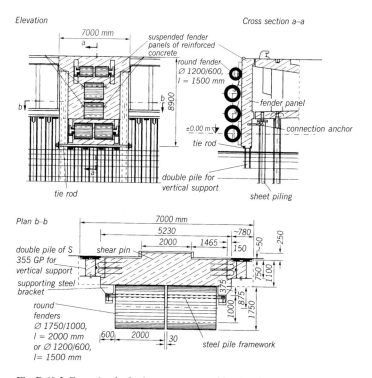

Fig. R 60-2. Example of a fender arrangement with cylindrical fenders

applicable, diagonal position. In the latter case they are frequently suspended as a protective "garland" in front of a quay wall, mole head, etc.

Cylindrical fenders are usually installed in a horizontal position. They may not be suspended directly on the quay wall with ropes or chains because of the risk of deflection and tearing under load. They are drawn over rigid steel tubes or steel trusses made from tubular sections, etc. These are then suspended from the quay wall with chains or steel cables, or placed on steel brackets located next to the fenders (fig. R 60-2).

Cylindrical fenders loaded axially and conical fenders
Cylindrical fenders can also be installed to carry loads in the longitudinal direction. However, owing to the risk of buckling, in this case only shorter cylinders are possible. If the reaction travel upon compression is not sufficient, several elements can be arranged adjacent to each other. To prevent buckling of such a row of elements, steel plates in suitable guides can be fitted between the individual elements, for instance (fig. R 60-3),

210

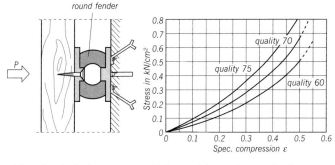

round fender

Stress in kN/cm²

quality 70
quality 75
quality 60

Spec. compression ε

a) Example under load   b) Characteristic stress/compression diagram

Fig. R 60-3. General data for round fenders loaded in longitudinal direction made of elastomer qualities with 60, 70 and 75 (ShA) to DIN 53505

As the fender is compressed, the reaction force rises sharply, but then drops again as the fender deforms.

The conical fender is a special form which is substantially less vulnerable to buckling. The energy and deformation characteristics are similar to those of an axially loaded cylindrical fender.

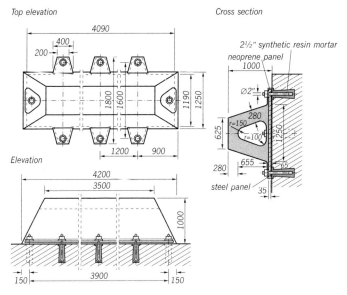

Top elevation   Cross section

2½" synthetic resin mortar
neoprene panel
Ø2"
steel panel

Elevation

Fig. R 60-4. Example of a trapezoidal fender

211

Trapezoidal fenders

In order to obtain a more favourable load–deflection curve, special shapes have been developed using special inlays, e.g. textiles, spring steels or steel plates vulcanised into the elements. Metal inlays must be blasted to a bright surface finish and must be completely dry before vulcanising. These elements are frequently made in a trapezoidal form with a height of approx. 0.2 to 1.3 m. They are attached to the quay wall with dowels and bolts (fig. R 60-4).

Floating fenders

The great advantage of floating fenders, primarily in tidal waters, is that ships are fendered practically exactly on the waterline and hence also roughly in line with their centre of gravity. Such fenders are available in foam- or air-filled versions.

Air-filled fenders are fitted with a blow-off valve which prevents the fender bursting if it is overloaded. This valve must be serviced regularly. Owing to their method of manufacture, foam-filled fenders can be produced in virtually any size and with virtually any characteristics. They have a core of closed-cell polyethylene foam and a jacket of polyurethane reinforced with a textile. The jacket is easily repaired. Special attention should be paid to the material properties of the jacket because the stresses during deformation are very high.

Fenders of rubber waste

Various seaports use old car tyres, mostly filled with rubber waste, as fenders, suspended flat against the face of quays. They have a cushioning effect, but do not possess any appreciable work capacity.

Frequently, several stuffed truck tyres – usually between 5 and 12 – are placed on a steel shaft with a pipe collar welded on at both ends for attaching the guy and holding ropes. The fender is placed with these ropes and held so that it can rotate on the face of the quay wall. The tyres are filled with elastomer slabs placed crosswise which brace the tyres against the steel shaft. The remaining voids are filled with elastomer material (fig. R 60-5). Fenders of this type – occasionally in an even simpler design with a wooden shaft – are economical. They have generally performed well in circumstances where the requirement to absorb the impact energy of berthing ships has not been severe, even though their working capacity and thus the anticipated berthing pressure cannot be reliably determined.

Not to be confused with these improvisations are the accurately designed fenders rotating freely on an axle. These fenders are fabricated mostly of very large special tyres which are either stuffed with rubber waste or inflated with compressed air. Fenders of this type are used successfully at exposed positions, such as the entrances to locks or dry docks, as well as at narrow harbour entrances in tidal areas. They are suspended

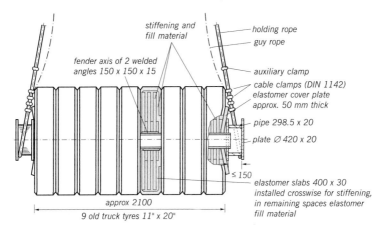

stiffening and fill material

holding rope

guy rope

fender axis of 2 welded angles 150 x 150 x 15

auxiliary clamp

cable clamps (DIN 1142)

elastomer cover plate approx. 50 mm thick

pipe 298.5 x 20

plate ⌀ 420 x 20

≤ 150

elastomer slabs 400 x 30 installed crosswise for stiffening, in remaining spaces elastomer fill material

approx 2100

9 old truck tyres 11" x 20"

Fig. R 60-5. Example of a fender made from truck tyres

horizontally and/or vertically as guidance for ships, which must always navigate with caution at such places.

Tyres from transport vehicles are occasionally used as fenders at ore loading/unloading facilities in the vicinity of open-cast mining operations. In such cases the energy absorption capacity should be determined in tests.

6.14.5.2 Fenders of natural materials

In countries in which suitable materials are available and/or funding is limited, suspended brushwood fenders can also be considered as fendering. However, if elastomer fenders are available, then brushwood fenders should be rejected because they involve a higher capital outlay and also higher maintenance costs than elastomer fenders. Brushwood fenders are subject to natural wear and tear due to ship operations, drifting ice, wave action, etc. The dimensions are adapted to the largest vessel berthed. Unless special circumstances require larger dimensions, fender sizes may be taken from table R 60-2:

Table R 60-2. Fender sizes

Size of vessel dwt	Fender length m	Fender diameter m
up to 10 000	3.0	1.5
up to 20 000	3.0	2.0
up to 50 000	4.0	2.5

6.14.6 Design advice

Fenders should be located at regular intervals along the side of the quay. The spacing of the fenders depends on the design of the fender system and the vessels anticipated. One important criterion here is the radius of the ship between bow and the flat side in the middle of the ship This radius defines at which fender spacing and projecting dimension of completely compressed fenders the bow makes contact with the quay wall between two fenders. The fender spacing must be adjusted to suit each particular situation.

As a rule, fenders should not be positioned more than 30 m apart.

The projection of the fender is not arbitrary. The maximum load moment of the cranes frequently influence the projection of the fenders.

It is difficult to design a fender system that is equally suitable for both large and small vessels. Whereas a fender designed for a large vessel is sufficiently "soft", this can be too inflexible for a small vessel. That might result in damage to the ship. Furthermore, the height of the fenders above the water level is more significant for small vessels than it is for large ones. In tidal waters floating fenders can offer considerable advantages.

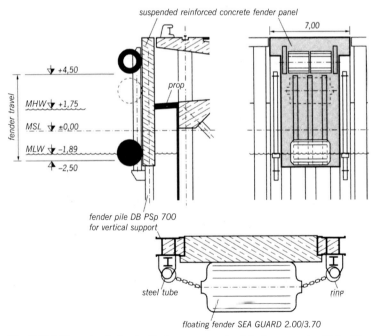

Fig. R 60-6. Example of a floating fender system at a berth for large vessels with berthing option for feeder and inland vessels

If container feeder ships or inland vessels are handled at berths for large ships, there is a danger of such vessels becoming caught beneath fixed fendering. In addition, the heeling of small ships during loading/unloading procedures at low water can lead to the fenders at the upper position damaging the superstructure and the cargo. In a new development for the container terminal in Bremerhaven, Germany, floating fenders have been installed in front of fender panels and can move up and down with the tide between lateral guide tubes. This solution is shown in fig. R 60-6. The fender construction here consists of a fixed upper cylindrical fender ($\varnothing$ 1.75 m) and a moving floating fender ($\varnothing$ 2.0 m) in front of a fixed fender panel. These diameters were chosen to ensure that sufficient heeling of smaller vessels is possible at low water levels.

6.14.7 Chains

Chains in fender systems should be designed for at least three to five times the theoretical force.

6.14.8 Guiding devices and edge protectors

6.14.8.1 General

Besides the actual fendering, which is specifically designed to absorb energy, there are many elements that are provided merely for constructional reasons, e.g. as guiding devices in channels and locks, as edge protectors, or as non-specific berthing facilities for smaller vessels. Such devices include a fender piles, timber fenders, rubbing strips and nosings.

6.14.8.2 Timber fenders and fender piles

Timbers used in seawater and brackish water can be attacked by the so-called shipworm (*Teredo navalis*). Such attacks can lead to total destruction of timbers in port facilities within just a few years. However, the destruction of the timber is practically invisible from the outside. The shipworm attacks mainly softwoods, but also certain European hardwoods, e.g. oak, and tropical hardwoods as well. The use of timber in loadbearing structures cannot be recommended in seawater and brackish water with more than 5‰ salt content. When, for example, timber fender piles are used, infestation by shipworm must be taken into account. If in doubt, products made from synthetic materials, e.g. recycled materials, are to be preferred.

6.14.8.3 Edge protectors

Edge protectors are made from solid sections or, frequently, from fender sections, e.g. cylindrical elastomer fenders. Owing to their small size or their shape, they have no noteworthy energy absorption capacity.

6.14.8.4 Rubbing strips of polyethylene

In addition to other friction components such as timber fenders, fender piles, etc., rubbing strips made of plastic, frequently polyethylene (PE), are used in order to reduce the friction stresses at waterfront structures when vessels are berthing and lying up. These components must absorb the loads arising from pressure and friction without fracture and be capable of transmitting them to the harbour structure via their mountings. In certain cases they must be supported by supplementary loadbearing members for this purpose. Polyethylene compounds of medium density to DIN EN ISO 1872 (HDPE) and high density to DIN 16 972 (UHMW-PE) have proven suitable for use as rubbing strips in hydraulic engineering and seaport construction. Standard forms are rectangular solid profiles with cross-sections of between 50×100 mm and 200×300 mm, and lengths of up to 6000 mm. Custom sections and lengths can also be supplied. HDPE sections are cast in moulds and are vulnerable to brittle fracture at low temperatures (below –6 °C). UHMW-PE sections are cut to suit the profile required and therefore have smooth edges.

In order to minimise friction forces, rubbing strips should be made of material that exhibits a very low friction coefficient together with low abrasion and wear rates, e.g. ultrahigh molecular mass polyethylene (UHMW-PE).

The shaped parts must always be free from voids and must be produced and processed in such a way that they are free from distortion and inherent stresses. The quality of processing can be checked by acceptance tests to verify the properties and by additional hot storage tests of samples cut from the sections.

Regenerated PE compounds of medium density may not be used because of the reduced material properties.

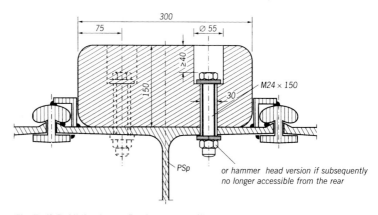

Fig. R 60-7. Sliding batten fixed to a sheet piling

216

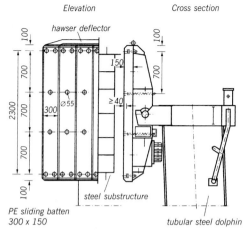

Elevation   Cross section

hawser deflector

150

300   Ø 55

≥ 40

2300

700
700
700
100
100

100
700

steel substructure

PE sliding batten
300 x 150

tubular steel dolphin

Fig. R 60-8. Equipping the fender apron of a tubular steel dolphin with sliding battens

Figs. R 60-7 and R 60-8 show fixing and construction examples. The heads of the fixing bolts should stop at least 40 mm below the contact surface of the rubbing strips. Replaceable bolts should be at least $\varnothing$ 22 mm and bolts embedded in concrete hot-dip galvanised and at least $\varnothing$ 24 mm.

6.15   Fenders in inland harbours (R 47)

The berthing areas of waterfront structures in inland harbours generally consist of concrete, sheet piling or faced natural stone. They are constructed either vertical or with a minimal batter to the land side (1 : 20 to 1 : 50).
To protect the waterfront structure and the hull of the ship, the ship's crew usually uses rubbing strips about 1 m long to prevent the hull making contact with the side of the quay.
The designer is recommended to refrain from equipping the waterfront structure with fender piles or timber fenders.

6.16   Foundations to craneways on waterfront structures (R 120)

6.16.1   General

The type of craneway foundation to be constructed along a waterfront structures depends mainly on the prevailing local subsoil conditions. For craneways with a wider track gauge, the subsoil is also to be investigated along the line of the inboard crane rail (R 1, section 1.2). In many cases, especially for heavy structures in seaports, structural requirements make it practical to construct a deep foundation for the outboard crane rail as an integral part of the quay wall, whereas the foundation for the land-side crane rail is generally independent of the waterfront structure, except at embankments below elevated structures, on pier slabs and the like.

By contrast, in inland ports the outboard crane rail is frequently also on a foundation separate from the waterfront structure. This facilitates later modification which could be required, for example, for changed operating conditions due to new cranes, or alterations to the waterfront structure. If the quay wall, craneway and crane are owned by different companies, it may be necessary to disperse ownership of the structure. The best overall solution should always be aimed at with respect to engineering and economic aspects.

Please refer to R 72, section 10.2.4, for details of designing long craneway beams without joints. Craneway beams over 1000 m long have been successfully constructed without joints.

6.16.2 Design of foundations, tolerances

The craneway foundations may be shallow or deep, depending on local subsoil conditions, the cranes' sensitivity to settlement and displacement, the crane loads, etc.

The permissible dimensional deviations of the craneway must be taken into account here, distinguishing between deviations during manufacture (assembly tolerances) and deviations during the course of operation (operational tolerances).

Whereas the assembly tolerances in harbour cranes mainly relate to displacement and fixing of the crane rails, the allowable operational tolerances must be taken into account when selecting the type of foundation depending on the subsoil.

Depending on the design of the crane portal, the following values can be taken as a guide for the operational tolerances:

- height of rail (gradient)   2‰ to 4‰
- height of rails in relation to each other   max 6‰ of track width
  (camber)
- inclination of rails in relation to each other   3‰ to 6‰
  (offset)

These operational tolerances include any assembly tolerances.

Considerably tighter tolerances apply, e.g. $\leq$ 1‰ for the height of a rail (gradient), when using special cargo handling equipment, e.g. container cranes. The designer is recommended to include the crane manufacturer in the planning work.

Please refer to [53] for details of the relationship between craneway and crane system.

6.16.2.1 Shallow craneway foundations

(1) Strip foundations of reinforced concrete

In soils not sensitive to settlement, the crane rail beams may be constructed as shallow strip foundations of reinforced concrete. The crane rail beam is then calculated as an elastic beam on a continuous, elastic support. In doing so, the maximum permissible soil pressures according to DIN 1054 for settlement-sensitive structures are to be observed. Furthermore, a soil settlement calculation must confirm that the maximum allowable non-uniform settlements for the crane stipulated by the crane manufacturer are not exceeded.

DIN 1045 is applicable for dimensioning the beam cross-section. The stresses from vertical and horizontal wheel loads are to be checked, including longitudinal loads due to braking.

In craneways with relatively narrow gauge widths, e.g. for gantry cranes spanning only one track, the design gauge of the tracks is maintained by tie or tie bars installed at intervals equal to about the gauge width. With wider gauges, both crane rails are designed separately on individual foundations. In this case the cranes must be equipped with a pinned leg on one side.

Please refer to R 85, section 6.17, for the design of the rail fastening.

Settlements of up to 3 cm can still be absorbed generally by installing rail bearing plates or special rail chairs. Where larger settlements are probable, a deep foundation will generally be more economical because later realignment and the consequent stoppage of cargo handling operations will cause much lost time and great expense.

(2) Sleeper foundations

Crane rails on sleepers in a ballast bed are comparatively easy to realign and so are used primarily in mining subsidence regions and where excessive settlement is expected. Even massive movements in the subsoil can be corrected quickly by realignment of elevation, lateral position and gauge. Sleepers, sleeper spacing and crane rails are calculated according to the theory of an elastic beam on a continuous, elastic support, and according to the standards for railway structures. Wooden sleepers, steel sleepers, reinforced concrete sleepers and prestressed concrete sleepers can be used. Wooden sleepers are preferred at facilities for loading lump ore, scrap and the like because of the reduced risk of damage from falling pieces.

6.16.2.2 Deep craneway foundations

Deep foundations are practical in settlement-sensitive soils or deep fills, insofar as soil improvement is not carried out by means of replacement, vibration and the like. An adequately deep foundation also reduces stresses on the waterfront structure.

Basically, all customary types of piles can be used for deep foundations to craneways. The piles under the outboard craneway in particular are loaded in bending due to the deflection of the quay wall. Likewise, asymmetric live loads can cause considerable horizontal, additional loads. All horizontal forces acting on the craneway are to be resisted either by the mobilised passive earth pressure in front of the craneway beam, by raking piles, or by effective anchoring.

For deep foundations on piles, the craneway beam is to be calculated as an elastic beam on a continuous, elastic support.

6.17 Fixing crane rails to concrete (R 85)

The crane rails are to be installed free from stresses but enabling longitudinal movement. Crane rail fixings tested for the respective type of use are available.

The following possibilities can be used for trouble-free installation of crane rails on concrete:

6.17.1 Bedding the crane rail on a continuous steel plate over a continuous concrete base

In this method the bedding plate is suitably grouted, or rests on slightly moist, compacted chip concrete. The crane rail fastenings are designed to permit the rail to move longitudinally, but to restrain it from vertical motion so securely that even negative bending stresses resulting from interaction between the concrete base and the rail will be satisfactorily accommodated. In calculating maximum moment, anchoring force and maximum concrete compression stress, the method of the subgrade reaction coefficient may be used.

Fig. R 85-1 shows an example for a heavy crane rail. Here, the concrete bedding is tamped in between the steel angles, levelled and given a levelling coat ≥ 1 mm of synthetic resin or a thin bituminous coating.

If an elastic intermediate course is laid between the concrete and the bedding plate, the rail and anchoring are to be calculated for this softer bedding, which can result in larger dimensions. The rails are to be welded to avoid joints as far as possible. Short rail bridges are to be used at expansion joints between quay wall sections.

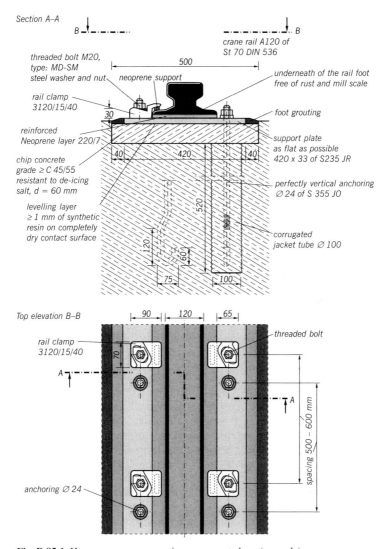

Fig. R 85-1. Heavy craneway on continuous concrete base (example)

221

6.17.2 Bridge-type arrangement with crane rail supported at intervals on bedding plates

In this case bedding plates of a special type are used which assure a well-centred introduction of the vertical forces into the points of support. They must also guide the rails, which have freedom of movement in the longitudinal direction. Furthermore, they must prevent tilting of the rail, which of necessity is quite high because it must act as a continuous beam. They must absorb the lifting forces resulting from both the negative bearing force as well as from the direct horizontal forces acting on the rail.

Lightweight crane rails of this type are used for normal general cargo cranes, preferably in inland ports, and for bulk cargo cranes. The heavy-duty design is to be recommended, above all, for the craneways of heavy lift cranes, very heavy short unloaders, unloader bridge cranes and the like. Rail sections S 49 and S 64 are used in light installations; in heavy installations PRI 85 or MRS 125 as per DIN 536 are used, or very heavy special rails made of St 70 or St 90.

A typical example of a light installation is shown in fig. R 85-2. Here, rail S 49 or S 64 according to the K-type of the railway structures of Deutsche Bahn AG rests on horizontal bedding plates. The rail, bedding plates, anchors and special dowels are placed completely assembled on the formwork or on a special adjustable steel support, which can be securely mounted. The concrete is then so placed with the aid of vibration that complete contact is made with the bedding plates. Occasionally, an intermediate course of approx. 4 mm thick plastic is also placed between bedding plates and upper surface of concrete (fig. R 85-2). With cambered bedding plates, design measures must ensure that the plastic intermediate layer cannot slip off.

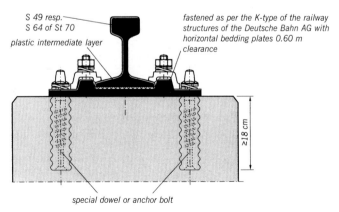

S 49 resp.
S 64 of St 70

plastic intermediate layer

fastened as per the K-type of the railway
structures of the Deutsche Bahn AG with
horizontal bedding plates 0.60 m
clearance

≥18 cm

special dowel or anchor bolt

Fig. R 85-2. Light craneway on individual supports

Fig. R 85-3 shows a heavy craneway in which the chair for the rail is cambered upwards so that the rail is supported tangentially on the camber. The bedding plate is supported or packed with a non-shrink material. The bedding plates are also provided with slotted holes in the transverse direction so that gauge changes can be corrected if need be. This bedding must be provided primarily for long-stem rails.

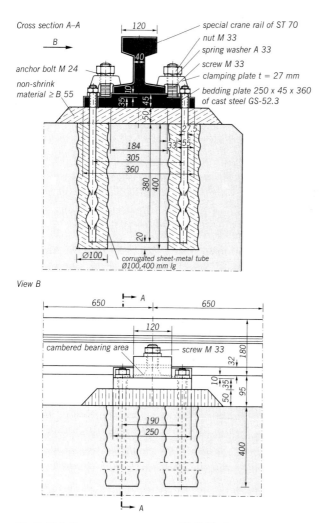

Fig. R 85-3. Heavy craneway on tamped individual supports

Intermittent support can also be provided with very high wheel loads. For crane rails with small section moduli, e.g. A 75 to A 120 or S 49, however, continuous support is recommended for loads above approx. 350 kN because otherwise the spacing between the panels or chairs becomes too small.

6.17.3 Bridge-type arrangement with crane rails supported on rail-bearing elements

When rail-bearing elements or so-called rail chairs are used, the rail is a continuous beam with an infinite number of supports. In order to make full use of the resilience of the rail, an elastic slab is inserted between rail and chair, e.g. for bearing pressures up to 12 N/mm², this could be made of neoprene or similar, or for higher pressures rubber fabric in

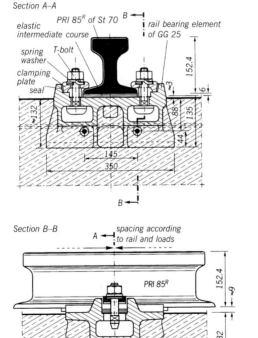

Fig. R 85-4. Example of a heavy craneway on rail bearing elements

thicknesses up to 8 mm. This also tends to cushion impacts and shocks on the crane wheels and chassis.

The top of the rail chair is cambered, thus causing the bearing force to be introduced centrally into the concrete. This cambered bearing lies some distance above the concrete, and a certain amount of yield occurs in the lock washers of the mounting hardware so that the rail will be free to move longitudinally. This enables temperature changes and rocking movements to be absorbed (fig. R 85-4). Through flexible shaping, the rail chairs can be adapted to any desired requirements. For example, the chairs offer the facility of subsequent rail alignment of:

$$\Delta s = \pm 20 \text{ mm (in transverse direction) and}$$
$$\Delta h = + 50 \text{ mm (in vertical direction)}$$

or of fitting lateral pockets to absorb edge protection angles for traversable rail sections.

The rail chairs are mounted together with the rails, whereby after adjustment and locking in position, additional longitudinal reinforcement is drawn through special openings in the chairs and connected with the projecting bars of the substructure (fig. R 85-4). The concrete grade depends on structural requirements. However, grade C 20/25 is required at least. Since the height of the rail cannot be adjusted easily by tamping, the design should be used only when any significant settlement can be ruled out.

If settlement and/or horizontal displacement of the crane rail necessitating re-alignment of the rail is anticipated, it must be taken into account right at the planning stage by designing an appropriate type of construction, e.g. with special chairs.

6.17.4   Traversable craneways

The demands of port operations frequently require the crane rails to be installed sunk into the quay surface so that they can be crossed without difficulty by vehicular traffic and port cargo handling gear. At the same time, all the other usual demands on craneways must also be met.

(1) Construction of a traversable heavy craneway
Fig. R 85-5 shows an example of a proven construction. The bedding of the rails on the craneway beam section completed first consists of chip concrete grade $> C \ 45/55$, levelled off horizontally by means of a flat steel bar (ladder template). For load distribution, the rail on a thin layer of bitumen rests on a bedding plate fixed on $a > 1$ mm synthetic resin levelling course. To prevent loads from the longitudinal movements of the rails and bedding plates being transferred to the bolt, this bedding plate is not connected to the fixing system. Subsequent installation of the bolts is preferable because this helps to position the bolts exactly.

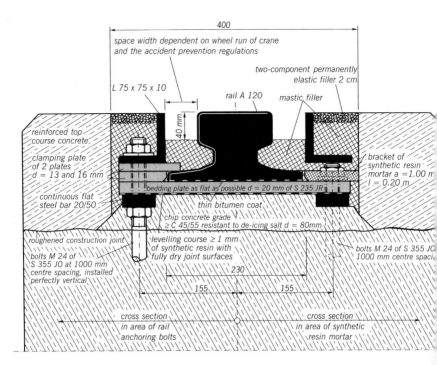

Fig. R 85-5. Execution example of traversible heavy craneway (the reinforcement is not shown)

However, this approach must be allowed for when reinforcing the craneway beam by leaving adequate space between the reinforcing bars for the sheet metal or plastic sleeve to be concreted in. If necessary, the bolt holes can also be drilled subsequently. In order to compensate for the transfer of horizontal forces and to hold the rail in an exact position, approx. 20 cm wide brackets of synthetic resin mortar are inserted between the foot of the rail construction and the bordering lateral edges of the top course concrete at approx. 1 m centre spacing.

It is expedient to use permanently elastic, two-part filler in the upper 2 cm of the head area of the reinforced top course concrete joined with stirrups to the remaining craneway beam. Further details can be taken from fig. R 85-5.

(2) Construction of a traversable light craneway
A proven example of this is shown in fig. R 85-6.
Horizontal ribbed slabs are fastened onto the flat, levelled reinforced concrete craneway beam with dowels and screw spikes, at a centre-to-centre spacing of approx. 60 cm. The crane rail, e.g. S 49, is connected

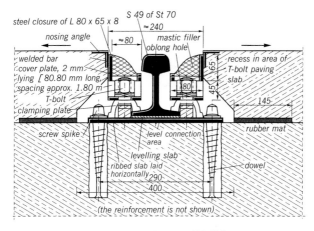

Fig. R 85-6. Execution example of a traversible light craneway

to the ribbed slabs with clamping plates and T-bolts in accordance with Deutsche Bahn AG specifications. Levelling slabs such as steel plates, plastic slabs and the like can be installed beneath the rail foot to correct slight differences in level in the concrete surface.

A continuous steel closure is installed as an abutment for any adjoining reinforced concrete raft foundations. This closure consists of an angle $80 \times 65 \times 8$ (grade S 235 JR) running parallel to the rail head, beneath which 80 mm long U 80 sections are welded at intervals of three ribbed slabs each. These sections have oblong holes on the bottom for fastening to the T-bolts. In the area of the intermediate ribbed slabs, the angle is stiffened by 8 mm thick steel plates. Recesses are cut in the horizontal leg above the fastening nuts and are covered by 2 mm thick plates after the nuts have been tightened. A bar is welded on along the rail head at the foot end of the angle to give the subsequent mastic filler an adequate hold below.

To pave the port ground surface, reinforced concrete slabs are placed loosely against the steel closure. In this case rubber mats should be laid underneath to prevent tilting, and at the same time to create a slope for drainage from the crane rails.

6.17.5 Note on rail wear

The wear to be expected for the foreseeable service life of all crane rails must be taken into account in the design. As a rule, a height reduction of 5 mm with good rail support is adequate. Furthermore, more or less frequent maintenance and, depending on the type, checks of the fixings, are recommended during operation to prolong the service life.

6.17.6 Eccentric concrete compression stress

In case that eccentric concrete compression stress beneath the bedding plate due to moving loads can not be prevented constructive, the eccentric compression stress has to be considered in the design of the supporting concrete. The influence of local impacts due to bolts beneath of the bedding plate and elastic deformations has to be investigated.

6.18 Connection of expansion joint seal in a reinforced concrete bottom to loadbearing external steel sheet piling (R 191)

Expansion joints in reinforced concrete bottoms, e.g. in a dry dock or similar, are protected against massive mutual displacement in the vertical direction by means of a joggle joint. Thus, only slight mutual vertical displacement is possible. The transition between the bottom and the vertical loadbearing steel sheet piling is via a relatively slim reinforced concrete beam connected securely to the sheet piling, to which the bottom plates separated by the expansion joint are flexibly connected, also by way of a joggle joint.

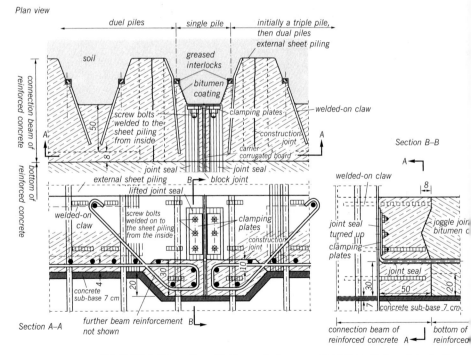

Fig. R 191-1. Connection of bottom seal of an expansion joint to U-shaped sheet piling, example

228

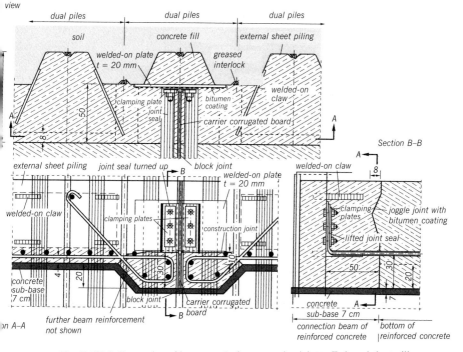

Fig. R 191-2. Connection of bottom seal of an expansion joint to Z-shaped sheet piling, example

The bearing plate joint with joggle joint is also constructed in the connection beams. The expansion joint of the bearing plate is sealed from below with a joint seal with loop. In U-section sheet piling in accordance with fig. R 191-1, this seal ends at a specially installed sheet piling pillar. In Z-section sheet piling, a connection plate is welded on to the sheet piling trough as shown in fig. R 191-2, and the joint seal is turned up and clamped to this.

The round interlocks of the connection piles (single piles for U-section, dual piles for Z-section) are to be generously greased with a lubricant before installation.

Further details are given in figs. R 191-1 and R 191-2.

6.19 Connecting steel sheet piling to a concrete structure (R 196)

The connection of steel sheet piling to a – possibly existing – concrete structure is always a one-off construction and must be adapted to suit the conditions and designed accordingly. The workmanship is especially important.

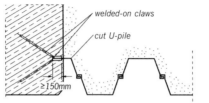

a) Connection to a subsequently produced concrete structure

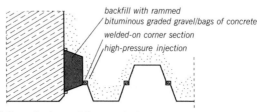

b) Connection to an existing concrete structure

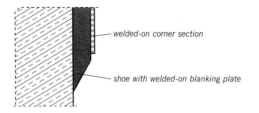

Fig. R 196-1. Connection of U-section sheet piling to a concrete structure

The connection should be as tight as possible and allow mutual vertical movement of the structures. The simplest possible solution should always be aimed for. Fig. R 196-1 shows construction examples for U-section sheet piling. Fig. R 196-1 a) shows connection to a concrete structure constructed previously. Here, the connection is achieved by an individual pile inserted through the formwork and then equipped with welded claws. It is embedded in the concrete structure during its construction to a depth sufficient for the necessary connection.

The connecting interlock must be treated previously in a suitable way (see R 117, section 8.1.20). Fig. R 196-1b) shows a recommended method for connecting steel sheet piling as closely as possible to an existing concrete structure. Instead of bituminous graded gravel, a backfill with small sacks of dry concrete (like for bagwork) has proved successful. For further stability it is advisable to stabilise the soil around the connection using a high-pressure injection method.

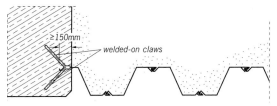

a) Connection to a subsequently produced concrete structure

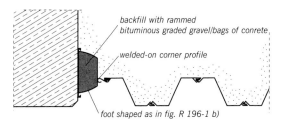

b) Connection to an existing concrete structure

Fig. R 196-2. Connection of Z-section sheet piling to a concrete structure

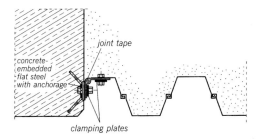

Fig. R 196-3. Connection of U-section sheet piling to a concrete structure with stringent requirements regarding the watertightness of the junction

Similar construction examples for Z-section steel piling are shown in fig. R 196-2.

If stringent requirements are made on the watertightness and/or flexibility of the connection, e.g. if seepage in dammed sections of waterways could prejudice stability, special joint designs with joint seals must be provided, these being connected with clamping plates to the sheet piling and to a fixed flange in the concrete construction (fig. R 196-3). The embedment depth of the sheet piling must be stipulated on the basis of the existence of low-permeability soil strata or of the allowable seepage path length.

231

The possible movements of the components with respect to each other must be checked carefully for this solution.

Please refer to DIN 18 195 parts 1, 2, 3, 4, 6, 8, 9 and 10.

6.20 Floating wharves in seaports (R 206)

The "Floating Wharves" specification of the German Federal Ministry of Transport applies to wharves on federal waterways. It can be applied correspondingly to seaports as well, taking into account the following advice:

6.20.1 General

In seaports, floating facilities are used for passenger ferry transport, as berths for port vessels and for pleasure craft. They consist of one or several pontoons and are connected to the shore by means of a bridge or permanent stairs. The pontoons are generally held by rammed guide piles, while the access bridge to the land is fastened to a fixed bearing on land and a movable bearing on the pontoon.

If the facility consists of several pontoons, interconnecting walkways ensure that it is possible to move from one pontoon to the next.

6.20.2 Design principles

Stipulation of the location for a floating facility must take account of current directions and velocities together with wave influence.

In tidal areas, HHW and LLW should be used as nominal water levels. The incline of the access jetty should not be steeper than 1 : 6 under normal tidal conditions and not steeper than 1 : 4 for extreme water levels. Especially when used by the public, the facility must comply with stringent requirements, even, for example, under icy conditions. Suitable structural and organisational measures must be provided here.

The bulkhead divisions of the pontoons must be chosen such that failure of one single cell through an accident or other circumstances will not cause the pontoon to sink. The cells should be vented individually, e.g. with swan-neck pipes. Cells with sounding pipes accessible from the deck are recommended for simplifying the checking of the watertightness. In certain cases it may be advisable to include an alarm system that warns of an undetected ingress of water. For industrial safety reasons, every cell should be accessible from the deck or through no more than one bulkhead.

Pore-free foam filling of the cells can also be considered.

A cambered beam (raised section in the pontoon surface) is required for water drainage.

It is advisable to provide a disconnecting option for the access bridge to guarantee that the pontoon can float away rapidly in the event of an accident, e.g. by two piles rammed in next to the bridge with suspended cross member.

The minimum required freeboard of the pontoon depends on the permissible heeling, anticipated wave height and anticipated use. For smaller facilities, e.g. for pleasure craft, a minimum freeboard of 0.20 m is adequate for one-sided use, whereas large steel pontoons have far greater freeboard heights. The following values serve as a guide:

- steel pontoon 30 m long,
  3 to 6 m wide   freeboard height 0.8 to 1.0 m
- steel pontoon 30 to 60 m long,
  12 m wide   freeboard height 1.2 to 1.45 m

The freeboard heights must be adjusted to suit the embarkation and disembarkation heights of the vessels, particularly when the facility is used by the public.

6.20.3   **Load assumptions and design**

As a basic rule, the position of the pontoon is to be verified with an even keel, with ballast balance being provided where necessary.

The following stipulations apply to principal and live loads:

A live load of 5 kN/m^2 is to be taken for floating stability and heeling calculations (one-sided load here).

The floating stability tests are also to be confirmed by a hydrodynamic analysis with tests where applicable. Causes of heeling, e.g. banking-up water pressure, current pressure and waves, are to be taken into consideration.

Verification is to be provided for heeling of the pontoon and the incline of the transition points between interconnected pontoons, together with the incline of the interconnecting walkways. Depending on the pontoon dimensions, heeling acceleration and mutual offset of several pontoons, heeling may not exceed 5°; the upper slewing limit is 0.25 to 0.30 m. Greater heeling angles are to be checked considering each individual case.

The ship's mooring impact as load from mooring ships is to be taken as 300 kN and 0.3 m/s, or 300 kN and 0.5 m/s for larger facilities (more than 30 m pontoon length).

A cushioning effect to reduce the ship's mooring impact for pontoons can be applied and verified by:

- fendering on the outer surface
- spring brackets and sliding battens and
- "soft guide dolphins" (e.g. coupled tubular piles)

7 Earthworks and dredging

7.1 Dredging in front of quay walls in seaports (R 80)

This section deals with the technical possibilities and conditions to be taken into account when planning and executing port dredging work in front of quay walls.

A distinction must always be made between new dredging and maintenance dredging.

Dredging down to the design depth as per R 36, section 6.7, should be carried out by grab dredgers, hydraulic dredgers, bucket-ladder dredgers, cutter-suction dredgers, cutter-wheel suction dredgers, ground suction dredgers or hopper suction dredgers and similar plant. When using cutter-suction dredgers, ground suction dredgers or hopper suction dredgers, they are to be equipped with devices which ensure exact adherence to the planned dredging depth. Cutter-suction dredgers with high capacity and high suction force are, however, unsuitable because of the danger of overdeepening and disturbing the soil below the cutter head. Dredging by suction dredgers without a cutter head must not be allowed in any case.

It should be noted that, even under favourable dredging conditions and with appropriate equipment, neither the bucket-ladder dredger nor the cutter-suction dredger nor the hopper suction dredger can create the exact theoretical nominal depth immediately in front of a vertical quay wall when dredging the last few metres because a wedge of earth 3 to 5 m wide remains unless the earth is able to slide down. The need to remove this residual wedge depends on the kind of fendering on the quay and on the hull forms of the ships which will berth there. The residual material can be removed only by grab or hydraulic dredgers. Under certain circumstances, the sheet piling troughs must be flushed free of cohesive soil.

When dredging a harbour with floating plant, the work is usually divided into cuts of between 2 and 5 m, depending on the type and size of the dredging plant. The intended utilisation of the soil excavated can also be relevant to the choice of dredging plant in the case of varying soil types. It is advisable to survey the front face of the quay accurately – not just after dredging is completed but also at intervals throughout the operation – to ensure early detection of any incipient, potential excessive movement towards the water side. If necessary, inspect the sheet piling after every dredging cut.

Please refer to R 73, section 7.4.4, regarding inspection by divers to detect any damage that may have occurred to sheet piling interlocks in areas on the underwater face of quays that have been exposed by dredging.

An approach such as that depicted in fig. R 80-1 can be more economical and less disruptive to port operations.

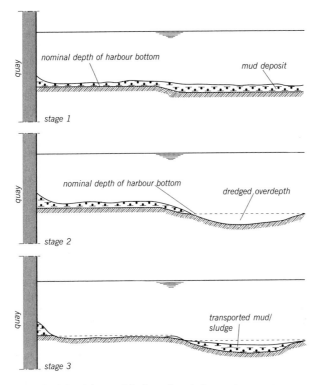

Fig. R 80-1. Dredging work in front of vertical quays in seaports
Stage 1: prevailing situation
Stage 2: situation after dredging with a hopper suction dredge
Stage 3: situation after work with a harrow or grab dredge

After overdeepening (stage 2) by a wet dredger, the mud/sludge lying in front of the quay is moved into the overdeepening of the harbour bottom with grab dredgers or a harrow (stage 3), which reduces the construction costs. The overdeepening should be created as far as possible during the new dredging stage.

Before every dredging operation that might exploit the theoretical total depth under the nominal depth of the harbour bottom, check the stability of the quay wall and restore this if necessary, particularly when the wall incorporates drainage systems. In addition, the behaviour of the quay wall must be monitored before, during and after dredging.

7.2 Dredging and hydraulic fill tolerances (R 139)

7.2.1 General

The specified depths and fill heights are to be produced within clearly defined permissible tolerances. If the difference in elevation between individual points of the excavation bottom or filled area exceed the permissible tolerances, then supplementary measures will be necessary. Specifying tolerances that are too tight can lead to disproportionately expensive extra dredging works at a later date. The stipulation of tolerances for dredging and hydraulic filling work is thus first and foremost a question of cost. The client must therefore carefully consider what value he places on a certain accuracy. In addition to deviations from the intended elevation (vertical tolerances), there are also horizontal tolerances to be considered when trenches are to be dug for such purposes as soil replacement, inverted siphons and tunnels. Here too, it is almost always necessary to reach an optimum balance between the extra costs involved in more extensive dredging and filling work, with more generous tolerances, and the extra costs due to reduced performance of the plant because of more precise working, together with the costs for any possible additional measures.

The accuracy of dredging work for inland waterways is generally better than in waterways for ocean-going vessels, where tides, waves, shoaling by sand and/or sludge deposits play an important role.

Maritime authorities normally demand minimum depths in waterways.

7.2.2 Dredging tolerances

Dredging tolerances are to be specified taking into account the following factors:

a) Quality demands regarding the accuracy of the depths to be produced resulting from the objective of the dredging work, e.g.:

   • regular, recurring maintenance dredging to eliminate sediments and maintain the navigability,
   • the creation or deepening of a berth in front of a quay wall or a navigable channel to improve the navigability,
   • the creation of a watercourse bottom to accept a structure (inverted siphon or tunnel constructions, bottom protection measures, etc.),
   • dredging to eliminate non-loadbearing subsoils within the scope of soil replacement measures,
   • dredging to remove contaminated soil sediments and thus improve the watercourse ecology.

   Each of these objectives calls for specific and distinct, different dredging accuracies. This affects the choice of plant and so has an influence on the cost of dredging work.

b) In addition, structural boundary conditions, the extent of dredging and the properties of the soil to be dredged must all be taken into account, e.g.:

- the stability of underwater embankments, moles, quay walls and similar constructions in the immediate vicinity,
- the depth below the planned harbour bottom to which the subsoil may be disturbed,
- horizontal or vertical dimensions of the dredging work, length and width of the excavation, thickness of the stratum to be dredged and the overall volume of the dredging works,
- soil type and properties, particle size and distribution, shear strength of soil to be dredged,
- the use of soil removed by dredging,
- any possible contamination with specific requirements for dealing with the soil removed by dredging.

This last point in particular is becoming increasingly significant due to the generally extremely high cost of handling and disposing of contaminated soils, and can therefore call for very tight tolerances.

c) Local circumstances continue to play a decisive role, influencing the use and control of dredging plant and the general strategy of the dredging works. Examples of this are:

- depth of the water,
- accessibility for the dredging plant,
- tide-related changes in water levels with changing currents,
- any seawater/freshwater changes,
- weather conditions (wind and current conditions),
- waves, sea conditions, swell,
- interruptions to the dredging works due to waterborne traffic,
- constraints due to the proximity of berths or ships at anchor,
- the extent of recurring sediments (sand or silt), possibly even during the dredging works.

d) And then the dredging plant itself and its equipment also play their part:

- the size and technology of the plant as well as its dredging accuracy depending on soil and depth,
- the instrumentation on board the plant (positioning device, depth measurement, performance measurement, quality of monitoring and recording technology for the entire dredging process),
- experience and skills of the dredging personnel,
- the magnitude of the specific loss of performance of dredging plant owing to the tolerances that have to be maintained, and the ensuing costs.

Table R 139-1. Guide values for vertical dredging tolerances in cm [55]

Dredger	Non-cohesive soils			Cohesive soils		Surcharges for		
	Sand	Gravel	Rock	Silt/sludge	Clay	Water depth 10–20 m	Current 0.5–0.1 m/s	Un-protected water-course
Grab dredger	40–50	40–50	–	30–45	50–60	10	10	20
Bucket-ladder dredger	20–30	20–30	–	20–30	20–30	5	10	10
Cutter-suction dredger	30–40	30–40	40–50	25–40	30–40	5	10	10
Cutter-wheel suction dredger	30–40	30–40	40–50	25–40	30–40	5	10	10
Suction dredger with reduced turbidity	10–20	–	–	10–20	–	5	5	–
Dipper dredger	25–50	25–50	40–60	20–40	35–50	10	10	10
Hopper suction dredger	40–50	40–50	–	30–40	50–60	10	10	10

Notes:

– Guide values for positive and negative deviations (in cm) for normal conditions (e.g. 50 = ±0.5 m).
– The lower value of each pair applies to work in which maximum accuracy is required.
– The upper value of each pair applies to work for which large plant would seem to be suitable.
– Normally the values given here will be under- or over-extended with a probability of 5%.
– Greater accuracy is possible with correspondingly greater effort (and cost).

238

Finally, it has to be taken into account that the result of dredging work and the accuracy achieved is generally checked and recorded by sounding. Therefore, the actual values obtained through sounding are always a combination of the dredging accuracy achieved and the accuracy of the sounding equipment used. So when specifying dredging tolerances, the limits of the sounding system employed to check the work must also be considered.

The specification of optimum and hence also economic dredging tolerances is therefore a complex problem. Before commencing major dredging operations, it is imperative to weigh up the many factors and their effects carefully. At the time of inviting bids, it is often unclear which dredger will be used and so it is wise to ask bidders to submit their price based on the specified accuracy along with their price for the accuracy they themselves propose and will guarantee (special proposal and special bid for dredging work). The client can then select the optimum solution based on the outcome of the submissions.

For general guidance, reasonable vertical dredging tolerances for various dredger types are given (in cm) in table R 139-1. These are based mostly on experience gained in the Netherlands [55]. Horizontal tolerances have been deliberately omitted because in the case of embankments (and that is the only case where horizontal tolerances are relevant, see R 138, section 7.5) these result from the vertical tolerance due to the required angle of the embankment. Furthermore, the horizontal tolerances should be specified for each individual case in conjunction with the requirements and the equipment available for determining position. The soundings should be made with instruments that measure the actual depth to the bottom and do not give a false reading to the top of an overlying suspended layer. A suitable frequency is to be selected for echo sounders. A two-frequency sounding is preferable.

7.2.3 Hydraulic fill tolerances

7.2.3.1 General remarks

For hydraulic fill, the tolerances depend largely on the accuracy with which the settlement of the subsoil and the settlement and subsidence of the fill material can be predicted. For this reason too, satisfactory subsoil and soil mechanics investigations are very important. Of course, final grading of the fill, which is usually performed by bulldozers in the case of sand, is always necessary. When a shallow hydraulic sand fill is to be placed over a yielding subsoil, the tolerance should be selected taking into consideration whether or not the fresh fill will be required to support site traffic immediately. A tolerance is relevant for practical purposes only for a hydraulic fill level intended for site traffic.

7.2.3.2 Tolerances taking settlement into account

If only minor settlement of the subsoil and the hydraulic fill is expected, a plus tolerance is generally required related to a specific fill height. When greater settlement is expected, the estimated amount of settlement should be indicated in the tender and taken into account in the hydraulic fill specification.

7.3 Hydraulic filling of port areas for planned waterfront structures (R 81)

7.3.1 General

R 73, section 7.4, deals with direct backfilling of waterfront structures. In order to build port areas with good loadbearing capacity behind planned waterfront structures, non-cohesive material, where possible with a wide range of grain sizes, should be pumped in. In hydraulic filling over water, a greater degree of density is generally achieved without additional measures than underwater (R 175, section 1.6), all other conditions being equal.

In all hydraulic filling work, but especially in tidal areas, measures must be taken to ensure good run-off of the filling water, as well as the water flowing in with the tide.

The fill sand should contain as little silt and clay as possible. The allowable volume depends not only on the intended waterfront structure and the required quality of the planned port area, but also on the time at which the site is to be built on or used at a later date. In this respect, the dredging and hydraulic filling processes are of vital significance. Further determining factors can be contamination in the filling material and the resulting impairment of the groundwater from leaking pore water. The upper 2 m of hydraulic fill must be ideal for compaction to ensure adequate distribution of loads under roads etc.

If the port area will include settlement-sensitive facilities, silt and clay deposits must be avoided. A content of fine particles < 0.06 mm of max. 10% should be allowed for the sand fill. For economic reasons it is frequently necessary to obtain the filling material locally, or to use material obtained, for example, from dredging work in the port. The dredger spoil is often loosened by means of cutter-suction or ground suction dredgers and pumped hydraulically directly to the planned port area. In this case in particular, soil investigations beforehand at the source are indispensable. Continuous, undisturbed soil profiles are to be taken in the dredging area by means of core borings, which helps to detect thin deposits of cohesive silt or clay strata. The use of static penetrometer tests and borehole dynamic penetrometer tests can provide a very good picture of the variations in silt and clay content (R 1, section 1.2). If the sand fill available contains larger silt and clay components, the flushing process must be coordinated accordingly. It is important to consider whether the sand fill is hydraulically pumped, dredged and hydraulically filled with a bucket-ladder or suction hopper dredger onto the future port area directly from the source, or obtained with a suction dredger and first loaded into scows. This method also enables some cleaning of the sand in the case of scow loading because mud and clay drain off when the scows overflow. If mud or clay has been deposited locally near the surface of the completed hydraulic fill, e.g. flushing field outlet, this material is to be removed to a depth of 1.5 to 2.0 m and replaced by sand (see R 175, section 1.6). This is necessary because if silt or clay inclusions are left in place, it may take a long time for the surplus pore water to drain away and permit the cohesive layers to become consolidated. Appropriate maintenance of the flushing field or suitable control of the flushing flow can prevent the creation of cohesive strata in the flushing field. Vertical drains can be used to accelerate consolidation (R 93, section 7.7). It is recommended that drainage ditches be dug as soon as possible after the hydraulic fill is completed.

Without special auxiliary means, it is not possible to produce certain types of slopes with hydraulic fill under water, nor truly horizontal surfaces. The natural slope of medium sand in still water appears to be 1 : 3 to 1 : 4, in depths ≥ 2 m below water table also up to 1 : 2 in some cases. The slope is even shallower with incoming currents.

7.3.2 Hydraulic filling of port areas on subsoil above the water table

Please refer to fig. R 81-1. The width and length of the flushing field and the layout of the outlets are of particular importance, particularly when the sand fill is contaminated by silt or clay.

Width, length and outlets must be defined so that hydraulic flush water carrying suspended solids and fines is drained as quickly as possible and, in particular, that no eddies ensue. To achieve this, hydraulic filling operations must continue without interruption if at all possible. If interruptions cannot be avoided, a check must be carried out after every unavoidable interruption, such as weekends, to establish whether a layout of fines has settled anywhere; if this has happened, the layer must be removed before filling is resumed. If flushing water containing suspended particles and fines is to be returned to the water, the corresponding regulations of the authorities are to be observed. If necessary, settlement basins are to be used and the separated silt particles disposed of separately. If the hydraulic fill dyke is to form the subsequent shoreline, a sand dyke with sheet covering is recommended.

In order that sand with the coarsest possible grain size settles in the immediate vicinity of the shoreline, it is recommended to place a shore pipe outlet on or in the immediate vicinity of the hydraulic fill dyke so

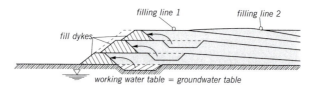

Fig. R 81-1. Hydraulic filling of port areas on a ground surface above the water table

that the dredger discharge will run along the dyke (fig. R 81-1). The risk of possible foundation failure must be taken into account.

7.3.3 Hydraulic filling of port areas on subsoil below the water table

7.3.3.1 Coarse-grained hydraulic fill sand (fig. R 81-2)

Coarse-grained sand can be used for filling without further measures. The inclination of the natural fill slope depends on the coarseness of the fill sand grains and on the prevailing currents. The fill material outside the theoretical underwater slope line will be dredged away later (R 138, section 7.5).

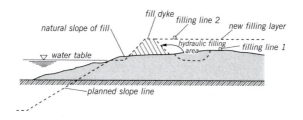

Fig. R 81-2. Hydraulic filling of port areas on a ground surface below water level

The non-cohesive soil filled in the first stage should reach a level of about 0.50 m above the relevant working water table for coarse sand and at least 1.0 m for coarse to medium sand. Above this, work continues between fill dykes. Filling in tidal areas may need to carried out to coincide with the tides.

7.3.3.2 Fine-grained hydraulic fill sand (fig. R 81-3)

Fine-grained fill sand is placed by filling or dumping between underwater dykes of rock fill material. This method can also be recommended when, for instance, insufficient space is available for a natural hydraulic fill slope because of waterborne traffic. Rock fill material is, however, not suitable when the final structure is to be a quay and the rock fill material would then have to be removed again.

242

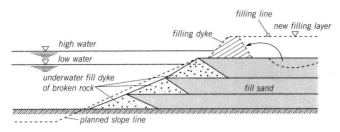

Fig. R 81-3. Underwater fill dike of broken rock. The fine-grained fill sand is flushed in or dumped

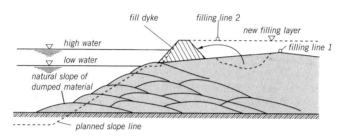

Fig. R 81-4. Underwater filling of dikes of coarse sand by dumping

It is also possible to build up the shore in advance with dumped sand (fig. R 81-4) and then fill behind this. The coarsest sand should be used for this method. Strong currents can nevertheless lead to drifting. The dumped sand outside the theoretical underwater slope line will be dredged away later (R 138, section 7.5). Some hopper suction dredgers are able to fill the sand in the so-called rainbow method. In this method, the fill material is deposited through a jet. Currents can cause considerable drifting, primarily with fine-grained material.

7.3.3.3 Land reclamation without large-scale removal of existing sediments or soft strata

If existing idle harbour basins are to be filled as part of restructuring work and the resulting surface used for harbour operations, it can be sensible and economical for larger amounts of existing sediments to be left in the port and covered by the fill material. In such cases, a local, restricted soil exchange down to the bearing bottom can be carried out to the extent required structurally as part of the planned final structure (embankment or quay wall). An underwater wall is created from bearing material as coarse as possible.

In the harbour basin behind the wall, the sediment remains on the bottom and is covered uniformly by the thinnest possible layers (max. 1 m) using the hydraulic filling method. The thickness of the layers should be such that local deformations and ground failure, also due to uneven surcharges, are avoided.

Special filling pontoons will be required for sand filling, which can be moved and positioned exactly in the vicinity of the filling surface.

The necessary consolidation periods between the placing of individual layers depend on the soil mechanics properties of the in situ sediments. As a general rule, and taking into account the bearing capacity of the in situ sediments, several layers are required, the load of which can consolidate the sediments.

This presumes careful soil investigations beforehand and accompanying measurements during the work on site. Settlement may need to be accelerated by additional measures in the form of vertical drains and preloading.

The land reclamation measures that are becoming ever more necessary in watercourses often encounter soft strata several metres thick with a low bearing capacity; however, the economic and ecological grounds for replacing such soft soil are often insufficient. In such cases an enclosing structure extending down to loadbearing soil strata must be built around the area to be reclaimed (polder). This can take the form of a sheet pile wall or a dam. Like with harbour basins, sand can then be introduced in layers within this peripheral structure using hydraulic fill techniques.

The plant as well as the control and positioning techniques available these days enable uniform layer thicknesses to an accuracy of 10 cm [243].

7.4   Backfilling of waterfront structures (R 73)

7.4.1   General
In order to provide a firm support for the backfill, and thereby prevent extensive future settlement and the imposition of heavy earth pressure on the structure, it may be advantageous – where practicable – to remove those layers of cohesive soil of poor bearing capacity occurring near enough to the structure to have an effect on it. This should be carried out before any pile driving or other work of similar importance. If such material is not removed, its effect on the structure must be taken into consideration. For thick layers of poor soil, the effect in the unconsolidated state must also be taken into consideration (R 109, section 7.9).

7.4.2   Backfilling in the dry
Waterfront structures erected in the dry shall also be backfilled in the dry whenever possible.

The backfill shall be placed in horizontal layers to suit the compaction plant used and well compacted. If possible, sand or gravel is to be used as fill material.

Non-cohesive backfill must have a minimum density of $D \geq 0.5$ if excessive maintenance of pavements, tracks and similar work is to be avoided. Density D is to be determined in accordance with R 71, section 1.5.

If the backfilling is executed with non-uniform sand in which the content of fines less than < 0.06 mm in diameter is less than 10% by weight, a toe resistance of at least 6 MN/m^2 shall be required at depths ≥ 0.6 m in static penetrometer tests. With satisfactory backfill and compaction, at least 10 MN/m^2 can generally be expected at depths ≥ 0.6 m. If possible, static penetrometer tests should be carried out regularly during the backfilling work.

For backfilling in the dry, cohesive types of soil such as boulder clay, sandy loam, loamy sand and, in exceptional cases, even stiff clay or silty clay may be considered. Cohesive backfill must be uniform throughout, placed in thin layers and especially well compacted to form a uniform solid mass without voids. This can be achieved without difficulty by using suitable compacting plant, e.g. vibratory rollers. An adequate cover of sand must be provided over the cohesive layers.

If the backfilled structure cannot deform sufficiently, a higher active earth pressure should be used in the calculations. Special investigations are required when there is any doubt.

7.4.3   Backfilling under water

Only sand and gravel or other suitable non-cohesive soil may be used as fill under water. A medium degree of density may be achieved if very non-uniform material is hydraulically pumped in such a manner that it is deposited like sedimentary material. But higher degrees of density can generally be achieved only through vibroflotation. When using uniform material as fill, hydraulic pumping alone can generally achieve a loose density only.

Please refer to R 81, section 7.3.1, and R 109, section 7.9.3, for details of the quality and procurement of the hydraulic fill sand.

The flushing water must be drained quickly. Otherwise, a severe increase in water pressure difference can cause greater loads and movements. Above all, when polluted material or silt and mud are to be expected, backfilling must proceed in such a manner that no sliding surfaces are formed. These lead to a decrease in the passive earth pressure or to an increase in the active earth pressure. Please refer to R 109, section 7.9.5. The drainage system of a waterfront structure may not be used to withdraw the dredger water when backfilling or it will become clogged and damaged.

In order to enable the backfill to settle uniformly and let the existing subsoil adapt itself to the increased surcharge, there should be a suitably long interval between completing backfilling and beginning the dredging.

7.4.4 Additional remarks

Sheet piling occasionally suffers driving damage at the interlocks, which results in a considerable flow of water at these points when there is a difference in water pressure. Backfill is then washed out and soil in front of the sheet piling piled up or carried off, resulting in voids behind the wall and shoaling or scour in front. Hydraulic backfilling considerably increases the risk of such damage. These defects can be recognised through settlement of the ground behind the wall. Even in non-cohesive soils, larger voids can be produced over time but go undetected in spite of careful inspection. Such voids often do not cave in until after many years of operation, and have often caused considerable damage to persons and property. In consideration of other actions, such as active earth pressure redistribution, consolidation of the backfill and the like, it is recommended at first to backfill the quay walls and then subsequently dredge them free in steps at adequate time intervals (R 80, section 7.1). During water-side dredging, divers should inspect the structure as early as possible for driving damage between water level and dredging pit bottom or harbour bottom. Otherwise, please note the inspection options during the driving of sheet piling according to R 105, section 8.1.13.

7.5 Dredging of underwater slopes (R 138)

7.5.1 General

In many cases underwater embankments are constructed as steep as stability requirements will allow. The inclination of the slope is stipulated, above alll on the basis of equilibrium investigations. Choppy seas and currents are considered as well as the dynamic effects of the dredging work itself. Experience has shown that slope failures frequently take place just at this stage. The high cost of resoring the planned slope justify extensive initial soil exploration and soil mechanics investigations as a basis for the planning and executing this kind of dredging.

Stability in non-cohesive soils can be increased during dredging by withdrawing groundwater by means of wells installed immediately behind the slope.

7.5.2 Effects of the soil conditions

The type and extent of the soil investigations must be suitable for determining those soil characteristics that influence the dredging operations.

The following soil mechanics parameters are of special significance:

Non-cohesive soils:

- grain-size curve
- weight density
- porosity
- critical degree of density
- permeability
- angle of internal friction
- cone resistances of penetro-meter tests or SPT values (No. of blows in borehole dynamic penetrometer test)
- geo-electric resistance measurements

Cohesive soils:

- grain structure
- weight density
- cohesion
- angle of internal friction
- undrained shear strength
- consistency index
- cone resistances of penetro-meter tests or SPT values (No. of blows in borehole dynamic penetrometer test)
- in situ density and moisture from nuclear geophysical measurements

Accurate knowledge of these soil parameters enables a suitable dredger to be selected, the specification of appropriate methods of working and a prognosis of the dredging performance achievable.

Adequate knowledge of the soil strata structure can be obtained by hose core borings. The records of soil findings obtained in this way can be supplemented and enhanced by colour photos taken immediately after removing the samples or after opening the hoses. Special problems may arise when dredging in loosely deposited sand if its density is less than the critical density. Large quantities of sand may become liquefied and "flow" due to minor effects such as vibration and local stress changes in the soil during dredging. Latent flow sensitivity of the in situ soil must be detected in good time in order to initiate countermeasures such as compacting the soil in the area influenced by the planned embankment, or at least making the slope shallower. The latter alone, however, is frequently not sufficient.

Even a comparatively thin layer of loosely deposited sand in the mass of soil to be dredged can lead to a flow failure during the dredging operation.

7.5.3 Dredging plant

Underwater slopes are excavated by dredgers whose type and capacity depend on:

- type, quantity and thickness of the layer of soil to be dredged,
- dredging depth and disposal of spoil.

The following dredger types may be considered for dredging the slopes:

- bucket-ladder dredger,
- cutter-suction dredger,
- cutter-wheel suction dredger,

- hopper suction dredger,
- grab dredger,
- dipper dredger.

The dredger must be selected according to the operating conditions. Availability of the dredger is also important.

The operating mode of suction dredgers means that they are apt to cause uncontrollable slope failures. They are therefore generally not considered qualified to meet the specifications for dredging underwater slopes. Undercutting must be avoided at all costs.

Slopes down to a depth of approx. 30 m can be successfully excavated with large cutter-suction dredgers; with large bucket-ladder dredgers, slopes down to approx. 34 m are feasible. Dipperdredgers excavate up to 20 m.

Dipper dredgers are preferably employed for dredging in heavy soils.

Grab dredgers are best suited to dredging only small quantities, or when dredging is to be carried out in accordance with R 80, section 7.1.

7.5.4   Execution of dredging work

7.5.4.1 Rough dredging

Dredging is preferably carried out above to just below the water level, where, for example, the profile of these parts of the slope is properly produced with a grab dredger. Before dredging the remainder of the underwater slope, dredging is carried out at such a distance from the embankment that the dredger can operate as close as possible to full capacity without causing any risk of a slope failure in the planned embankment.

Indications as to the safe distance to be maintained between the dredger and the planned embankment are gained by observing soil slips and the ensuing slopes which occur during the rough dredging work.

After conclusion of the rough dredging work, a strip of soil remains clear of the top of the underwater slope. This must be removed by a suitable method (figs. R 138-1 and 138-2).

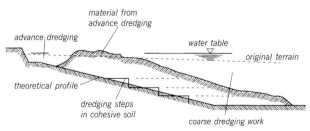

Fig. R 138-1. Dredging an underwater slope with bucket-ladder dredges

7.5.4.2 Slope dredging

Dredging along underwater slopes must be carried out carefully so that slope failures are kept within bounds and under control.

(1) Bucket-ladder dredging

Formerly, bucket-ladder and grab dredgers were employed exclusively both for coarser dredging and for dredging slopes. Dredging can be accomplished with small bucket-ladder dredgers, starting at a depth of approx. 3 m below the water level.

For practical reasons, the bucket-ladder dredger operates parallel to the slope, generally dredging layer by layer. Fully automatic or semi-automatic control of movements of the dredger ladder is possible and is to be recommended.

The slope is dredged in steps. The type of soil determines the extent to which the steps may intrude into the theoretical slope line (fig. R 138-1). In cohesive soils, the steps are generally dredged on the intended slope line. In non-cohesive soils, however, intrusion into the intended slope line is not allowed. The possible removal of the protruding soil depends on the tolerances which are to be stipulated contingent on the soil conditions and the marginal conditions listed under R 139, section 7.2.2. The depth of the steps depends on the soil conditions and is generally between 1 and 2.5 m.

The precision with which slopes can be built in this manner depends, among other things, on the planned slope inclination, the type of soil and the capabilities and experience of the dredger crew.

With slope inclinations of 1 : 3 to 1 : 4 in cohesive soils, a tolerance of ± 0.5 m measured vertical to the theoretical slope line may be accepted. The tolerance in non-cohesive soils must be +0.25 to +0.75 m, depending on the dredging depth.

(2) Cutter suction or cutter-wheel suction dredging

Apart from bucket-ladder dredgers, cutter-suction and cutter-wheel suction dredgers are also suitable for building underwater slopes, and often do so better and at a lower cost.

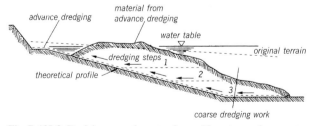

Fig. R 138-2. Dredging an underwater slope with cutter-suction or cutter-wheel suction dredges

If a spoil area is not available within a reasonable distance, dredged sand can be loaded into barges with the help of supplementary plant. Coarser sand will settle and remain in the barges. But finer material will flow out with the overflowing water.

When dredging, the cutter-suction dredger preferably moves along the slope, dredging layer by layer like the bucket-ladder dredger. Computerised control of the dredger and dredger ladder is recommended.

Fig. R 138-2 shows how the cutter-head works upward parallel to the theoretical slope line after having made a horizontal cut. Underwater slopes of the greatest precision can be produced in this way. In the case of computerised dredger control, tolerances of +0.25 m measured at right angles to the slope are satisfactory for small cutter-suction dredgers, and +0.5 m for larger dredgers. If dredging is done without special control, the same tolerances as recommended for bucket-ladder dredging apply. A prerequisite is that the soil should exhibit no flow tendencies.

7.6   Scour and scour protection at waterfront structures (R 83)

7.6.1   General

Scour can be caused by natural currents or erosion effects caused by ship movements, such as

- natural gradients or drift currents,
- ship-waves, backflow currents and screw currents caused by ships.

In port engineering in the vicinity of the ship berths, the action of the ship screws is a prime eroding element, with speeds of up to 4 to 8 m/s possible near the harbour bottom. This contrasts with the current speeds of natural river or drift currents and backflow currents caused by the ships of only 1 to 2 m/s. Erosion resulting from waves caused by ship action has an influence on the slopes. A corresponding safeguard is to be included in the principles for coastal construction, see R 186, section 5.9.

Scour caused by the natural currents of the water occurs mainly where erosive currents can occur at headlands, narrow passage openings, etc.

Special account must be taken of the ship's screw action and its effect on the harbour bottom, particularly where ships berth and depart again without tug assistance. This includes, in particular, ferries, Ro-Ro vessels and container ships.

Scour caused by the drive screw or bow thruster affects only a limited area, but the high and turbulent velocities involved do particularly endanger water structures in the vicinity, such as ferry beds, quay structures, locks, etc.

7.6.2   Scour caused by ships

7.6.2.1   Scour caused by jet formation from the stern screw

The jet velocity caused by the rotating screw, so-called induced jet speed (occurs directly behind the screw), can be calculated after [170]:

$$v_0 = 1.6 \cdot n \cdot D \cdot \sqrt{k_T} \qquad (1.1)$$

n = speed of screw [1/s]
D = diameter of screw [m]
k_T = thrust coefficient of screw, $k_T = 0.25 \ldots 0.50$

The simplified formula for a mean value of the thrust coefficient is:

$$v_0 = 0.95 \cdot n \cdot D \qquad (1.2)$$

If the output of the screw P is known instead of the speed, the induced jet speed can be calculated according to the following assumption:

$$v_0 = C_P \left[\frac{P}{\rho_0 \cdot D^2} \right]^{1/3} \qquad (1.3)$$

P = screw output [kW]
ρ_0 = density of the water [t/m³]
C_P = $0.87 \cdot \left[\left(\dfrac{k_T}{k_Q} \right) \cdot \sqrt{k_T} \right]^{1/3}$ = 1.48 for free screw
  (without nozzle) [1]
  = 1.17 for screw
  in nozzle [1]
k_Q = moment magnification factor for a screw
k_T/k_Q = 7.5 (approximate value)

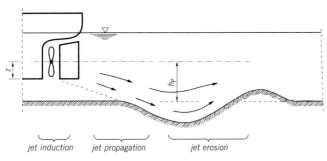

jet induction   jet propagation   jet erosion

Fig. R 83-1. Jet formation caused by the stern screw

251

As it progresses further, the jet generated expands cone-like due to the turbulent exchange and mixing processes and loses speed with increasing distance from the screw.

The maximum speed occurring near the bottom which is essentially responsible for scouring can be calculated as follows, see also [170]:

$$\frac{\max v_{\text{bottom}}}{v_0} = E \cdot \left(\frac{h_P}{D}\right)^a \tag{2}$$

E = 0.71 for single-screw vessels with central rudder, after [170]
  = 0.42 for single-screw vessels without central rudder, after [170]

  = 0.42 for twin-screw vessels with central rudder, after [170]
    valid for $0.9 < h_P/D < 3.0$ after [170]
  = 0.52 for twin-screw vessels with twin rudders
    located after the screws
    valid for $0.9 < h_P/D < 3.0$ after [170]

a = −1.00 for single-screw vessels
  = −0.28 for twin-screw vessels

h_P = height of screw shaft above bottom [m]
  = $z + (h - t)$

z = $\left(\dfrac{D}{2}\right) + 0.10...0.15$ [m]

h = water depth [m]
t = draught

The speed of the screw which is relevant to the water jet velocity depends on the power plant output used for berthing and departing. Practical experience has shown that this screw r.p.m. for port manoeuvres is as follows:

- approx. 30% of the rated speed for "slow ahead", and
- approx. 65–80% of the rated speed for "half-speed ahead".

According to German experience hitherto [233] and taking account of safety aspects, a speed corresponding to 75% of the rated speed should be selected for designing the bottom protection measures. This corresponds to 42% of the rated power (see [234]).

The rated speed or, where applicable, increased speeds at maximum power plant output must be assumed for particularly critical local conditions (high wind and flow loads on the ship, nautically unfavourable channels, manoeuvres in icy conditions), or for basin trials in shipyards. These conditions must be clarified with the future operator of the port facilities, particularly with the port authority.

7.6.2.2 Water jet generation by the bow thruster

The bow thruster consists of a screw which works in a pipe and is located at 90° to the longitudinal axis of the ship. It is used for manoeuvring out of a stationary position and is therefore installed at the bow – more rarely at the stern. When the bow thruster is used near to the quay, the water jet generated strikes the quay wall directly and is deflected in all directions from there. The critical element for the quay wall is the part of the water jet directed at the harbour bottom, which causes scour in the immediate vicinity of the wall upon striking the bottom, see fig. R 83-2.

The velocity at the bow thruster outlet $V_{0,B}$ can be calculated according to [170] as per equation (3):

$$v_{0,B} = 1.04 \left[\frac{P_B}{\rho_0 \cdot D_B^2} \right]^{1/3} \tag{3}$$

P_B = output of bow thruster [kW]
D_B = inside diameter of bow thruster opening [m]
ρ_0 = density of water [t/m³]

Jet velocities of 6.5 to 7.0 m/s must be expected for the bow thrusters of large container ships ($P_B = 2500$ kW and $D_B = 3.00$ m).

The velocity of the part of the water jet striking the bottom max v_{bottom}, which is responsible for erosion, is calculated as follows:

$$\frac{\max v_{bottom}}{v_{0,B}} = 2.0 \cdot \left(\frac{L}{D_B} \right)^{-1.0} \tag{4}$$

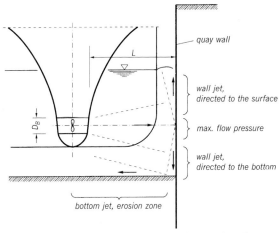

Fig. R 83-2. Jet load on the harbour bottom from the bow thruster

L = distance between bow thruster opening and quay wall [m] (see fig. R 83-2)

The bow or stern thruster usually operates at full power.

7.6.3 Scour protection

The following measures can be considered for averting the dangers to waterfront structures due to scour:

(1) Scour surcharge to the structures.
(2) Covering the bottom with a loose or grouted stone fill.
(3) Covering the bottom with flexible composite systems.
(4) Monolithic concrete slabs, e.g. in ferry beds.
(5) Jet-deflecting design of quay walls.

Notes on 1):

In this case a protective layer is not applied and the scour formation is accepted. The structure is secured by calculating the theoretical foundation bottom at a depth which takes account of the corresponding scouring depth (scour surcharge).

The scouring depth to be expected in front of quay walls as a result of ship manoeuvres was investigated in [233] on the basis of model tests. Like with other scouring processes, it was shown that once the maximum loading capacity of the relevant grain in the harbour bottom has been exceeded, assuming erodible material, the scouring depth at first increases very rapidly as the jet speed rises but then slows down.

Scouring due to a bow thruster can be estimated, e.g. after [233], with the following equation:

$$\frac{T_K}{d_{85}} = C_M \cdot 3.05 \cdot \left(\frac{v_{bottom}}{\sqrt{d_{85} \cdot g \cdot \Delta'}} \right)^{2.25} \tag{5}$$

T_K = scouring depth [m]
d_{85} = relevant grain size of harbour bottom [m]
v_{bottom} = bottom velocity from eq. (4) [m/s]
Δ' = relative density of bottom material under water [1]
    = $(\rho_s - \rho_0)/\rho_0$
C_M = 1.0 for stationary jet load
    = 0.34 for jet load during berthing manoeuvres [1]

Owing to the complexity of the scouring processes, this approach should only be regarded as a rough guide.

Notes on 2):

A loose stone cover (natural stone and waste materials such as slag) constitutes one of the common protection systems. The following requirements must be met:

- Adequate stability when exposed to screw action.
- Installation of the stones so as to cover the bottom reliably. This means installing the loose stone in two or three layers.
- Installation on a grain or textile filter rated for the relevant subsoil, see [128] and [149].
- Connection to the waterfront structure underneath the current to be safe from erosion, especially with a sheet pile quay wall.

Verification of flow stability is provided after [170] according to the following assumption:

$$d_{req} \geq \frac{v_{bottom}^2}{B^2 \cdot g \cdot \Delta'} \tag{6}$$

d_{req} = required diameter of the stones [m] (upper layer)
v_{bottom} = bottom velocity as per eq. (4) [m/s]
B = stability coefficient [1] according to [170]
  = 0.90 for stern screw without central rudder
  = 1.25 for stern screw with central rudder
  = 1.20 for bow thruster
g = 9.81 (gravitational acceleration) [m/s^2]
Δ' = relative density of bottom material under water [1]
  = $(\rho_s - \rho_0) / \rho_0$
ρ_s, ρ_0 = density of loose stone material or water [t/m^3]

For bottom speeds of 4 to 5 m/s, stone fills create problems because the corresponding diameters d_{req} = 0.7–1.0 m are so large that they cannot be handled easily.

At high bottom speeds stone fills in grouted form represent a good alternative. We distinguish here between full and partial grouting.

In full grouting the grouting material fills all the voids between the stones and this results in a cover layer not unlike plain (i.e. unreinforced) concrete. Normally, the grouting is placed in such a way that the tips of the stones still project in order to help dissipate the energy of the currents. If there is a water pressure difference below the cover layer, the weight of the grouting must take account of this. A geotextile membrane is laid between the fully grouted cover layer and the subsoil in order to prevent soil mixing with the stones and to guarantee that the voids within the stone matrix can be filled completely with the grouting material.

In partial grouting the amount of grouting material introduced into the stone matrix is just enough to hold the stones in position. However, the stone fill still retains sufficient permeability to prevent a water pressure difference below the cover layer. Like with loose stone fill, a filter is required between subsoil and cover layer.

Thanks to its interlacing effect, partially grouted stone fill remains stable in velocities up to 6 to 8 m/s, see [235].

Generally, stone classes II and III [97] should be used for stone fill in conjunction with partial grouting. In the majority of cases it is sufficient to grout the uppermost 50 to 60 cm of the stone fill.

The partial grouting of the stone fill should be carried out evenly over the area of the fill in an attempt to include every stone in the interlacing effect. The amount of grouting material required should be selected such that a minimum pore volume of 10 to 20% remains after grouting to prevent water pressure differences below the grouted stone fill.

Partial grouting can be placed properly in a layer thickness of up to about 60 cm in one operation. If the total thickness of grouting required is greater than this, this must be built up in layers.

Sufficiently hardwearing scour protection can generally be achieved with 150 to 200 l/m^2 of grouting material (for a layer thickness of 60 cm). See [235] for a proposed design for scour protection.

At the junction with the quay wall, a strip approx. 1 m wide must be fully grouted.

The grout must be made from mortars or concretes with erosion stabilisers or colloidal mortars with good bonding properties and suitable for placing under water. However, bitumen can be used as well when a fully grouted solution is required. Methods and the materials must comply with the requirements of ZTV-W LB 210 [132], and the contractor must be ISO-accredited. Suitability tests under the local conditions are to be carried out for every project.

The junction between structure or structural components and grouted cover layers must be designed to prevent gaps through which soil could be washed away.

A grouted cover layer exhibits only limited flexibility. Therefore, to avoid scour, the quantity of grouting material must be reduced to approximately 60–80 l/m^2 at the edges, or rather the transition to the unreinforced harbour bottom, or flexible protective elements must be used to adapt the protection to the soil formations.

Notes on 3):

Composite systems are systems in which various basic elements are combined to create a planar covering. The most important principle to be observed here is that the elements are to be connected in a flexible way for good adaptation to and stabilisation of edge scouring. The following engineering solutions are known:

- concrete elements connected by ropes or chains
- interlocking concrete blocks
- mesh containers filled with broken rocks
  (stone or ballast mats, gabions)

- geotextile mats filled with mortar
- geotextile mats filled with interlocking concrete blocks

Adequately sized systems systems provide excellent stability. However, a general fluid mechanics design assumption applied only to certain cases because of the individual variety of the systems available, see [170], so that these systems are frequently dimensioned according to the manufacturer's experience.

When the system connections are flexible enough, such systems exhibit good edge scouring behaviour, i.e. any edge scour is stabilised automatically by the system, thus preventing regressive erosion.

The disadvantage of rubble-filled mesh mats (although equally good in terms of stability and edge scour properties) is that the wire mesh is vulnerable to corrosion, sand wear and mechanical damage. If the gabion mesh is damaged beyond repair, the gabions lose their stability. The mats or gabions must be joined together with tension-resistant connections.

Notes on 4):

An underwater concrete bottom offers very effective erosion protection for a limited area (ferry beds, trials basins in shipyards) because its depth can be controlled with far greater precision than a stone fill. The homogeneous structure of carefully produced concrete enables the thrust force transferred locally to the bottom by the screw to be distributed across a wide area so that even with very high jet loads the bottom remains stable.

The disadvantage is that the rigid concrete bottom cannot adapt to differential settlements, and fractures can occur. Furthermore, the bottom cannot automatically stabilise any edge scouring, meaning that special solutions are required. In ferry beds cut-off walls have proved successful as a termination to the underwater concrete bottom. Underwater concrete bottoms – depending on current load and installation methods – are built in thicknesses of 0.3 to 1.0 m. The proper installation of underwater concrete is a technically complicated and costly process (divers, poor visibility, submerged formwork, prevention of segregation in the concrete, etc.). A rigid concrete bottom as erosion protection is economic when it can be built in the dry, e.g. within the protection of a cofferdam.

Notes on 5):

Giving the sheet pile wall a batter in order to create a cushion of water between front edge of quay and side of ship, if necessary combined with an arrangement of jet deflectors, provides efficient options for avoiding or at least minimising the load on the bottom. Such measures are particularly suitable for reducing scour due to jet erosion caused by a bow thruster (see fig. R 83-3). According to [236] the following reduction

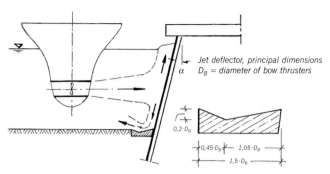

Jet deflector, principal dimensions
D_B = diameter of bow thrusters

Fig. R 83-3. Jet-deflecting measures at a quay wall to reduce scour, minimum dimensions [236]

in scour can be expected for an arrangement involving an inclined sheet pile wall combined with a jet deflector at the base of the wall (valid for $L/D_B \sim 4$ and $h_{P,B}/D_B \sim 2$):

$$T_{K,\alpha,m,SL} = C_{SL} \cdot (1 + 0.005 \cdot \alpha) \cdot C_\alpha \cdot T_K \qquad (7)$$

T_K = scouring depth due to bow thruster without jet-deflecting measures

$T_{K,\alpha,m,SL}$ = scouring depth due to bow thruster with jet-deflecting measures

= angle of sheet pile wall

C_α = jet-splitting ratio for angle of wall

C_{SL} = scouring reduction when using a jet deflector

	$\alpha = 0°$	$\alpha = 10°$	$\alpha = 20°$	$\alpha = 30°$
C_α	1.00	0.78	0.58	0.38
C_{SL}	0.25	0.20	0.10	0.05

A reduction in scour according to eq. (7) amounting to about 16% of the scour on the unprotected bottom can be expected when using suitable measures to deflect the jet, i.e. for example, wall batter $\alpha = 10°$ plus jet deflector at the bottom. Bottom protection is in this case unnecessary.

7.6.4   Dimensions of strengthening measures

The dimensions of strengthening measures should be chosen taking into account fluid mechanics aspects such that the jet velocities at the edges are reduced to such an extent that there is no danger of edge scouring leading to undermining of the strengthening measures. Depending on the specific reduction functions, see eq. (2) for stern screws and eq. (4)

258

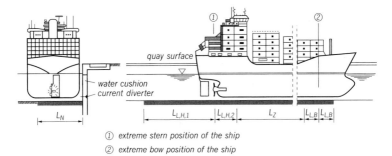

① extreme stern position of the ship
② extreme bow position of the ship

Fig. R 83-4. Dimensions of scour protection areas alongside a quay

for bow thrusters, such an approach leads to very large areas of strengthening, which are linked to considerable capital outlay.

Considering the economic aspects and the fact that it is not the bottom but rather the structure (quay wall etc.) that requires protection, the areas of strengthening should be arranged in such a way that the strengthening measures can withstand at least intensive flow loads. Moreover, the minimum dimensions of the strengthening measures should be chosen such that edge scouring does not reduce the area of the structurally effective passive earth pressure wedge at the base of the quay wall to an unacceptable extent. The following values are recommended as a first approximation and as minimum dimensions (see fig. R 83-4), although it should be remembered that 70…80% of the maximum bottom speed is to be expected at the edges. The strengthening should be terminated with suitable edge protection on harbour bottoms vulnerable to erosion; this protection should be able to adapt flexibly to any edge scouring and thus stabilise this.

For single-screw vessels:

- perpendicular to the quay
  L_N = $(3 ... 4 \cdot D) + \Delta RS$
  D = screw diameter

- parallel to the quay
  $L_{L,H,1}$ = $(6 ... 8 \cdot D) + \Delta RS$
  $L_{L,H,2}$ = $3 \cdot D + \Delta RS$
  $L_{L,B}$ = $(3 ... 4 \cdot D_B) + \Delta RS$
  ΔRS = allowance for edge protection, approx. 3 … 5 m

For twin-screw vessels:

Owing to the arrangement of two screws adjacent to one another, the above values for diameter are to be doubled.

The total extent of the strengthening measures parallel to the quay depends on the expected variations in the berthing positions. The intermediate length L_Z may be left unstrengthened on berths with precisely defined ship positions.

Berths for frequent traffic, e.g. ferries, and quay structures particularly sensitive to settlement require the extent of the strengthening measures to be investigated in greater detail than as required by the above recommendations (minimum dimensions) by analysing the propagation behaviour of the screw jet.

7.7 Vertical drains to accelerate the consolidation of soft cohesive soils (R 93)

7.7.1 General

The consolidation (primary settlement) of soft cohesive, relatively impermeable strata, can generally be accelerated by means of vertical drains. This cannot be used as solution for the secondary settlements that occur in some soils, caused by creep of the soil with no change in pore pressure.

Vertical drains are particularly successful in soft cohesive soils. Since the horizontal permeability of these soils is generally higher than the vertical permeability, the drainage paths are considerably shorter. In stratified soils with alternating permeability (e.g. layers of clay and mud-flats sand), the pore water in layers of low permeability is drained through adjacent layers of higher permeability. As a result, the consolidation is further accelerated.

When secondary settlement accounts for a substantial portion of the total settlement, or – as in peat – settlement is to be expected due to material conversion, vertical drains are usually inadvisable.

7.7.2 Applications

Vertical drains are used where bulk goods, dykes, dams or fill are dumped on soft cohesive soils. This shortens the period of consolidation and the in situ soil attains the loadbearing capacity required for the intended use at an earlier date. Vertical drains are also used to prevent sliding of slopes or terraces, and to limit lateral flow movements due to fill.

When using vertical drains, the designer must also check whether their use promotes the spread of any harmful contaminants that might be present in the soil.

7.7.3 Design

When designing a vertical drain system, take the following factors into account.

260

- The application of a surcharge greater than the total of all future loads can even compensate for later secondary settlement to a certain extent.
- Primary settlement can be estimated according to the theory of consolidation (TERZAGHI), secondary settlement according to KEVERLING BUISMAN and the combined settlement formula after KOPPEJAN [56]. However, owing to the simplifications contained in these methods and the inhomogeneity of the subsoil, these estimates should be applied to the probable bandwidths of the soil mechanics parameters. The progress of the consolidation must always be checked by measuring the settlement and the excess pore water pressure in situ. Only by means of these measurements is it possible to deduce reliably the conclusion of the settlement and the shear strength available.
- In many cases the settlement observed in vertically drained soil is greater than that found under similar conditions in undrained soil because the stagnation gradient of the consolidation is smaller with vertical drainage than it is without.
- When using vertical drains, the consequences for the hydrological boundary conditions should be carefully assessed.
- If there is a confined water table below the stratum to be consolidated, the drains must terminate at least about 1 m above the lower layer.

The structure of the subsoil must be very carefully investigated beforehand in order to specify the optimum spacing of the drains. However, the water permeability of the in situ soil types can only be determined reliably by means of tests loads.

The following criteria are relevant for designing the drains:

(1) The drain inlet area must be large enough.
(2) The drain must possess sufficient discharge capacity.
(3) Thickness and permeability (horizontal and vertical) of the cohesive soil strata to be dewatered.
(4) Permeability of neighbouring soil strata.
(5) Desired acceleration of the consolidation period.
(6) Costs involved.

Criterion (1) governs in most cases. The available consolidation time is also of particular importance.
Early systematic preliminary investigations with settlement, water level and pore water pressure measurements can be used to find the most practicable and economical type, layout and construction of the drains. The effects of later construction work must also be taken into consideration; soiling of the drain walls caused by construction work can substantially reduce the water entering the drain and hence its efficiency. This also applies to run-off down into a water-bearing layer, or to lateral discharge in a sand or gravel layer placed on top. In general, $\varnothing$ 25 cm

sand drains can be installed at a centre-to-centre spacing of 2.5 to 4.0 m, but plastic drains are normally installed closer together.

Plastic drains are supplied in widths of approx. 10 cm. The design should also take account of the manufacturer's specific product information.

The following factors are critical for the choice of drain type (sand or plastic):

- The soil/drain contact surface may become smeared during the installation of a plastic drain. This will increase the resistance to water entering the drain. The favourable action of thin, horizontal sand layers cannot take effect.
- Vertical drains experience the same settlements as the soil strata in which they are installed. Large settlements may cause detrimental twists, which seriously impair the function of the drain. Tests on twisted vertical drains can provide information about the reduction in the flow capacity of the drain in the case of considerable compression of the drained stratum.

7.7.4 Construction

Boring, jetting, vibrating and driving methods can be used to install drains. The most common types are the flat or round plastic drains, most of which are installed by jacking or vibration. In order to facilitate the movement of plant and the transportation of sand over the (mostly) soft terrain, and to prevent the bore spoil from fouling the ground surface, a layer of sand of good permeability at least 0.5 m thick is placed on the ground before work starts. This also acts as a collector and discharge area for the water being drained. If the sand layer is too thick and compacted, it is much harder to install the plastic drains. The watery bore spoil is carried off in ditches to locations where deposits can cause no harm.

In the event of contaminated subsoil, avoid installation procedures that produce bored or jetted material. Furthermore, any existing compacting stratum should not be penetrated. In order to gain time for the utilisation of the drained ground, the drains should be installed at an early stage and the subsoil preconsolidated with soil fill on the ground.

7.7.4.1 Bored sand drains

For bored sand drains, depending on the type of soil, cased or uncased boreholes are sunk with diameters of about 15 to 30 cm, and filled with sand. The casing is pulled if necessary. Bored sand drains have the advantage that they cause minimum disturbance of the subsoil, which is especially important in sensitive soils. In addition, they involve the least reduction in permeability of the soil in the horizontal direction, which is vital for their success. They should be bored swiftly where there is a water pressure difference. When no casing is used, they are to be flushed free of any obstructions so as to provide a perfectly clear hole; when a water pressure difference exists, the holes should be filled without interruption to minimise the danger that the sand will be fouled by crumbling of the walls into the hole. Special care must be taken to ensure that the drains are not narrowed by soil penetrating from the sides.

The fill material must be chosen so that consolidation water can flow in and out without hindrance. The proportion of fine sand (≤ 0.2 mm) should not exceed 20%.

7.7.4.2 Jetted sand drains

Section 7.7.4.1 applies correspondingly to jetted sand drains. They are inexpensive and quick to install, but have the disadvantage that deposits of fines on the bottom and on the drain walls can cause a marginal reduction in the performance of the drain. Special care must be paid to this during their construction. Large quantities of jetting water are required even when penetrating sand strata because the jetting water can escape via the walls. After they are jetted clear and the gear removed, the sand filling must commence immediately and continue swiftly without interruption.

7.7.4.3 Driven sand drains

If displacement of the soil during driving has no harmful effect on the bearing capacity of the subsoil (sensitivity) and its horizontal permeability, driven sand drains may be used. Thus, for example, a pipe 0.3 to 0.4 m in diameter is plugged with gravel or sand at its lower end and driven into the soft subsoil until the desired depth is reached using a ram working inside the pipe. After the bottom plug has been knocked out, the pipe is filled with suitable graded sand (as with bored drains) and driven into the subsoil, while the pipe is simultaneously withdrawn. In this manner, a sand drain is created whose volume is about two to three times that of the volume of the pipe, and the soil around the pipe is permanently disturbed. Any existing structural strength in the in situ soil is thereby lost. Zones which may have become blocked when the pipe was being driven will be opened again, at least partially, by the driving of the sand and the resulting increase in the wall area of the drain. With this method it is perfectly feasible to direct the drainage both into a lower water-bearing stratum and into a sand fill on the surface.

7.7.4.4 Plastic drains

Plastic drains are introduced into the soft soil by means of a needle unit. They can be inserted in the soft soil without water, which essentially prevents any soiling of the drain walls.

7.7.5 Checks during construction

The effectiveness of vertical drainage depends largely on the care taken during construction. In order to avoid failures, continuous inspection of the sand filling is necessary, and the effectiveness of the drains must be checked in good time by means of fill tests with water, water level measurements in individual drains, pore pressure measurements between the drains, pressure soundings or water pressure measurements and observations of settlement of the ground surface.

7.8 Subsidence of non-cohesive soils (R 168)

7.8.1 General

Subsidence is a certain component in the volume reduction of a non-cohesive mass of soil, e.g. sand.

Altogether, volume reductions are caused by:

(1) Grain readjustments resulting in an increase in density.
(2) Deformation of the grain structure.
(3) Grain failure and grain splintering in areas with large local stress peaks at grain contact points.

These effects can even be started simply through load increases. The grain readjustments according to (1) occur chiefly during vibrations, reductions in the structure resistances and/or the friction effects between the grains.

The volume reductions due to the latter two actions are designated as subsidence, which takes place when during very heavy wetting or drying out of non-cohesive soils the so-called capillary cohesion between the grains, i.e. the friction effect due to the surface tension of the residual water in the pore grooves (at the points of grain contact), decreases substantially or disappears completely. If this occurs, for example, in connection with a rising groundwater level, effects due to the changes of the menisci between the grains at the respective groundwater level also act on the grains.

7.8.2 Degree of subsidence

If moist sand is loaded and fully wetted, e.g. by rising groundwater, the result is a load–settlement diagram as per fig. R 168-1.

The subsidence of non-cohesive soils is contingent on the granulometric composition, grain shape, grain roughness, initial moisture content, state of stress in the ground and, above all, on the initial density. The more loosely the soil is deposited, the greater is the subsidence. In very loosely deposited, uniform, fine sands it reaches 8%. But even after a high degree of compaction it can still reach 1–2% of the layer thickness.

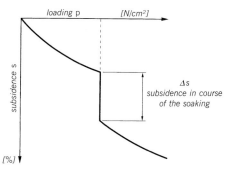

Fig. R 168-1. Load–settlement diagram of moist sand at soaking

In general, subsidence is greater in round-grained than in angular-grained soils. Uniform sands show a greater degree of subsidence than non-uniform sands. The difference, however, is only recognisable in very loose and loose deposits. In this connection, the sudden subsidence of coarse silt when heavily wetted has been known for some time.

The danger of subsidence in sand decreases rapidly with the increase in the proportion of cohesive particles.

Tests as per [57] have shown that sands with the same initial moisture content and same initial density experience settlements and subsidence caused by loading and wetting, whose total may be assumed with adequate accuracy irrespective of the magnitude of the loading. Accordingly, under small loads, settlement is small and subsidence large.

7.8.3   Effects of subsidence on structures, and countermeasures

Subsidence occurs essentially during the first soaking of non-cohesive soils. At further soaking, after repeated groundwater lowering, subsidence is minimal.

In order to avoid or reduce subsidence as a result of flooding or wetting during rising groundwater or when completely dry, the sand must be extensively compacted beforehand. The density achieved can be checked by taking samples or with the help of dynamic, static or radiometric penetrometer tests (see R 71, section 1.6).

If necessary, subsidence, which may cause damage, may be effected in advance by adding large quantities of water during the installation or during soil replacement groundwater lowering protection by allowing the groundwater level to rise temporarily.

7.9   Soil replacement procedure for waterfront structures (R 109)

7.9.1   General

Soil replacement for waterfront structures in areas with thick, soft cohesive soil strata is only economic when the required quantity of sand fill is readily available at a favourable price. Soil replacement is also recommended when the soil investigations show that driving obstacles must be expected, which could lead to damage to the sheet piling. The depth of the dredged excavation must be sufficient to ensure the stability of the intended waterfront structure at all times. In this respect, it may be necessary to excavate all soft cohesive strata down to loadbearing soil. When excavating below water, the soil lost during excavating is deposited as a layer of silt on the bottom of the excavation. Added to this is the silt from the water itself. If this silt were allowed to remain on the bottom of the excavation and was then covered by the replacement soil, it would form a layer of large settlement and low strength within the subsoil. Therefore, the silt on the bottom of the excavation must be removed as completely as possible before installing the replacement soil.

The complete excavation of in situ soft cohesive strata is also necessary when the waterfront structure can accept only minimum settlement.

Comprehensive soil investigations and soil mechanics tests are necessary at the preliminary design stage in order to determine the cost of excavation and soil. Only then can the dimensions and the depth of the excavation, the plant required and its capacity be calculated with any reliability.

The cost estimate requires the anticipated thickness of the layer of silt lost from the soil during excavation work and from the movement of bed material and sediment to be investigated as carefully as possible (materials and amounts depending on the flow velocities in the course of the various tides and depending on seasonal variations). Based on such work, the installation of the replacement soil can be planned and carried out systematically in order to minimise deposits of silt.

Before a decision is made in favour of the soil replacement procedure for major structures, a trial pit of appropriate size should be excavated beforehand and kept under constant observation for an extended period. Attention is drawn to the vulnerability of sand fill to erosion and scour in flowing watercourses. It follows that temporary scaffolding and similar structures placed on the sand must be deeply embedded unless a protective covering has been installed.

7.9.2 Excavation

7.9.2.1 Selecting the dredger to be used

In general, only bucket-ladder dredgers, cutter-suction dredgers, cutter-wheel suction dredgers or dipper dredgers can be used for excavating cohesive soil. If a layer with buried obstacles must be dredged (e.g. soil interlaced with rubble), a suction dredger may be employed only if it is equipped with special pumps with sufficiently large apertures because otherwise the buried obstacles will remain in place and form a layer that is almost impossible to penetrate during later driving work.

With all types of dredger, soil losses cannot be fully avoided, and these lead to the formation of an intrusive layer on the dredger pit bottom (fig. R 109-2). When excavating with bucket-ladder dredgers (fig. R 109-1) in particular, it must be accepted that the dredging operation itself (overfilled buckets, incomplete emptying of buckets in the dumping area, overflowing of barges) will lead to a thick sediment. Therefore, to ensure that the layer of silt/sludge is removed as thoroughly as possible, a shallower cut must be used when reaching the dredger pit bottom. For this, a slack lower bucket chain must be used together with low bucket and cutting speeds. The scows may be fully loaded, but overflowing with soil losses must be absolutely avoided.

Use of cutter-suction and cutter-wheel suction dredgers results in an undulating dredger pit bottom as per fig. R 109-2, and the intrusive layer is thicker than that produced by a bucket-ladder dredger.

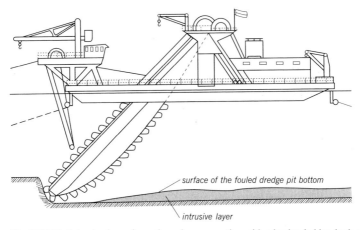

Fig. R 109-1. Intrusive layer formation when excavating with a bucket-ladder dredge

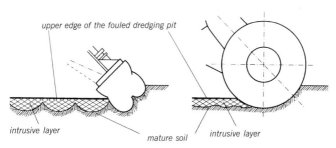

Fig. R 109-2. Intrusive layer formation when excavating with a cutter-suction or cutter-wheel suction dredge

267

The intrusive layer thickness can be reduced by using a special cutting head shape, a slower rotation, short thrusts and slow cutting speed. The intrusive layer must be removed immediately before applying the replacement soil.

7.9.2.2 Executing and checking dredging work

In order to guarantee proper dredging, the dredger pit must be generously dimensioned according to the size of the dredger selected, and marked appropriately. Suitable survey markers and positioning devices must be clearly and unmistakably recognisable by the dredger crew at all times of day or night. Marking the dredger cutting width solely on the side lines of the dredger is not adequate.

Excavation is performed in steps corresponding to the mean profile inclination at the edge of the dredger pit. The depth of the cuts depends on the type and size of the plant and on the character of the soil. Strict control over the cut widths must be maintained because cuts that are too wide can cause an excessively steep slope in places, with resulting slope slips. Orderly progress of dredging can be controlled quite effectively by modern surveying methods (e.g. depth sounder in combination with the global positioning system – GPS) which can also be used for early detection of any profile changes caused by slips in the underwater slope. Measurements with inclinometers on the edge of the excavation have also proven successful for monitoring slips on underwater slopes.

The last sounding should be taken immediately before the sand fill is placed. In order to obtain information on the characteristics of the dredger pit bottom, soil samples are taken for testing. For this, a hinged sounding tube with a minimum diameter of 100 mm and a gripping device (spring closure) has proven effective. Depending on conditions, this tube is driven 0.5 to 1.0 m or even deeper into the dredger pit bottom. After retrieval and opening, the core in the tube permits an assessment of the soil strata at the dredger pit bottom.

7.9.3 Quality and procurement of the fill sand

When planning soil replacement works, it is advisable to investigate the sand procurement areas by means of boreholes and penetrometer tests before starting the actual work on site. The fill sand should contain only very little silt and clay, and no major accumulations of stones.

If the fill sand available is severely contaminated and/or stony, but otherwise usable, to avoid local accumulations of fines and stones hydraulic filling methods may not be used but instead the material must be dumped.

In the interests of continuous, rapid and economical filling, adequately large deposits of suitable sand must be available within a reasonable distance. In determining the amount of fill required, soil drift is to be taken into account. The finer the sand, the stronger the current over and in the dredger pit, the slower the rate at which the fill is placed and the greater the depth to the dredger pit bottom, the greater will be the loss of sand due to drifting.

Powerful suction dredgers are recommended for sand procurement because, along with their high output, they clean the sand at the same time. The cleaning action can be intensified by proper loading of the barges and longer overflow times. Samples of the fill sand are to be taken frequently from the barges and tested for compliance with the requirements in the design, especially for the maximum allowable silt content. Dredging of sand with suction dreggers and barges generally improves the quality of the sand since the fine graines are washed out.

7.9.4   Cleaning the dredger pit bottom before filling with sand

The dredger pit bottom must be adequately cleaned immediately before starting the filling work. For this purpose, the use of silt suction plant is possible if the deposits are not too firm. If, however, a longer period of time has elapsed between conclusion of the dredging work and the start of suction work on the silt, the silt may already be so solid that another cleaning cut may be necessary.

The water jet or water injection method has proven effective here. Large quantities of water are pumped at approx. 1 bar through nozzles to the dredger pit bottom, using a cross-member suspended under a floating machine. The clearance to the bottom is kept to a minimum of between 0.3 m and 0.5 m. This method turns any settled silt back into a full suspension.

The water jet method must be carried out before placing the sand fill until the bottom is proven to be free of silt.

The cleanliness of the dredger pit bottom must be checked regularly. The sounding tube described in section 7.9.2.2 can be used for this. If only soft deposits are to be expected, a properly designed grab can also be used for taking the samples; a hand grab may also be considered. A combination of silt soundings and depth soundings with differing frequencies is a good control option.

If an adequately clean bottom cannot be ensured, the keying between the natural bearing soil and the fill sand is to be established by other suitable measures. This can best be achieved in cohesive bearing subsoil by a layer of crushed rock of adequate thickness, which must be placed very quickly.

Such a safety measure can be of special importance on the passive earth pressure side. As there is generally no pile driving work in this area, rubble can be used to even better effect than crushed rock.

With non-cohesive soils and a dredging pit of minimal depth, keying between the fill soil and the subsoil can also be achieved by the "dowelling" effect of multiple-vibratory cores (vibroflotation with a unit of 2 to 4 internal vibrators).

7.9.5 Placing the sand fill

The dredger pit can be filled hydraulically or by dumping, or even both methods at the same time. Particularly in waters heavily burdened with sediment, this operation must be carried out without interruption around the clock, with carefully coordinated equipment, at all times adhering to a plan worked out in every detail from the very beginning. Winter operations with loss of working days to be expected due to cold weather, drifting ice, storms and fog should be avoided.

Placing the sand fill should follow dredging of the poor soil as soon as possible to minimise the unavoidable deposits of sediment (silt) accumulating in the meantime. On the other hand, however, mixing of the soil to be replaced with that being filled may not be permitted due to insufficient time between dredger and fill operations. This danger exists particularly in waters with strong reversing currents (tidal area) where it must be guarded against with special care.

A certain amount of contamination of the sand fill by continuous depositing of silt cannot be avoided. However, it can be minimised if the fill is installed rapidly. The action of the expected contamination on the soil mechanics characteristics of the fill sand must be taken into account accordingly. The sand must be deposited in a manner that will prevent – as far as possible – the build-up of continuous silt layers. In case of heavy silt deposits, this can only be achieved by continuous, efficient operations, not even interrupted at weekends.

It has proven effective to process the sand surface with the water injection method while placing the sand fill. This makes it possible to avoid silt contamination of the sand to a great extent. Sand losses occurring in this way must be taken into account.

Should interruptions, and thus larger silt deposits, occur despite all efforts, the silt must be removed before further sand filling takes place, or it must be rendered harmless later by suitable measures. During any interruptions, a check must be made as to whether and where the surface elevation of the fill has changed.

In order to avoid causing active earth pressure on the waterfront structure in excess of the design load, the dredger pit must be filled in such a way that silted-up slopes occurring during the filling will have an inclination opposite to that of the failure plane of the active earth pressure slip wedge that will later act on the waterfront structure. The same applies similarly to the passive earth pressure side.

Please refer to the PIANC report of the 3rd Wave Commission, part A, Harbour Protective Structures, for explanations of the construction phases [171].

7.9.6 Checking the sand fill

Soundings are to be taken constantly during the sand filling, and the results recorded. In this way, changes in the fill surface caused by the filling processes itself and those due to the effects of the tidal currents can be determined to a certain extent. At the same time, these records show clearly, for example, how long a surface has remained unchanged, or if it has undergone prolonged exposure to the deposition of settling solids, so that measures can be initiated to eliminate any intrusive layers which may have formed.

The taking of surface samples from the fill area can be dispensed with only when there is speedy uninterrupted hydraulic filling and/or barge dumping.

After completion of the fill, and also during the work if required, the fill must be investigated and tested by means of core borings or other equivalent methods. These borings are to be sunk to the in situ soil at the dredger pit bottom.

An acceptance certificate forms the official basis for final design of the waterfront structure and any measures which may be required to adapt the design to conditions at the site.

7.10 Calculation and design of rubble mound moles and breakwaters (R 137)

7.10.1 General

Moles differ from breakwaters particularly as regards the type of use. Moles can support vehicular or at least pedestrian traffic. The crown is therefore generally higher than that of a breakwater, which may even lie submerged below the still water table. Breakwaters do not always have a shore connection.

When moles and breakwaters are constructed by a rubble mound method, reliable results from subsoil investigations are indispensable, as are careful analyses of the wind and wave conditions, currents and presence of littoral drift. For the sake of simplicity, the remaining discourse here refers only to rubble mound breakwaters.

The layout and cross-section of large rubble mound breakwaters are determined not only by their purpose but also by the method of construction.

7.10.2 Soil investigations, stability analyses, settlement and subsidence, plus advice on construction

The structure of the subsoil below proposed rubble mound breakwaters must be thoroughly investigated by means of soil mechanics field tests plus penetrometer tests and boreholes. In addition, comprehensive geophysical investigations could prove useful in certain instances.

271

Together with the results of soil mechanics laboratory tests, these form the foundation for the structural design and the stability calculations.

Loosely deposited non-cohesive soils in the area below the proposed breakwater must be compacted beforehand, soils of low bearing capacity must be replaced. The displacement of soft cohesive layers through deliberately overloading the soil so that the rubble penetrates into the subsoil, or blasting below the proposed rubble to displace the soil, can also be considered. However, both these methods can lead to larger differential settlements in the finished structure because the displacement achieved is never uniform.

Layers of silt/sludge are displaced by heaping the rubble and the resulting mud roll is removed because it can otherwise penetrate into the rubble and have a permanent detrimental effect on its properties.

Rubble mound breakwaters are to be checked for ground and slope failure. In doing so, the action of the waves is taken into account using the characteristic value of the "design wave". In seismic zones the risk of soil liquefaction must also be assessed.

The total of settlement due to rubble surcharge, subsidence due to wave action and penetration of the rubble into the subsoil, or subsoil into the rubble, can amount to several metres and when determining the height of the finished breakwater must be compensated for by specifying a higher cross-section. Rubble mound breakwaters are permeable, allowing the water to flow through them. Adequate filtering in the adjoining layers must be guaranteed in the case of an inhomogeneous structure.

7.10.3 Specifying the geometry of the structure

The primary initial parameters for specifying the cross-section of the breakwater are:

- design water levels
- significant wave heights, wave periods and the approach direction of the waves
- subsoil conditions
- materials available

The elevation of the crown is specified such that after settlement has concluded, overtopping of the waves is reduced to a minimum. It is proposed to set the breakwater crown at the following elevation:

$$R_c = 1.2 \cdot H_S + s$$

where:

R_c = freeboard height
(elevation of crown above still water level)

H_S =   significant wave height $H_{1/3}$ of design sea conditions
s =   total of expected final settlement, subsidence and penetration

When $R_c < H_S$, a considerable overtopping of the waves must be accepted.
When $R_c = 1.5 \cdot H_S$, overtopping of waves is almost ruled out.
The overtopping q can be calculated using [206] and [207].

Calculate the size of stones in the cover layer using tried-and-tested empirical calculations, e.g. after HUDSON, whose method is described below.
The size of the stones that can be obtained economically from quarries is often insufficient for the cover layer. In such cases, shaped concrete blocks, e.g. as listed in table R 137-1 and illustrated in fig. R 137-1, may be used instead.
Table R 137-1 also includes guide values for the sea-side embankment.

Minimum crown width: $B_{min} = (3 \text{ to } 4) \cdot D_m$

B_{min} =   minimum width of crown of breakwater [m]
D_m =   average diameter of single stone or block in cover layer [m]

$$D_m = \sqrt[3]{\frac{W}{\rho_s}} \text{ or } D_m = \sqrt[3]{\frac{W_{50}}{\rho_s}}$$

As in most cases fine-grain material is considerably cheaper than the coarse cover layer material, the majority of breakwaters have the following, traditional structure:
• core,
• filter layer, and
• cover layer
in accordance with fig. R 137-2.

However, it may also be the case, e.g. when transport routes are long, that the difference in price between core and cover layer materials is no longer significant. In such cases, particularly when using floating plant, building the breakwater using blocks of uniform size is another option. A base filter is particularly important with coarse-grain cores.
If a roadway is required along the top of a mole, the crown stones are often covered by a crown structure of concrete.
Crown walls on rubble mound breakwaters are very frequently used for deflecting overtopping and splashing water and for gaining access to moles. However, they form an intrusion which reveals the considerable settlement and differential settlement. Deformation and cracks in crown walls are therefore common.

7.10.4 Designing the cover layer

The stability of the cover layer, given the prevailing wave conditions, depends on the size, weight and shape of the armour blocks, as well as on the slope of the outer covering. In a series of tests over a period of many years, HUDSON developed the following equation for the required block weights [21, 172] and [174]. It has proven itself in practice and the equation is as follows:

$$W = \frac{\rho_s \cdot H_{des,d}^3}{K_D \cdot \left(\frac{\rho_s}{\rho_w} - 1\right)^3 \cdot \cot \alpha}$$

where:

W = block mass [t]
ρ_s = mass density of block material [t/m^3]
ρ_w = mass density of water [t/m^3]
$H_{des,d}$ = height of "design wave" [m], the characteristic value being multiplied by the partial safety factor
α = slope angle of cover layer [°]
K_D = shape and stability coefficient [1]

The preceding equation is valid for a cover layer composed of rocks of roughly uniform weight. The most common shape and stability coefficients K_D of rubble and shaped blocks for a sloped cover layer according to [21] have been compiled in table R 137-1. Fig. R 137-1 shows examples of common shaped blocks.

If settlement or subsidence movements in accordance with section 7.10.2 are possible, when selecting the type of cover layer components it should be borne in mind that additional tensile, bending, shear and torsion stresses can occur as a result of the form of the component. Owing to the high impact loads, the K_D values must be halved if the Dolos block is used. The following modified equation is recommended as per [21] for the design of an outer layer composed of graded rock, for "design wave" heights up to approx. 1.5 m:

$$W_{50} = \frac{\rho_s \cdot H_{des,d}^3}{K_{RR} \cdot \left(\frac{\rho_s}{\rho_w - 1}\right)^3 \cdot \cot \alpha}$$

where:

W_{50} = mass of a rock of average size [t]
K_{RR} = shape and stability coefficient
$\quad\quad K_{RR} = 2.2$ for breaking waves
$\quad\quad K_{RR} = 2.5$ for non-breaking waves

Table R 137-1. Recommended K_D values for the design of the cover layer at an allowed destruction up to 5% and only insignificant wave overflow, excerpt partially from [21].

Type of outer covering elements	Number of layers	Type of placing	Breakwater side K_D[1]		Breakwater end K_D		Slope
			Breaking waves[5]	Non-breaking waves[5]	Breaking waves	Non-breaking waves	
Smooth, rounded racks	2	random	1.2	2.4	1.1	1.9	1 : 1.5 bis 1 : 3
	3	random	1.6	3.2	1.4	2.3	1 : 1.5 bis 1 : 3
Angular rubble	2	random	2.0	4.0	1.9	3.2	1 : 1.5
	3	random	2.2	4.5	1.6	2.8	1 : 2
	2	carefully placed[2]	5.8	7.0	1.3	2.3	1 : 3
					2.1	4.2	1 : 1.5 bis 1 : 3
					5.3	6.4	1 : 1.5 bis 1 : 3
Tetrapode	2	random	7.0	8.0	5.0	6.0	1 : 1.5
					4.5	5.5	1 : 2
					3.5	4.0	1 : 3
Antifer Block	2	random	8.0	–	–	–	1 : 2
Accropode	1		12.0	15.0	9.5	11.5	bis 1 : 1.33
Coreloc	1		16.0	16.0	13.0	13.0	bis 1 : 1.33
Tribar	2	random	9.0	10.0	8.3	9.0	1 : 1.5
					7.8	8.5	1 : 2
					6.0	6.5	1 : 3
Tribar	1	uniformly placed	12.0	15.0	7.5	9.5	1 : 1.5 bis 1 : 3
Dolos	2	random	15.8[3]	31.8[3]	8.0	16.0	1 : 2[4]
					7.0	14.0	1 : 3

[1] For slope of 1 : 1.5 to 1 : 5.
[2] Longitudinal axis of rocks perpendicular to the surface.
[3] K_D values confirmed experimentally only for slope 1 : 2.
[4] If requirements are higher (destruction < 2%), the KD values must be halved.
[5] Slopes steeper than 1 : 2 are not recommended.
Breaking waves occur more often when still water depth in front of the breakwater decreases the wave height.

The mass of the largest rocks should be $3.5 \cdot W_{50}$ and that of the smallest at least $0.22 \cdot W_{50}$. Owing to the complex processes involved, the block masses according to [21] should generally not be reduced if the structure will be exposed to waves approaching at an oblique angle.

According to [172] it is recommended that the "design wave" assumed in the HUDSON equation should be at least $H_{des,d} = H_s$ where this value is generally extrapolated with the aid of peak value statistics over a fairly long period (e.g. 100-year return). Adequate wave measurement data must be available for the extrapolation. The significance of the design wave is obvious in that the required mass of individual blocks W increases proportionally with the cube of the wave height.

In the planning of a rubble mound breakwater, economic considerations may make it advisable to depart from the criterion for maximum protection of the outer covering against damage if extreme loads from wave action occur only very infrequently, or if rapid, extensive land accretion on the sea-side of a stretch of the breakwater towards the shore end makes the outer covering unnecessary there. The most economical design should be chosen if the capitalised maintenance costs together with the anticipated cost of repairing other damage which may occur in the port area are less than the increase in capital cost which would occur if the block masses were designed to resist an unusually high wave which occurs only infrequently. In this case, however, the general local maintenance possibilities and the expected time required to make repairs should always be taken into account.

Further calculation methods are to be found in [40] and [172]. [208] contains fundamental information on how the size, installation thickness and dry mass density of the stones used influence the stability of a bonded cover layer with respect to flow and wave loads as well as proposals for calculating technically equivalent cover layers.

The report of the PIANC working group 12 of the Permanent Technical Committee II for Coastal and Maritime Zones [172] makes use of, besides the HUDSON equation, the VAN DER MEER equation for the calculation of cover layers for rubble mound breakwaters.

These equations take into account the breaking form of the waves (plunging and surging/collapsing breakers) which according to the Irribaren index is calculated from the height and period of the waves. However, they also take into account storm duration, degree of destruction and breakwater porosity. The equations were derived from model tests with waves whose distribution with respect to wave height and wavelength corresponds to a natural wave spectrum. HUDSON, on the other hand, used only regular (uniform) waves for his tests in 1958 [202] and [203]. The method of calculation according to VAN DER MEER presumes a knowledge of many detail relationships, as can be found in [172]. The result of calculations for the cover layer size according to

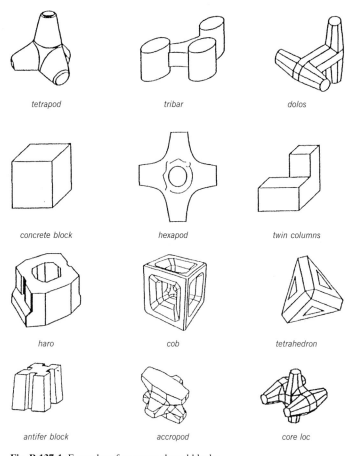

tetrapod	tribar	dolos
concrete block	hexapod	twin columns
haro	cob	tetrahedron
antifer block	accropod	core loc

Fig. R 137-1. Examples of common shaped blocks

HUDSON and VAN DER MEER differ considerably at the limits of their applicability.

The designer is therefore recommended to investigate the chosen cross-section as a whole by means of hydraulic model tests at an acknowledged hydraulics institute when large mole or breakwater structures are planned. The effect of the crown wall on the overall stability of the breakwater can be assessed at the same time.

7.10.5 Breakwater construction

In actual practice, as recommended in [21] and shown in fig. R 137-2, breakwaters constructed in three graded layers have proven effective.

277

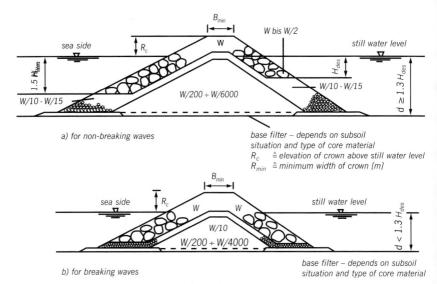

a) for non-breaking waves

base filter – depends on subsoil
situation and type of core material
R_c ≙ elevation of crown above still water level
R_{min} ≙ minimum width of crown [m]

b) for breaking waves

base filter – depends on subsoil
situation and type of core material

Fig. R 137-2. Filter-type breakwater construction in three stages

Symbols used in the figures:

W = mass of individual block [t]
H_{des} = height of "design wave" [m]

Rubble mound breakwaters should not be constructed as a single layer of rocks.

It is recommended that the sea-side slope be generally no steeper than 1 : 1.5.

Particular attention should be paid to the support for the cover layer, especially if it is not taken down to the base of the slope on the sea side. According to the requirements for the stability of the slope, an adequate retreat must be provided (fig. R 137-2a).

The filter rules are also to be obeyed vis-à-vis the subsoil. It is frequently practical to achieve this, especially at the base under the outer layers of large blocks, by using a special filter layer (gravel filter, geotextile, sand mat, geotextile filter-type container) because these are more stable when being installed.

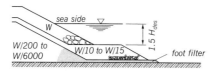

Fig. R 137-2a. Securing the base of the sea side of the breakwater

7.10.6 Execution of work and use of plant

7.10.6.1 General

The construction of rubble mound moles and breakwaters often requires the placing of large quantities of material in a comparatively short time under difficult local conditions due to weather, tide, sea conditions and currents. The interdependency of the individual operations under such conditions demands especially careful planning of the individual construction steps and use of plant.

The design engineer and the contractor should obtain precise information about the wave heights to be expected during the construction period. In order to do this, they require information about the prevailing sea conditions during the construction period, not only details of rare wave events. The duration of the wave heights H_s and H_{max} occurring over 12 months, for example, can be estimated according to fig. R 137-3 [175]. A reliable description of the wave conditions by means of a wave height duration line requires long observation periods.

The design of a breakwater must allow for construction that avoids major damage even when storms suddenly occur, e.g. layer construction with few steps.

When determining the capacity of the site installation or selecting the capacity of the equipment, realistic assumptions must be taken into consideration for possible interruptions due to bad weather.

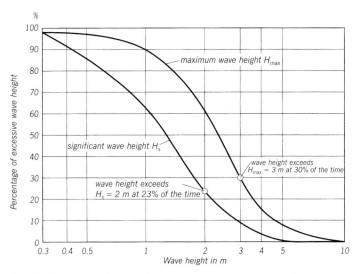

Fig. R 137-3. Wave height duration line. Duration of waves exceeding a certain wave height in a twelve month period, e.g. $H_s = 2$ m; $H_{max} = 3$ m

Depending on the local functions, the fill work is carried out:

(1) with floating plant,
(2) with land-based plant when the structure is advanced from the shore,
(3) with fixed scaffolding, mobile jack-up platforms, etc.,
(4) with cableways,
(5) with a combination of (1) to (4).

Construction methods with fixed scaffolding, mobile jack-up platforms and the like are employed preferably at especially exposed locations subject to the aggravating influence of wind, tide, wave action and currents. This applies to an even greater degree if there is no harbour of refuge available at or near the construction site.

7.10.6.2 Provision of fill and other materials

The provision of rubble and other fill materials requires prudent planning and depends on the sources and transport options, which often represent the biggest problems when procuring coarse materials.

7.10.6.3 Placing material with floating plant

When placing the fill with floating plant, the cross-section of the breakwater must be rated according to the plant. Dumping scows always require an adequate depth of water. Deck scows can be used for lateral displacement with lesser water depths. Today's computerised position-finding methods mean that floating plant is now able to achieve the precision which was previously only possible with land-based plant.

7.10.6.4 Placing fill material with land-based plant

The working level of the land-based plant should as a rule lie above the action of normal wave action and surf. The minimum width of this working plane is to be adapted to the requirements of the construction plant being used. When placing the fill with land-based plant, the material is brought to the actual site for advance filling using rear dumpers.

This construction method usually requires a core protruding out of the water with an overwide crown. The core, which acts as roadway, frequently has to be removed again to a certain thickness before the cover layer can be applied to ensure adequate keying and hydraulic homogeneity.

When the working level is narrow, it is frequently advantageous to use a portal crane for placing the material, as the material for the ongoing work can be passed through below the crane.

Rip-rap intended for installation by crane is brought to the site mostly in rock skips on platform trailers, trucks with special tailgates or low-bed trailers. Where narrower roadways are concerned, trailers are used that can be revered without turning.

Large rocks and prefabricated concrete elements are placed with multi-blade grabs or special tongs.

Electrical or electronic instruments in the operator's cab of the placing crane also facilitate true-to-profile filling under water.

Core fill and cover layer should follow each other as rapidly as possible in order to avoid unprotected core material from being washed away, or to at least minimise this danger, especially when the breakwater is constructed by advancing its head into the sea.

For further information, please refer to [173].

7.10.6.5 Placing material with fixed scaffolding, mobile jack-up platforms and the like

Placing the fill from a fixed scaffolding, mobile jack-up platform and the like or also with a cableway may be considered, above all, for bridging a zone with constantly heavy surf.

Otherwise, section 7.10.6.4 applies accordingly.

When using a jack-up platform, the placing progress generally depends on the efficiency of the crane. Therefore, the crane used should have a high lifting capacity in addition to the required reach.

The design should clarify which parts of the breakwater cross-section should be filled with a very calm sea and which can still be placed at a certain wave action. This applies both to the core material and to the precast concrete elements of the cover layer. When lowering precast concrete elements into position, even a slight swell can cause impact movements underwater because of the large weight of the parts, which in turn causes cracks and failure.

7.10.6.6 Settlement and subsidence

Uniform and minor settlement of rubble mound breakwaters are compensated for by increasing the height of the breakwater. Only after settlement has concluded, which should always be assessed by checking settlement levels, should the concrete on the crown be constructed, and then not in excessively long sections.

If significant differential settlement is expected, crown walls should not be constructed because their settlement later leads to a visually unsatisfactory overall appearance of the breakwater, even though its function and stability are not at risk.

7.10.6.7 Accounts for the materials used

The settlement and subsidence behaviour of such structures is difficult to predict. Therefore, when accounts are to be based on the drawings, it is recommended to stipulate easily maintained tolerances (±) depending on the shape of the mole and its individual layers right at the very start to compensate for the settlement and, if applicable, penetration volume (see section 7.10.2, paragraph 6) irrespective of the engineering methods chosen.

Furthermore, tenders should clearly stipulate whether accounts are to be based on the quantities used or measurements of the completed works. Settlement levels should be included in the tender when basing accounts on the quantities used.

If the subsoil as revealed by the soil investigations leads the designer to expect particular difficulties when accounting for the quantities used or in the case of the planned displacement of cohesive players (see section 7.10.2, paragraph 3), it is advisable to base the accounts on weight, provided no other solution specially suited to these subsoil conditions is possible [173]. Measurement of the completed works (highest point of a stone layer or the use of a sphere/hemisphere at the base of a levelling staff) must be specified.

7.11 Lightweight backfilling to sheet piling structures (R 187)

If light, durable filling material that is insoluble in water can be procured economically, or if suitable, non-decomposable, non-toxic industrial by-products or similar are available, they can occasionally be used as backfill material. They may make a sheet piling structure more economical if, in addition to the weight, the compressibility and consolidation values are also low, and the permeability will not decrease over time. Depending on the shear strength, the momentary stresses and anchoring strength can be reduced as a result of the low weight when installed.

If a construction with light backfill is under consideration, conventional solutions should always be examined by way of comparison and the effects of light backfilling should be examined in both positive and negative respects for the structure as a whole (e.g. reduced stability against slippage and slope failure, settlement of the surface, fire risk, etc.).

7.12 Soil compaction using heavy drop weights (R 188)

It is primarily soils with good water permeability that can be effectively compacted with heavy drop weights. This method can also be used on weakly cohesive and non-water-saturated cohesive soils because the compressible pore fluid tears open the soil structure due to the impacts and so creates additional drainage possibilities. Under especially heavy impacts, successful compaction has been proved even in water-saturated cohesive soils.

In order to assess the success of compaction measures in advance, the soil to be compacted must be carefully examined using soil mechanics methods. In zones that can be compacted only with difficulty or not at all, soil replacement is recommended. The effects of compaction should be ascertained beforehand in tests in order to establish the effects deeper in the ground as well as the economic feasibility and, if applicable, the environmental compatibility of the measures (noise, vibration of neighbouring structures or components such as canals, drains, etc.). Values for the target degree of density to be achieved can be established on the basis of the results of tests. Please refer to R 175, section 1.6.4, for verification of the degree of density attained.

Experience has also been gained in the use of heavy drop weights for compacting soils below the outer water level.

7.13 Consolidation of soft cohesive soils by preloading (R 179)

7.13.1 General

Frequently, only areas with soft cohesive soil strata with inadequate bearing capacity are available for extensions to harbours. However, in many cases the soft soil strata can be improved by preloading such that subsequent settlement resulting from use of the harbour facilities does not exceed the permissible tolerances. The shear strength of the soft strata is permanently improved in this way [112–114].

Under certain conditions preloading also compensates for later horizontal deformations so that, for example, pile foundations are no longer subject to unacceptable lateral pressures and deformations [115, 116].

Apart from preloading, later settlement can also be anticipated by vacuum (see section 7.13.7). To do this, vertical drains are installed within a defined volume of soil from a stable working level. These do not penetrate as far as a low-lying water-bearing stratum, but instead terminate approx. 0.5 to 1.0 m above this. Afterwards, the body of soil at the surface is covered by a synthetic sealing membrane on top of a layer of sand and a negative pressure is applied to the enclosed block of soil. This causes pore water and air to be extracted via the vertical drains and initiates a consolidation of the soft strata. The effect can be increased by building a mineral diaphragm wall at the sides. It is also possible to position the vertical drains in slits that terminate in a horizontal drain. After filling the slit with material of low permeability, a negative pressure can be applied to the horizontal drain.

7.13.2 Application

The aim of preloading is anticipate the settlement of soft soil strata that would otherwise take place during the later utilisation of the area (fig. R 179-1). The timescale for this depends on the thickness of the soft strata, their water permeability and the level of preload. However, preloading is effective only when there is sufficient time for the consolidation to take place. It is advisable to use the maximum possible preload and to apply this at such an early stage that all primary settlement

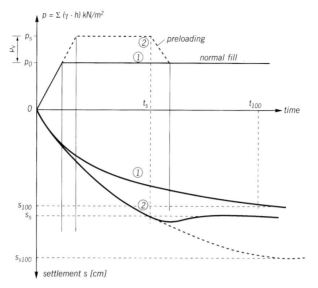

Fig. R 179-1. Relationship between settlement, time and surcharge (principle)

is completed before commencing any construction work for the actual waterfront structure. However, this is not always feasible.

Preloading in accordance with fig. R 179-1 is divided into:

a) The part of the preload fill that is to be constructed in the form of earthworks (permanent filling). It generates the surcharge stress p_0.

b) The excess preload fill, which temporarily acts as an additional surcharge with the preload stress p_v.

c) The sum of both fills (total fill), which yields the overall stress $p_s = p_0 + p_v$.

7.13.3 Bearing capacity of the in situ soil

The magnitude of the preload is limited by the bearing capacity of the in situ soft subsoil. In many cases it is at first only possible to add a thin layer and to wait for the consolidation due to this surcharge. A further preload layer can be added in the next step.

The maximum height of fill h of a preload can be estimated with

$$h = \frac{4\,c_u}{\gamma}$$

The shear strength c_u, and thus the permissible height h of further preload fill, increases with the consolidation.

284

Occasionally, however, especially when filling underwater, the covering results in a desirable or acceptable displacement of the in situ soft cohesive soil underneath. In this case it is accepted that a mud roll consisting of in situ soil is created ahead of the fill. This must be removed by dredging. Experience has shown that this method is only successful for very soft sediments; it involves hardly foreseeable risks during construction regarding the unintended covering of sediments.

7.13.4 Fill material

The permanent fill material must exhibit filter stability with respect to the soft subsoil. If applicable, filter layers or geotextiles should be installed before the permanent fill is added. Otherwise, the required quality of the permanent fill material is governed by the intended use.

7.13.5 Determining the height of preload fill

7.13.5.1 Principle

The requirements with respect to the preload fill arise principally from the construction time available. The dimensioning is based on the consolidation coefficient c_v. However, reliable figures for the consolidation coefficient c_v can only be obtained through test fills. If c_v values are derived from the time–settlement curves of compression tests, experience shows that only a rough estimate of the consolidation time is possible. Therefore, such values should be used for preliminary appraisals only and the values must be checked by measuring the actual settlement during filling operations. If the mean stiffness module E_s and the mean permeability k of the in situ soil are known, c_v can also be calculated from:

$$c_v = \frac{k \cdot E_s}{\gamma_w}$$

If this equation is used, it must be borne in mind that the permeability coefficient k, too, may embody considerable scatter, meaning that this procedure can be recommended only with reservations, and then only for preliminary appraisals.

The most reliable method is to determine the consolidation coefficient c_v in advance from a trial fill, during which the settlement and also, if possible, the pore water pressure progression can be measured. c_v is then calculated using fig. R 179-2 according to the following equation:

$$c_v = \frac{H^2 \cdot T_v}{t}$$

where:

T_v = specific consolidation time
t = time
H = drainage length of soft soil stratum

For $U = 95\%$, i.e. virtually complete consolidation, $T_v = 1$ and thus:

$$c_v = \frac{H^2}{t_{100}}$$

7.13.5.2 Dimensioning of preload fill

For the dimensioning of the preload fill, the drainage length of the soft soil stratum and the c_v value must be known. Additionally, the consolidation time t_s (fig. R 179-1) must be specified (construction schedule). $t_s/t_{100} = T_v$ is determined with $t_{100} = H^2/c_v$ and the required degree of consolidation $U = s/s_{100}$ under preloading is determined with the aid of fig. R 179-2. The 100% settlement s_{100} of the permanent fill p_0 is determined with the aid of a settlement calculation in accordance with DIN 4019.

The magnitude of the preload p_v (fig. R 179-1) is then derived in accordance with [114] as:

$$p_v = p_0 \cdot \left(\frac{A}{U} - 1 \right)$$

A is the relationship of the settlement s_s after removing the preload to the settlement s_{100} of the permanent fill: $A = s_s/s_{100}$.

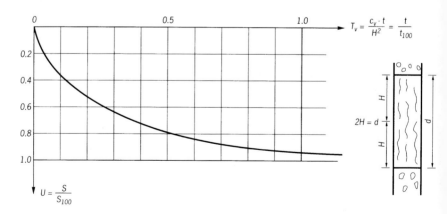

Fig. R 179-2. Relationship between time factor T_v and degree of consolidation U

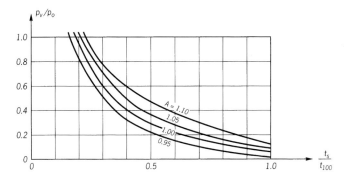

Fig. R 179-3. Determination of preload p_v as a function of time t_s

A must be equal to or greater than 1 if complete elemination of the settlement is to be achieved (fig. R 179-3).

Thus, for example, with a stratum drained on both sides, a depth $d = 2 \cdot H = 6$ m, a c_v value $= 3$ m^2/year and a specified consolidation time of $t_s = 1$ year:

$$t_{100} = \frac{3^2}{3} = 3 \text{ years}$$

$$t_s/t_{100} = 0.33$$

$$U = 0.66 \, [1]$$

As shown in fig. R 179-3, with $A = 1.05$:

$$p_v = p_0 \cdot \left(\frac{1.05}{0.66} - 1 \right) = 0.6 \cdot p_0$$

Please note that p_v is limited by the bearing capacity of the in situ soil (see section 7.13.3).

7.13.6 Minimum extent of preload fill on the surface

In order to save preload material, the soil stratum to be stabilised is generally preloaded in sections. The sequence of preloading is governed by the construction schedule. To achieve effective dissemination of stress in the soft strata which is as uniformly distributed as possible, the area of the fill may not be too small. As a guide, the smallest side length of the preload fill should equal two to three times the sum of the depths of the soft stratum and the permanent fill.

287

7.13.7 Design of vacuum consolidation with vertical drains

When designing ground improvement works using the vacuum consolidation technique, please take into account the following factors:

- With conventional preloads, the increase in the shear strength is achieved through an additional load (total stresses). In contrast to this, the vacuum method achieves the consolidation by reducing the pore water pressure, the total stresses remain unchanged.
- Installing the working level (simultaneously with the drain layer) can lead to problems if the bearing capacity of the in situ soil does not permit a layer thickness of at least about 0.8 m.
- The water within the drain layer is collected and drained away with the help of a horizontal drainage system.

Special pumps that can pump, or rather extract, water and air simultaneously are connected to the drain layer; these generally achieve a maximum vacuum of 75% of atmospheric pressure (approx. 0.75 bar). When calculating the total, effective and neutral stresses in the soil, the atmospheric pressure P_a must be considered as follows:

$$\sigma = z \cdot \gamma + h \cdot \gamma_r + P_a$$
$$u = z \cdot \gamma_w + P_a$$
$$\sigma' = \sigma - u = z \cdot \gamma' + h \cdot \gamma_r$$

Once the vacuum block has been established and is in operation, the neutral stress u decreases by a value equal to the atmospheric pressure P_a and the total stress σ remains unchanged. As a result, the effective stress σ' in the soil increases during the vacuum and the following relationship applies:

$$\sigma'_{vacuum} = \sigma' + P_a$$

In doing so, the effective stress in the soil increases isotropically by a value equal to the atmospheric pressure.

The pumping capacity n reduces the atmospheric pressure P_a by the value

$$\Delta\sigma = n \cdot P_a$$

During the consolidation, the increase in the shear strength $\Delta\tau$ depends on the degree of consolidation and increases by the value

$$\Delta\tau = U_t \cdot (\tan\varphi' \cdot \Delta\sigma)$$

where:

$\Delta\sigma$ = negative pressure applied

$$U_t = 1 - \frac{\Delta u_t}{\Delta \sigma} \quad \text{degree of consolidation at time } t$$

where Δu_t = pore water pressure difference at time t

The final settlement and the progression of the settlement can be calculated using the consolidation theory of TERZAGHI and BARRON. The shear strength of cohesive soils after consolidation corresponds to that due to an equivalent sand surcharge. The temporary strength of non-cohesive soils due to the application of the negative pressure is lost again when the vacuum pump is switched off.

7.13.8   Implementation of vacuum consolidation with vertical drains

Firstly, a working level (thickness > approx. 0.8 m) is installed. Vertical drains (a size equivalent to $\varnothing$ 5 cm is usual but flat and round drains can be used in combination) are then installed from this layer extending down to about 0.5 to 1.0 m above the underlying stratum.

To maintain the vacuum, it is vital to avoid or prevent hydraulic contact with intermediate water-bearing sand strata in the subsoil (e.g. by means of deep trenches or diaphragm walls). Therefore, in inhomogeneous subsoil conditions indirect investigations (e.g. static penetrometer tests) during the work are necessary in order to establish the depth quickly beforehand.

In subsoil conditions with major variations, test installations without drain material can be carried out on a wide grid (e.g. 10×10 m) from which the depth can be determined together with the indirect investigations. The test installations enable the plant operator to detect increased penetration resistance of intermediate sand strata or the topside of in situ sands.

Any defects in the covering impervious membrane are difficult to locate and repair. Therefore, stony, angular material should not be used for the drainage layer/fill. Flooding with water can be used to protect the membrane.

In special cases the impermeable soft stratum can itself function as the membrane maintaining the vacuum. To do this, the vertical drains are interconnected hydraulically by a horizontal drain approx 1.0 m below the top of the soft stratum.

The settlement, the negative pore water pressure in the soil and the horizontal deformations on the edges of the vacuum block must be observed for the entire time because settlement and degree of consolidation are generally difficult to forecast with any precision. By observing the settlement and stresses, the numerical models can be calibrated by the real conditions in order to calculate reliable residual settlement values. The horizontal deformations of the total system are approximately zero in the centre of the block. Deformations in highly compressible soils are very difficult to observe and inclinometers are unreliable. The thickness of the vacuum block is limited by economic considerations on the one hand and the feasible depth of vertical drains (usually 40–50 m) on the other.

7.13.9 Checking the consolidation

The consolidation can be checked by measuring the settlement and the pore water pressure. At the edges of the fill inclinometers can indicate whether the bearing capacity of the subsoil has been exceeded.

A limiting value should be specified for the settlement rate at which the preload fill can be removed, e.g. in mm per day or cm per month.

7.13.10 Secondary settlement

It should be noted that preloading can anticipate secondary settlement to only a very small extent (in highly plastic clays, for example). If secondary settlement is expected on a fairly large scale, special supplementary investigations are necessary.

7.14 Improving the bearing capacity of soft cohesive soils by using vertical elements (R 210)

7.14.1 General

Grouted or ungrouted ballast or sand columns, resting on low-lying loadbearing strata, are frequently constructed for the foundations to earthworks on soft cohesive soils.

The columns carry the vertical loads and transfer them to the loadbearing subsoil, and in doing so they are supported by the surrounding soil. Therefore, the soil must exhibit a strength of at least $c_u > 15$ kN/m^2. Grouted vibro-replacement stone columns need such support during installation only.

Very soft, organic cohesive soils cannot guarantee the necessary degree of support and therefore such foundation systems can be built in this soil type only when the lateral support is achieved by other means, e.g. geotextiles. Ballast and sand columns exhibit drainage properties similar to those of vertical drains (see R 93, section 7.7.1) and hence increase the initial shear strength of the in situ soil.

7.14.2 Designing pile-type foundation systems

Vibroflotation is a method of improving the subsoil that has been used and accepted for a long time. However, its use in non-cohesive, compactable soils is limited. Even just small proportions of silt can prevent compaction by vibration because the fine-grain soil particles cannot be separated from each other by vibration.

Vibro displacement compaction was developed for these soils. Soil improvement by this means is achieved, on the one hand, by displacing the in situ soil and, on the other, by introducing columns of compacted, coarse-grain material into the in situ soil. But even vibro displacement compaction requires the support of the surrounding soil. As a guide to the applicability of this method, the undrained shear strength c_u should generally be greater than 15 kN/m^2.

The loadbearing elements arranged on a grid are embedded in the low-lying loadbearing soil strata. The load transfer into the loadbearing elements takes place via a layer of sand over the tops of the elements. This layer of sand is hence intrinsic to the system and is generally known as the loadbearing layer.

The principle of this loadbearing system is based on the fact that a stiff support is generated at the head of each loadbearing element. This concentrates the stress over the loadbearing elements and relieves the surrounding cohesive soil (vault formation).

The loadbearing effect of the loadbearing layer can be enhanced by including geotextile reinforcement above the loadbearing elements, spanning like a membrane over the soft strata. In order to achieve a sufficient margin of safety against punching through the loadbearing layer, the inclusion of a geotextile layer is recommended as minimum reinforcement.

When designing geotextile-reinforced earthworks on pile-type foundation elements, take the following factors into account:

- The pile diameter is generally 0.6 m and the spacing between the piles arranged on a regular grid is approx. 1.0 to 2.5 m.

- The geotextile reinforcement is generally placed 0.2 to 0.5 m above the heads of the columns. Additional layers of reinforcement are placed at a distance of 0.2 to 0.3 m above the first layer. This prevents the failure mode "sliding of one geotextile on another geotextile".

- The analytical verification of the stability for the construction phases and the final condition can be carried out on the basis of DIN 4084 using curved failure planes or rigid body failure mechanisms. In doing so, the layout of the pile-type foundation should be converted into a planar equivalent system with wall-like plates but maintaining the area ratio. The additional resistances due to the "truncated" piles and the geotextile reinforcement may be taken into account.

- The load on the pile elements and the surrounding soft soil sensitive to settlement depends on the degree of load redistribution E in the loadbearing layer. The force F_P due to the load redistribution E that must be carried by one pile as a result of the vault effect, taking into account the area of influence A_E, is calculated as follows:

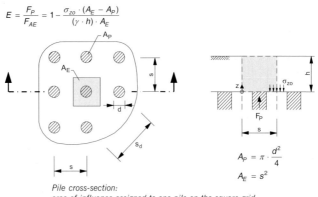

$$E = \frac{F_P}{F_{AE}} = 1 - \frac{\sigma_{zo} \cdot (A_E - A_P)}{(\gamma \cdot h) \cdot A_E}$$

Pile cross-section:
area of influence assigned to one pile on the square grid

Fig. R 210-1. Loading on the piles [241]

$$E = \frac{F_P}{F_{AE}} = 1 - \frac{\sigma_{zo} \cdot (A_E - A_P)}{(\gamma \cdot h) \cdot A_E}$$

This is valid for:

rectangular grid $\qquad\qquad$ triangular grid

$F_P = E \cdot (\gamma \cdot h + p) \cdot s_x \cdot s_y \qquad F_P = E \cdot (\gamma \cdot h + p) \cdot 1/2\, s_x \cdot s_y$

The load redistribution E can be determined theoretically with the help of vaulting in which it is assumed that a redistribution of stress takes place only within a limited zone above the soft stratum. For more details please consult [241]. The load redistribution in the loadbearing layer here depends directly on the shear strength of the material of the loadbearing layer.

- The membrane effect of the geotextile layer can also further relieve the soft strata.
- Please refer to [241] and [242] for more details of the calculation.
- The membrane effect of the geotextile reinforcement is improved by laying the reinforcement as close as possible over the almost rigid piles. To prevent shearing of the geotextile, provide a levelling layer over the heads of the piles so that the material does not rest directly on the heads of the piles.

7.14.3 **Construction of pile-type foundation systems**
For constructing the pile elements, a working level with adequate loadbearing capacity for the necessary plant is essential. This can be achieved through improving the soil at the surface or by replacing the topmost layer of soil.

The following forms are possible as pile elements:

(1) Bored piles to DIN EN 1536 or displacement piles to DIN EN 12 699
(2) Vibrated stone columns and grouted or partly grouted vibro-replacement stone columns

Vibro-replacement stone columns are produced with a bottom-feed vibrator (a vibrating probe suspended from a crawler crane). After reaching the design depth or the loadbearing subsoil, the bottom-feed vibrator is raised by a few decimetres and coarse-grain material is driven out of a chamber in the vibrator by compressed air or water. Afterwards, the vibrator is lowered again, which compacts the material just added. The in situ soil is thereby displaced and also compacted. This process is repeated in several stages until a compacted pile element is created from the bottom upwards.

To prevent the pile material escaping into the surrounding soil, vibro-replacement stone columns are recommended only for cohesive soils with an undrained shear strength $c_u \geq 15$ kN/m^2.

Grouted or partly grouted vibrated stone columns make use of the same method as the vibro-replacement stone column. However, the coarse-grain material is mixed with a cement suspension as it is installed; alternatively, it can be entirely of concrete. The same rule applies here: an undrained shear strength $c_u \geq 15$ kN/m^2 is recommended. However, $8 < c_u < 15$ kN/m^2 is permissible for intermediate layers with a soft consistency, provided the thickness of the intermediate layer does not exceed 1.0 m.

The soil displacement gives rise to a pore water pressure difference which, however, is quickly dispersed owing to the drainage effect of the load-bearing elements. In grouted or partly grouted vibrated stone columns or vibrated concrete columns, the drainage effect is at best very limited.

7.14.4 **Design of geotextile-encased columns**

The above foundation systems are not suitable for very soft cohesive soils with $c_u < 15$ kN/m^2. In such cases the lateral support provided by the soil must be achieved with additional measures, e.g. by means of a geotextile sleeve in which the sand is placed ("sand-filled geotextile-encased column").

When designing earthworks using this system, please take into account the following factors:

- The geotextile-encased column is a flexible loadbearing element which can adapt to horizontal deformations.
- Practical experience on site and in the laboratory has shown that there is no risk of punching because the settlement behaviour of the geotextile-encased column is equal to that of the surrounding soil.

- Residual settlement can be accelerated by applying an overload greater than the sum of all the later loads (excess fill).
- The sand columns are arranged on a uniform grid (usually triangular). The spacing between the columns is approx. 1.5m to approx. 2.0 metres depending on the grid.
- The diameter of the columns is generally 0.8 m.
- It is recommended to install a layer of horizontal geotextile reinforcement above the geotextile-encased columns. This increases the stability during critical phases of construction and reduces the expansion deformations.
- The stress concentration over the columns leads to an increase in the overall shear strength in the columns and to a decrease in the surrounding soft strata. The shear strength effect should be taken into account by including so-called equivalent shear parameters after [238] and the pore water pressure due to the surcharge.
- The design of the geotextile sleeve can be based on [239] in such a way that the short-term strength F_K of the geotextile is reduced by various factors Ai and the factor of safety γ, and compared with the calculated tensile forces in the geotextile.

$$ F_d = \frac{F_K}{A_i \cdot A_{i+1} \cdot \gamma} $$

The calculation of the respective tensile force F_d in the geotextile is based on the compression, or rather settlement, of an individual column as a result of the effective stress concentration over the column and the horizontal support of the segment of surrounding soft soil. There ensues a volume-based bulging of the column over the depth, which in turn causes stretching of the geotextile. Further details and methods of calculation can be found in [238] and [240].

- Measurements of the pore water pressure difference and stresses in and above the soft strata during and after construction work allow the true state of consolidation to be checked if so required. In all other respects the consolidation characteristics of the column are similar to a large-diameter vertical drain.

7.14.5   Construction of geotextile-encased columns

A soil displacement or soil excavation method is used to install the columns. Whereas with soil excavation an open pipe down to the level of loadbearing subsoil is introduced through vibration and then the soil is excavated from within the pipe, here the soil is displaced using a steel pipe with a conical, closable flap which is vibrated into the ground. After installing the pipe, a pre-assembled geotextile sleeve (either woven directly as a tube or factory-sown to form a tube) is suspended in position.

Afterwards, the column is first filled loosely with a drainage-compatible material (e.g. sand, gravel, ballast) and the pipe extracted while been vibrated. This compacts the column material.

The initial shear strength of the soft soil increases when using the more economic soil displacement method. This may need to be checked by measuring the undrained shear strength before and after installing the columns.

Soil displacement is recommended for contaminated soils because this avoids the need to excavate and dispose of the soil. A brief increase in the pore water pressure after establishing the column must be expected when using soil displacement; however, this is quickly dispersed by the drainage effect of the column. In addition, the soil displacement causes a temporary lifting of the top level of the soft stratum. The raising of the surface of the site increases with the number of rows of columns [244]. The columns are constructed either from a working level installed previously, or from a floating pontoon.

7.15   Installation of mineral bottom seals under water and their connection to waterfront structures (R 204)

7.15.1   Definition

A mineral underwater seal comprises natural, fine-grained soil composed of or prepared in such a way that it either possesses very low permeability without the addition of other substances to achieve the sealing effect, or achieves the desired properties through the inclusion of additives (seals of grouted rubble stone layers are covered in section 12.1.3).

7.15.2   Installation in the dry

Mineral seals installed in the dry are dealt with in detail in [155]. Recent additions are geosynthetic clay liners [245, 237].

7.15.3   Underwater installation

7.15.3.1   General

When watertight harbour basins or waterways are to be deepened or expanded, it is often necessary to install bottom seals underwater, sometimes while waterborne traffic is operating. Measures are to be taken against the prejudicial effects in respect of the stability of the structures involved and the effects on ground water quality and level of the water penetrating the subsoil for the event of resulting temporary seal failure. Special requirements are to be placed on the seal material to be installed, depending on the installation procedure.

7.15.3.2 Requirements

Mineral seals installed underwater can be mechanically sealed to at best only a limited extent. They must therefore be prepared homogeneously and installed in a consistency such that a uniform sealing action is guaranteed from the outset, the material installed evens out unevenness of the level plane without tearing, withstands the erosion forces of waterborne traffic even during installation, and is capable of guaranteeing the watertightness of the connections to the waterfront structures, even if deformation of these structures occurs.

If the seal is to be created on slopes, the installation strength must also be sufficient to guarantee the stability on the slope.

Sufficient resistance against the following must be verified for mineral seals:

- the danger of disintegration underwater of the newly installed sealing material,
- erosion from the backflow of the waterborne traffic adapted to the site conditions,
- the breaching of the seal in the form of thin channels with coarsely grained subsoil (piping),
- slip on slopes inclined at up to 1 : 3, and
- the stresses when the seal is covered with filters and hydraulic structure blocks.

Seals made from natural soil materials without additives generally satisfy these requirements, provided the sealing material fulfils the following conditions (geosynthetic clay liners must be considered separately):

– proportion of sand (0.063 mm $\leq d$)	< 20%
– proportion of clay ($d \leq 0.002$ mm)	> 30%
– permeability	$k \leq 10^{-9}$ m/s
– undrained shear strength	15 kN/m$^2 \leq c_u \leq 25$ kN/m^2
– depth (with 4 m water depth)	$d \geq 0.20$ m

With mixtures produced with certain additives and a proportion of cement, which solidify after installation, the flexibility of the seal in the final state must not be impaired, this being checked by means of tests, e.g. in accordance with [156].

Special investigations are necessary if mineral seals are installed at greater water depth, or on gravel-like soils, or with large subsoil pore sizes or slopes steeper than 1 : 3, and also in the case of dimensioning the sealing material in respect of self-closure upon gap formation and the sealing effect of butt joints (see [157] and [158]).

See [132] with regard to suitability and monitoring tests.

Several procedures, some protected by patents, are currently available for the soft mineral seals, which are today only installed in a single layer

[237]. In all the processes it is recommended to operate the installation plant from elevated platforms.

7.15.4   Connections

The connection of mineral bottom seals to structures is generally by means of a butt joint, the sealing material generally being pressed with suitable devices adapted to the shape of the joint (e.g. sheet piling profile). An amount of sealing agent corresponding to the line of the joints is applied in advance with suitable equipment. Since the sealing effect is generated by the direct contact stress between the sealing agent and the joint (see [157] and [158] in this respect), the application process must be undertaken with great care.

The length of contact between a mineral seal and a sheet pile wall or structural component should be at least 0.5 m for a sealing material with an undrained shear strength $c_u = 25$ kN/m^2, and at least 0.8 m in the case of a higher strength. The strength of a mineral seal should not exceed $c_u = 50$ kN/m^2 at the junction with a wall. The connection of a geotextile sealing membrane makes use of a sealing fillet made from a suitable sealing material with the contact length between this and the sealing membrane being at least 0.8 m.

8 Sheet piling structures

8.1 Material and construction

8.1.1 Design and driving of timber sheeting (R 22)

8.1.1.1 Range of applications
Timber sheeting is advisable only when the existing soil is favourable for driving and when the required section moduli are not too large. In permanent structures, the pile tops must be at an elevation which is continuously wet, so that they will not decay. The risk of infestation from marine borers must be prevented. Timber sheeting is only advisable if no other materials can be used because of local conditions.

DIN 1052 – timber structures – is to be applied accordingly. Connecting elements of steel must be hot galvanised at least or have an equivalent corrosion protection.

8.1.1.2 Dimensions
Timber piles are usually made of highly resinous pine wood, but also of spruce and fir. Standard dimensions and design of the piles and the joints are indicated in fig. R 22-1. For straight or chamfer-tongued joints in general the tongue is made several millimetres longer than the depth of the groove, so that a snug fit develops as the pile is driven.

So-called "corner piles" are placed at the corners. These are thick squared timbers in which the grooves for the adjoining piles are cut corresponding to the corner angle.

8.1.1.3 Driving
Mostly double piles joined by pointed dogs are driven. Staggered driving or driving in panels is always required (see R 118; section 8.1.11) to protect the piles and increase the watertightness of the wall. The point of each sheet pile is bevelled on the tongue side so that the pile is pressed against the wall already standing. The cutting edge has a shoe which is reinforced by a 3 mm thick steel plate in soil which is difficult to drive into. The top of the steel pile is always protected from splitting by a conical ring of flat forced steel about 20 mm thick.

8.1.1.4 Watertightness
Timber sheet piling becomes somewhat tight as the result of swelling of the wood. As with other sheet piling, the watertightness of excavation enclosures in free water can be improved with ashes, fine slag, wood shavings and other environmentally compatible materials which are sprinkled in the water on the outside of the sheet piling while pumping out the excavation. Large leaky areas can be temporarily repaired by

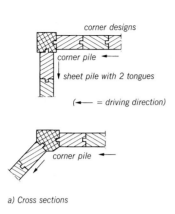

corner designs

corner pile ←

sheet pile with 2 tongues

(← = driving direction)

corner pile ←

a) Cross sections

Rule of the thumb for pile thickness t
t (cm) = 2 x l (m)
Pile width b = approx. 25 cm
Pile length l ≤ 15 m

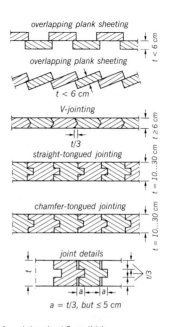

overlapping plank sheeting

t < 6 cm

overlapping plank sheeting

t < 6 cm

V-jointing

t/3

straight-tongued jointing

chamfer-tongued jointing

joint details

a = t/3, but ≤ 5 cm

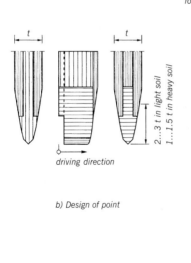

driving direction

b) Design of point

2...3 t in light soil
1....1.5 t in heavy soil

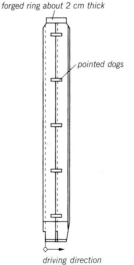

forged ring about 2 cm thick

pointed dogs

driving direction

c) Double sheet pile

Fig. R 22-1. Timber sheet piling

Table R 22-1. Characteristic values of tropical hardwoods

Common name of wood	Botanical name	Mean density	Mois-ture	Abso-lute com-pressive strength	Elas-ticity module	Abso-lute bending strength	Shear strength	Durability as per TNO*		Teredo resistance
								in moist soils, in water or water changing zone	exposed to the elements	
		kN/m³		MN/m²	MN/m²	MN/m²	MN/m²	years	years	
Demerara Greenheart	Ocotea rodiaei	10.5	dry / wet	92 / 72	21 500 / 20 000	185 / 107	21 / 12	25	50	Yes, but somewhat less than Basralocus
Opepe (Belinga)	Sarcoce-phalus	7.5	dry / wet	63 / 50	13 400 / 12 900	103 / 92	14 / 12	25	50	
Azobe (Ekki Bongossi)	Lophira procera	10.5	dry / wet	94 / 60	19 000 / 15 000	178 / 119	21 / 11	25	50	Yes, but limited
Manbarklak (Kakoralli)	Eschweilera long pipes	11.0	dry / wet	72 / 52	20 000 / 18 900	160 / 120	13 / 11	15–25	40–50	Yes
Basralocus Angelique	Dicorynia paraensis	8.0	dry / wet	62 / 39	15 500 / 12 900	122 / 80	11.5 / 7	25	50	Yes
Jarrah	Eucalyptus marginata	10.0	dry / wet	57 / 35	13 400 / 9 900	103 / 66	13 / 9	15–25	40–50	Yes, but limited
Yang	Dipterocarpus Afzelia	8.5	dry / wet	54 / 39	14 600 / 12 300	109 / 80	11 / 10	10–15	25–40	No
Afzelia (Apa Doussie)	Afzelia africana	7.5	dry / wet	66 / 30	13 000 / 9 900	106 / 66	13 / 9	15–25	40–50	No

*) TNO = Nijverheisorganisatie voor Toegepast Natuurwetenschappelijk Onderzoek

placing canvas over them, but must be permanently sealed by divers with wooden laths and caulking for permanency.

8.1.1.5 Protecting the wood

Since native wood is usually well protected against rotting only under water, timber sheet piling subject to permanent loads must lie below the groundwater table and below low water in waterways. In tidal areas, they may extend to mean water level. In other cases, timber must be protected by means of an environment-friendly impregnating agent. This also applies to areas at risk of attack by shipworm (*Teredo navalis*) where there are marine borers, i.e. generally in water with a salt content exceeding 9‰. Tropical hardwoods are more resistant under such circumstances. The data for tropical hardwoods are listed in table R 22-1.

8.1.2 Design and driving of reinforced concrete sheet piling (R 21)

8.1.2.1 Range of applications

Reinforced concrete sheet piling may be used only with the assurance that the sheet piles can be driven into the soil without damage and with tight joints. Their use should however be restricted to structures in which requirements in respect of watertightness are not high, e.g. groynes etc.

8.1.2.2 Concrete

When selecting a type of concrete for reinforced concrete sheet piling, take into account the respective exposure classes in relation to the ambient conditions (DIN 1045-1, section 6.2).

8.1.2.3 Reinforcement

The cover to the main structural reinforcing bars in freshwater and seawater shall be at least $c_{min} = 50$ mm, with a nominal dimension of $c_{nom} = 60$ mm, and hence larger than the values given in DIN 1045. In other respects, reinforced concrete sheet piling is designed according to DIN 1045, also taking into account the loading case of lifting the piles out of the precasting forms and lifting them before driving. Sheet piles generally have main longitudinal reinforcement of grade BSt 500 S. In addition, the sheet piles have a helical reinforcement of grade BSt 500 S or M or of wire rod ∅ 5 mm. Details of the layout are shown in fig. R 21-1.

8.1.2.4 Dimensions

Piles to be driven must have a minimum thickness of 14 cm but should in general not be thicker than 40 cm for reasons of weight. Apart from the driving conditions, the thickness depends on construction details and structural requirements. The normal pile width is 50 cm, but the

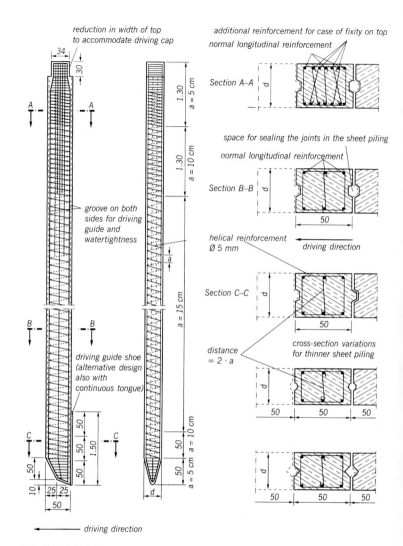

Fig. R 21-1. Reinforced concrete sheet piling

width of the pile top is reduced to 34 cm if possible so that it fits into the normal driving cap. The piles are made up to 15 m long, or in exceptional cases 20 m long.

Normal groove shapes are shown in fig. R 21-1. The width of the grooves is up to $^1/_3$ of the sheet pile thickness, but not bigger than 10 cm. The groove runs continuously to the lower end of the pile on the leading side. On the opposite side, the point has a tongue about 1.50 m long which fits the groove. A groove runs upward from this tongue (fig. R 21-1). The tongue guides the points as the pile is driven. It may run from the point to the top of the pile and help create tightness, but only in non-cohesive soil if the soil is of such a nature that a filter is automatically created behind every joint after minor leaching has taken place, to prevent any further leaching.

8.1.2.5 Driving

The point of the pile is bevelled to about 2 : 1 on the leading side, so that the pile presses against the sheeting already driven. This design is also retained for jetted piles. Driving is facilitated if the piles have a wedge-shaped point. The piles are always driven as individual piles, and in the tongue and groove design with the groove side forward. A driving cap should be used if the driving is done with a pile hammer. Piles should be driven with the heaviest (slow-stroke) hammers possible and with a small drop. A similar effect is achieved when using an appropriate hydraulic hammer whose impact energy can be carefully adjusted to suit the respective driving energy needed.

In soils of fine sand and silt, better progress will be made if jetting is used.

8.1.2.6 Watertightness to prevent loss of soil

If the sheet pile has only a short tongue, sufficient space for the insertion of joint leakproofing is available. Before this is placed, the grooves should always be cleaned with a jet of water. The space is then filled with a good concrete mixture using the contractor method. With large grooves, a jute sack filled with freshly mixed concrete can be lowered into place. A seal of bituminised sand and fine gravel can also be used. In any case, the leakproofing is to be so inserted that it fills the entire space without leaving any gaps.

The seal is best achieved with a C-shaped special section, with which a cavity of the appropriate size is obtained and which can also be lightly prestressed.

However, especially in surrounding non-cohesive, finely grained soil, above all with tidally-induced water level fluctuations, the achievable seal is restricted, so that the requirements relating to the seal cannot be too high, see section 8.1.2.1. Subsequent resealing of reinforced concrete sheet piling is possible only at great expense.

8.1.3 Steel sheet piling (R 34)

8.1.3.1 General

Steel sheet piling is frequently an ideal solution in regarding both, structural considerations and driving conditions (see R 106, R 119, R 200, R 201, sections 6.4, 6.5, 6.8 and 6.9), which is also capable to absorbe localised overloading without endangering overall stability. Damage caused by collisions and accidents often can be repaired simply. The sheet piling can be made watertight using measures according to recommendation R 117, section 8.1.20.

Corrosion and corrosion protection measures are dealt with in recommendation R 35, section 8.1.8.

8.1.3.2 Selecting of profile and steel grade

In addition to structural considerations and economic aspects, the choice of type and section depends on the driving conditions existing at the site, the stresses when being installed and in operation, the acceptable deflection, the watertightness of the interlocks and the allowable minimum wall thickness, whereby especially possible mechanical loads on the sheet piling from berthing manoeuvres of ships and ship groups and from sand abrasion effects are to be taken into account (R 23, section 8.1.9).

Impeccable driving of the piling and, in permanent structures, adequate durability must be assured.

Combined steel sheet piling (R 7, section 8.1.4) is often economical for larger section moduli. Section strengthening with welded-on plates or connectors can be taken into consideration, just so the selection of a higher-strength, weldable steel grade with minimum yield point above the values stated in table R 67-1, section 8.1.6.2. The choice of higher strength steels is to be agreed with the manufacturers on placing the order. Otherwise, recommendation R 67, section 8.1.6 is to be observed.

8.1.4 Combined steel sheet piling (R 7)

8.1.4.1 General

Combined steel sheet piling is constructed by the alternate placing of different types of sections or elements. Long and heavy sections designated as 'primary elements' alternate with shorter and lighter 'secondary elements'. The most customary wall shapes and wall elements are described in detail in R 104, section 8.1.12.

8.1.4.2 Structural system

Usually, only the primary elements can be assumed to carry vertical loads.

The horizontal loads that act directly on the secondary elements must be transferred to the primary elements.

Experience has shown that unwelded secondary elements of Z-section with a wall thickness of 10 mm are stable with a clear spacing between the primary elements of 1.50 m, and those of U-section are stable with a clear spacing between the primary elements of 1.80 m up to a hydrostatic pressure difference load of 40 kN/m^2. The prerequisite for this is an extensive relieving of the secondary elements from earth pressure, for which a sufficiently compactly layered full backfill is necessary.

In the event of clearances and/or loads above this, the stresses must be checked. In such cases, horizontal intermediate walings can be used as supplementary support components or, as simplification, test results according to EN 1993-5, section 5.5.2 (5) and 5.5.3 (2), which have been derived under the regulations of EN 1990, appendix D, can be used.

If a clear primary element spacing of maximum 1.80 m and a minimum embedment depth of 5.00 m is adhered to, the full passive earth pressure can be assumed for simplicity in the passive earth pressure area even if the secondary elements are not driven as deep as the primary elements. In the case of larger clear primary elements spacing shorter embedment lengths, check whether the continuous, plane passive earth pressure according to section 5.2.2 of DIN 4085 (Feb 1987) instead of the volumetric passive earth pressure in front of the narrow compression areas of the primary elements in accordance with section 5.13.2 of DIN 4085. In combined sheet piling, allowance for the out-of-centre secondary elements as part of a composite cross-section is appropriate (R 103, section 8.1.5) only if displacement of the axis through the centre of gravity is largely prevented by reinforcement of the primary elements on the opposite side.

8.1.4.3   Design advice

The material grades and manufacture of the primary elements are to be certified with acceptance test certificate 3.1 to DIN EN 10 204. Here, commercially available general structural steels according to DIN EN 10 025 works test certificates are generally adequate. For special steel grades, certificates to DIN EN 10 204 (acceptance test certificate 3.1).are to be provided.

If required, slight deviations in shape and size from the standard are to be the subject of special agreements. Furthermore, existing contraction supervisory permits have to be taken into account with regard to the intended purpose when it comes to pipes of fine-grained structural steels to DIN EN 10 219 and of thermo-mechanically treated steels.

Tubes as primary elements must have the corresponding interlock sections welded on, to guarantee adequate connection of the secondary elements. For this purpose, the interlock connection must comply with the tolerances in R 67, section 8.1.6, and be able to transmit reliably the loads from the secondary elements to the tubes. At the intersection points with girth and helical weldings, interlock sections must fit closely against the skin of the tube.

Tube seams and interlock sections must be executed correspondingly at the intersection point. The welding contractor must perform a method test to DIN EN 288-3 for interlock welds on tubes made from fine-grained steels.

Tubes as primary elements with interior interlocks are useful, if a rotary drill system is used to reduce noise level and vibration impact and if obstacles have to be removed.

The steel grade of the tubes should comply with DIN EN 10 025 and fulfil all requirements otherwise made of sheet piling steels.

Tubes as primary elements of the tubular sheet piling are manufactured in the works in full length spirally welded, or as individual lengths longitudinally welded and connected by girth welds. Differing piling thicknesses stepped on the inside are usual for longitudinally welded tubes. Fully or semi-automatic machines can be used for joining on longitudinal, spiral and girth welds.

Longitudinal and helical line welds must be subjected to ultrasonic testing, in which colour-coded failure areas are repaired manually during the test procedure. Repaired weld seams can be documented under a new ultrasonic test by a connected printer log.

Girth welds between the individual pipe lengths and transverse welds between the coil ends of the SN pipes must be checked by X-ray [127, p. 132–134].

The secondary elements consist generally of Z-section double piles or U-section triple piles. In the case of tubular primary elements, they are usually arranged in the wall axis, and for box- or H-piles in the water-side interlock connection; it has to be mentioned, that particular in the case of tubular primary elements no smooth berthing surface will be achieved.

8.1.4.4 Installation

A combined sheet piling must be driven with particular care. The primary elements (box piles, H-piles or tubular piles) are usually driven in. When driving the primary elements, the recommendations E 104, section 8.1.12 and R 202, section 8.1.22 are to be observed. This is the only way to ensure that the primary elements stand parallel to each other within the permissible tolerances at the planned spacing and without any distortion, so that the secondary elements can be installed with the minimum possible risk of damage to the interlocks.

When driving in pipes, any obstacles can be removed by excavating inside of the pipe. However, the prerequisite here is that there are no structural elements inside the pipe, that protrudes more than the interior lock chambers (see section 8.1.4.3). In addition, the inner diameter of the tube must be at least 1200 mm to allow operation of suitable dredging equipment. When there are many obstacles in the soil, it is advisable to drive the elements in using the vibration method to obtain early information about them.

8.1.4.5 Structural analyses

In the case of tubes as primary elements the check of safety against bulges is not required, if the primary element tubes are concreted in or filled with non-cohesive material to the very top.

Please refer to R 33, section 8.2.11 for the transmission of the axial load from the tube into the subsoil. Where large axial loads and large pipe diameters are concerned, it may be necessary to weld a sheet metal cross or similar into the foot of the pipe to activate the necessary plug for the point resistance, which is to be verified by trial stress where applicable. The prerequisite for flawless formation of the plug is in any case adequate readiness for compaction of the soil in the area of the pile foot (see R 16, section 9.5). In these cases, any obstacles are to be removed in advance, by soil replacement for example.

Furthermore it is also possible to use inside dog-bars adjusted to the pipe walls and supported on a synthetically inserted plug (concrete plug) to generate the required point resistance.

8.1.5 Shear-resistant interlock connections for steel sheet piling (Jagged Walls) (R 103)

8.1.5.1 General

From the structural viewpoint, we distinguish between walls without shear force transmission in the interlocks and those with shear force transmission, the so called "Jagged Walls".

When determining the cross-sectional values of connected sheet piling, all sheet piles are taken into consideration. The jagged wall-section however may be calculated as a uniform cross-section only if full shear force absorption in the interlocks is certain.

Jagged walls in wave-like form consist of "U"-shaped sheet piles in which half the wave length consists at least of one individual pile. In this case, the uniform cross-section is achieved already by transmission of shear forces in every second interlock. In case of two individual piles per half wave length, the interlocks are alternately placed on the wall axis (neutral axis) and outside in the flanges. Here the uniform cross-section is only achieved when all interlocks on the wall axis are linked shear-resistant. The interlocks in the flanges are the threaded locks in construction. The interlocks located on the wall axis can be drawn together in the workshop and prepared accordingly for the transmission of the shear forces, namely:

- by welding the interlocks together in accordance with sections 8.1.5.2 and 8.1.5.3, or
- by crimping the interlocks; however, only a partial connection can be achieved because the interlocks at the crimping points are displaced by several millimetres to take up the loads. The degree of partial connection depends on the number of crimping points per member, which has a critical influence on the deformation behaviour, respectively the degree of displacement.

8.1.5.2 Bases of calculation for welding

The shear stresses in the welded interlocks originating from the main loadbearing system, and from the loading and supporting influences acting on the system, are determined by the equation:

$$T_d = V_d \cdot \frac{S}{I}$$

where:

V_d = design value of the shear force. For sheet piling where the moment component from active earth pressure may be reduced as per R 77, section 8.2.2, calculation may be simplified by using the shear forces figure based on the unreduced bending moment figure. In the case of redistributed earth pressure the calculation is done by using the resulting shear force figure based on that earth pressure figure [MN].

S = static moment of the cross-section portion to be connected, referred to the centroidal axis of the connected sheet pile wall [m³]

I = moment of inertia of the connected sheet piling [m⁴]

For interrupted welds, the shear stress to section 4.9 of EN 1993-1-8 should be set correspondingly higher.

The verification of the welds is to be carried out according to section 4.5.4 of EN 1993-1-8, in which the plastic analysis – assuming a uniform shear stress – according to (1) of section 4.9 of EN 1993-1-8 is permitted. For steel grades with yield stresses not covered by table 4.1 of EN 1993, the β_w value may be obtained through linear interpolation.

8.1.5.3 Layout and execution of the welding seams

The interlock welds shall be so laid out and executed as to achieve optimum continuous absorption of the shear forces. A continuous seam is best for this. If an interrupted seam is used, the minimum length is to be 200 mm, provided that the static shear stresses mentioned in section 8.1.5.2 do not require a longer weld. In order to keep the secondary stresses within limits, the interruptions in the seam should be ≤ 800 mm.

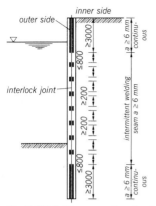

Fig. R 103-1. Diagram to show interlock welding at walls of killed, brittle-fracture resistant steel, easy driving and only slight corrosion in harbour and groundwater (6 mm = 15/64")

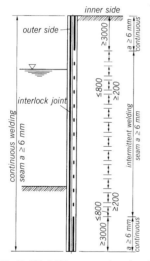

Fig. R 103-2. Diagram to show interlock welding at walls of killed, brittle-fracture resistant steel, difficult driving or stronger corrosion from the outside, in the harbour area (6 mm = 15/64")

Continuous seams shall always be used in areas where the sheet piling is subject to heavy concentration of stresses, especially near anchor connections and at the point where the equivalent force C is introduced at the foot of the wall (fig. R 103-1).

In addition to the requirements introduced by static forces, the effects of driving stresses and corrosion must be considered. In order to be able to cope with the driving stresses, the following measures are necessary:

309

(1) The interlocks are to be welded on both sides at the tip and toe ends.
(2) The length of the weld depends on the length of the sheet pile and on the difficulty which may be encountered in driving.
(3) These seam lengths are to be ≥ 3000 mm in sheet piling walls for waterfront structures.
(4) Moreover, additional seams as per fig. R 103-1 are necessary for light driving, and seams as per fig. R 103-2 for difficult driving.

In areas where the harbour water is aggressively corrosive, a continuous welding seam of $a \geq 6$ mm (fig. R 103-2) is run on the outer side down to the sheet pile toe.

If there is greater corrosion both in the harbour water and in the groundwater, a continuous seam of $a \geq 6$ mm must also be welded on the inner side of the wall.

8.1.5.4 Choice of steel grade

As the amount of welding work on connected sheet piling is comparatively extensive, the sheet piles are to be manufactured of steel grades with full suitability for fusion welding. In view of the starting points of the welds not only at discontinuous welding, killed steels without tendency to brittle fracture shall be used as per R 99, section 8.1.18.2.

8.1.5.5 Basis of calculation for crimping of interlocks

The design value of the shear stress at the crimping points is arising from statical system. The loads and supporting actions are determined according to section 8.1.5.2. The analysis of the pressing points is carried out according to sections of 5.2.2 and 6.4 of EN 1993-5, with the resistances of the pressing points determined according to DIN EN 10 248.

8.1.5.6 Layout and execution of crimping points

A spacing of up to 400 mm is usual for double crimping points, and according to EN 1993-5, section 6.4, should not exceed 700 mm. Check in each case whether the number of crimping points per unit of width of pile is sufficient for the resulting shear force of the shear force areas with same direction. The usual crimping point spacing can be reduced to suit the shear force. This aspect should be discussed and agreed with the manufacturer prior to fabricating the piles.

8.1.5.7 Welding on reinforcing plates

In order to avoid rust forming on the inner contact surfaces of reinforcing plates, these must always be welded to the sheet piles around their full circumference. They should be tapered to reduce the leap in the moment of inertia. The seam thickness a should be at least 5 mm without corrosion and at least 6 mm where corrosion is severe. If a reinforcing plate spans an interlock located on the flange of the sheet pile, both sides of this interlock must be given a continuous weld in the area of the plate, plus an extension of at least 500 mm on each side of the plate. Even on the side opposite that on which the plate is located, the welds should be $a \geq 6$ mm thick, while the weld under the plate should be of a thickness which will permit seating the plate without subsequent machining. If this is not done, the welds attaching the plate may be seriously damaged during driving.

If welding of the interlock is not desired, the plates are to be cut in half and welded on each part of the pile according to the procedure above with a surrounding seam.

8.1.6 Quality requirements for steels and interlock dimension tolerances for steel sheet piles (R 67)

This recommendation applies to steel sheet piles, trench sheeting and driven steel piles, which are all called steel sheet piles in the following. DIN EN 10 248-1 and -2 as well as DIN EN 10 249-1 and -2 apply.

If steel sheets are stressed in the depth direction, e.g. in the case of special piles or junctions for circular and diaphragm cells (see R 100, section 8.3.1.2), steel grades with appropriate properties must be ordered from the sheet pile manufacturer in order to avoid terracing failure, see EN 1993-1-10.

8.1.6.1 Designation of steel grades

Grades of steel with the designation S 240 GP, S 270 GP and S 355 GP are used for hot-rolled steel sheet piles in normal cases, as indicated in sections 8.1.6.2 and 8.1.6.3.

The grades of steels with yield stresses up to 355 N/mm^2 should be verified with a acceptance test certificate 3.1 to DIN EN 10 204 – and higher quality steels with acceptance test certificate 3.1 stating the 14 alloy elements (like, for example, for S 355 J3 G3 as per DIN EN 10 025). In special cases, e.g. for accommodating greater bending moments, steel grades with higher minimum yield points up to 500 N/mm^2 can be used, regarding recommendation R 34, section 8.1.3. The higher steel grade is then to be verified by a mill certificate, acceptance test certificate or acceptance test protocol according to DIN EN 10 204. When using steel grades with a minimum yield stress exceeding 355 N/mm^2, in Germany a general building authority approval for sheet piles of this grade of steel should be provided.

Steel grades S 235 JRC, S 275 JRC and S 355 JRC are possible for cold-rolled steel sheet piling.

In special cases, e.g. as stated in 8.1.6.4, fully killed steels to DIN EN 10 025 are used.

8.1.6.2 Characteristic mechanical and technological properties

Table R 67-1. Characteristic mechanical properties for steel grades for hot-rolled steel sheet piling

Steel grade	Minimum tensile strength N/mm^2	Minimum yield point N/mm^2	Minimum % elongation for measuring length of $L_0 = 5.56 \cdot \sqrt{S_0}$	Former name
S 240 GP	340	240	26	St Sp 37
S 270 GP	410	270	24	St Sp 45
S 320 GP	440	320	23	–
S 355 GP	480	355	22	St Sp S
S 390 GP	490	390	20	–
S 430 GP	510	430	19	

The mechanical properties for steels S 235 JRC, S 275 JRC and S 355 JRC are described in DIN EN 10 025.

8.1.6.3 Chemical composition

The ladle analysis is binding for verification of the chemical composition. If a verification of the values of single bars is wanted, a separate agreement has to be made for the testing. This analysis can be used as an additional test in case of doubts.

Table R 67-2. Chemical composition of the ladle/bar analysis for hot-rolled steel sheet piling

Steel grade	Chemical composition % max. for ladle/bar				
	C	Mn	Si	P and S	N [*] [**]
S 240 GP	0.20/0.25	–/–	–/–	0.040/0.050	0.009/0.011
S 270 GP	0.24/0.27	–/–	–/–	0.040/0.050	0.009/0.011
S 320 GP	0.24/0.27	1.60/1.70	0.55/0.60	0.040/0.050	0.009/0.011
S 355 GP	0.24/0.27	1.60/1.70	0.55/0.60	0.040/0.050	0.009/0.011
S 390 GP	0.24/0.27	1.60/1.70	0.55/0.60	0.040/0.050	0.009/0.011
S 430 GP	0.24/0.27	1.60/1.70	0.55/0.60	0.040/0.050	0.009/0.011

[*] The stipulated values may be exceeded on condition that for every increase by 0.001 % N, the max. P level is decreased by 0.005 %; but the N-level of the ladle analysis may not be higher than 0.012 %.

[**] The maximum nitrogen value does not apply when the chemical composition has a minimum total level of aluminium of 0.020 %, or when there are sufficient N-binding elements. The N-binding elements are to be stated in the test certificate.

8.1.6.4 Weldability, special cases

An unlimited suitability of steels for welding cannot be presumed, since the behaviour of a steel during and after welding depends not only on the material but also on the dimensions and the shape, as well as on the manufacturing and service requirements of the structural member. Generally, killed steels are preferred (R 99, section 8.1.18.2 (1)).

All sheet piling steel grades can be assumed to be suitable for arc welding, provided the general welding standards are observed. When selecting higher strength steels S 390 GP and S 430 GP, the welding stipulations of the building authority approval must be adhered to. To allow for weldability, the carbon equivalent CEV should not exceed the values for steel grade S 355 as per DIN EN 10 025 table 4, in the interests of weldability. Rimming steels should not be used.

In special cases with unfavourable welding conditions because of external influences (e.g. plastic deformation due to heavy driving, at low temperatures) or the nature of the structure, spatial stresses and in the case of not predominantly static loads as per R 20, section 8.2.6.1 (2), fully killed steels are to be used as per DIN EN 10 025 in grade groups J2 G3 or K2 G3 with regard to the required brittle fracture resistance and ageing resistance.

The welding material is to be selected according to DIN EN 499, DIN EN 756 and DIN EN 440, or according to the data provided by the manufacturer (R 99, section 8.1.18.2 (2)).

8.1.6.5 Types of interlocks and coupling connections

Examples of proven types of interlocks for steel sheet piling are shown in fig. R 67-1. The nominal dimensions a and b, which can be obtained from the manufacturers, are measured at right-angles to the least favourable direction of displacement. The minimum interlock hook connection, calculated from a minus b, must correspond to the values in the pictures. In short partial sections, the values may not fall more than 1 mm below these minimum values. In forms 1, 3, 5 and 6, the required coupling must be present on both sides of the interlock.

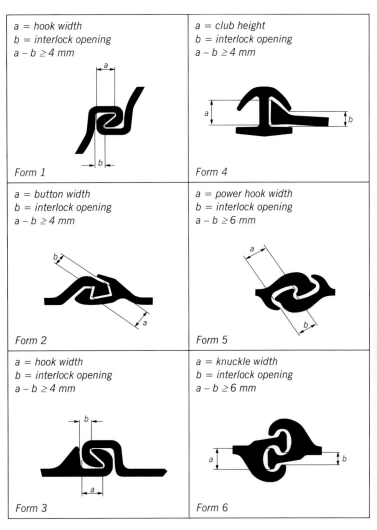

a = hook width b = interlock opening a − b ≥ 4 mm Form 1	a = club height b = interlock opening a − b ≥ 4 mm Form 4
a = button width b = interlock opening a − b ≥ 4 mm Form 2	a = power hook width b = interlock opening a − b ≥ 6 mm Form 5
a = hook width b = interlock opening a − b ≥ 4 mm Form 3	a = knuckle width b = interlock opening a − b ≥ 6 mm Form 6

Fig. R 67-1. Established types of interlock and coupling for steel sheet piles

8.1.6.6 Permissible dimensional deviations for interlocks

Deviations from the design dimensions occur during rolling of sheet piles and connectors. The tolerable deviations are summarised in table R 67-3.

Table R 67-3. Allowed interlock tolerances

Form	Rated dimensions (to section drawings)	Tolerances of rated dimensions		
		Designation	plus [mm]	minus [mm]
1	Hook width a Interlock opening b	Δa Δb	2.5 2	2.5 2
2	Button width a Interlock opening b	Δa Δb	1 3	3 1
3	Button width a Interlock opening b	Δa Δb	1.5 ... 2.5[1] 4	0.5 0.5
4	Club height a Interlock opening b	Δa Δb	1 2	3 1
5	Power hook width a Interlock opening b	Δa Δb	1.5 3	3.5 1.5
6	Knuckle width a Interlock opening b	Δa Δb	2 3	3 2

[1] depending on the section

8.1.7 Acceptance conditions for steel sheet piles and steel piles on site (R 98)

Although careful and workmanlike construction methods are of the greatest importance whenever steel sheet piles or steel piles are used as part of a structure, it is essential that the material is in impeccable condition at delivery on site. In order to achieve this, specific inspection of the material at the site is necessary. As a supplement to the manufacturer's own shop inspection, a works acceptance can be agreed in each case. For shipments to overseas, inspection is frequently carried out before the piles are loaded on board.

The acceptance procedure at the site should specify that every unsuitable pile will be rejected until it has been refinished in a usable condition, unless it is rejected out right. Acceptance on the building site is based on:

DIN EN 10 248-1 and -2 for hot-rolled sheet piling
DIN EN 10 249-1 and -2 for cold-formed sheet piling
DIN EN 10 219-1 and -2 for cold-formed welded hollow sections

In addition, R 67, section 8.1.6.6, applies to the interlock tolerances. And with regard to the limit deviation for straightness of combined sheet pile walls, R 104, section 8.1.12.4, paragraph 1, applies as well for primary elements also for tubes as primary elements.

Please refer to DIN EN 12 063, section 8.3, for more detailed information regarding the storage of components on site.

8.1.8 Corrosion of steel sheet piling, and countermeasures (R 35)

8.1.8.1 General

Steel in contact with water undergoes the natural process of corrosion, which is influenced by numerous chemical, physical and, occasionally, biological parameters. Different corrosion zones (fig. R 35-1) along the exposed height of the sheet pile wall, which are characterised by the type of corrosion (surface, pitting or tuberculation) and its intensity.

To measure the degree of corrosion we can assess the loss in wall thickness [mm] or – related to the time in use – the rate of corrosion [mm/a]. Typical mean and maximum values for wall thickness reduction are shown in figs. R 35-3 and R 35-4. These diagrams are based on numerous wall thickness measurements on sheet pile walls and piles or dolphins in the North Sea and Baltic Sea as well as inland and can be assigned to the corrosion zones shown in fig. R 35-1. Due to the multitude of factors

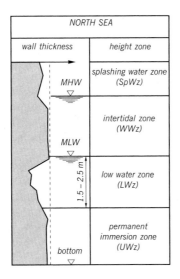
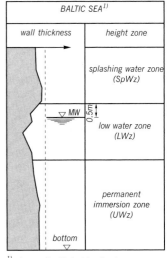

[1] also applicable to inland waters

Fig. R 35-1. Qualitative diagram of the corrosion zones for steel sheet piling with examples of the North Sea and Baltic Sea

that can influence corrosion, the loss of wall thickness is subject to a very wide scatter.

8.1.8.2   How corrosion influences loadbearing capacity, serviceability and durability

Corrosion influences the loadbearing capacity, serviceability and durability of unprotected steel sheet piling and this should be examined as follows:

(1) Corrosion decreases the design value of the component resistance corresponding to the different reductions in wall thickness in the individual corrosion zones (fig. R 35-1). Depending on the shape of the bending moment diagram, this can reduce the loadbearing capacity and the serviceability of the structure (EN 1993-5, sections 4 to 6).

In analyses of loadbearing capacity and serviceability according to EN 1993-5, the section modulus and the cross-sectional area of the sheet piles are to be reduced in proportion to the mean values of the wall thickness losses according to fig. R 35-3a (freshwater) or fig. R 35-4a (seawater).

Unprotected sheet piling should be designed in such a way that the maximum bending moment lies outside the zone of maximum corrosion.

(2) In the light of experience gained over the past 10 to 20 years, it would seem that corrosion can limit the durability of sheet pile walls (EN 1993-5, section 4) to a useful life of 20-30 years, especially in the seawater of the North Sea and Baltic Sea [246]. Once the steel is rusted through, the fill behind the wall can be washed out, thus leaving voids and probably leading to soil settlement. This is frequently linked with considerable safety risks and restrictions on use. U-sections frequently rust through in the middle of the flanges on the water side and Z-sections frequently at the flange–web to junctions on the water side (fig. R 35-2).

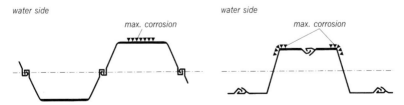

Fig. R 35-2. Zones of possible rusting through U- and Z-piles in the low water zone (sea-water)

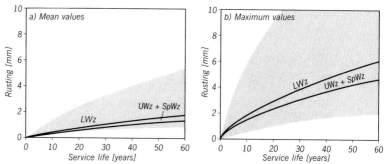

Fig. R 35-3. Decrease in thickness as a result of corrosion in the fresh water zone

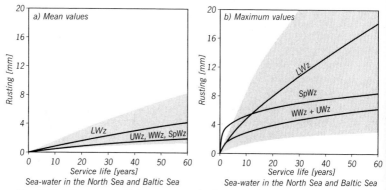

Sea-water in the North Sea and Baltic Sea   Sea-water in the North Sea and Baltic Sea

Fig. R 35-4. Decrease in thickness as a result of corrosion in the sea-water zone

The basis for assessing the durability (estimating the useful life until the occurrence of the first holes caused by corrosion) are the maximum values for wall thickness losses according to fig. R 35-3b (freshwater) or fig. R 35-4b (seawater).

New, unprotected sheet pile walls should be planned and designed for the design values for mean and maximum wall thickness losses according to the simplified diagrams (figs. R 35-3 and R 35-4), unless other wall thickness measurements on adjacent structures are available. To avoid uneconomic designs, it is recommended to use the values above the regression curves only if local experience suggests this is necessary.

For older, unprotected sheet pile walls, the design values should always be determined from ultrasound measurements of the wall thickness in order to assess the local corrosion influences affecting the mean and maximum values for wall thickness reduction. Information on performing, troubleshooting and analysing such measurements can be found in [176] and [247].

318

8.1.8.3 Design values for loss of wall thickness in various media

The information below can be used in the sense of design values forming the basis for the design of new sheet pile structures and for checking existing sheet pile walls, provided no local experience is available.

(1) Freshwater

The design values for the reduction in wall thickness for sheet pile walls in freshwater can be taken from the regression curves of fig. R 35-3 depending on the age of the structure. The area shaded grey represents the scatter of those structures investigated.

(2) Seawater of the North Sea and Baltic Sea

The design values for the reduction in wall thickness for sheet pile walls in the seawater of the North Sea and Baltic Sea are shown in fig. R 35-4 depending on the age of the structure. The area shaded grey represents the scatter.

The corrosion rates for freshwater and seawater resulting from the diagrams are comparable with internationally accepted values [151].

(3) Briny water

Brackish water zones are characterised by a mixture of freshwater from inland areas with the seawater of the North Sea or Baltic Sea. The design values for the reduction in wall thickness can be interpolated from the figures for seawater (fig. R 35-4) and freshwater (fig. R 35-3), depending on the location of the structure within the briny water zone.

(4) Atmospheric corrosion

Atmospheric corrosion, i.e. corrosion above the splash water zone (fig. R 35-1), on waterfront structures is low, amounting to a corrosion rate of about 0.01 mm/a. Higher values can be expected where de-icing salts are in use or in areas involving the storage and handling of substances known to attack steel.

(5) Corrosion in the soil

The anticipated corrosion rate on both sides of steel sheet piling embedded in natural soils is very low at 0.01 mm/a.

The same corrosion rate applies when the sand filling behind the sheet pile wall is such that the troughs of the piling are also completely embedded.

Whenever possible, the surface of the sheet piling should be isolated from aggressive soils or aggressive groundwater. Such soils include humus and carbonaceous soils, e.g. waste washings. The aggressiveness of soils and groundwater can be assessed using DIN 50 929. In soils the corrosion can be further aggravated by the activity of bacteria that attack steel [249] and [250]. Such bacterial activity can be expected when organic substances reach the rear face of the sheet pile wall, either through circulating water (e.g. in the vicinity of landfill sites for domestic waste or in irrigation areas) or through soils with a high organic content. In such cases higher corrosion rates and, typically, non-uniform degradation can be expected.

8.1.8.4 Corrosion protection

At the early design stage the following factors should be taken into account as corrosion protection measures:

- purpose and design life of the structure
- general and specific corrosion rate at the location of the structure
- experience with corrosion phenomena in adjacent structures
- possibilities for including corrosion protection in the design
- profitability calculations [105] for unprotected sheet piling should always take account of the costs for premature repairs to unprotected sheet pile walls (e.g. preplating).

Subsequent protection measures or complete renewal are extremely difficult so that particular care is required in the planning and execution of the protection system.

Specifically adjusted protection measures may be required, depending on the nature and intensity of the corrosion load.

(1) Corrosion protection with coatings

According to available experience, coatings can delay the start of corrosion by more than 20 years.

The prerequisite is sandblasting to a standard degree of purity Sa 2½ and the selection of a suitable coating system. Advice on selecting coating systems depending on the local conditions and also surface preparation, laboratory tests, workmanship and supervision of coating work plus maintenance of coating systems can be found in DIN EN ISO 12 944. ZTV-W 218 [177] contains additional advice on contractual matters.

With a view to health and environmental protection, bituminous systems will be replaced in future by coating systems with little or no solvents on the basis of epoxy resins or polyurethane resins, or corresponding combinations of tar substitutes and hydrocarbon resins. The coating should be applied completely in the workshop so that only transport and installation damage needs to be touched up on the construction site.

The design of the coating system should make allowances for the possibility of subsequent installation of cathodic corrosion protection. Attention must be paid to corresponding compatibility of the coating materials. If steel sheet piling is to be protected by coatings from sand abrasion, the abrasion value (A_w) of the coating material must be taken into account [153].

(2) Cathodic corrosion protection

Corrosion under the water line can be substantially eliminated by electrolytic means by the installation of cathodic protection using impressed current or sacrificial anodes.

Additional coating or partial coating is an economical measure and usually indispensable in the interests of good current distribution and lower power requirements later on.

Cathodic protection systems are ideal for protecting those sections of sheet piling, e.g. the tidal low water zone, in which a renewal of protective coatings or the repair of corrosion damage in unprotected sheet piling is not possible or only at high financial and technical expense. Sheet piling structures with cathodic corrosion protection require special construction measures and must therefore be taken into account in the design phase [178].

(3) Alloy additives

On the basis of available experience, the addition of copper to the steel does not increase its service life in the submerged zone. However, an alloy of copper in connection with nickel and chromium as well as with phosphorous and silicon increases the service life in and above the splash zone, especially in tropical regions with salt-laden moving air.

No differences in resistance to corrosion could be determined among the different types of sheet pile steels according to DIN EN 10 248 and the steels mentioned in DIN EN 10 025 (structural steels) and DIN EN 10 028 together with DIN EN 10 113 (higher strength fine grain structural steels). If higher strength is achieved by the addition of niobate, titanium and vanadium, this also has a positive effect on the corrosion behaviour.

(4) Corrosion protection by oversizing

In order to extend the service life, with regard to exceeding the bearing capacity it is possible to select profiles with greater section modulus or higher steel grade according to E 67, section 8.1.6.1. Profiles with thicker backs and webs provide better protection from rusting through.

(5) Constructional measures

- Sheet piling structures with cathodic corrosion protection require special structural measures [59].
- As far as corrosion attack is concerned, constructions in which the back of the sheet piling is only partially backfilled or not at all have certain disadvantages.
- Surface water should be collected and drained clear of the rear face of the sheet pile wall. This applies particularly to quays handling aggressive substances (fertilisers, cereals, salts, etc.).
- Free-standing, open piles are exposed to corrosion on the whole periphery, closed piles e.g. box piles on the other hand essentially just

on the outer surface. The inner surface of the pile is protected when the pile is filled with sand.

- In the case of free-standing sheet piling, e.g. flood protection walls, the coating of the sheet piling must be extended deep enough into the ground area; settlement must be taken into account.
- A sand bed is recommended for placing (round steel) tie rods, inhomogeneous fill material must be avoided. A coating or other protection is essentially superfluous. The design according to R 20, section 8.2.6.3, permits a certain degree of corrosion. Once a coating has been applied, it should not be damaged because such damage favours pitting. The anchor heads should be sealed carefully.
- At ship berths, timber fenders, fender piles and permanent fender systems should always be used to protect sheet piling against constant chafing and scuffing caused by pontoons or ships or their fenders. This protection should ensure that there is never direct contact between the surface of the sheet pile wall and the ship or pontoon. Otherwise, greater decreases in thickness and further corrosion must be expected which clearly exceed the data indicated in figs. R 35-3 and R 35-4.
- Corrosion should always be expected at the transition from steel to reinforced concrete (e.g. at concrete capping beams). See [248] for details. R 95, section 8.4.4, is to be taken into account for protecting steel capping profiles.

8.1.9 Danger of sand abrasion on sheet piling (R 23)

Heavy sand abrasion affects particularly walls of reinforced or prestressed concrete steel piles.

When steel sheet piling is used, it must be coated with a system which can permanently withstand the sand abrasion at the site of installation. According to DIN EN ISO 12 944 part 2 appendix B, the mechanical stress from sand abrasion can be divided into three groups, according to the quantity of transported sand and current.

Assessment of the necessary abrasion-resistance of the coating is performed in accordance with a procedure described in the "Directive for testing coating agents for corrosion protection in steel hydraulic engineering" [153].

8.1.10 Driving assistance for steel sheet piling by means of shock blasting (R 183)

8.1.10.1 General

If difficult driving is anticipated, it is necessary to check which driving aids can be used in order to prepare the subsoil in such a way that driving progress is economic, whilst at the same time avoiding overloading the

equipment or overstressing the elements, and reducing power requirements. The latter will also result in less noise and vibration. It should
also be ensured that the necessary embedment is reached.

Shock blasting is frequently used in rocky soils.

The blasting fragments the rock along the planned line of the sheet pile
wall in such a way that a vertical, ballast-filled trench arises into which
the sheet piles can be, preferably, vibrated. The loosening should extend
as deep as the planned base of the sheet pile wall and should be wide
enough to accommodate the sheet pile wall profile. On either side of the
trench the rock remains stable.

8.1.10.2 Blasting method

Trench blasting with short-period detonators and inclined blastholes as
shown in fig. R 183-1, has proved to be a successful way of achieving
the aims of blasting while causing minimal vibration. Trenches up to 1
m wide can be produced by this method [80].

The blasting sequence begins with the stab-holes V-cut (shown on the
right of the longitudinal section in fig. R 183-1). This relieves the stress
in the rock by creating a second free face in addition to the surface of the
trench, against which the rock can be thrown. This improves the effect
of the blasting and leads to lower vibration than is the case with vertical
blastholes.

Following the V-cut, the other charges detonate in succession at intervals
of, usually, 25 ms, which allows the earlier detonations to create space
for the later ones. In addition, the detonations bounce off one another so
that the rock along the line of the blastholes in the trench is thrown
backwards and forwards several times and thus reduces the size of the
fragments.

The explosive acts in a V-form (included angle 90°) in the direction of
the free face (see cross-section in fig. R 183-1). In order that the designed
sheet pile wall can be driven to the intended depth despite the narrow tip
of the cone of debris, the blastholes for the explosives must extend below
the planned toe of the sheet pile wall. Above the cone of debris there
arises a ballast-filled trench into which the sheet piles can be vibrated
without damage afterwards.

The spacing of the blastholes roughly corresponds to the average width
of the trench, which should be a few hundred millimetres wider than the
sheet pile profile to be used. Common blasthole spacings for inclined
drilling are 0.5 to 1.0 m.

In changing rock conditions the individual explosive charges should be
placed in the hard rock segments in order to achieve an optimum blasting
effect. In principle, every type of rock can be blasted. But critical for
achieving the aims of the blasting are the choice of a suitable blasting
method, the detonation sequence, the layout of the charges, the type of

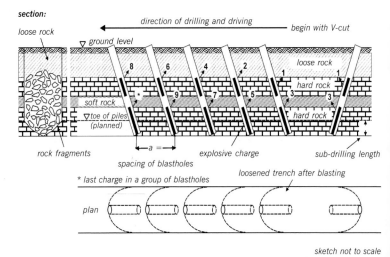

Fig. R 183-1. Principle of trench blasting with inclined blastholes

explosive (waterproof, highly explosive) and, in particular, accurate positioning, spacing and angle of the blastholes.

8.1.10.3 Advice on blasting

The following basic advice should be taken into account when carry out blasting work:

(1) Before beginning work it is necessary to test the subsoil by means of trial blastholes and trial blasting in order to obtain information about the optimum blasthole spacing and the explosives required. In doing so, the propagation curves of ultrasound measurements prior to and after blasting can be used, for example, to estimate the volume loosened.

As vibration of sheet piles causes less damage than impact driving, the latter should be permitted only in exceptional circumstances. In doing so, the limiting frequency or the penetrating frequency must be taken into account. If driving difficulties are encountered, further blasting should be carried out without delay after raising the sheet piles. Shock blasting creates a trench into which sheet piles can be vibrated without causing a compacting effect. Simply splitting the rock allows only impact driving. Predrilling and replacing the soil with gravel fill can lead to a compacting effect during vibration.

(2) The driving of the piles must follow quickly after blasting because the loosening effect in the trench gradually decreases due to the load from above and, possibly flowing water (hydraulic compaction).

(3) The loosening of the area already blasted can be assessed prior to driving the sheet pile wall with the help of the dynamic penetrometer test. A high number of blows ($N_{10} > 100$) indicates that difficulties will be experienced when driving the sheet pile wall.

(4) As evidence, the vibration due to blasting occurring at the nearest objects should be comprehensively measured with vibration measuring instruments to DIN 45 669. This can be carried out by the contractor within the scope of his own monitoring procedures.

(5) Blasthole logbooks should be kept. These should record the boundaries of the strata by means of the drilling progress, colour and sieve residue for every blasthole. Examples of such logbooks can be found in [254]. Further, the contact pressure, flushing losses and water-bearing strata should be noted in order to obtain information on clefts, voids and narrowing of the blasthole. Finally, record the angle and depth of every blasthole. The logbooks should be made available to the chief blasting engineer prior to placing the charges in the holes so that he can achieve the best results.

(6) Spot-checks to assess the accuracy of the blastholes should be carried out, especially at the start of the drilling. Precise blasthole measurement systems (e.g. BoreTrack) are available. A drilling accuracy of 2% is feasible. The axis of the blastholes and the sheet pile wall must lie in one plane throughout the intended depth so that the sheet piles are always placed into the stone loosened by the blasting.

(7) Only water should be used for flushing. Flushing with water allows a changing colour and the drilling debris to be identified, whereas flushing with air produces only a uniform cloud of dirty dust. Another problem is that the air introduced under high pressure is forced into softer strata and existing clefts, which may create undesirable paths. This can, in turn, lead to blowouts during blasting that are remote from the desired blasting point and thus diminish the success of the blasting.

(8) Soft cohesive soils and non-cohesive soils should be drilled with casings because there is a risk of material caving in to the inclined blastholes.

(9) The blasting parameters should be systematically optimised on the basis of the records for the first blasts. Separate blasting trials are then unnecessary. Strict coordination of the drilling and driving

works and a constant exchange of information on the work carried out so far is necessary and improves the success of the operations. In some circumstances subsequent blasting beneath partly driven piles may be necessary.

(10) The maximum amount of explosive per detonation stage should be established on the basis of the vibration calculations prior to any blasting. On no account exceed this amount.

(11) The structural calculations are based on a sheet pile toe that is fixed in solid rock. Therefore, the trench may not be too wide.

8.1.11 Driving corrugated steel sheet piles (R 118)

8.1.11.1 General

Driving of sheet steel piles represents a widely used and proven construction method. Although driving of steel sheet piles may seem relatively simple, it cannot be tolerated that driving work is carried out without sufficient special knowledge and in a negligent manner. The resident engineer and the driving team should always adhere to the target of the assigned task that steel sheet piles are so driven as not to adversely affect the intended purpose of the wall, and to achieve a closed sheet piling wall of maximum safety.

The more difficult the soil conditions, the longer the piles, the greater the embedment depth and the deeper the later dredging in front of the sheet piling, the greater will be the need to insist on a high order of competence in the construction work. Poor results can be expected if long driven elements are driven one after the other to final penetration because the exposed interlock is then too short and therefore proper initial guidance to the next pile is not given.

A working basis for judging the behaviour of the ground with respect to pile driving is obtained from borings and soil mechanics investigations, as well as from load and penetration tests, see R 154, section 1.9. In critical cases however, driving tests are required. If it is necessary to adhere closely to the designed position of the sheet piles, the behaviour of these test piles at selected locations should be observed with the aid of measurement facilities to determine the amount of deflection which may occur.

The success and quality of the sheet piling installation depend largely on the correct manner of driving. This demands as a prerequisite that in addition to suitable reliable construction equipment, the contractor himself must possess broad experience and therefore be capable of making the best use of the skills of qualified technical and supervisory personnel. All plant and methods for installing sheet piles must comply with DIN EN 996.

8.1.11.2 Driven elements

Steel sheet piling in waveform of U- or Z-sections is generally driven in pairs. Triple or quadruple piles may also have technical and economical advantages in specific cases.

Sheet piles joined in pairs should, as far as possible, be securely connected to form a unit by crimping or welding the middle interlocks. This facilitates handling of the driving elements, as well as the driving so that elements already in place are hardly dragged down. The driving of single piles should be avoided if possible.

Due to technical requirements, it can be necessary from the very first to use sheet piles with heavier wall thickness when difficult subsoil and/or deeper embedment is concerned, or to use a higher steel grade than required for purely structural reasons. The pile tip and, if necessary, also the pile head may have to be reinforced occasionally, which is particularly recommended for driving in soils with stone interstratification and rocky soils.

Please refer to R 98, section 8.1.7 for the acceptance conditions for steel sheet piles at the site.

8.1.11.3 Driving equipment

The size and efficiency of the required driving equipment are determined by the driving elements, their steel grade, dimensions and weights, the embedment depth, the subsoil conditions and the driving method. The equipment must be so constructed that the driving elements can be driven with the necessary safety and careful handling, and at the same time be guided adequately, which is above all essential in the case of long piles and deep embedment.

Drop hammers, diesel hammers, hydraulic hammers and rapid-stroke hammers may be used for pile driving. Vibratory hammers (see R 154, section 1.8.3.2, and R 202, section 8.1.22) and press-in-systems are also used (see R 154, section 1.8.3.3). The efficiency of a drop hammer depends on the ratio of hammer weight to weight of piling element plus driving cap. The target should be a ratio of 1 : 1.

Rapid-stroke hammers tend not to damage the piling element and are especially well-suited for driving in non-cohesive soils. Slow-stroke, heavy hammers are used universally, particularly in cohesive soils.

Take the following factors into account when choosing a type of hammer:

- total weight of hammer
- weight of piston, energy of single blow, type of acceleration
- energy transmission, force transmission (driving cap and guide)
- driven elements: weight, length, angle, cross-section, form
- subsoil (see R 154)

A good degree of efficiency for a hammer-type installation is achieved through optimum coordination of these factors. Hydraulic hammers tend to be universal in application; their energy per blow can be carefully controlled to suit the respective driving energy requirements and the subsoil (including rock).

For a hammer-type installation, a driving cap to match the requirements of driving equipment and piling elements is essential.

Suitable preliminary investigations (e.g. the use of prognosis models) are recommended for estimating the effects of driving vibrations and noise. If necessary, driving tests may be performed. DIN 4150 parts 2 and 3 contain guidance on the values that cause a nuisance to persons and damage to buildings (see also [7]).

Please refer to R 149, section 8.1.14, with regard to noise protection.

8.1.11.4 Driving sheet piles

The following must be observed when driving:

The driving blow should generally be introduced centrally in the axial direction of the driving element. The effect of the interlock friction which acts only on one side, can be countered if required, by a suitable adjustment of the point of impact.

The driven elements must be guided with regard to their stiffness and driving stresses, so that their design position is achieved in the final state. For this, the pile driver itself must be adequately stable, firmly positioned and the leads must always be parallel to the inclination of the driving element. The driven elements should be guided at least at two points which are spaced as far apart as possible. A strong lower guide and spacer blocks for the driving element in this guide are specially important. The leading interlock of the pile being driven must also be well guided. When driving without leads, care must be taken to ensure tight contact between the hammer and the driving element by the use of well-fitting leg grips. When a free hanging pile driver is employed, it must be securely moored to restrain its movements as fully as possible.

The first driven element must be positioned with special care at the intended inclination of the wall. In this way, good interlock engagement is ensured when driving the remaining elements in deep water. In case of difficult subsoil conditions and for deeper embedment, a driving method with two-sided interlock guidance of the driven elements is recommended, if not already required for other reasons. This refers to a situation when normal continuous driving does not produce the desired result because of increased driving resistance, and because it becomes impossible to hold the pile in the design position.

In such cases, driving should be staggered (e.g., initial driving with a light hammer and redriving with a heavy one) or in panels whereby several driven elements are pitched and then driven in the following sequence: $1 - 3 - 5 - 4 - 2$.

When constructing sheet piling enclosures, staggered driving is also recommended.

U-shaped piles tend to lean forwards in driving direction with the pile head, whereas Z-section piles tend to lean backwards.

This can be prevented by driving staggered or in panels. For normal continuous driving of U-shaped sheet piles, the web extending in driving direction can be bent out by several millimetres so that the theoretical pile width is somewhat enlarged. In the case of Z-section sheet piles, the leading web can be slightly pressed in toward the trough. If leaning over cannot be prevented by these methods, taper piles must be used. Care must be taken with these that they are properly designed for driving, so that the flanges do not plough up the soil. For this, the corrugation of the taper pile must be similar at both ends, and the connecting flange with a welded-in wedge must lie in the driving direction (see fig. R 118-1a).

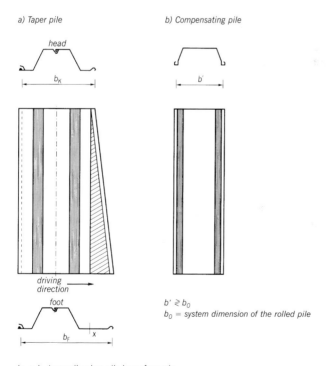

a) Taper pile

b) Compensating pile

head

b_K

driving direction

foot

b_F

x

$b' \gtrless b_0$

b_0 = system dimension of the rolled pile

$b_K < b_F$ taper pile when pile leans forwards

$b_K > b_F$ taper pile when pile leans backwards

Fig. R 118-1. Taper and compensating piles

The chamfering of the toes of either U- or Z-section sheet piles can lead to damage to the interlocks and is therefore not permitted.

If the unit spacing of certain stretches of sheeting must be maintained with great accuracy, the width tolerance must be observed. If necessary, compensating piles (fig. R 118-1b) must be inserted.

Driving can be assisted by shock blasting to R 183, section 8.1.10, predrilling, soil replacement or jetting to R 203, section 8.1.23. Rocky subsoils can be perforated by drilling boreholes and thus loosened to such an extent that the sheet piling can be driven in.

The energy required for driving is less and the driving progress greater, the greater the care used to position and guide the driven elements, and the better the hammer and driving procedure are in line with local conditions. The minimum penetration per blow for a drop-type hammer should be in accordance with the manufacturer's specification.

8.1.12 Driving of combined steel sheet piling (R 104)

8.1.12.1 General

In view of the long lengths, particularly for primary elements, the greatest possible care must be used when driving. This is the only way to ensure satisfactory success with the desired penetration and undamaged interlock connections.

8.1.12.2 Wall forms

Combined steel sheet piling (R 7, section 8.1.4) consists of primary and secondary elements.

Rolled or welded I-beams or I-shaped steel sheet piles are especially well suited for primary elements. In order to increase the section modulus, additional plates or connectors can be welded on. I-beams welded together to form a box pile or double piles can also be used. Special constructions, such as primary elements as welded box piles of U- or Z-section, which are connected to each other with web plates, may also be considered.

LN or SN-welded pipes are suitable primary elements with welded corner sections or single piles (see R 7, section 8.1.4). In special cases, interlock chambers are welded flush to the outer edge of the pipes in the pipe wall.

Secondary elements are generally corrugated steel sheet piles in the form of double or triple piles. Triple piles may require adequate stiffeners for structural and constructional reasons. Other suitable constructions may also be considered if they can properly transmit the acting forces to the primary elements and can be installed undamaged.

8.1.12.3 Types of piling elements

If secondary elements with interlock types 1, 2, 3, 5 or 6 as per R 67, section 8.1.6 or DIN EN 10 248-2 are used, matching interlocks or sheet pile cuts are to be attached to the primary elements by shear resistant welded joints. The outer and inner welding seams of these joints should be at least 6 mm thick. The single units of the intermediate piles are to be secured against displacement by welding or crimping their interlocks together.

If secondary elements with interlock type 4 as per R 67 section 8.1.6 are used, primary elements with conforming interlock design are to be chosen. Type 4 connectors are as a rule mounted on the secondary elements, but occasionally also on the primary elements.

When piles are driven to greater embedment depth, the connectors which are laterally mounted on the secondary elements are welded to the latter only on their upper end to maintain the rotation flexibility of the connection and reduce interlock friction during driving. The length of these welds depends on the length of the piles, embedment depth, soil conditions and such driving difficulties as will probably be encountered. Generally, the length of the weld will be between 200 and 500 mm. With especially long piles and/or heavy driving, an additional safety weld at the head of the pile is recommended. At lesser embedment depth, a shorter safety weld at the top of the piles is generally sufficient. Care must be taken that the driving cap covers the outer connectors, but not completely, so that enough play is left between primary elements in case it should be necessary to drive the secondary elements so that the elevation of their tops is below that of the primary elements.

If the connectors are mounted to the primary elements they are welded-on with shear resistant joints ($a \geq 6$ mm), if a higher moment of inertia and section modulus are desired. However, reduced rotational flexibility of the connection must be accepted.

If the primary elements consist of U- or Z-section sections connected to each other with web plates, the plates are to be welded continuously to the U- or Z-section sections on the outside, and inside at the ends of the primary elements for at least 1000 mm. The welding seams must be at least 8 mm thick. Furthermore, the primary elements must be stiffened at tip and head with wide plates between the web plates so the driving energy can be transmitted without damaging the primary elements.

When deep penetration and/or especially long primary elements are required, these should consist of box piles or double piles having wide-flange or box sheet pile sections because they provide the desired greater rigidity referred to the z-axis and greater torsional rigidity. The resulting increased driving costs must be taken into account.

8.1.12.4 General requirements placed on piling elements

In addition to the otherwise customary requirements in accordance with R 98, section 8.1.7, the primary elements must be straight with the gage as a rule no more than $\leq 1\,‰$ of the pile length. They must be free of warp and, in the case of long piles or deep penetration, must have adequate flexural and torsional rigidity.

The top of the primary elements must be finished plane and at right angles to the pile axis, and must be so shaped that the hammer blow is introduced and transmitted over the entire cross-section by means of a well-fitting driving cap. If reinforcement is placed at the toe of the pile, e.g. vanes for increasing axial bearing capacity, these must have a symmetrical layout so that the resultant force of the resistance to penetration will act along the centre of gravity of the pile, eliminating the tendency of the pile to creep out of alignment. Furthermore, the vanes should terminate far enough above the pile point in order to provide some guidance of the pile during driving.

The intermediate piles are to be so designed that they have the greatest adaptability to changes in position, and can thus follow acceptable deviations of the primary elements from the design position. In secondary elements with exterior interlocks to R 67, section 8.1.6 (fig. R 67-1) (Z-section), better adaptation to greater positional changes of the primary elements from the design position is possible with free rotation. In intermediate piles with interlocks on the axis (U-section), adaptation is possible only by deformation of the section. They must at the same time be so designed that horizontal sagging under later loading also remains within reasonable limits.

The interlocks must engage in a snug fit and be capable of absorbing such lateral tensile stresses as may occur (see R 97, section 8.1.6). Special care must be taken that matching interlocks fit properly and are not distorted.

8.1.12.5 Driving procedure

Driving must be carried out that the primary elements are embedded straight, vertically or at the prescribed batter, parallel to each other and at the designed spacing. Prerequisites for this are good guidance of the piles during positioning and driving; guides at two levels (top and bottom) are helpful. Furthermore, a suitable, heavy, adequately rigid and straight pile template, adapted to the length and weight of the piles is necessary. All equipments and methods for installing sheet piles must comply with DIN EN 996.

On building sites with extreme weather conditions and high waves, jack-up platforms have proven ideal. The position must be constantly checked because the jack-up platform can shift as a result of driving vibrations. Adequately large working pontoons suitable for the weather conditions with guide piles and heavy driving equipment are also suitable. Due to the weight of the primary elements and the hammers required, the primary elements are first positioned with a vibrator and subsequently driven to their final depth with heavy equipment. The secondary elements can be positioned and installed afterwards.

When driving tubes as primary elements, torsional movements can be prevented by guiding the interlocks.

Moreover, a satisfactory result will be more likely if a fixed guide waling is placed at the lowest practicable elevation. The spacing of the primary elements is indicated on the waling by means of welded-on frames, taking any width tolerances into account. Furthermore, the pile head shall be guided by means of the driving cap by the leads during driving to ensure that the top of the pile is always held in the position required by the design. Care must be taken to assure that the play between the pile and the cap and between the cap and the leads is always as small as possible. The driving sequence of the primary elements must be determined so that the pile toe encounters compacted soil uniformly on its total circumference, and never on only one side. This is achieved by driving in the following sequence:

$$1 - 7 - 5 - 3 - 2 - 4 - 6 \text{ (large driving step)}.$$

At least the following sequence should be observed:

$$1 - 3 - 2 - 5 - 4 - 7 - 6 \text{ (small driving step)}.$$

In general, all of the primary elements should be driven to full penetration without interruption. Subsequently, the secondary elements can be successively placed and driven.

In unusually deep water, or if the free portion of the sheet pile wall is very high, vertical guidance may be employed in driving the primary elements. For this purpose, use may be made of structural steel guide frames, which can be adjusted as necessary to fit exactly to the heights and widths concerned. The sides of these frames are fitted with connectors which match the primary elements. In addition, in order that the completed wall is straight, an arrangement for horizontal guidance must be provided above the water level.

In soil which is suitable for jetting, and which is free of rocks, the primary elements, and if necessary, also the secondary elements, can be driven with the aid of water jetting. In this case, the jetting devices are to be installed symmetrically and properly guided. Drifting of the piles from the design position can be prevented by careful handling.

The certainty of constructing a faultless, tight wall is increased, if a trench is excavated, as deep as practicable, before driving begins, thus increasing the guided depth of the wall, while at the same time decreasing the driving depth.

In boulders and hard soil layers replacement of soil is recommended. For building sites on land, the diaphragm wall method with adjusted sheet piling is feasible.

8.1.13 Observations during the installation of steel sheet piles, tolerances (R 105)

8.1.13.1 General

During the installation of steel sheet piling, the position, setting and condition of the elements must be constantly controlled and suitable measurements carried out to check when the design position is reached. Together with the correct starting position, intermediate phases are also to be checked, particularly after the first few metres of penetration. This should make it possible to detect even the slightest deviations from the design position (slant, out-of-line, distortion) or deformations of the pile head so that early corrections can be made and, if necessary, suitable counter-measures initiated.

The penetration, line and setting of the elements are to be observed frequently and with particular care when the subsoil is full of obstacles. If a driving element no longer moves, i.e. unusually slow penetration, the driving should be stopped straight away. In the case of continuous driving, the subsequent piles can be inserted first. Later a second attempt can be made to drive the protruding pile in deeper. In the case of unusual penetration of primary elements they have to be pulled and driven again under special precautions. Driven elements which move only with great difficulty just before reaching the design depth so that there is a risk of damages to the toe area, should not be driven further. Individual, shorter but undamaged elements are to be given preference to sheet piling which is true to specification but possibly damaged. Shorter embedment than final depth should not impair the total concept (e.g. undistribute groundwater flow) and stability.

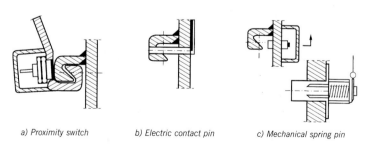

a) Proximity switch   b) Electric contact pin   c) Mechanical spring pin

Fig. R 105-1. Signal transmitters

If observed peculiarities such as extreme distortion or skewed driven elements give rise to the assumption that they have been damaged, an attempt should be made to inspect the piles by dredging them partially free and pulling them to investigate the subsoil for driving obstacles.

8.1.13.2 Interlock damage, signal transmitters

When a driving element runs out of the interlock connection (declutching), it is usually not possible for this to be detected by observations, particularly in the case of increasing penetration resistance.

Interlock damage cannot be completely ruled out even when driving is performed carefully, so checks are sensible to increase the safety. Such checks can take the form of visual inspections of the interlocks where these are visible, or may involve the use of interlock declutching detectors. Various systems are available which indicate the condition of the interlock either locally or over its full length (see fig. R 105-1).

Method a) measures continuously whether the interlock connection is still intact during driving. In method b), once the interlock of the threading pile reaches the signal transmitter, the contact is registered and indicated electrically, whereas in method c) a spring pin is tripped to provide a single mechanical indication.

8.1.13.3 Driving deviations and tolerances

As the driving depth increases, so do the deviations from the vertical. The following tolerances should be included in the calculations at the planning stage for deviations of the piles in compliance with DIN EN 12 063:

± 1.0% for normal soil conditions and driving on land,
± 1.5% of the driving depth for driving in water,
± 2.0% of the driving depth with difficult subsoil.

Verticality is to be measured at the top meter of the driving element. The deviation of the sheet piling top perpendicularly to the wall axis may not exceed 75 mm for driving on land and 100 mm for driving in water. For combined piling, the precise position of the primary elements has to fulfil strict requirements, therefore the tolerances are always to be fixed in any special case. An example of dimensional deviations in a combined sheet pile wall as shown in DIN EN 12 063, fig. 6, is helpful when specifying tolerances.

8.1.13.4 Measurements, equipment

The correct starting position and also intermediate positions of the driven elements can be checked easily by using two theodolites each checking the position in the y- or z-axis. This method should generally be prescribed for driving primary elements in combined piling walls. At the end of the driving procedure and after removal of the guides each primary element is to be surveyed in the installed position, to draw any necessary conclusions for driving the filling piles.

When working with spirit levels, these must be long enough (at least 2.0 m), if necessary with ruler. The use of a theodolite is however preferable. The checks are to be repeated at different points to compensate local irregularities.

8.1.13.5 Records

Records of the driving observations are to be kept according to DIN 12 699, section 10; the list in this standard corresponds to the preprinted forms of DIN 4026:1975 which has been withdrawn. Under difficult driving conditions, the driving curve for the whole driving procedure should also be recorded for the first three elements and then for every 20th element.

Modern pile drivers record the driving results on data carriers so that the information about the driving results can be made available by computers and suitable software directly. This is recommended particularly under difficult driving conditions in changing soils.

8.1.14 Noise protection, low-noise driving (R 149)

8.1.14.1 General remarks on sound level and sound propagation

The noise emissions of a machine are characterised by the acoustic power or the sound pressure level measured at a defined distance. The acoustic power is the sound energy emitted by the machine per unit of time. It does not depend on the ambient conditions and hence is a machine parameter that is assessed according to subjective criteria (e.g. human hearing). The principal method of assessment is the A-weighted sound power level, which is 10 times the logarithm of the ratio of the acoustic power to the reference power ($P_0 = 1$ pW $= 1 \cdot 10^{-12}$ W). The A-weighting reflects a filter for the frequency response of the human ear. The type of assessment is identified either by using the index L_{WA} or by a suffix to the unit of measurement dB(A).

The sound power level, being 10 times the logarithm of the ratio of the sound pressure of the machine to a reference pressure (in air: $p_0 = 20$ μPa), is a variable that depends on the measuring distance and the ambient acoustic conditions.

Other important parameters that can be used for estimating the noise emissions from a machine are the sound distribution over various frequency bands (third-octave, octave, narrowband), maybe also chronological fluctuations and directional characteristics.

Due to the logarithmic calculation in acoustics, it is easy to appreciate that a doubling of the number of sound sources raises the sound pressure

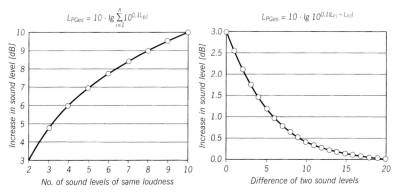

$$L_{PGes} = 10 \cdot \lg \sum_{i=1}^{n} 10^{0.1 L_{pi}}$$

$$L_{PGes} = 10 \cdot \lg 10^{0.1(L_{p1} - L_{p2})}$$

Fig. R 149-1a. Increase in sound level when several sound levels of the same loudness come together

Fig. R 149-1b. Increase in sound level for two levels of different loudness

level by 3 dB(A). However, investigations have revealed that the sound pressure level must rise by 10 dB(A) in order to be perceived by the human ear as "twice as loud". Fig. R 149-1a shows the increase in the total sound pressure level upon superimposing a number of equal sources. Sources with different sound levels have a totally different influence on the overall sound level. If the difference between the sound pressure levels of two sources is greater than 10 dB(A), the quieter source actually has no influence on the overall sound level (see fig. R 149-1b).

Consequently, measures to counteract noise can only be effective when initially the loudest individual noise levels are reduced. The elimination of weaker individual noise levels only makes a slight contribution to noise reduction.

Given ideal free-field propagation in the infinite semi-spherical room, the sound pressure of a specific sound source is reduced with every doubling of the distance by 6 dB(A) on account of the geometric propagation of sound energy on the four-fold surface.

$$\Delta L_p = -20 \cdot \log \frac{S}{S_0}$$

S = distance 1 to sound source [m]
S_0 = distance 2 to sound source [m]
ΔL_p = change in sound pressure [dB]

In addition, the sound is attenuated by up to 5 dB(A) in larger distances over grown uneven ground on account of air and ground absorption and, if existing, vegetation or buildings.

337

On the other hand, it must be taken into account that the simple sound reflection on a structure in the vicinity of the sound source or at concrete or asphalt surfaces can cause an increase of up to 3 dB(A) in the sound level, depending on the absorption and propagation degree of the surface. When there are several reflecting surfaces, each can be substituted by a theoretical mirror sound source with the same loudness as the original sound source and the resulting increase in level calculated taking account of the calculation rules for the interaction of several sound sources (see fig. R 149-1a).

With sound propagation over larger distances, it must also be taken into account that the decrease in sound level from meteorological influences, such as wind currents and temperature stratification, can have both a positive effect, i.e. in the sense of a larger decrease in sound level, and a negative effect. For example, a positive temperature gradient (increase in air temperature at altitude = ground inversion) results in an amplification of the sound level because of the deflection of the sound rays back to the ground at locations approx. 200 m away from the sound source. This effect is to be found in particular over water surfaces, which are generally colder than the ambient air which heats up more quickly, thus resulting in a positive temperature gradient, similar to the earth surface cooling down rapidly after sundown.

In interaction with the ground reflection, the curvature of the sound rays also means that the propagation of the sound remains limited to a corridor between the ground and the inversion layer, so that the geometric propagation attenuation is reduced to half.

The influence of the wind can be compared to that of temperature. Here again, the lower reduction in sound level in the wind direction is caused by a change in the horizontal wind speed with increasing height and the resulting deflection of the sound rays downwards. This effect is particularly noticeable on cloudy or foggy days, when the wind can be observed with a speed of up to 5 m/s as laminar air current. On the other hand, turbulence and vertical air circulation caused particularly during the day by sunshine can result in a higher reduction in sound level from scatter and refraction of the sound rays.

8.1.14.2 Regulations and directives

Special attention must be paid to the following:

- General Administrative Regulation for Protection against Construction Noise – Emission of Noise. The Federal and State Regulations for Reduction of Construction Noise, CARL HEYMANNS Verlag KG, Cologne, 1971.

- General Administrative Regulation for Protection against Construction Noise – Emission Measuring Methods. CARL HEYMANNS Verlag KG, Cologne, 1971 [180].

- Directive 79/113/EEC of the European Council dated December 19, 1978, for Harmonisation of the Legal Regulations of the Member States Referring to Ascertaining the Noise Emission Level of Construction Machinery and Equipment (ABl. EU 1979 No. L 33 p. 15) [181].
- 15th Regulation for Enforcement of the *Bundesimmissionsschutzgesetz* (Immissions Protection Act) of 10 November 1986 (Construction Plant Noise Act) [182].
- Directive 2000/14/EC of the European Parliament and Council of 8 May 2000, "Noise emission in the environment by equipment for use outdoors" (ABl. No. L 162 of 3 July 2000, p. 1; ber. ABl. No. L 311 of 12 December 2000, p. 50).
- ISO 9613-1, June 1993 edition: Acoustics – attenuation of sound during propagation outdoors – Part 1: Calculation of the absorption of sound by atmosphere.
- DIN ISO 9613-2, October 1999 edition: Acoustics – attenuation of sound during propagation outdoors – Part 2: General method of calculation (ISO 9613-2, 1996).
- 3rd Regulation to the *Gerätesicherheitsgesetz* (Device Safety Act) – Machine Noise Information Act of 18 January 1991 (BGBl. 15. 146; 1992, p. 1564; 1993, p. 704).
- VDI Directive 2714 [Jan 1988] Sound propagation outside – calculation method.

Furthermore, the higher regulations from the state legislation regarding the quiet periods to be observed at night and on public holidays must be complied with, together with the laws and regulations regarding occupational safety and protection (accident prevention regulations UVV, GDG i.V.m. 3., GSGV).

The tolerable noise immissions in the influence area of a noise source are stipulated graduated according to the need to protect the surrounding areas from construction noise. The need for protection results from the structural use of the areas as stipulated in the local development plan. If no local development plan exist or the actual use deviates considerably from the use intended in the local development plan, the need for protection results from the actual structural use of the areas.

According to the AVV publication on construction noise, the effective level generated by the construction machine at the emission site may be reduced by 5 or 10 dB(A) when the average daily operating period is less than 8 respectively 2 ½ hours. On the other hand, a nuisance surcharge of 5 dB(A) is to be added when clearly audible sounds such as whistling, singing, whining or screeching can be heard from the noise. If the evaluation level of the noise caused by the construction machine evaluated in this way exceeds the tolerable emission indicative value by more than 5 dB(A), measures should be initiated to reduce the noise.

This however is not necessary when the operation of the construction machines does not cause any additional danger, disadvantages or nuisances as a result of not merely occasionally acting outside noises.

The emission measuring procedure serves to register and compare the noises of construction machines. For this purpose, the construction machines are subjected to a minutely described measuring procedure during various operating procedures under defined overall conditions. As part of the standardisation of EU regulations, the noise emissions of a construction machine is now stated as sound power level L_{WA} referred to a semi-spherical surface of 1 m^2. The sound pressure level L_{PA} is still frequently used, referred to a radius of 10 m around the centre of the sound noise or at the operator's workstation in combination with occupational safety regulations.

Emission levels have been defined for the certification or use of various construction machinery which can be observed easily using state-of-the-art technology. Up to now, no binding indicative values have been stipulated for pile drivers.

8.1.14.3 Passive noise protection measures

In the case of passive noise protection, suitable measures prevent the sound waves from propagating unhindered. Baffles prevent the spreading of sound in certain directions. The baffle may not have any leaks or open joints. In addition, the baffle must be lined with sound-proofing material on the side facing the noise source, otherwise reflections and so-called standing waves will reduce the effect of the baffle. The effectiveness of a sound baffle depends on the effective height and width of the sound baffle and the distance from the sound source being baffled. Basically the baffle should be erected as close as possible to the sound source.

So-called encapsulated solutions with sound jackets, sound aprons or sound chimneys surround the sound source completely with sound-proofing material. A soundproofing jacket enclosing the pile driver and pile can reduce the noise level. Unfortunately, encapsulated solutions to reduce driving noise make working conditions much more difficult. As encapsulation makes it impossible to watch the driving procedure and thus increases the risk of accidents, the use of this type of passive noise protection is very limited. Encapsulation is expensive, it places considerable extra weight on the plant and seems to need constant repair. If some form of a screen cannot be avoided, a U-shaped mat of textured sheeting suspended over the hammer and piling element is preferable and still enables the pile driver operator to watch the driving procedure. A reduction of up to 8 dB(A) can be achieved in the screened direction. Up to now, sound chimneys have been used only for smaller driving units.

8.1.14.4 Active noise protection measures

The most effective and, as a rule, cheapest measure to reduce noise levels both at the workstation and in the neighbourhood is to use construction machines with low noise emissions. For example, compared to the striking hammer, the use of vibration hammers can reduce the sound level considerably. Hydraulic pressing-in of sheet piles and the insertion of primary tube elements using the pile drilling method can be classified as low-noise levels in any case. But the scope for using these low-noise driving methods depends essentially on the quality of the existing subsoil. Active measures also include the construction methods which make it easier to drive the sheet piling or piles into the subsoil, thus reducing the energy required for the driving procedure. Together with loosening borings or shock blasting and jets, these also include limited soil replacement in the area of the driven elements, as well as setting the sheet piling in suspension-supported slots. However, the feasible noise protection measures can be used only if they are permitted by the soil and construction conditions.

8.1.14.5 Planning a driving site

In planning the construction procedure, an attempt should be made to keep the anticipated environmental nuisance of the alter building site to an absolute minimum. Necessary baffling measures are cost-intensive and frequently not taken into account in the calculation. The assurance and observance of only short construction periods with severe disturbances followed by adequately long periods of low or no noise, e.g. over the midday period – should be striven for. If a certain reduction in capacity has to be taken into account in the calculations from the very beginning, then this must be mentioned separately and clearly in the tender.

8.1.15 Driving of steel sheet piles and steel piles at low temperatures (R 90)

At temperatures above 0 °C and with normal driving conditions, steel sheet piles of all steel grades can be driven without hesitation.

If driving has to be done at lower temperatures, special care is required in the handling of the driven elements as well as during driving.

In soils favourable for driving, driving is still possible down to temperatures of about –10 °C, particularly when using S 355 GP and higher steel grades.

Fully killed steels as per DIN EN 10 025 are however to be used when difficult driving with high energy expenditure is expected, and when working with thick-walled sections or welded driven elements.

If even lower temperatures prevail, steel grades with special cold workability are to be used.

8.1.16 Repairing interlock damage on driven steel sheet piling (R 167)

8.1.16.1 General

Interlock damage can occur during driving of steel sheet piles or from other external actions. However, the risk is the less, the better the recommendations dealing with the design and construction of sheet piling structures are observed with care and diligence. Please refer especially to the following recommendations; R 34, section 8.1.3, R 73, section 7.4, R 67, section 8.1.6, R 98, section 8.1.7, R 104, section 8.1.12, and R 118, section 8.1.11.

These recommendations also contain numerous possibilities of precautions to limit the risk.

If damages still occur, one advantage of structural steel is that it can be repaired with comparatively simple means, and the repair possibilities are extremely adaptable.

8.1.16.2 Repairing interlock damage

If the driving behaviour shows that interlock damage may be expected over an extended area and the sheet piling cannot be extracted again, e.g. due to pressure of time, repairs are considered consisting chiefly of large-area grouting of the soil behind the wall. Particular success was achieved when repairing combined steel sheet piling with the high-pressure injection method (HDI) (see fig. R 167-5). For permanent structures however, the interlock damage should be additionally secured in the following manner.

Individual interlock damage is subsequently repaired on the spot. Solutions are required that not only seal the interlock but also restore the necessary load-carrying capacity. This is especially important for corrugated sheet pile walls.

The type of sealing of interlock breaks depends, above all, on the size of the interlock opening and on the sheet pile section. Various methods were employed in the past. Repair work is generally carried out on the outboard side. Smaller interlock openings can be closed with wooden wedges. More extensive damage can be temporarily sealed, e.g. with a rapid-setting material, such as lightning cement or two-component mortar, which will be placed in sacks. For permanent safety however, the opening must be fully bridged with steel parts. This applies especially to the parts of a damaged sheet piling which protrude beyond the quay face. Thereby an adequate and secure fastening of the forward placed steel parts at the sheet piling must be provided. In addition, the damaged area must be concreted in order to prevent the soil from being washed out. The concrete installed may need to be reinforced in order to make it less vulnerable to the impact of ships. It is recommended that an additional protective layer, e.g. of ballast, be installed in the bottom area to prevent scouring.

The work must be executed mostly below water and therefore always with diver assistance. Very high demands are therefore to be placed on the technical ability and reliability of the divers.

The sealing elements installed must extend at least 0.5 m, but better 1.0 m below the design bottom (R 36, section 6.7) in front of the waterfront structure. Scouring expected from currents in the harbour water or screw action must be taken into account.

The smoothest possible sheet piling surface on the water side is to be striven for. Protruding bolts, for example, are to be burnt off after the damaged area has been concreted, once they have fulfilled their function as formwork element. The steel plates and the sheet piling are to be joined to the concrete with anchor elements, so-called rock claws. When repairing sheet piling, the waling frequently located on the land side has to be taken into account, as well as possibly existing sheet piling drainage, including gravel filter.

The solutions shown in figs. R 167-1 to R 167-5 for the repair of underwater areas have proven successful in practice, but lay no claim to completeness. Interlock damage on combined sheet piling is repaired if still possible by rear driving of secondary elements (fig. R 167-4) and otherwise from the water side according to fig. R 167-3.

Repairs in accordance with figs. R 167-1 to R 167-3 presume that there are no obstructions to the repair work from leaking soil, either because either there is no prevailing pressure difference behind the wall or the temporary sealing measures described above have a sufficiently stabilising effect.

If for reasons of soil mechanics or for technical, economic or other reasons, such temporary sealing measures are not possible, driven elements combining all the necessary functions can be inserted as shown in fig. R 167-4.

In *rear driving*, e.g. according to fig. R 167-4, the pressure difference behind the wall presses the driving element against the intact parts of the sheet piling during dredging on the water side. This applies to both combined and corrugated sheet piling.

In *forward driving*, suitable measures must be taken to ensure that the base of the driving element is always pressed against the primary element, in other words, it has to be located as near to the primary element as possible. Before dredging takes place, the head of the driving element must be secured to the primary element, using hammerhead screws, for example. The elevation of the first dredging cut must be coordinated with the bearing capacity of the driving element between this upper fixed support and the lower, more flexible earth support. After this has been done, the driving element must be secured to the primary element in the exposed area, etc.

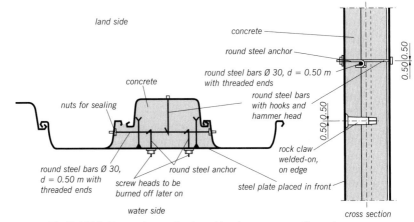

Fig. R 167-1. Executed example of repairing damage at a small opening

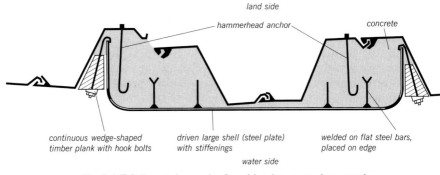

Fig. R 167-2. Executed example of repairing damage at a large opening

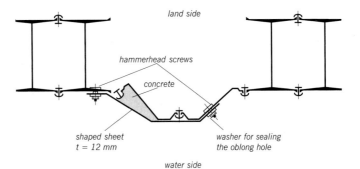

Fig. R 167-3. Executed example of repairing damage at an opening in combined sheet piling

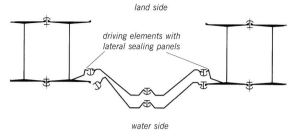

land side

driving elements with
lateral sealing panels

water side

Fig. R 167-4. Executed example of repairing damage by rear driving with a driving element at an opening in combined sheet piling

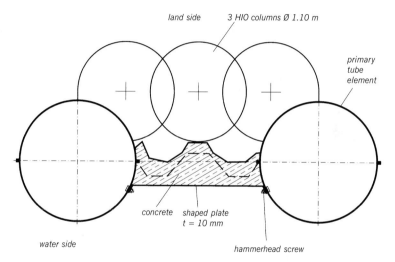

land side   3 HIO columns Ø 1.10 m

primary
tube
element

concrete   shaped plate
t = 10 mm

water side

hammerhead screw

Fig. R 167-5. Executed example of repairing damage to declutching in combined tubular sheet piling

During dredging operations, constant investigations of the repaired area by divers are necessary. Any local leaks can be sealed by injections over an appropriate area.

8.1.17   Design of pile driving templates (R 140)

8.1.17.1   General
Insofar as driving operations are not possible from existing or hydraulically filled land, the following possibilities may be considered:

(1) Driving from a pile driving templates.
(2) Driving from a mobile working platform.

8.1.17.2 Design of pile driving templates

The pile driving templates can be built of either steel, timber or reinforced concrete piles. The design of the template must make allowance for the following, especially for reasons of economy:

(1) The trestle piles can be driven from a pontoon or a barge. The inaccuracies resulting from this method must be taken into account in the construction. Occasionally, it may be possible to drive some of the piles from existing ground or from an existing template.

(2) The length of the template depends on the progress made by the pile driver, and on whether the trestle will be needed for work later on. Parts which are no longer needed are reclaimed and used to extend the template. Templates which are no longer required, are extracted. If this cannot be done, even with jetting and vibrating, they should be cut off below the elevation of the existing and planned harbour bottom. In this connection, dredging tolerances and possible future deepening of the harbour should be taken into consideration. When extracting the templates, any side effects on the structure must be taken into consideration.

(3) The design and construction of the template should be simple, keeping in mind the need to salvage and reuse the structural elements, with minimum waste.

(4) The danger of scour should be given special attention in the case of uniform fine sand on the harbour bottom. To this end, the piles should be of a generous length as a matter of course. Moreover, the area in the vicinity of the trestle should be constantly observed during the construction period, especially if strong currents and sand fill are present. An example of the latter is a foundation which has been improved by soil replacement. If scour occurs, the affected area should be immediately filled with graded gravel.

(5) The trestle piles should be driven well clear of the permanent structure, so that they will not be dragged down by the driving of the structural piles. The separation required depends on the type of soil. Open holes left in the bottom after extraction of piles from cohesive soils should, if necessary, be filled with suitable material.

(6) Pile driving templates built close to the shore can consist of a row of piles in the water and a bearing on shore for the rail beam. Where such a shore connection does not exist, the trestle is supported by two or more rows of piles. Structure piles can also be used, if initially designed for that purpose.

8.1.17.3 Loading assumptions

The design of the foundation of the driving template should take account of the inherent loads and the operating conditions etc. corresponding to DIN EN 996, section 4.1.

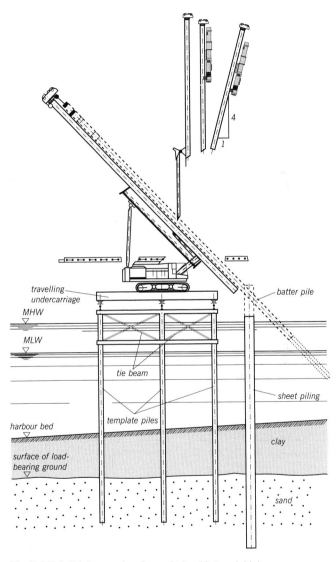

Fig. R 140-1. Driving template for vertical and battered driving

Together with the loads from the pile driver and the undercarriage, and, when used, the tower crane, the current pressure load, wave action and ice pressure load are to be taken into account for the driving template (template piles and tie braces). Insofar as the driving trestle is not safeguarded by additional measures (safety dolphins) from ship contact – e.g. pontoons or similar equipment delivering the driven elements to the site, vessel impact and, where applicable, also line pull forces of up to 100 kN each are to be taken into account in rating the driving trestle and its piles, even in the most unfavourable possible position.
Please refer to DIN EN 996.

8.1.18 Design of welded joints in steel sheet piles and driven steel piles (R 99)
This recommendation applies to welded joints in steel sheet piles and steel piles of every type.

8.1.18.1 General requirements
The design of the joints is carried out according to EN 1993-1-8. Design and fabrication must comply with the requirements of DIN EN 12 063 or 12 699 at least; in the case of more demanding requirements please refer to DIN 18 800 part 7, and DIN EN 729-3. Shelters to protect against wind and weather must be provided for on-site welding. The joints to be welded are to be thoroughly cleaned, kept dry and, if necessary, preheated.

8.1.18.2 Materials
(1) Parent metals
Sheet piling steel grades according to R 67 and steels according to DIN EN 10 025 can be used with welding suitability described in R 67, section 8.1.6.4.
The steel grades used must always be verified by an acceptance test certificate 3.1 B to DIN EN 10 204, which indicates both mechanical and technological properties as well as the chemical composition (R 67, section 8.1.6.1).

(2) Filler metals
The filler metals are to be selected by the welding engineer of the welding contractor licensed to perform the work, taking into account the proposals of the sheet piling and steel pile manufacturer.
Basic electrodes or welding materials with a high degree of basicity may generally be used (filler wire, powder).

8.1.18.3 Classification of the welded joints
(1) General
The butt joint is intended to replace as fully as possible, i.e. to 100%, the steel cross-section of the sheet piles and driven piles.

The percentage of the effective butt weld sectional area in relation to the full steel cross-section, called effective butt coverage in the following, is however contingent on the construction type of the elements, the offset of edges at the joint ends and on the prevailing conditions on site (table R 99-1).

If the steel cross-section of the sheet piles or piles is not achieved by the butt weld cross-section, and if a full joint is required for structural reasons, splice plates or additional sections are to be used.

(2) Effective butt coverage
The effective butt coverage is expressed as a percentage and is the ratio between the butt weld cross-section and the steel cross-section of sheet piles or piles.

Table R 99-1. Effective butt coverage expressed in %

Construction type of sheet piles or piles	Effective butt coverage as %	
	in the work-shop	under the pile driver
a) Pipe sections, calibrated joint ends, root welded through	100	100
b) Piles of I-shaped sections, box sheet piles Cross section reduction with material removed from the throats	80–90	80–90
c) Sheet piles — Single piles	100	100
c) Sheet piles — Double piles Interlock area only with one-sided welding U-shaped sheet piles Z-shaped sheet piles	90 80	~ 80 ~ 70
d) Box piles consisting of individual sections Individual sections to be jointed then assembled Box pile to be jointed	100 70–80	50–70

8.1.18.4 Making weld joints

(1) Preparation of the joint ends
The cut of the section to be welded should lie in a plane at right-angles to the axis; an offset in the joint is to be avoided.

Special attention is to be paid to ensuring a good fit between the cross-sections and, in the case of steel sheet piling, to preservation of a fair passage in the interlocks as well. Differences in width and height between pieces to be welded should not exceed ± 2 mm so that offsets in the welding edges will not exceed 4 mm.

It is recommended that hollow piles composed of several sections first be manufactured in the required full length, and then cut into working lengths after having been suitably match-marked (e.g. for transportation, driving, etc.).

The ends intended for the butt joint are to be checked for doublings for a length of about 500 mm.

(2) Edge preparation for welding

In the shop, butt welding is generally performed with V- or Y-groove welds. Both edges of the butt joint are to be suitably prepared.

If a butt joint must be made in the field on driven steel sheet piles or steel piles, the top of the driven element must first be trimmed as required by R 91, section 8.1.19. The extension piece is to be prepared for a butt weld, with or without a root pass.

(3) Welding procedure

All accessible surfaces of the butted section are to be fully connected. Where possible, the roots are gouged and sealed with root passes.

Root positions which are not accessible require a high degree of accuracy in fitting the sections to be joined and careful edge preparation.

The proper welding sequence depends on various factors. Special care must be taken that residual tensile stresses from the welding process will not be superimposed on tensile stresses which will occur when the structure is in service.

8.1.18.5 Special details

(1) Joints are to be placed as far as possible into a cross-section where stresses are low, and staggered by at least 1 m.

(2) In preparing H-shaped sections for welding, the throat areas of the web are to be drilled out in such a way that the drilled openings form a semi-circle with a diameter of 35 to 40 mm open to the flanges and sufficient to ensure full penetration of the flanges, using a root pass. The surface of these openings must be machined to remove any notches after completion of welding. Run-on and run-off plates must be provided for the flange welds in the vicinity of the butt joint in order to achieve a clean termination at the flange. After removing the plates, grind the edge of the flange to remove any notches.

(3) If flange splice plates are required for the butt coverage for structural reasons, the following rules are to be observed:

a) The thickness of splice plates shall not exceed the thickness of the spliced section members by more than 20%. The maximum thickness of the splice plates is 25 mm.

b) The width of the plates shall be dimensioned in a way that they can be welded all-round on the flanges without end craters

c) The ends of the plates shall be tapered to $^1/_3$ b with each side converging at a rate of 1 : 3.

d) Before the plate is placed, the butt weld is to be ground down flush.

e) Non-destructive tests must be completed before the splice plates are placed.

(4) If butt joints in service are not subjected to predominantly static load within the meaning of R 20, section 8.2.6, splice plates over the joints are to be avoided wherever possible.

(5) If butt joints are scheduled e.g. as a result of transportation or driving practice, only killed steels should be used.

(6) Butt joints under the pile driver are to be avoided as far as possible for economic reasons and because unfavourable weather conditions could have a negative effect on the welding.

(7) If welded joints have unavoidable openings through which the soil could leach out, such openings must be sealed in a suitable manner (see section R 117, section 8.1.20).

8.1.19 Burning off the tops of driven steel sections for loadbearing welded connections (R 91)

If the tops of driven steel sheet piling or steel piles are to be fitted with welded loadbearing connections (e.g. welded joints, bearing fittings or the like), these may not be placed in areas with driving deformation. In such cases, the top ends of the piles must be cut off to a point below the limit of the deformation, or all welding seams must be run beyond the deformed area.

This measure is to remove embrittlement which would have a detrimental effect on the loadbearing welded connections.

8.1.20 Watertightness of steel sheet piling (R 117)

8.1.20.1 General

Because of the required play in the interlocks, steel sheet piling walls are not absolutely watertight, which is however not necessary in general. The degree of watertightness of interlocks threaded at the shop (W interlocks) is mostly less than for the driven threaded interlocks (B interlocks) which are partially filled with soil in the area below ground surface.

Progressive self-sealing (natural sealing), due to corrosion with incrustation as well as the accumulation of fine particles in sediment-laden water, can usually be expected over the course of time. Appendix E of DIN EN 12 063 can be used to estimate the permeability of sheet pile wall joints. The suppliers of steel sheet piles can provide seepage resistances for the various synthetic seals according to section 8.1.20.3. If very strict requirements have been specified for the watertightness, realistic tests must be carried out. With a wide scatter of results, a safety factor should be applied to the mean value.

8.1.20.2 Assisting the natural sealing process

With hydrostatic pressure difference acting on one side and at walls standing free in water, e.g. at an excavation pit sheeting, the natural sealing process can be assisted if necessary by pouring in sealing material, e.g. boiler ash etc., provided the interlocks are constantly under water and the sealing materials are poured close to the interlocks into the water. When dewatering excavations, especially high pumping rates are necessary at the beginning, so that the greatest possible differential head is quickly built up between inner and outer water.

Thereby, the interlocks join together well. Furthermore, this causes an adequately strong flow of water to the excavation interior and thus effectively flushes the sealing materials into the interlocks. The costs for sealing and pumping are to be adjusted to each other optimally. However, this sealing method produces no permanent success under fluctuating hydrostatic pressure difference from the sides and with movements of the sheet piling in free water through wave impact or swells, etc.

8.1.20.3 Synthetic seals

Sheet piling interlocks may be artificially sealed both before as well as after installation.

(1) Sealing method before the sheet piles are driven
a)   Filling the interlocks with a durable, environmentally compatible, sufficiently plastic compound, namely the W-interlocks at the shop and the B-interlocks at the plant or at the site.
b)   A noticeable improvement in watertightness is achieved by fitting an extruded polyurethane seal to the B-interlocks in the factory.
     When employing this method, also those B interlocks can be sealed which are no longer accessible after driving, e.g. the areas below the bottom of excavation or waters. Please refer to c) concerning the position of the sealed joint.
     With both sealing methods, the achievable watertightness of the interlocks depends on the hydrostatic pressure difference and the

embedment method. Driving places little stress on the seal, since the movement of the pile in the interlock takes place in one direction only. In the case of vibration, the stress is greater, depending on the driving speed. It cannot be ruled out here that the seal will be damaged as a result of friction and temperature increases.

c) Interlock joints of W interlocks are welded tight, either at the plant or at the site. In order to avoid cracks in the watertight seam during the installation procedure, additional seams are required, e.g. on both sides at the head and foot of the driven element, as well as counter-seams in the area of the tight seam. The tight seam must lie on the correct side of the sheet piling, e.g. on the air/water side for dry docks and navigation locks.

d) The method known from the refurbishment of existing sites, i.e. "sheet pile wall with small bored piles" can be used around the B interlocks specifically for seals to dams in waterways with the highest demands on watertightness. W interlocks are welded tight prior to installation. The bored piles with diameters of approx. 0.1 to 0.3 m are installed with a casing. The sheet piles are then driven before the concrete of the piles has fully hardened. This method can be used for depths down to about 15 m. A combination of B interlocks with the method according to b) above can prove beneficial.

(2) Sealing methods after installation of the sheet piling

a) Caulking the interlock joints with wooden wedges (swelling effect), with rubber or plastic cords, round or profiled, with a caulking compound capable of swelling and setting, e.g. fibrous material mixed with cement.

The cords are tamped in with a blunt chisel. Light pneumatic hammers have also proven themselves.

The caulking work can also be carried out for water-bearing interlocks. B interlocks can generally be made watertight better than squeezed W interlocks.

The interlock joint must be cleaned of clinging particles of soil before caulking is begun.

b) The interlock joints are welded tight. As a rule, these are only the B interlocks, as the W interlocks were already welded tight before the piles were driven, see section 8.1.20.3 (1) c).

Direct welding of the joint is possible for dry and properly cleaned joints. Water-bearing joints should be covered with steel plates or sections, which are welded to the sheet piling by two fillet welds. With this method, a fully watertight sheet piling may be achieved.

c) At the completed structure, plastic sealing compounds may be placed into the accessible joints above the water level at any time, or PU foam injected into the interlock chamber by impact or screw nipples.

Care must be taken in such cases to ensure that the flanks of the plastic compound are applied on a dry surface. This can be achieved following previous temporary sealing of the joints.

In the case of box sheet piling with double locking bars, sealing can also be achieved by filling the emptied cells with a suitable sealing material, e.g. with underwater concrete.

Special emphasis is drawn to the fact that interlocks which are not sealed behave like vertical drains in strata of low permeability.

At greater water table differences and especially where wave action may possibly prevail, the interlocks must be sealed with extra care if fine sand or coarse silt is present behind the sheet piling, as this non-cohesive material can be easily washed out through the voids of the interlocks.

8.1.20.4 Sealing of penetration points

Aside from the watertightness of the interlocks, special attention must be paid to adequate sealing of the points of penetration of anchors, waling bolts and the like.

Lead or rubber washers are to be placed between sheet piling and base plates, as well as between base plates and nuts. In order not to damage the sealing washers, the anchor must be tensioned by means of a turnbuckle, and the waling bolts with the nuts on the waling side.

The holes in the sheet piling for the waling bolts and, if necessary also for the anchor, are to have ridges cleanly removed so that the base plate rests fully.

8.1.21 Waterfront structures in regions subject to mining subsidence (R 121)

8.1.21.1 General

Predictable ground movements and their changes in the course of time are to be taken into account in the design work. A distinction must be made here between:

(1) movements in the vertical direction, subsidence, and
(2) movements in the horizontal direction.

Since the movements follow generally in different time intervals, subsidence, tilting, torsion, tearing and squeezing may occur in changing sequence.

Where local subsidence occurs, the elevation of the groundwater table usually remains unchanged. This is also true of the water level in navigation channels.

Insofar as any information might be available on planned or current underground mining operations, the company carrying out this work should be informed of proposed waterfront construction as early as possible. The plans must be submitted to these companies, and it must be left to their discretion whether to propose safety measures and have them carried out, or assume responsibility for repairing good any damages caused by the mining operations.

It is not usually possible to predict the locations or extent of failures. If at the outset the responsible mining company is not willing to undertake measures to prevent mining damages, or does not consider such measures necessary, the owner cannot be advised to undertake any extra expenditure for safety measures in advance. However, it would be wrong to adopt types of construction which are especially susceptible to damage or failure due to mining operations, or which are especially difficult to repair, if it appears likely that harmful effects from mining operations may occur later. It must be mentioned in this connection that solid waterfront structures have frequently been heavily damaged by being torn or squeezed, as well as by torsion. In contrast, no appreciable damage to structures of U- or Z-section steel sheet piles has been confirmed up to now. Such structures can therefore generally be recommended at waterfront regions where mining subsidence may occur. The following suggestions must be given particularly careful consideration in planning, design, calculation and construction.

8.1.21.2 Design advice

The magnitude of the earth movements to be expected must be ascertained from the responsible mining company. The ground elevations and load assumptions are determined on the basis of this data.

Vertical movements can make it necessary to establish the elevation of the top of the waterfront structure higher to allow for the probable magnitude of subsidence; this is generally more economical than increasing the height of the wall following subsidence. If subsidence of varying depth is expected along the length of the waterfront structure, about which the mining surveyor can make quite reliable and correct predictions, and if it is not planned to increase the height of the structure later on, the elevation of the top of the structure, when placed, must be increased by varying amounts to compensate for the anticipated local subsidence. Thus in the final state, the top of the wall will reach a horizontal position throughout most of its length. It is frequently more practicable however to increase the height of the waterfront structure only in later years. However, the resulting load increases on sheet piling and anchoring should be taken into account in the original design, in order to avoid the need for later reinforcing, which is as a rule quite expensive. The effect of the hydrostatic pressure difference is also to be determined exactly for all stages of the height increase and taken into account in the calculations.

Longitudinal pulling and squeezing of a waterfront structure, built with U- or Z-section sheet piling generally do not suffer from damages because the accordion effect enables the structure to accommodate ground movements.

A transverse push against the waterfront structure creates a negligibly slight displacement of the wall toward the water. A pull transverse to the water structure may severely overstress the anchor rods only if unnecessarily good stability exists in the lower failure plane due to overlength anchor rods. This can also be the case with firmly embedded overlength anchor piles. If dredging should become necessary in front of the waterfront structure to meet a temporary requirement, the harbour bottom shall not be dredged deeper than is absolutely necessary to expose as little as possible of the sheet piling.

8.1.21.3 Advice on design, calculations and construction

Beyond taking into consideration the interim stages and the final stage, the waterfront sheet piling generally does not require any overdesigning, unless this is requested by and paid for by the mining company. This also applies to a reinforced concrete capping beam and its reinforcement, provided that it steel remains above water after the mining subsidence takes place. Any damaged places can easily be demolished and repaired afterwards.

In order to minimise the susceptibility of the structure to harmful mining effects, the portion of the sheet piling above the anchoring should be as little as possible, so that the anchor rods and waling should be placed as close as possible below the top of the sheet piling; this is why the sheet piling moment should not be reduced for active earth pressure redistribution due above all to expected pulling.

The steel grades for the sheet piling can be chosen according to R 67, section 8.1.6. For the capping beam and the waling, the steel grades S 235 J2G3, S 235 J2G4 and S 355 J2G3 should be selected in accordance with DIN EN 10 025.

The latter also applies to anchor rods. If tie rod rods are used, upsetting in the thread area of the anchor rod is permissible, if requirements as per R 20, section 8.2.6.3 are fulfilled. Upset tie rods offer the advantage of a grater elongation path and higher flexibility than tie rods without upsetting in the thread area; besides this, they are easier to install and cheaper.

When the sheet piles are delivered, special attention should be paid to inspecting the interlock tolerances for compliance with R 67, section 8.1.6.

The motion possibilities of the sheet piling are not impaired in horizontal direction when the interlock joints are welded together, e.g. because of watertightness. However, the interlock welding does hinder the vertical possibilities of motion of the sheet piling. This can still be maintained to an adequate degree however, if the interlock joints are only welded together at certain intervals. In order to also ensure the watertightness in these joints, they are filled with an elastic, profile sealing compound before being joined at the shop, which does not hinder the vertical motions.

Accessible interlock joints can also be sealed at the site, e.g. by installing an elastic sealing compound in front of the joint. This compound is supported by a plate construction, which will not deter vertical movements. For the other structural members, welded constructions are to be avoided if possible, if they impair the motion possibilities of the sheet piling.

The foregoing remarks apply similarly to the interaction of reinforced concrete structural members and the sheet piling. It is especially important that the flexibility of the sheet piling is not limited by the presence of solid structural members. The waterfront structure and the craneway are to be designed on separate foundations to give opportunities for independent settlement and realignment. The same applies to a reinforced concrete capping beam, whose expansion joints should be spaced at about 8 to 12 m, depending on the magnitude of the expected varying subsidence. If the craneway is not laid on ties, see R 120, section 6.16.2.1 (2), but rather on reinforced concrete, the reinforced concrete beams should be connected to each other by sturdy tie beams to maintain a true gauge.

It is advisable to omit electric contact line channels and depend on trailing cables instead.

Walings consisting of two U-channels are preferable to other arrangements because the waling bolts can more easily adapt themselves to deformations. They are to be generously designed and to be so manufactured that waling reinforcing will not become necessary later on.

Oblong holes should be provided in the joints of the walings and the steel capping beams to allow longitudinal movement of the wall. As a substitute, circular holes may be enlarged carefully with circle oxycutters. They are to be finished as far as necessary in order to avoid or eliminate notches which can produce cracks in the steel.

If a sheet piling wall must later be increased in height, this should be anticipated and taken into account in the original design of the capping beam (simple dismantling).

Anchor connections in a capping waling are to be avoided.

Horizontal or gently sloped anchor rods are recommended for the lowest possible additional stresses produced by differential settlement between anchorage and wall. The anchor connections are to be made completely flexible. The end hinges are to be placed if possible in the waterside trough of the sheet pile section, so that they are accessible and can be easily observed.

357

8.1.21.4 Observations of structures

Waterfront structures in regions subject to mining subsidence require periodic inspection and checks on measurements. Even if the mining company is liable for any damage, the owner of the facility still remains responsible for its safety.

8.1.22 Vibration of U- and Z-section steel sheet piles (R 202)

8.1.22.1 General

Vibratory hammers are used to install piles by means of vibration. They generate vertically directed vibrations through unbalances rotating synchronously in opposition. The vibrator must be connected rigidly with clamping jaws to the material being driven, whereby the soil is induced to co-vibrate. This can considerably reduce driving resistances in the soil such as skin friction and toe resistance.

Good knowledge of the interactions of the vibrating hammer, the material being driven and the soil is an important prerequisite for planning an operation.

Please refer to the influences of soil and driving material as mentioned in R 118, section 8.1.11.3, and R 154, section 1.8.3.2.

In terms of bearing capacity, settlement behaviour and soil properties, the effects of vibration when installing foundation elements is not identical with the effects due to installation with impact hammers. In particular loading tests are recommended when combined tension and compression forces are expected.

8.1.22.2 Terms and parameters for vibratory hammers

Important terms and parameters are as follows:

(1) Type of drive
- electrical
- hydraulic
- electro-hydraulic

(2) Driving power P [kW]
This ultimately determines the efficiency of the hammer. At least 2 kW should be available per 10 kN centrifugal force.

(3) Effective moment M [kg m]
This is the product of the total mass m of the unbalances, multiplied by the spacing r of the centre of gravity of the individual unbalance from tits axis of rotation.

$$M - m \quad r \,[\text{kg m}]$$

The effective moment also plays a part in determining the vibration width or amplitude.

(4) Revolutions per minute n [rpm]
The rpm of the unbalanced shafts have a quadratic effect on the centrifugal force. Electric vibration hammers operate with constant speed and hydraulic hammers with fully variable speed adjustment.

(5) Centrifugal force (exciting force) F [kN]
This is the product of the effective moment and the square of the angular velocity

$$F = M \cdot 10^{-3} \cdot \omega^2 \text{ [kN]} \quad \text{where} \quad \omega = \frac{2 \cdot \pi \cdot n}{60} \text{ [s}^{-1}\text{]}$$

In practice the centrifugal force is a comparative variable of different machines. However, the revolutions per minute and the effective moment at which the optimum centrifugal force is reached must also be taken into account here.

The latest generation of vibrators includes the option of a stepless variable speed and an adjustable static moment during operation. The advantage of such vibrators is that they can be started with an amplitude of zero, free from all resonance. Only upon reaching the preselected speed are the weights extended and adjusted. This avoids the undesirable peaks during start-up and slow-down.

(6) Vibration width S, amplitude $\bar{x}$ [m]
The vibration width S is the total vertical shift of the vibrating unit in the course of one revolution of the unbalances. The amplitude $\bar{x}$ is half the vibration width. In the manufacturer's equipment lists, the quoted amplitude – the value of S is frequently mistakenly quoted for $\bar{x}$ – is the quotient of the effective moment (kg m) and the mass (kg) of the oscillating vibrator.

$$\bar{x} = \frac{M}{m_{\text{Ham,dyn}}} \text{ [m]}$$

On the other hand, the "working amplitude" $\bar{x}_A$ necessary in practice is an unknown variable. The divisor hereby is the total co-vibrating mass.

$$\bar{x}_A = \frac{M}{m_{\text{dyn}}} \text{ [m]}$$

where $m_{\text{dyn}} = m_{\text{Ham,dyn}} + m_{\text{Pile mat}} + m_{\text{Soil}}$. In prognoses $m_{\text{Soil}} \geq 0.7 \cdot (m_{\text{Ham,dyn}} + m_{\text{Pile mat}})$ should be assumed.
A theoretical "working amplitude" of $\bar{x}_A \geq 0.003$ m should be aimed for.

(7) Acceleration a [m/s²]

The acceleration of the pile material acts on the grain structure of the soil encountered. The grain structure stratification should be constantly moved during vibration in order to approach the "pseudo-liquid" state in the ideal case.

The product of the "working amplitude" and the square of the angular velocity yields the acceleration "a" of the driving material.

$$a = \overline{x}_A \cdot \omega^2 \ [\text{m/s}^2] \text{ where } \omega = \frac{2 \cdot \pi \cdot n}{60} \ [\text{s}^{-1}]$$

Experience shows that $a \geq 100$ m/s² is desirable.

8.1.22.3 Connection between equipment and driving element

An essentially rigid connection must be made between the hammer and the driving material with the clamping jaws. Hydraulic clamping jaws are mostly used for this. Since the hammer should be located on the centroidal axis of the driving material during vibration as during driving, double clamp is recommended for double piles. The following factors should be considered in order to achieve the optimum transfer of the energy into the pile material:

- profile
- method of clamping
- driving resistance, especially interlock friction

The clamping force as well as the number and position of the clamping jaws are to be chosen accordingly.

8.1.22.4 Criteria for selecting a vibrator

For idealised (uniform, re-arrangeable and water-saturated) soils, a hammer should be selected with 15 kN centrifugal force as the minimum for each m driving depth and with 30 kN centrifugal force as the minimum for each 100 kg driving material mass.

$$F = 15 \cdot \left(t + \frac{2 \cdot m_{\text{Pile,mat}}}{100} \right) \ [\text{kN}]$$

where:

t = embedment depth [m]
$m_{\text{Pile,mat}}$ = driving material mass [kg]

Alternatively, refer to the details of the driving equipment manufacturer or to computer-assisted prognosis models in order to estimate the penetration behaviour (see [251, 252, 253]). A calibration test is recommended for larger construction projects.

8.1.22.5 General experiences

(1) The action and effects of the vibration are virtually impossible to predict.

If the vibration is effective and penetration speeds ≥ 1 m/min are achieved, damaging effects are unlikely. Penetration speeds ≤ 0.5 m/min should be accompanied by measurements. Brief slower penetration speeds (e.g. through consolidated strata) can occur.

If there are structures within the area affected by the vibration, a forecast should be drawn up in order to select plant and methods that guarantee that the vibrations do not exceed the guidance values of DIN 4150 parts 2 and 3. However, this does not rule out damage caused by settlement.

(2) In soils which are not very amenable to rearrangement or in dry soils, jetting can be used (R 203, section 8.1.23).

Predrilling with small advance intervals or soil exchange should also be considered as aids.

(3) The compaction effect mentioned in R 154, section 1.8.3.2, is more likely to occur at high revolutions per minute. It can be expedient in such cases to continue the work with an equivalent vibrator, but one which operates at a lower speed. Determination of a in accordance with section 8.1.22.2 (7) is useful here for orientation.

(4) Reference is made particularly to R 117, section 8.1.20.3 (1) b) regarding the achievable watertightness of artificially pre-sealed interlocks.

(5) R 118, section 18.1.11, applies accordingly for the work on site.

Vibration logs should contain as a minimum the time of each 0.5 m penetration. Alternatively, the log can be produced by means of electronic data acquisition.

(6) Vibration is generally a low-noise method of embedment. High noise levels can occur with defective vibratory action as a result of co-vibration of the piling and driving jaw, through hitting against one another. Co-vibration can be intensive in the case of high piling and staggered or compartmental embedment. The use of an embedment aid in accordance with section 8.1.22.5 (2) or padded driving jaws can provide a remedy.

(7) The risk of settlement in the vicinity of existing structures must be taken into account even when using modern high-frequency vibrators with variable eccentric weights.

(8) It should be noted that slow penetration rates can lead to heating and hence also to welding of the interlocks. In the case of a brief slower penetration rate, water cooling of the pile, especially around the interlocks, can prevent overheating.

(9) Dual piles should be vibrated by use of dual clamping jaws. The piles should have no punch holes in the clamping zone.

8.1.23 Jetting when installing steel sheet piles (R 203)

8.1.23.1 General

Please refer to the driving assistance with "water jet" in recommendations R 104, section 8.1.12.5, R 118, section 8.1.11.4, R 149, section 8.1.14.4, R 154, section 1.9.3.4, and R 202, section 8.1.22. To summarise, jetting can be used with impact driving, vibrating and pressing, in order to:

a) generally facilitate the embedment,
b) prevent overloading of the equipment and overstressing of the driven elements,
c) achieve the structurally required embedment depth,
d) reduce soil shocks,
e) reduce costs through shortening installation times and reducing power requirements and/or enabling lighter equipment to be used.

Jetting can be carried out with various pressures depending on the soil structure and strength.

8.1.23.2 Low-pressure jetting

A water jet is directed at the base of the installation elements through flushing pipes. The subsoil is loosened by the water injected under pressure and the loosened material is carried away in the jetting flow. Essentially, the toe resistance is reduced hereby. Depending on the soil structure, the skin friction and the interlock friction are also reduced by the flowing, rising water. The method is limited by the strength of the subsoil, the number of jetting lances, the magnitude of the hydrostatic pressure and the volume of water introduced. In order to establish the necessary parameters, it is recommended that trial driving operations are carried out before the method is used.

(1) Ratings
- Jetting lance $\varnothing$ 45–60 mm
- Jetting pressure at pump 5 to 20 bar
- Required nozzle action can be obtained by constricting the tip of the lance or with special jetting heads
- Water required: up to approx. 1000 l/min, delivered by centrifugal pumps.

The low-pressure jetting method is used for loose, closely-stratified soils, especially dry mono-grained soils or in sandy soils mixed with gravel. Depending on the difficulty of embedment, the jetting lances are flushed in next to the driving elements, or fixed directly to the driving elements. Reductions in the soil properties and settlement may occur as a result of the introduction of relatively large volumes of water.

(2) More recent experience

A special low-pressure jetting method (15–40 bar), combining the jetting process with placing the driving elements by vibration, has been used with success for some time. It enables sheet piles to be driven into very compact soils, which would be very difficult without this aid. Because of its environmental compatibility, the low-pressure jetting method is also used in residential areas and inner cities.

Success depends on the correct adjustment of the jetting, and selection of the vibrator for the soil encountered.

The vibrator should be selected in accordance with R 202, section 8.1.22, and equipped with adjustable effective moment and stepless speed regulations.

Usually, two to four lances of 1/2 to 1 inch diameter are secured to the driving elements (double pile). The lance tip ends flush with the base of the pile. The optimum arrangement is the use of one pump per lance.

Jetting begins simultaneously with vibration in order to prevent the opening of the lance from becoming clogged by the ingress of soil material.

If penetration rates ≥ 1 m/min are achieved, jetting can be retained until the structurally required embedment is reached. The soil properties previously determined for the sheet piling calculation generally then also applies. However, the angle should be limited to $\delta_a = +^1/_2\ \varphi$ or $\delta_p = -^1/_2\ \varphi$. The transmission of high vertical forces requires test loading.

8.1.23.3 High-pressure jetting

Ratings:
- Jetting lances (precision tubes), e.g. $\varnothing$ 30 × 5 mm,
- Jetting pressure 250–500 bar (at pump)
- Special nozzles in screw-in nozzle container (generally circular-section jet nozzles, dia. 1.5–3 mm; occasionally, fan-jet nozzles may also be appropriate),
- Water required: 30 to 60 l/min per nozzle, supplied by reciprocating pump.

High-pressure jetting can enable sheet piles to be driven in rock of varying solidity. The relatively low volumes of water mean that high-pressure jetting is also an expedient driving aid under unfavourable circumstances, e.g. in areas subject to settlement risk. It is ideal for firmly stratified, highly pre-loaded cohesive soils, e.g. in silt and clay rocks and mellow sandstone. Economic use is achieved only if construction operations are geared toward reuse of the jetting lances. The lances are fed into pipe clamps welded to the pile and the lance jetting heads are fixed on the sheet pile in such a way that the nozzle lies approx. 5 to 10 mm above the bottom elevation of the pile.

High-pressure jetting can be adapted particularly well to local conditions. During the course of construction operations, intensive observations are required for fine tuning. For example, it may be necessary to change the nozzles frequently because of unusual wear, or to re-adjust the lance arrangement or the nozzle diameter to prevailing conditions resulting from changes in the soil. To limit the angle of the active and passive earth pressures when installing piles with the aid of jetting, see section 8.1.23.2.

8.1.24 Pressing of U- and Z-section steel sheet piles (R 212)

8.1.24.1 General

The increasing demands for low noise levels and no vibrations during the driving of steel sheet piles can be met by static jacking of the piles. This method is often the only solution when the economic sheet piling forms of construction are to be erected in sensitive environmental conditions and adjacent to structures sensitive to settlement.

Currently, the cost for the installation itself is frequently still higher than methods employing hammers or vibrators. However, the cost seen in relation to the total cost is dropping gradually thanks to the omission of additional measures according to R 149, section 8.1.14.

8.1.24.2 Pressing equipment

We distinguish between three types of pressing:

Firstly, those set on the panel of piles by means of a crane. Secondly those which were guided by a leader and thirdly those which move free from any support on the head of the piles. The first one needs a two level guiding template as the second one only needs a guidance near the ground surface. The first and the second press are equipped with several adjacent rams in the direction of the wall axis and press in the piles by staggered installation.

Due to its small required installation area the first method is frequently used. This so called single-pile-press installs a single pile continuously to the final depth. It operates directley over the edge of the terrain, fitted on the piles already installed, and need no further guiding devices or supports. The sheet pile to be pressed in is both aligned and pressed by the ram in the head of the jack. This type of press activates the reaction forces via the adjacent piles already driven. The first two types of presses activate the necessary reaction forces via the partially installed piles and the equipment loads.

Some types of presses can be fitted and operated with driving aids, e.g. jetting equipment (R 203, section 8.1.23.2), or drilling gear.

8.1.24.3 Elements for pressing

The majority of the machines available on the market today can press in only single U- or Z-section sheet piles due to the arrangement of the rams. Presses suspended from a crane and those with a leader guidance can operate with double, triple or quadruple piles that are to be driven as a single pile, whereas the single-pile presses – as their name suggests – can install only single piles.

In order to minimise the interlock friction resistance in non-cohesive soils, the open interlocks of the sheet piles should be filled with a waterproofing material, e.g. hot bitumen or similar.

When selecting sheet piles, the stresses and strains of pressing should be considered as well as the structural requirements.

8.1.24.4 Pressing-in sheet piles

The sensitivity of this method with respect to friction resistances and the stresses on the piles calls for special care during installation.

The maximum pressure force available is about 1500 kN. As in the case of single-pile presses the reaction forces are activated via adjacent piles already driven. The permanent pressure should not exceed approx. 800 kN in order to minimise the movements in the part of the wall already installed.

The use of driving aids such as low-pressure jetting (see R 203) or drilling to loosen the soil enables sheet piles to be pressed into difficult soils as well. To limit the angle of inclination of the active and passive earth pressures when using jetting, see section 8.1.23.2. It is always necessary to remove or break up obstacles in the soil.

8.2 Calculation and design of sheet piling

8.2.0 General

The characteristic actions and resistances for calculating the stresses and strains on a sheet pile wall are applied according to limit state LS 1B to DIN 1054. The actions due to hydrostatic pressure are to be determined in line with recommendations R 19, section 4.2, R 113, section 4.7, and R 114, section 2.9.

Please refer to R 214, section 8.2.0.1 for information regarding the use of modified partial safety factors.

To verify the loadbearing capacity, examine limit state LS 1B, taking into account the analyses of the following failure modes for the retaining structure:

- ground failure in the passive earth pressure area due to horizontal action effects from the soil support $B_{h,d}$.
  (For analysis format $B_{h,d} \leq E_{ph,d}$ see DIN 1054, section 10.6.3.)

- Axial sinking of the sheet pile wall in the subsoil due to vertical action effects V_d.
  (For analysis format $V_d \leq R_{1,d}$ see R 4, section 8.2.4.6.)

The verification of stability for the deep failure plane is also carried out for LS 1B, verification of safety against slope failure, on the other hand, for LS 1C.

If the structural system changes in the course of building work on site or due to subsequent building measures, all the sheet pile wall analyses required must be re-examined. If, for instance, during a certain building phase a concrete bottom is cast in front of the sheet pile wall, this becomes a support for subsequent loading cases and represents a fundamental change to the structural system and the nature and effective direction of the soil support.

Verification of loadbearing capacity for the elements in steel sheet pile structures is to be carried out according to R 20, section 8.2.6.

Verification of loadbearing capacity for the elements in reinforced concrete or timber sheet pile structures is to be carried out according to DIN 1045-1 or DIN EN 1995-1.

The serviceability for limit state LS 2 covers conditions that make the structure unusable but do not bring about collapse. This analysis must be carried out in detail for waterfront structures to ensure that any wall deformation – if necessary taking into account anchor elongation and the resulting settlement behind the wall – does not damage the structure and its surroundings.

8.2.0.1 Partial safety factors for actions and effects and resistances (R 214)

When calculating sheet piling structures as well as the anchor walls and anchor plates for tie rods, the following partial safety factors apply for analyses of LS 1B:

- γ_G and γ_Q for actions according to table R 0-1
- $\gamma_{G,red}$ for hydrostatic pressure effects according to R 216, section 8.2.0.3
- γ_{Ep} and γ_{Gl} for resistances according to table R 0-2
- $\gamma_{Ep,red}$ for the passive earth pressure according to R 215, section 8.2.0.2

8.2.0.2 Determining the design values for the bending moment (R 215)

A reduced partial safety factor $\gamma_{Ep,red}$ may be used for the passive earth pressure according to table R 215-1 when determining the bending moments provided certain boundary conditions are complied with:

Table 215-1. Reduced partial safety factor $\gamma_{Ep,red}$ for passive earth pressure for determining the bending moments

LS 1B	LC 1	LC 2	LC 3
$\gamma_{Ep,red}$	1.20	1.15	1.10

We distinguish between the following cases:

- Below the theoretical ground surface in front of the retaining structure – hereinafter designated the "calculation bottom" – there are soils to which the following classification features apply:

  (1) Non-cohesive soil must exhibit a mean strength at least equal to table A3 of DIN 1055-2 so that reduced partial safety factors $\gamma_{Ep,red}$ may be used:

Designation	Degree of density D		Cone resistance
	$U \le 3$	$U > 3$	q_c [MN/m^2]
• low strength	$0.15 \le D < 0.30$	$0.20 \le D < 0.45$	$5.0 \le q_c < 7.5$
• medium strength	$0.30 \le D < 0.50$	$0.45 \le D < 0.65$	$7.5 \le q_c < 15$
• high strength	$0.50 \le D < 0.75$	$0.65 \le D < 0.90$	$15 \le q_c < 25$

  (2) Cohesive soil must exhibit a stiff state according to DIN 18 122 part 1 so that reduced partial safety factors $\gamma_{Ep,red}$ may be used:

State	Consistency index I_C
soft	$0.50 \le I_C < 0.75$
stiff	$0.75 \le I_C < 1.00$
semi-firm to firm	$1.00 \le I_C < 1.25$

  The redistribution of the active earth pressure to R 77, section 8.2.2.1, is carried out down to the calculation bottom.

- Below a level lower than the calculation bottom, soils are present with at least a medium strength, or rather, stiff consistency. The reduced partial safety factors may only be used below this low level – hereinafter called "separating plane". The soft, or rather, less solid, soil strata between calculation bottom and separating plane may only be applied as a surcharge p_0 on the separating plane. The redistribution of the active earth pressure to R 77, section 8.2.2.1, is carried out in this case down as far as the separating plane instead of as far as the calculation bottom (fig. R 215-1).

- If the reduced partial safety factor $\gamma_{Ep,red}$ is not used, the earth pressure redistribution diagram corresponding to fig. R 215-2 must be continued down to the level of the calculation bottom.

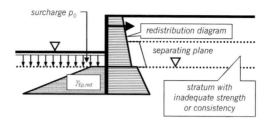

Fig. R 215-1. Loading diagram for determining the bending moments with reduced partial safety factors in soils with inadequate strength or consistency between calculation bottom and separating plane

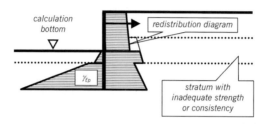

Fig. R 215-2. Loading diagram for determining the bending moments without reduced partial safety factors in soils with inadequate strength or consistency below the calculation bottom

- If below the calculation bottom only soils of lower strength or consistency are available than is necessary for the application of $\gamma_{Ep,red}$, the bending moments must be calculated without reduced partial safety factors γ_{Ep}. The redistribution of the active earth pressure according to R 77, section 8.2.2.1, is in this case is carried out down as far as the calculation bottom.

8.2.0.3 Partial safety factord for hydrostatic pressure (R 216)

In the case of the boundary conditions given below, a reduced partial safety factor γ_G may be used for the hydrostatic pressure (permanent action in LS 1B to table R 0-1) in accordance with DIN 1054, section 6.4.1 (7), for loading cases LC 1 and LC 2. The partial safety factors $\gamma_{G,red}$ are given in table R 216-1:

Table R 216-1. Reduced partial safety factors $\gamma_{G,red}$ for hydrostatic pressure actions

LS 1B	LC 1	LC 2	LC 3
$\gamma_{G,red}$	1.20	1.10	1.00

The partial safety factors for hydrostatic pressure actions may be reduced only when at least one of the three following conditions is satisfied:

- Verified measurements are available regarding the elevations and chronological dependencies between groundwater and outer water levels to guarantee the hydrostatic pressure used in the calculations and also as a basis for classification into loading cases LC 1 to LC 3.
- Numerical models of bandwidth and frequency of occurrence of the true water levels and hence the hydrostatic pressure lie on the safe side. These forecasts are to be checked through observations, beginning with the construction of the sheet pile wall. In the case of larger measurements than those predicted, the values on which the design was based must be guaranteed by measures such as drainage, pumping systems, etc.
- The geometrical boundary conditions present limit the water level to a maximum value, as is the case, for example, with the top edges of sheet pile walls designed as flood protection walls by limiting the height of raised water.
  Drainage systems incorporated behind the sheet pile wall do not represent a clear geometrical limit to the water level in the meaning of this stipulation.

8.2.1 Sheet piling structures without anchors (R 161)

8.2.1.1 General

Contingent on the bending resistance of the wall, sheet piling fully fixed in the ground and without an anchorage can be economical if there is comparatively little difference in ground surface elevations. Such an arrangement can also be used for larger differences in ground level if the installation of an anchor or another head prop would be very costly and if relatively large displacements of the pile head can be regarded as harmless in terms of serviceability.

8.2.1.2 Design, calculations and construction

In order to attain the necessary stability of sheet pile walls without anchors, i.e. with 100% fixity in the ground, the design, calculations and construction must satisfy the following requirements. If in doubt, employ sheet piling with anchors, in which the stability does not depend exclusively on the fixity in the ground.

- Ascertain all the actions as accurately as possible, e.g. also compaction earth pressure when backfilling in accordance with DIN 4085. This applies especially to those actions applied to the upper section of the sheet piling because these can substantially affect the design bending moment and the embedment depth required.

- It must be possible to establish an exact classification into loading cases LC 1 to LC 3 taking into account, for example, unusually deep scouring and special water pressure differences.
- The design bottom level may in no case be exceeded in the passive earth resistance area. Therefore, the design bottom level used in the calculations must include the additional depth required for any scouring and dredging work.
- The structural calculations for sheet pile walls without anchors, i.e. fully fixed in the ground, may be carried out according to BLUM [61], with the active earth pressure applied in a classic distribution, taking into consideration R 4, section 8.2.4, because redistribution of the earth pressure is not possible in this system.
- The theoretical embedment depth required, determined according to R 56, section 8.2.9, must be reached in the construction work.
- In the serviceability state – i.e. with characteristic actions – the deformation of the free-standing sheet pile wall is to be calculated in addition to the internal forces. The deformation values that occur must be investigated to ensure their compatibility with the structure and the subsoil, e.g. with respect to the formation of gaps in cohesive soils on the active earth pressure side which could fill up with water) and with the other project requirements. This is especially applicable when overcoming larger differences in ground levels.
  Free-standing, backfilled sheet piling behaves more favourably because displacement and deflection of the wall occur in the construction state and are therefore generally harmless.
- The displacement of the sheet piling depends on the extent to which the passive earth pressure was utilised. The elastic deflections of the wall fully fixed in the ground play a major role here.
- The inclination of the sheet piling to the vertical during installation is generally selected so that a visibly unfavourable overhang of the wall at the top is avoided when applying the critical actions and thus maximum deflection.
- The top of a sheet pile wall without anchorage, at least in a permanent structure, shall be provided with a capping beam or waling of steel or reinforced concrete to distribute the actions and prevent non-uniform deformations as far as possible.

8.2.2 Calculations for sheet piling structures with fixity in the ground and a single anchor (R 77)

8.2.2.1 Active earth pressure

The use of the active earth pressure has proved worthwhile for the relatively elastic anchorage of retaining structures without prestress common in harbour engineering.

The earth pressure force calculated with the so-called classic distribution – but reduced by the cohesion complement – must be redistributed over the height H_E for the analysis of the sheet pile wall. Redistribution of the earth pressure is not permitted if when subtracting the cohesion component the stresses in the anchorage are lower than without redistribution of the earth pressure. This is possible, for example, with the "trenching" method of construction in the case of varying soil strata, if the total pressure to be redistributed is considerably reduced by deducting the cohesion component for cohesive strata in the area of the redistribution depth H_E and therefore the redistributed earth pressure ordinates in the anchorage area are smaller than without redistribution.

The ratio of the position of the anchor head a to the redistribution depth H_E serves as a criterion for deciding the case when selecting the redistribution diagrams (R 77, section 8.2.2.3). The distances H_E and a as well as the non-redistributed earth pressure distribution $e_{a,k}$ as a result of the dead load of the soil, plus cohesion if necessary, and extensive surcharges on the ground are defined in figs. R 77-1 and R 77-2. The latter shows an example of a construction with depth $H_{\ddot{U}}$ and a reinforced concrete slab to screen the earth pressure.

The following definitions apply:
- H_G Total difference in ground levels
- $H_{\ddot{U}}$ Depth of construction from ground level to underside of screening slab
- H_E Depth of earth pressure redistribution above calculation bottom or separating plane to R 215, section 8.2.0.2. The depth H_E begins at the underside of the screening slab in the case of a construction with such a slab.
- a Distance of anchor head A from top edge of redistribution depth H_E

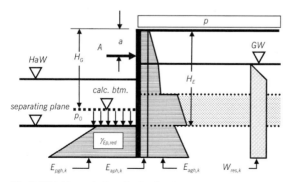

Fig. R 77-1. Example 1: redistribution height H_E and anchor position a for determining bending moment with $\gamma_{Ep,red}$

371

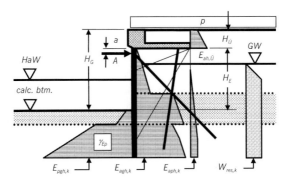

Fig. R 77-2. Example 2: redistribution height H_E and anchor position a for determining bending moment with γ_{Ep}

On the action side the non-redistributed active earth pressure is applied beneath the calculation bottom or separating plane to R 215, section 8.2.0.2.

8.2.2.2 Passive earth pressure

When analysing a sheet pile wall, the soil reaction anticipated is entered into the calculation using the BLUM [61] method with a linear increase opposite to the true progression. At the same time, the equivalent force C required is applied to maintain equilibrium.

The characteristic soil support $B_{h,k}$ required for determining the embedment length is formed here by the mobilised passive earth pressure $E_{ph,mob}$, which must exhibit a progression related to the characteristic passive earth pressure $E_{ph,k}$ and may not be redistributed.

8.2.2.3 Earth pressure redistribution

The earth pressure redistribution is selected depending on the method of construction:

- trenching in front of the wall (cases 1 to 3, fig. R 77-3)
- backfilling behind the wall (cases 4 to 6, fig. R 77-4)

We distinguish here between three ranges for the anchor head distance a:

- $0 \leq a \leq 0.1 \cdot H_E$
- $0.1 \cdot H_E < a \leq 0.2 \cdot H_E$
- $0.2 \cdot H_E < a \leq 0.3 \cdot H_E$

Besides the designations shown in figs. R 77-1 and R 77-2, in figs. R 77-3 and R 77-4 the magnitude of the mean value em of the earth pressure distribution over the redistribution depth H_E is as follows:

$$e_m = e_{ahm,k} = \frac{E_{ah,k}}{H_E}$$

The loading diagrams of figs. R 77-3 and R 77-4 include all the anchor head positions a in the range $a \leq 0.3 \cdot H_E$. These redistribution diagrams do not apply to anchorages at lower levels and in such cases the appropriate earth pressure diagrams must be determined separately.

If the ground surface is a short distance below the anchor, the earth pressure may be redistributed according to the value $a = 0$.

"Trenching in front of the wall" method of construction

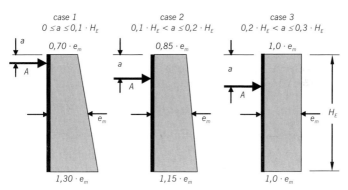

Fig. R 77-3. Earth pressure redistribution for the "trenching in front of the wall" method of construction

"Backfilling behind the wall" method of construction

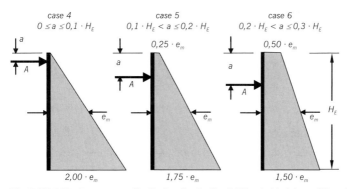

Fig. R 77-4. Earth pressure redistribution for the "backfilling behind the wall" method of construction

373

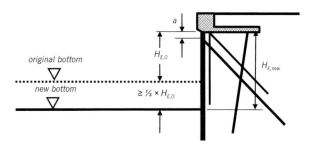

Fig. R 77-5. Additional excavation depth required for earth pressure redistribution according to the "trenching" method

The loading diagrams for cases 1 to 3 are only valid for the condition that the earth pressure can redistribute to the stiffer support areas as a result of adequate wall deformation. A "vertical earth pressure vault" therefore forms between the anchor and the soil support. Consequently, cases 1 to 3 may not be used when

- the sheet pile wall is backfilled to a large extent between bottom of watercourse and anchorage, and subsequent excavations in front of the wall are not so deep that adequate additional deflection takes place (guide value for adequate excavation depth: approximately one-third of redistribution depth $H_{E,0}$ of original system corresponding to fig. R 77-5);
- there is cohesive soil behind the sheet pile wall which is not yet sufficiently consolidated;
- the retaining wall with increasing rigidity does not exhibit the wall deflections necessary for forming a vault like, for example, in reinforced concrete diaphragm walls. In this case the designer must check whether the displacement of the support at the toe as a result of mobilisation of the passive earth pressure is adequate for an earth pressure redistribution according to the "trenching" method cases 1 to 3.

If loading diagrams cases 1 to 3 are not permissible, cases 4 to 6 applying to the a/H_E value for the "backfilling behind the wall" method of construction may be used.

8.2.2.4 Bedding

A sheet pile wall with a single anchor can also be designed using a horizontal bedding as a soil support [67–71]. It should be noted that the soil reaction stresses $\sigma_{h,k}$ in the calculation bottom as a result of characteristic actions may not be greater than the characteristic – i.e. maximum possible – passive earth pressure stress $e_{ph,k}$ (DIN 1054, eqs. (47) and (48)).

8.2.3 Calculation of sheet pile walls with double anchors (R 134)

In contrast to R 133, section 8.4.7, which deals with issues regarding auxiliary anchoring, R 134 covers sheet piling structures with double anchors, i.e. there are two anchor positions at different elevations.

The total actions applied to the sheet pile wall due to earth and hydrostatic pressures are assigned to the two anchor positions A_1 and A_2 as well as the soil support B. Owing to the distribution of the actions over the existing structural system, the majority of the total anchorage force is taken by the lower anchor A_2.

If the anchors consist of round steel bars, it is advisable to connect both anchor positions to a common anchor wall at the same elevation and to apply the direction of the resultant of the anchor forces A_1 and A_2 as the direction of the anchor when analysing the stability for the deep slip plane.

Anchorages not connected to a common anchor wall (e.g. grouted anchors to DIN EN 1537) require the two anchors to be analysed independently when checking stability.

8.2.3.1 Active and passive earth pressures

Take into account the active and passive earth pressures in the same way as for a wall with a single anchor.

8.2.3.2 Loading diagrams

The loading diagrams shown in R 77, section 8.2.2, for a wall with a single anchor are valid in principle for determining the internal forces, support reactions and embedment length of a sheet pile wall with double anchors. The elevation of the anchor head a required for determining the a/H_E value and for specifying the loading diagram is here taken to be the average elevation between the two anchor positions A_1 and A_2. The earth pressure is then redistributed over the depth H_E down to the calculation bottom or separating plane in a similar fashion to the sheet pile wall with a single anchor.

8.2.3.3 Considering deformations due to previous excavations

As earlier deflections of sheet pile walls due to slippage of the soil on the earth pressure slip plane can be only partly reversed, the effects of temporary conditions on the stresses in the final condition must be taken into account when they are relevant for verifying serviceability. This might be the case, for example, when considering the wall deflection at the level of anchor A_2, which for a sheet pile wall temporarily anchored only at point A_1 should be considered as a yielding support for the structural system in the final condition.

8.2.3.4 Bedding

Like the sheet pile wall with a single anchor, a wall with double anchors can be designed using horizontal bedding forces as a soil support (R 77, section 8.2.2.4).

8.2.3.5 Comparative calculations

As a comparison a calculation according to R 133, section 8.4.7, "Auxiliary anchoring at the top of steel sheet piling structures" should be carried out when designing the top of the wall and positioning the upper anchor A_1. If this calculation results in higher stresses, this shall govern the design.

8.2.4 Applying the angle of earth pressure and the sheet pile wall analysis in the vertical direction (R 4)

Realistic earth pressure angles must be used when determining the active and passive earth pressures and analysing sheet pile walls for the structural calculations (see R 56, section 8.2.9, fig. R 56, for the rules covering +/– signs).

The magnitude of the angle of earth pressure selected, or permissible, depends on the maximum angle of friction between building material and subsoil (wall friction angle) physically possible and the relative displacements of the sheet pile wall with respect to the soil.

The angle of earth pressure has a major influence on the analysis of the sheet pile wall in the vertical direction, but which the following equilibrium and limit state conditions must be satisfied:

- vertical equilibrium   to section 8.2.4.3
- vertical loadbearing capacity   to section 8.2.4.6

The angle of earth pressure for untreated sheet pile surfaces can be selected within the limits given below. In the case of treated surfaces, lubricating layers present in the subsoil, piles installed with the help of jetting (R 203, section 8.1.23) etc., the upper and lower limits may need to be reduced accordingly.

For the angle of earth pressure to be used for other types of retaining structure, e.g. diaphragm or bored pile walls, see also DIN 4085.

8.2.4.1 Angle of inclination $\delta_{a,k}$ of active earth pressure

Assuming straight slip planes, the angle of inclination $\delta_{a,k}$ of active earth pressure may lie between the limits

$$-\tfrac{2}{3} \cdot \varphi'_k \leq \delta_{a,k} \leq +\tfrac{2}{3} \cdot \varphi'_k$$

8.2.4.2 Angle of inclination $\delta_{p,k}$ of passive earth pressure

Assuming straight slip planes, the angle of inclination $\delta_{p,k}$ of passive earth pressure for an angle of friction $\varphi'_k \leq 35°$ and for horizontal terrain may lie between the limits

$$-\tfrac{2}{3} \cdot \varphi'_k \leq \delta_{p,k} \leq +\tfrac{2}{3} \cdot \varphi'_k$$

Otherwise, curved slip planes must be assumed.
When assuming curved slip planes, the angle of inclination $\delta_{p,k}$ of passive earth pressure may lie between the limits

$$-\varphi'_k \leq \delta_{p,k} \leq +\varphi'_k$$

The approach using straight slip planes with the aforementioned limits is permissible to simplify the equation because the K_{ph} values used when determining the characteristic passive earth pressure $E_{ph,k}$ are not substantially different for the respective limit value min $\delta_{p,k}$ for the methods with street and curved slip planes.

8.2.4.3 Vertical equilibrium as the basis for using min $\delta_{p,k}$

(1) Analysis format
The following analysis ensures that the angle of inclination $\delta_{p,k}$ of passive earth pressure selected for the calculation will actually occur in practice. Therefore, the value of the angle of inclination $\delta_{p,k}$ must be one that satisfies the equilibrium condition $\Sigma\, V = 0$ to DIN 1054, section 10.6.3 (5). According to this the minimum characteristic total vertical action effect V_k for every combination of actions must be at least equal to the upward vertical component $B_{v,k}$ of the characteristic soil support B_k to be mobilised. The analysis format for this equilibrium condition is:

$$V_k = \sum V_{k,i} \geq B_{v,k}$$

Carry out this analysis with the same angles of inclination used previously in the calculations for the active and passive earth pressures.

(2) Varying vertical force action effects $V_{Q,k}$
Vertical components $V_{Q,k}$ of the action effects due to varying actions Q may not be entered into the equilibrium condition when they have beneficial effects, i.e. do not cause any significant soil support components $B_{v,k}$. This applies, on the one hand, to action effects that occur directly at the top of the wall, e.g. the water-side support reaction $F_{Qv,k}$ of the superstructure as a result of the actions due to cranes and stacked loads; on the other hand, this also applies to the vertical component $A_{Qv,k}$ of those anchor force components that occur as a result of horizontal,

377

varying actions in the region of the top of a wall or above the anchor position, e.g.

- lateral crane impacts and storm fastenings,
- line pull forces, and
- active earth pressure due to varying actions.

(3) Magnitude of equivalent force C to be used for the analysis
For the analyses in the vertical direction, please note that the BLUM method [61] results in an equivalent force C_{BLUM} that is too large because the passive earth pressure mobilised for the soil support B_k is assumed to act with its full magnitude down to the theoretical base of the sheet pile wall *TF*. When the true progression of the soil reaction B_k is taken into account, the equivalent force C is only about half its theoretical magnitude. At the same time, the associated soil support B_k is reduced by exactly this value (see R 56, section 8.2.9, fig. R 56-1).
In order to compensate for this error, the horizontal components of the equivalent force and the soil support ($C_{\mathrm{h,k,BLUM}}$ and $B_{\mathrm{h,k}}$) are each reduced by $\frac{1}{2} \cdot C_{\mathrm{h,k,BLUM}}$ when calculating the associated vertical components. The remaining characteristic equivalent force $\frac{1}{2} \cdot C_{\mathrm{h,k,BLUM}}$ is designated $C_{\mathrm{h,k}}$. This also applies in principle to section 8.2.4.6, "Vertical loadbearing capacity" in which the vertical components of the remaining characteristic equivalent force components due to permanent and varying actions are designated with $C_{\mathrm{Gv,k}}$ and $C_{\mathrm{Qv,k}}$ respectively.

(4) Characteristic vertical force component $V_{k,i}$ due to dead loads
The direction of the component corresponds to the angle of inclination δ, positive = down, negative = up:

- $V_{\mathrm{G,k}} = \Sigma\, F_{\mathrm{G,k}}$
  as a result of *permanent* axial actions F at the top of the wall

- $V_{\mathrm{Av,k}} = A_{\mathrm{v,k,MIN}}$
  as a result of the anchor force $A_{\mathrm{v,k,MIN}} = A_{\mathrm{Gv,k}} - A_{\mathrm{Qv,k}}$

- $V_{\mathrm{Eav,k}} = \Sigma\, (E_{\mathrm{ah,k,n}} \cdot \tan \delta_{\mathrm{a,k,n}})$
  as a result of the active earth pressure E_{ah} with n layers down to the depth of the theoretical toe *TF*

- $V_{\mathrm{Cv,k}} = C_{\mathrm{h,k}} \cdot \tan \delta_{\mathrm{C,k}}$
  as a result of the equivalent force $C_{\mathrm{h,k}}$

(5) Characteristic upward component $B_{\mathrm{v,k}}$ of the soil support reaction B_k
- $B_{\mathrm{v,k}} = |\Sigma(B_{\mathrm{h,k,r}} \cdot \tan \delta_{\mathrm{p,k,r}}) - C_{\mathrm{h,k}} \cdot \tan (\delta_{\mathrm{p,k\ at\ theor.\ toe}}) |$
  as a result of the soil support B_k with r layers down to the depth of the theoretical toe *TF*, including a correction term

378

8.2.4.4 Angle of inclination $\delta_{C,k}$ of equivalent force C_k

According to the BLUM method [61] for a sheet pile wall fully fixed in the ground, the soil reaction below the theoretical toe TF on the action side is used to resist the equivalent force C. The extra depth required to resist this reaction force must be calculated according to R 56, section 8.2.9, as an addition Δt_1 to the embedment depth t_1. The direction of action of equivalent force C used in this calculation is inclined at the angle $\delta_{C,k}$ to the horizontal.

The angle of inclination $\delta_{C,k}$ must be determined for every loading combination when checking the vertical equilibrium according to section 8.2.4.3.

In the usual notion of a parallel displacement of the wall or overturning about the top or bottom, the active and passive sliding wedges move relative to the wall, whereas the appearance of equivalent force C is not linked with the formation of a failure body. Therefore, the angle of inclination of the equivalent force is generally taken to be $\delta_{C,k} = 0$. In the case of other relative displacements, the angle of inclination $\delta_{C,k}$ of equivalent force C_k may be assumed to lie between the limits

$$ -\tfrac{2}{3} \cdot \varphi'_k \le \delta_{C,k} \le +\tfrac{1}{3} \cdot \varphi'_k $$

8.2.4.5 How large vertical forces influence the angle of inclination

Downward vertical forces acting on the retaining wall cause a relative displacement between the sheet pile wall and the soil, the magnitude of which depends on the magnitude of the vertical force and the strength of the subsoil below the base of the sheet pile wall. If at the same time another relative displacement between subsoil and sheet pile wall occurs, different to the one that causes the normal formation of the sliding bodies for active and passive earth pressure, this can lead to the need to assume a different angle of inclination. This must be taken into account in the analyses according to sections 8.2.4.3 and 8.2.4.6:

- The characteristic angle of inclination $\delta_{p,k}$ of the passive earth pressure can have negative values up to the limit specified in section 8.2.4.2. As the values of $\delta_{p,k}$ decrease, the result is higher K_{ph} values and hence also higher values for the mobilisable soil support B_k.
- The characteristic angle of inclination $\delta_{C,k}$ of the equivalent force C_k can have negative values up to the limit specified in section 8.2.4.4 if the downward displacement of the sheet pile wall with respect to the subsoil is significant. The change in the direction of action has no effect on the theoretical value $C_{h,k,BLUM}$ of the equivalent force from the sheet pile wall calculation.
- If relative displacements between subsoil and sheet pile wall are to be expected owing to the subsoil properties around the base of the wall such that the earth pressure angle $\delta_{a,k}$ can have a negative value, this

379

must be specified within the scope of the subsoil investigation. A reduction in the angle of inclination $\delta_{a,k}$ leads to an increase in the earth pressure force.

- If it is necessary for the loads on the wall due to large vertical forces to assume another resisting vertical force component $E_{av,k}$ and the associated relative displacement represents no risk for the serviceability of the structure, then a negative earth pressure angle $\delta_{a,k}$ up to the limit specified in section 8.2.4.1 may be used.

Upward vertical forces in a sheet pile wall – e.g. as a result of compression raking piles, tie rods inclined upwards or inclined line pull forces – change the relative displacement between wall and soil. In doing so, the active earth pressure force $E_{ah,k}$ is reduced until it reaches the positive limit for $\delta_{a,k}$ according to section 8.2.4.1.

It is also necessary to vary the angle of inclination $\delta_{p,k}$ in order to achieve vertical equilibrium. The effect of this is a distinctly lower value for the passive earth pressure force $E_{ph,k}$ until it reaches the positive limit for $\delta_{p,k}$ according to section 8.2.4.2. The corresponding effects should be taken into account in the analyses to sections 8.2.4.3 and 8.2.4.6.

8.2.4.6 Vertical loadbearing capacity

(1) Analysis format
When analysing the safety against failure of the sheet pile wall due to sinking into the subsoil (to DIN 1054, section 10.6.6), all the downward loads and the axial resistance must be taken into account with their design values: for LS 1B the total load V_d may be equal to the axial resistance $R_{1,d}$. The analysis format for the limit state condition is:

$$V_d = \sum V_{d,i} \leq R_{1,d}$$

(2) Design value V_d for vertical force due to dead loads
The design value of the downward vertical force Vd includes the individual components of the various types of action according to section 8.2.4.3 (4). However, the characteristic part-loads are increased – separately according to cause within each loading combination – by the associated partial safety factors according to table R 0-1 due to the permanent (G) and varying (Q) actions:

- $V_{F,d} = \sum (V_{F,G,k} \cdot \gamma_G + V_{F,Q,k} \cdot \gamma_Q)$
  as a result of *maximum* vertical actions F at the top of the wall

- $V_{Av,d} = \sum (V_{Av,G,k} \cdot \gamma_G + V_{Av,Q,k} \cdot \gamma_Q)$
  as a result of *maximum* anchor force component A_v

- $V_{Eav,d} = \sum (V_{Eav,n,G,k} \cdot \gamma_G + V_{Eav,n,Q,k} \cdot \gamma_Q)$
  as a result of the active earth pressure F_a with n layers down to the depth of the theoretical toe *TF*. The vertical component $E_{av,d}$ of the

active earth pressure is assigned to the loads $V_{d,i}$ as a wall friction force when the active earth pressure force $E_{a,d}$ acts downwards with a positive angle of inclination $\delta_{a,k}$. When $\delta_{a,k}$ is negative, $E_{av,n,G,k}$ and $E_{av,n,Q,k}$ are relieving forces subtracted in full from the vertical loads.

- $V_{Cv,d} = \Sigma (V_{Cv,G,k} \cdot \gamma_G + V_{Cv,Q,k} \cdot \gamma_Q)$
  as a result of the equivalent force C. The vertical force component $C_{v,d}$ as a result of permanent and varying actions is assigned to the loads $V_{d,i}$ as a wall friction force when the equivalent force C acts downwards as the result of a positive angle of inclination $\delta_{C,k}$.

(3) Design value of axial resistance
The design value $R_{1,d}$ of the axial resistance consists of the sum of the characteristic part-resistances divided by the partial safety factor γ_P:

$$R_{1,d} = \frac{\Sigma R_{1k,i}}{\gamma_P}$$

The partial safety factor γ_P according to table R 0-2 for the resistance components to be determined by using empirical values according to action (4) below is $\gamma_P = 1.40$ for all loading cases.

By contrast, loading tests for the loadbearing members of the sheet pile wall (R 33, section 8.2.11.2) permit sufficiently accurate statements about the magnitude of the toe resistance $R_{1b,k}$, the skin resistance R1s,k and hence the effective total resistance $R_{1,k}$ that is available in the axial direction for a sheet pile wall loaded horizontally according to the BLUM method. The partial safety factor $\gamma_{Pc} = 1.20$ applies to the resistance R1,k determined with loading trials.

(4) Characteristic values of part-resistances R_{1i} for LS 1B
The geometrical and constructional conditions according to R 33, section 8.2.11.2, must be satisfied in order to use the toe resistance force R_{1b}. The axial sheet pile wall displacement required to mobilise a toe resistance force $R_{1b,k}$ is greater than that required to mobilise a skin resistance force $R_{1s,k}$. When using a toe resistance $q_{b,k}$ the soil reaction force B_k and the equivalent force C_k may be considered with upward, i.e. negative, angles of inclination δ. The vertical components $B_{v,k}$ and $V_{Cv,k}$ of these forces are therefore assigned to the axial part-resistances $R_{1,i}$.

- $R_{1b,k}$ toe resistance force as a result of toe resistance $q_{b,k}$
- $R_{1Bv,k} = |B_{v,k}|$ wall friction resistance force as a result of the mobilised soil reaction B_k to section 8.2.4.3 (5)
- $R_{1Cv,k} = |V_{Cv,k}|$ wall friction resistance force as a result of the equivalent force C_k for a negative angle of inclination $\delta_{C,k}$
- $R_{1s,k}$ skin resistance force as a result of skin friction $q_{s,k}$

Skin resistance force $R_{1s,k}$ is only possible with a value below the theoretical bottom edge exceeding the required length of the sheet pile wall, which is determined with the help of the embedment length to R 56, section 8.2.9. It is not permitted to use the additional skin friction on both sides of the wall down to its theoretical bottom edge to carry vertical forces. The wall friction resistance forces $B_{v,k}$ and $C_{v,k}$ plus the active earth pressure component $E_{av,k}$ are already acting in this area.

8.2.5 Taking account of inclined embankments in front of sheet piling and unfavourable groundwater flows in the passive earth pressure area of non-cohesive soil (R 199)

8.2.5.1 How sloping embankments affect the passive earth pressure
The magnitude of the passive earth pressure for an inclined embankment depends on the negative angle of inclination β of the surface of the embankment. Even the smallest negative inclinations produce a substantial reduction in passive earth pressure.

8.2.5.2 How unfavourable groundwater flows affect the passive earth pressure
The influence of a flow around the sheet pile wall as a result of different water levels in front of and behind the wall must be taken into account in the design (R 114, section 2.9.3.2).

Irrespective of this, it is necessary to verify the stability of the bottom against a heave failure for LS 1A according to R 115, section 3.2.

Any risk to the stability caused by subsurface erosion of the bottom as a result of flows is to be investigated according to R 116, section 3.3. If necessary, the measures given in that section are to be initiated.

8.2.6 Bearing stability verification for the elements of sheet piling structures (R 20)

8.2.6.1 Quay wall

(1) Predominantly static stresses
Bearing capacity verification for all types of sheet pile wall construction is to be carried out according to DIN EN 1993-5, section 5. According to this, the analysis format for safety against failure of the sheet pile wall section with the design value S_d for internal forces and design value Rd for the resistance of the section is:

$$S_d \leq R_d$$

DIN EN 1993-5 refers to DIN EN 1997-1 for the method of calculation. The procedure according to DIN 1054 and hence also the EAU procedure correspond to DIN EN 1997-1.

References in DIN EN 1993-5 to DIN EN 1993-1 are to be implemented with corresponding methods according to DIN 18 800.

Raking piles and all parts of the construction at the top of the sheet pile wall and the heads of the piles for the connection of walings, capping beams or reinforced concrete superstructures are designed according to DIN 18 800.

The design values to be used in DIN 18 800 are determined with the help of the assumptions, principles and calculations of sections 8.2.0 to 8.2.5.

(2) Predominantly alternating stresses
Non-backfilled sheet pile walls standing isolated in the water are loaded primarily cyclic by the impact of waves. In doing so, a large number of load cycles occur over the lifetime of the wall so that verification of fatigue strength to DIN 19 704-1 is required. In addition, please refer to DIN 18 800, part 1, El. (741).

In order to prevent adverse action from the notch effect, such as from structural welding seams, tack welds, unavoidable irregularities on the surface due to the rolling process, pitting and the like, killed steels as per DIN EN 10 025 should be used in such cases (see R 67, section 8.1.6.4).

8.2.6.2 Anchor wall, walings, capping beams and anchor head plates

(1) Predominantly static stresses
Section 8.2.6.1 (1) applies for the verification of bearing capacity for anchor plates and anchored sheet pile walls fixed in the ground. Walings, capping beams, bracing and anchor head plates are calculated according to DIN 18 800. In doing so, it may be necessary to increase the partial safety factors for resistances of walings and capping beams according to R 30, section 8.4.2.3. The loadbearing capacity of the sheet pile wall sections with regard to the transfer of anchor forces must be verified according to DIN EN 1993-5, section 6.4.3.

See section 8.2.6.1 (1) with respect to the references to other DIN EN standards given in section 6.4.3.

(2) Predominantly alternating stresses
Section 8.2.6.1 (2) applies to verification of bearing capacity. Turned structural bolts of at least strength class 4.6 are to be used for bolted connections in walings and capping beams. Fatigue assessment procedure is to be provided according to DIN 18 800 part 1, section 8.2.1.5, El. (811).

8.2.6.3　Tie rods and waling bolts

DIN EN 1993-5, section 6.2, forms the basis for the design of tie rods and waling bolts, but modified with the use of k_t^* instead of k_t and A_{core} instead of the stressed cross-sectional area A_s (and therefore the calculated design value of the section resistance lies on the safe side).

See section 8.2.6.1 (1) with respect to the references to other DIN EN standards given in section 6.2.

(1) Predominantly static loads
The materials for tie rods and waling bolts are given in R 67, section 8.1.6.4.

The analysis format for the limit state condition of loadbearing capacity according to DIN EN 1993-5 is:

$$Z_d \leq R_d$$

Calculate the design values using the following variables:

Z_d	design value for anchor force $Z_d = Z_{G,k} \cdot \gamma_G + Z_{Q,k} \cdot \gamma_Q$
R_d	design resistance of anchor $R_d = \min [F_{tg,Rd}; F_{tt,Rd}^*]$
$F_{tg,Rd}$	$A_{shaft} \cdot f_{y,k} / \gamma_{M0} = A_{shaft} \cdot f_{y,k} / 1.10$
$F_{tt,Rd}^*$	$k_t^* \cdot A_{core} \cdot f_{ua,k} / \gamma_{Mb} = 0.55 \cdot A_{core} \cdot f_{ua,k} / 1.25$
A_{shaft}	cross-sectional area at shaft
A_{core}	cross-sectional area of core in thread
$f_{y,k}$	yield stress
$f_{ua,k}$	tensile stress
γ_{M0}	partial safety factor to DIN EN 1993-5 for anchor shaft
γ_{Mb}	ditto, but for threaded segment
k_t^*	notch factor

The notch factor $k_t = 0.80$ given in DIN EN 1993-5 is reduced to $k_t^* = 0.55$ when determining the design value for the resistance of the threaded segment. This takes into account any additional stresses due to the installation of the anchor under the less-than-ideal conditions on a building site and any ensuing, unavoidable bending stresses in the threaded segment. Regardless of this, it is also necessary to provide constructional measures to ensure that the anchor head can rotate sufficiently.

The additional serviceability analyses called for in DIN EN 1993-5 are already implied in the limit state condition $Z_d \leq R_d$ owing to the value selected for the notch factor k_t^* and the customary upsetting relationships between shaft and thread diameters; the additional analyses are therefore unnecessary. Tie rods can have cut, rolled or hot-rolled threads as per R 184, section 8.4.8.

Prerequisite for a proper design is a suitable detail for the anchor connection. This means the anchors must be connected via a form of articulated joint. The anchors must be installed higher so that any settlement or subsidence does not cause any additional stresses.

Upsetting the ends of the anchor bars for the threaded segments and T-shaped heads as well as tie rods with hinged eyes are permissible

- when using steel grades J2 G3, J2 G4, K2 G3 and K2 G4, but not thermo-mechanically rolled steels of grades J2 G4 and K2 G4, (R 67, section 8.1.6.1 must be taken into account);
- when other steel grades, e.g. S 355 J0, are used and accompanying tests ensure that the strength values in DIN EN 10 025 are upheld after the normalising procedure of the forging process;
- when the upsetting, T-shaped heads and hinged eyes are provided by specialist fabricators and it is ensured that the mechanical and technological values in all parts of the tie rod are in accordance with the steel grade selected, and that the alignment of the fibres is not impaired during the machining process and detrimental microstructure disruptions are reliably avoided.

The loading case "failure of an anchor" called for in DIN EN 1993-5, section 6.2.2, does not need to be considered for tie rods because the aforementioned verification of loadbearing capacity is carried out with the reduced notch factor k_t^* and the tie rods hence exhibit sufficient loadbearing reserves in order to avoid any failures due to additional stresses caused during installation.

Tie rods may be installed without an anti-corrosion coating when the groundwater and subsoil contains – at worst – only minimal substances that could attack steel.

After installation, tie rods must be bedded in a sufficiently thick layer of sand in filled soil on all sides and over their full length.

If it is necessary to coat the tie rods with an anti-corrosion treatment, measures will be necessary on site to ensure that the coating is not damaged. If, despite this, damage does occur, the coating must be repaired so that the original quality is restored.

The aforementioned measures reduce the risk of anodic areas on the tie rods and any ensuing pitting.

The design and installation of sheet pile wall anchorages with grouted anchors is covered by DIN 1054 and DIN EN 1537.

(2) Predominantly dynamic stresses

Anchors are generally subjected to static loads. Primarily dynamic stresses only occur in anchors in rare, exceptional cases (section 8.2.6.1 (2)), but more frequently in waling bolts. Only specially killed steels to DIN EN 10 025 may be used where dynamic stresses occur.

Verification of the fatigue strength is to be provided according to DIN 18 800 part 1, section 8.2.1.5, with the criterion of section 7.5.1.

If the basic static load is less than or equal to the cyclic stress amplitude, it is recommended to apply a permanent, controlled prestress exceeding the stress amplitude to the anchors or waling bolts. This ensures that the anchors or waling bolts remain under stress and do not fail abruptly when the when the stress increases again.

A prestress, which is not defined exactly, is applied to anchors and waling bolts, in many cases during the installation procedure. In such cases without controlled prestressing, only a stress $\sigma_{Rd} = 80$ N/mm^2 may be assumed for the thread of the anchors or waling bolts, regardless of loading class and steel grade, neglecting the prestress.

Always ensure that the nuts of the waling bolts cannot loosen due to repeated changes in stress.

8.2.6.4 Steel cable anchors

Steel cable anchors are used only for predominantly static loads. They are to be designed in such a way that the design value N_d for the anchor tensile force satisfies the condition $N_d = N_{failure}$ / 1.50 for LC 1 and LC 2, and the condition $N_d = N_{failure}$ / 1.30 for LC 3.

The mean modulus of elasticity of patented steel cable anchors should not be less than $E = 150\ 000$ MN/m^2 and must be guaranteed by the manufacturer to a tolerance of $\pm 5\%$.

8.2.7 Consideration of axial loads in sheet piling (R 44)

In general, sheet piling is stressed chiefly by bending. If an additional compressive force acts in the axis of the wall, appropriate verification of bearing capacity is to be provided according to DIN EN 1993-5.

The bending stress in the sheet piling can be influenced both favourably but also unfavourably by an intentional or unintentional eccentric application of the vertical loads at the top of the wall.

When verifying the bearing capacity considering longitudinal forces and stresses, it is generally adequate to take the buckling length as the spacing between the zero points of the bending moment diagram for the span.

If severe corrosion is expected in the areas of the bending moments critical for sizing the section, then reduced cross-section values are to be considered at those positions.

8.2.8 Selection of embedment depth for sheet piling (R 55)

Technical, structional, operational and economic requirements may be relevant to the embedment depth for sheet piling in addition to the corresponding structural calculations and the supplement required by R 56, section 8.2.9. Any foreseeable future deepening of the harbour and any danger of scour below the calculation bottom should be considered to the same extent as safety against slope and foundation failure, failure by heave and erosion influences.

These latter requirements usually result in such a large minimum embedment depth that partial fixity is available at least, apart from the special case of foundations in rock. Even if a simple support would be adequate theoretically, it is often advisable to increase the embedment depth because this can have economic advantages as well. The resistance of the section is utilised more uniformly over the length of the sheet pile wall and therefore at least partial fixity of the sheet pile wall is advisable when using the BLUM method [61].

If the sheet pile wall also has to transfer vertical loads into the subsoil, not all the piles have to continue as far as the loadbearing stratum. It can be sufficient for the embedment length of only some of the piles to be long enough that they are effective as vertical loadbearing piles if with this number of piles it can be shown that the loadbearing capacity against sinking into the subsoil is adequate.

8.2.9   **Determining the embedment depth for sheet pile walls with full or partial fixity in the soil (R 56)**

If sheet piling is calculated according to the BLUM method [61], with full fixity in the soil (degree of fixity $\tau_1 = 100\%$) the entire embedment length below the calculation bottom consists of the embedment length t_1 down to the theoretical toe plus the extra length Δt_1 (driving allowance). The extra length Δt_1 is necessary in order to accommodate the design value of the horizontal component $C_{h,d}$ of the equivalent force C actually effective at the theoretical toe TF as a soil reaction force distributed over the depth Δt_1. If the more accurate analysis to determine Δt_1 (shown below) is not carried out, then for a sheet pile wall with full fixity in the soil the extra length may be considered in a simplified form with

$$\Delta t_1 = \frac{t_1}{5}$$

However, this approach is only valid when the actions include no substantial hydrostatic pressure component.

The design value $C_{h,d}$ of the residual equivalent force according to the BLUM method [61], corrected according to R 4, section 8.2.4.3 (3), is:

$$C_{h,d} = \sum (C_{Gh,k} \cdot \gamma_G + C_{Qh,k} \cdot \gamma_Q)$$

or, with a reduced partial safety factor for the hydrostatic pressure component and separated according to the components of the equivalent force:

$$C_{h,d} = \sum (C_{Gh,k} \cdot \gamma_G + C_{Gh,W,k} \cdot \gamma_{G,red} + C_{Qh,k} \cdot \gamma_Q)$$

387

The equivalent force components to be used are:

$C_{Gh,k}$ as a result of permanent actions G,
$C_{Gh,W,k}$ as a result of the permanent action hydrostatic pressure, and
$C_{Qh,k}$ as a result of varying actions Q.

The associated partial safety factors are:

γ_G for permanent actions,
$\gamma_{G,red}$ for hydrostatic pressure with a permissible reduction, and
γ_Q for varying actions.

The characteristic value $E_{phC,k}$ for the soil support for the equivalent force $C_{h,d}$ at failure is the magnitude of the passive earth pressure force on the equivalent force side below the theoretical toe TF is:

$$E_{phC,k} = \Delta t_1 \cdot e_{phC,k}$$

The characteristic value for the passive earth pressure stress $e_{phC,k}$ on the equivalent force side at depth TF is:

$$\begin{aligned} e_{phC,k} &= \sigma_{z,C} \cdot K_{phg,C} & \text{for non-cohesive soils, or} \\ e_{phC,k} &= \sigma_{z,C} \cdot K_{pgh,C} + c'_k \cdot K_{phc,C} & \text{for cohesive soils} \end{aligned}$$

(taking into account the respective consolidation state as a result of the shear parameters $c_{u,k}$ or φ'_k and c'_k).
Calculate the vertical soil stress $\sigma_{z,C}$ at depth TF on the equivalent force side.
The design value $E_{phC,d}$ for the soil support to accommodate the equivalent force $C_{h,d}$ is calculated with the partial safety factor γ_{Ep} for the passive earth pressure as follows:

$$E_{phC,d} = \frac{E_{phC,k}}{\gamma_{Ep}}$$

The analysis format for upholding the limit state condition when accommodating the equivalent force $C_{h,d}$ as a soil reaction is:

$$C_{h,d} \leq E_{phC,d}$$

In a further development of the approach by LACKNER [73], the limit state condition provides the size of the extra length Δt_1 required below the theoretical toe TF for walls fully fixed in the ground:

$$\Delta t_1 \geq \frac{C_{h,d} \cdot \gamma_{Ep}}{e_{phC,k}}$$

The above equation for the extra length in the case of full fixity in the ground (degree of fixity $\tau_1 = 100\%$) is also used to determine the extra length for sheet pile walls with only partial fixity in the ground, i.e. for any degree or fixity over the possible bandwidth of $\tau_1 = 100\%$ to $\tau_0 = 0\%$ (simply supported in the soil). The degree of fixity designated here with τ_{1-0} for a partially fixed sheet pile wall is $\tau_{1-0} = 100 \cdot (1 - \varepsilon / \max \varepsilon)$ [%] with the end tangent angle ε of the line of bending for the theoretical toe TF selected and the end tangent angle $\max \varepsilon$ for a simple support in the soil. The embedment depth associated with the degree of fixity τ_{1-0} is designated t_{1-0} and the extra length Δt_{1-0}.

Partial fixity in the soil is associated with – compared with a full fixity – smaller values for the equivalent force component $C_{h,d}$ and hence also extra lengths $\Delta t_{1-0} < \Delta t_1$. In the case of a simple support for the sheet pile wall in the soil ($\tau_0 = 0\%$), $C_{h,d} = 0$ and $\Delta t_0 = 0$. A necessary minimum value Δt_{MIN} must be maintained for the additional embedment length, which is defined depending on the degree of fixity present ($100\% \geq \tau_{1-0} \geq 0\%$):

$$\Delta t_{MIN} = \frac{\dfrac{\tau_{1-0}}{100} \cdot t_{1-0}}{10}$$

Definitions of the symbols used in fig. R 56-1:

t total embedment depth required $t = t_1 + \Delta t_1$ for the sheet pile wall fully fixed in the soil [m]

TF theoretical toe of sheet pile wall (load application point for equivalent force C) [m MSL]

t_1 distance between TF and calculation bottom [m]

Δt_1 extra length for accommodating equivalent force $C_{h,d}$ via a soil reaction force below TF [m]

$\sigma_{z,C}$ vertical soil stress at TF on equivalent force side [kN/m²]

$\delta_{p,k}$ angle of inclination of passive earth pressure [°]

$K_{pgh,C}$ passive earth pressure coefficient at TF on equivalent force side for angle of inclination $\delta_{C,k}$ [1]

$\delta_{C,k}$ angle of inclination of equivalent force C [°]
     $\delta_{C,k}$ is specified within the limits according to R 4, section 8.2.4.4, based on verification of the vertical equilibrium to R 4, section 8.2.4.3.

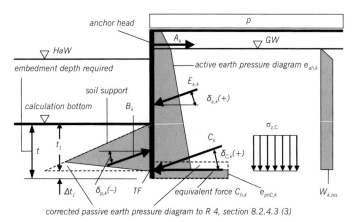

corrected passive earth pressure diagram to R 4, section 8.2.4.3 (3)

Fig. R 56-1. Actions, support and soil reactions of a sheet pile wall fully fixed in the ground for determining the additional embedment length Δt_1

8.2.10 Staggered embedment depth for steel sheet piling (R 41)

8.2.10.1 Application

Piles (generally dual piles) are frequently driven to different depths for technical and – for fully fixed walls– also for economic reasons. The permissible extent of this staggering (i.e. difference in embedment length) depends on the stresses in the region of the base of the longer piles and on construction considerations. For driving reasons, staggering within a driving unit is not recommended for corrugated sheet piles.

At failure a uniform, continuous passive earth pressure failure zone forms in the region of the soil support of staggered sheet piling (similar to closely spaced anchor plates), taking into account the geometrical boundary conditions according to R 7, section 8.1.4.2. Therefore, the full soil reaction down to the bottom of the deeper piles can be used when determining the stresses without taking into account the staggering. The bending moment present at the bottom edge of the shorter piles must be withstood by the longer piles alone. In corrugated steel sheet piling therefore, only adjacent piles (double piles at least) are staggered (figs. R 41-1 and R 41-2) in order to limit the stresses in the deeper piles.

A length of 1.0 m is usual for the staggering. In practice, it has been found that a structural check of the longer piles is unnecessary. For a greater staggering, it is necessary to verify the loadbearing capacity of the deeper piles with respect to multiple stresses due to bending moment combined with longitudinal and shear forces.

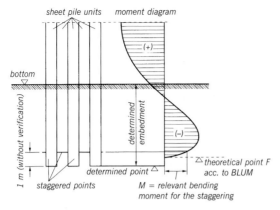

Fig. R 41-1. Staggered toe of sheet piling for a sheet pile wall fully fixed in the soil

8.2.10.2 Sheet pile walls fixed in the soil

- In the BLUM method [61] sheet pile walls fully fixed in the soil ($\tau_1 = 100\%$) may exploit the entire stagger dimension s to save steel: the longer sheet piles are driven to the theoretical depth determined according to R 56, section 8.2.9 (fig. R 41-1), the shorter piles terminate at a higher level (shorter by the stagger dimension s).

- In the case of walls partially fixed in the soil ($100\% > \tau_{1-0} > 0\%$) the steel saving depends on the degree of fixity present. A corresponding saving in steel is achieved by driving the longer sheet piles below the level of the theoretical toe of the wall by a certain fraction of the stagger dimension s_U, whereas the shorter piles terminate at the higher level (shorter by the stagger dimension s).

The dimension s_U depends on the degree of fixity τ_{1-0} [%] and is calculated from $s_U = (100 \cdot \tau_{1-0}) \cdot s / (2 \cdot 100)$.

8.2.10.3 Sheet pile walls free supported in the soil

When the sheet piling is free supported in the earth, the stagger dimension s no longer leads to a saving in steel (owing to the equation for the dimension s_U from section 8.2.10.2 which also applies to the degree of fixity $\tau_0 = 0\%$), but rather to an increase in the mobilisable soil support B, but this cannot be applied in the calculations.

In this case the longer sheet piles (see fig. R 41-2) must be driven below the theoretical underside of the wall by a distance $s_U = s/2$.

If the stagger dimension s is > 1.0 m, the loadbearing capacity of the longer piles must be verified according to fig. R 41-2.

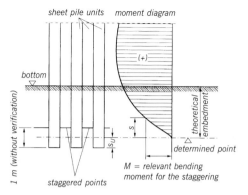

sheet pile units   *moment diagram*

(+)

bottom

1 m (without verification)

s

s_u

theoretical embedment

determined point

M = relevant bending moment for the staggering

staggered points

Fig. R 41-2. Staggered toe of sheet piling for a sheet pile wall free supported in the soil

The same applies to reinforced concrete or timber sheet piling, providing the pile joints have sufficient strength to guarantee that the shorter and the longer sheet piles will act together.

8.2.10.4 Combined sheet piling

Sheet piling composed of primary and secondary piles (R 7, section 8.1.4) must take into account the water pressure difference present for flows around the wall such that the necessary safety against hydraulic heave failure (R 115, section 3.2) is guaranteed in front of the shorter secondary piles. Longer piles should be used when scour represents a risk.

If soft or very soft layers prevail in the harbour bottom area, the embedment depth of the short or of the secondary piles respectively should be determined by special investigations.

8.2.11 Vertical loads on sheet piling (R 33)

8.2.11.1 General

Besides its function as a retaining structure in bending designed to accommodate horizontal actions, a sheet pile wall can also accommodate vertical loads in its axial direction provided it is driven deep enough into loadbearing subsoil, and provided it can be regarded as having adequate loadbearing capacity owing to the driving energy required during its installation.

The vertical loadbearing capacity is to be assessed according to R 4, section 8.2.4.6. For the sheet pile walls loaded in bending and dealt with here, a loadbearing effect similar to piles using the skin resistance $R_{1s,k}$ is only permissible below the theoretical toe of the wall, i.e. when the selected embedment depth t is greater than the embedment depth required reqd $t = t_{1-0} + \Delta t_{1-0}$.

In steel sheet piling in general, the axial resistance to sinking into the subsoil increases over time as a result of the accumulation of incrustations on the steel surface.

8.2.11.2 Toe and skin resistance of sheet piling against sinking as a result of vertical loads

The axial resistance of the sheet pile wall, or rather the sheet pile wall loadbearing elements, that enables it to accommodate vertical loads in the soil consists of the toe resistance $R_{1b,k}$ as a result of an end-bearing pressure $q_{b,k}$ and – if according to R 4, section 8.2.4.6 (4) the embedment depth selected is greater than the embedment depth required – a skin resistance $R_{1s,k}$ as a result of a skin friction $q_{s,k}$.

The characteristic resistance $R_{1,k}$ is to be determined according to DIN 1054, section 10.5.2 or 8.4, and the design value $R_{1,d}$ is obtained from the following equation:

$$R_{1,d} = \frac{R_{1,k}}{\gamma_P}$$

where γ_P is the partial safety factor.

When mobilising axial resistances it should be noted that the skin resistance R_{1s} becomes effective after just minimal relative displacements, but the toe resistance R_{1b}, on the other hand, basically requires large displacements, unless local experience enables the piling elements to be classed as sufficiently stable directly after installation. If the geometry of the section selected does not permit a plug to form within the loadbearing element, the soils specialist must provide information on the embedment depth required in the loadbearing subsoil and the effective cross-sectional area for the toe resistance in addition to details of the end-bearing pressure $q_{b,k}$ to be assumed.

Suitable flat or profiled steel sections can be welded together at the toe in such a way that the formation of a plug assumed in the calculations is also guaranteed in soil mechanics terms. Arrange the steel sections at the toe in such a way that they create the minimum hindrance during installation, but so that they can be driven to the required depth without damage in order to transfer the vertical loads into the subsoil with a sufficient margin of safety.

The end-bearing pressure $q_{b,k}$ on the area enclosed by the wall cross-section may be assumed for box sections. Reduce the bearing area for sheet pile walls made of corrugated sections with an average web spacing > 400 mm. The toe resistance of such corrugated sheet pile walls can be determined in some circumstances using the WEISSENBACH method [256], which is based on investigations carried out by RADOMSKI. In this method a shape factor based on the geometry is used to take into account the different included angles of the various corrugated section types. The loadbearing capacity determined with this method was not achieved in some loading trials. Therefore, when determining the toe resistance of corrugated sheet pile walls, the method using the end-bearing pressure applied to the n-times enlarged steel cross-section is the safer approach, i.e. $A_b = n \cdot A_s$, where $n = 6$–8.

In combined sheet pile walls with primary piles made from rolled I-sections, the flat or profiled steel sections mentioned above can be welded to the toe in order to achieve the plug formation if required when the clear spacing between the flanges of the loadbearing piles is greater than 400 mm.

8.2.11.3 Resistance of the subsoil to axial tension loads

With regard to the wall friction resistances, the same conditions apply here in principle as for compression loads. As a result of the unfavourable effects of tension loads corresponding to R 4, section 8.2.4.5, tension loads on sheet pile quay walls should be avoided whenever possible.

8.2.12 Horizontal actions parallel to the quay in steel sheet pile walls (R 132)

8.2.12.1 General

Combined and corrugated steel sheet piling reacts comparatively elastically to horizontal forces acting parallel to the shore. When such actions occur, a check must be carried out to verify that the resulting horizontal action effects parallel to the quay can be accommodated by the sheet piling, or whether additional measures are required.

In many cases in-plane stresses in sheet piling structures due to active earth pressure and hydrostatic pressure can be avoided if an appropriate design is chosen. For example, crossing anchors at quay wall corners according to R 31, section 8.4.11; or radial arrangements of tie rods, with an anchor plate arranged at the centre of the curve, at curving sections of quay walls or radiused pier heads. Further tie rods then continue back from this central plate to an anchor wall of sheet pile sections. The anchor force resultant of the anchors connected here must act in the same direction as the anchor force resultant of the radial anchors so that the central anchor plate remains in equilibrium.

8.2.12.2 Transmission of horizontal forces into the plane of the sheet piling

The transmission can take place via the available construction members such as capping beam and waling if these are suitably designed. Otherwise, additional measures are required, e.g. the installation of diagonal bracing behind the wall. Welding the interlocks in the upper section will suffice in the case of lower longitudinal forces.

The action components parallel to the quay due to line pulls act at the mooring gear, the maximum actions due to wind at the crane wheel arresting points and those due to ship friction at the fenders. The load application point of these friction forces can occur at any place along the wall. This also applies to horizontal loads as a result of crane braking which must be transferred from the superstructure into the top of the wall. The construction elements which transmit longitudinal forces to the sheet piling can be designed so that this transmission will be distributed over a considerable length of the wall instead of concentrating it in a short section.

For this purpose, the flanges of steel walings should be bolted or welded to the land-side sheet piling flanges (fig. R 132-1).

The transmission of longitudinal forces can also be effected by steel strips which are welded onto the waling and braced against the sheet piling webs (fig. R 132-2).

When the waling consists of two channels sections, the waling bolts can be used to transmit longitudinal forces only when both waling channels are joined by a vertical welded plate with holes, at the land-side sheet piling flanges. The force from the waling bolts is accommodated by the plate via bearing stresses, whereas the bolts are subject to shearing stresses (fig. R 132-3).

When loads parallel to the plane of the sheet pile wall occur, the capping beams and walings, including their joints, should be designed for bending, with both axial and shear stresses.

In order to transfer horizontal actions in the direction of the quay from a reinforced concrete capping beam to the top of the sheet pile wall, the latter must be adequately connected to the beam. The design of the reinforced concrete cross-section must take account of all the global and local stresses and strains that occur in this area.

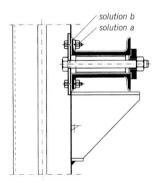

Fig. R 132-1. Transmission of longitudinal forces by means of close-tolerance structural bolts in the waling flanges (solution a) or by welding (solution b)

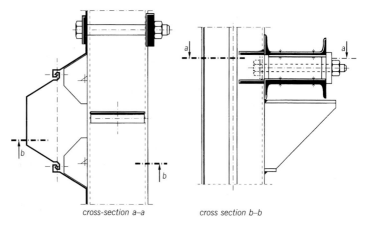

cross-section a–a   cross section b–b

Fig. R 132-2. Transmission of longitudinal forces by means of steel cleats welded to the waling

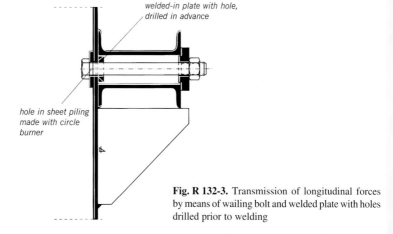

welded-in plate with hole,
drilled in advance

hole in sheet piling
made with circle
burner

Fig. R 132-3. Transmission of longitudinal forces by means of wailing bolt and welded plate with holes drilled prior to welding

8.2.12.3 Transmission of horizontal forces acting parallel to the line of the sheet piling from the plane of the wall into the subsoil

Horizontal longitudinal forces in the plane of the sheet pile wall are transmitted into the ground by friction on the land-sided sheet piling flanges and by resistance in front of the sheet piling webs. The latter however may not be of greater magnitude than the friction in the ground for the length of the sheet piling trough.

In non-cohesive soils the force can therefore be accommodated entirely by friction, for which a reasonable mean value of the friction coefficient between earth and steel as well as between earth and earth is assumed to be the friction coefficient. In non-cohesive soils the effectiveness of this force transmission increases with the angle of internal friction and the strength of the backfill; in cohesive soils the force transmission improves as the shear strength and the consistency of the soil increase.

When transferring horizontal forces acting parallel to the plane of the sheet pile wall from the capping beam or waling into the subsoil, additional bending moments occur transverse to the principal loadbearing direction of the sheet pile wall. The bending moments can be calculated with a numerical model of a fixed or simply supported anchor wall. Instead of the resisting soil reaction as a result of the mobilised passive earth pressure, however, the aforementioned average wall friction force or a corresponding shear resistance are assumed in this case.

As a rule, only shear resistant welded double piles should be considered as bearing elements for the absorption of these additional stresses. Unwelded piles may be considered only as single piles.

When absorbing horizontal forces parallel to the shore, the sheet piles are stressed in two planes by bending. When superimposing the resulting stresses, according to DIN 18 800 only the design value for the equivalent stress $\sigma_{v,d}$ may be assumed, which – under the boundary conditions for individual corner stresses given in the standard – may exceed the maximum permissible axial stress $\sigma_{R,d}$ by 10%, i.e. just reach the yield stress $f_{y,k}$.

By taking into account wall friction resistances in the horizontal direction, in the sheet piling calculation only a reduced angle of earth pressure $\delta_{a,k,red}$ may be applied. In doing so, the magnitude of the resultant of the vector addition of the two friction resistances orthogonal to one another may not exceed the maximum possible wall friction resistance value of the sheet pile wall with respect to the soil.

8.2.13 Calculation of anchor walls fixed in the ground (R 152)

Obstacles in the subsoil, such as ducts, services and the like, sometimes mean that it is not possible to connect the tie rods to the centre of the anchor wall. In such cases the tie rods must then be positioned at a higher level and connected in the upper area of the anchor wall.

The calculations for the anchor wall in such cases are to be carried out as for a non-anchored sheet pile wall for limit state LS 1B. The anchor force A_k of the front sheet pile wall to be anchored is entered into the calculation as the characteristic value F_k of the tensile force acting at the top of the anchor wall.

The partial safety factors for the earth pressure and, if applicable, hydrostatic pressure actions are applied according to R 214, section 8.2.0.1, the factor for the tensile force F_k is the quotient of the design value A_d and the characteristic value A_k of the anchor force, both of which are taken from the calculations for the sheet pile wall.

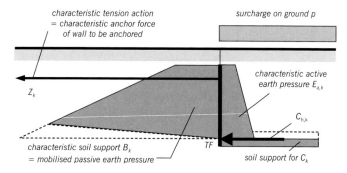

Fig. R 152-1. Actions, soil reaction and equivalent force for an anchor wall fully fixed in the ground as required for the stability verification for LS 1B

The extra embedment depth Δt required is calculated according to R 56, section 8.2.9.

In this form of construction the sheet piles of the anchor wall are considerably longer than an anchor plate loaded centrally and also have a larger cross-section.

Staggering of the anchor wall according to R 42, section 8.2.14, is allowable only at the lower end, but can be carried out here up to 1.0 m, without any special verification, for deep anchor walls.

In the case of a predominantly horizontal groundwater flow, a number of weepholes must be provided in the anchor wall if the hydrostatic pressure acting on the wall is to be reduced. The resulting hydrostatic pressure must be taken into account in the design of the sheet piles for the anchor wall.

8.2.14 Staggered arrangement of anchor walls (R 42)

In order to save materials, anchor walls may be staggered in the same manner as the waterfront sheet piling. Both ends may be staggered in the same wall. In general, the stagger should not be more than 0.5 m. When the wall is staggered top and bottom, all the double piles can be of the same length, i.e. 0.5 m shorter than the height of the anchor wall. The dual piles are driven alternately so that of each pair of dual piles one extends to be lower bottom edge of the anchor wall, with the other at the higher level. A stagger greater than 0.5 m is permissible only for deep anchor walls when the bearing capacity of the soil support and the loadbearing capacity of the section regarding bending moment, shear force and axial force have been verified. Such verification is also required for a stagger of 0.5 m if the overall height of the anchor wall is less than 2.5 m. In such a case it must be verified that the bending moments can be transferred from the lower ends into the neighbouring piles.

A similar procedure may be used for reinforced concrete and timber piles if the pile joints have sufficient strength to ensure that all the piles act together.

8.2.15 Steel sheet piling driven into bedrock or rock-like soils (R 57)

8.2.15.1 When bedrock shows a fairly thick decomposed transition zone, with solidity increasing with depth, or when the rock is soft, experience has shown that steel sheet piles can be driven so far into the rock to achieve at least sufficient free support.

8.2.15.2 In order to make it possible to drive sheet piles into bedrock, the piles must be modified and strengthened at the toe, and if need be, also at the top, depending on pile section and type of rock. It is recommended that the sheet piling be of steel grade S 355 GP (R 67, section 8.1.6), in consideration of the high driving energy required. Heavy hammers and a correspondingly smaller drop height are very effective for this work. A similar effect can be achieved by use of hydraulic hammers, the impact energies of which can be regulated in a controlled manner in line with the particular driving energy requirement (R 118, section 8.1.11.3).
If there is strong, hard rock up to the upper surface, test piles and rock investigations are indispensable. If necessary, special measures must be taken for protection of the pile point and for ensuring proper alignment. Borings of diameters 105 mm to 300 mm are sunk at a spacing corresponding to the pile width of the sheet piling wall to perforate and destress the subsoil in such a manner as to facilitate driving of the sheet piles.
The same effect can be achieved by high-pressure jetting in the case of rock with changing hardness and the like. (See R 203, section 8.1.23.3).

8.2.15.3 If deeper driving embedment into bedrock is required, precision blasting can be used to loosen the bedrock in the area of the sheet piling and facilitate driving. In selecting the section and the steel grade, possible irregularities in the foundation soil and resulting driving stresses must be taken into account. For loosening blasting, please refer to R 183, section 8.1.10. The advantage of pre-drilling is that the properties of the rock in its undisturbed state remain intact. Compared to blasting, this has positive effects on the lower support reaction of the sheet piling. The pre-drilling depth is also less than the depth required for blasting.

8.2.16 Waterfront sheet piling in unconsolidated, soft cohesive soils, especially in connection with undisplaceable structures (R 43)
For various reasons, harbours and industrial facilities with associated waterfront structures today must sometimes be constructed in areas with poor foundation soil. Existing alluvial cohesive soils, sometimes with

layers of peat, are surcharged by a layer of fill and thus altered into an unconsolidated state.

The resulting settlement and horizontal displacement require special construction features and a structural design treatment that is best suited to the particular site.

In unconsolidated, soft cohesive soils, sheet piling structures may be built as "floating" only when neither the serviceability nor the stability of the entire structure and its parts will be endangered by the resulting settlement and horizontal displacement and/or their differences. In order to evaluate the conditions and to adopt the required measures, the expected settlement and displacement must be calculated.

If a quay wall is constructed in unconsolidated, soft cohesive soils, in connection with a structure on a foundation with practically no settlement to be expected, such as a rigidly founded structure on piles, the following solutions may be used:

8.2.16.1 The sheet piling may be anchored or supported so that it is free to move in a vertical direction so that the stability and serviceability of the connection to the structure remains effective even at the maximum theoretical displacement.

The solution is quite straightforward apart from the settlement and displacement calculations. For operational reasons however, it can generally be used at structures on piles only for sheet piling lying behind them. The vertical friction force occurring at the support must be taken into consideration in the design of the piles supporting the structure. Slotted holes are not sufficient at the anchor connections of front sheet piling. In fact, freely sliding anchoring is then required.

8.2.16.2 The sheet piling is supported against vertical displacements by driving a sufficient number of driving units deep enough to penetrate into the bearing, deep-lying foundation soil. In this case the loadbearing capacity of the sheet pile wall with respect to the following vertical loads must be guaranteed by the deeper piles alone:

(1) dead load of the wall,
(2) soil clinging to the sheet piling due to negative skin friction and adhesion, and,
(3) axial actions on the wall.

The solution is technically and operationally practicable in case of forward-positioned piling. Since the cohesive soil clings to the sheet piling during the settlement process, the active earth pressure decreases. If the supporting soil in front of the toe of the sheet piling also settles, the characteristic passive earth pressure and hence the potential soil support also decreases as a result of negative skin friction. This must be taken into consideration in the sheet piling calculation.

400

In calculating the vertical load on the sheet piling arising from soil settlement, negative skin friction and adhesion for the initial and final state are taken into consideration.

8.2.16.3 Apart from the anchoring or support against horizontal forces, the sheet piling should be so suspended from the structure that the actions mentioned in section 8.2.16.2 are transmitted to the structure and then into the loadbearing soil.

In this solution, the sheet piling and its structural suspension elements are calculated in accordance with the information in section 8.2.16.2.

8.2.16.4 If the loadbearing soil is not at an excessive depth, the entire wall is driven down into the loadbearing soil. The passive earth pressure in this stratum is calculated with the usual angles of earth pressure and the partial safety factors as per table R 0-2. When calculating the soil reaction of the overlying soil with a lower strength or consistency, only a reduced characteristic passive earth pressure may be assumed. A serviceability calculation for LS 2 is required.

8.2.17 Effects of earthquakes on the design and dimensioning of waterfront structures (R 124)
See section 2.13.

8.2.18 Design and dimensioning of single-anchor sheet piling structures in earthquake zones (R 125)

8.2.18.1 General
Careful checks must first be made on the basis of the soil findings and the soil mechanics investigations, as to the effect which the vibrations occurring during a probable earthquake may have on the shear strength of the subsoil.

The results of these investigations can be the decisive criteria for the design of the structure. For example, when soil conditions are such that liquefaction must be expected as per R 124, section 2.13, no high-lying anchor wall or anchor plates may be used unless the mass of earth supporting the anchoring is adequately compacted in the course of the construction work, and the danger of liquefaction thus eliminated. Please refer to R 124, section 2.13, regarding the magnitude of the seismic coefficient k_h and other actions as well as the design values for loads and resistances plus the safety factors required.

8.2.18.2 Sheet piling calculations
Taking into consideration the sheet piling loads and support reactions determined according to R 124, sections 2.13.3, 2.13.4 and 2.13.5, the

calculation can be carried out as per R 77, section 8.2.2, but without redistribution of the active earth pressure.

The characteristic active and passive earth pressures determined with the imaginary angles of inclination for the reference and slope surfaces are generally used as a basis for all calculations and analyses, although tests have shown that the rise in the active earth pressure from an earthquake does not increase linearly with the depth, but rather is higher near the ground surface. The anchoring is therefore to be generously dimensioned.

8.2.18.3 Sheet piling anchoring

Verification of stability of anchoring for the deep failure plane is to be carried out according to R 10, section 8.4.9, taking account of the additional horizontal forces occurring at reduced live loads due to acceleration of the mass of earth to be anchored and the pore water contained therein.

8.3 Calculation and design of cofferdams

8.3.1 Cellular cofferdams as excavation enclosures and waterfront structures (R 100)

8.3.1.1 General

Cellular cofferdams are constructed of straight web piling with high interlock tensile strength ranging from 2 to 5,5 MN/m, depending on type of steel and type of section. Cellular cofferdams offer the advantage that they can be designed as stable gravity walls without waling and anchoring, even if embedment of the steel walls is not possible caused by rocky subsoil.

Cellular cofferdams can be economical for higher water depths, i.e. large differences in ground levels, where longer structures intersect, and when bracing or anchoring is impossible or uneconomic. In some circumstances the additional requirement in sheet piling surface area is thus compensated by the weight-savings of the lighter and shorter sheet pile sections, and by the omission of walings and anchors.

8.3.1.2 Cell construction for cofferdams

We distinguish between
- cellular cofferdams with circular cells (fig. R 100-1a),
- cellular cofferdams with diaphragm cells (fig. R 100-1b),
- cellular cofferdams with special fixed point cells within the cofferdam (e.g. "clover leaf" cell), and
- mono-cell cofferdams.

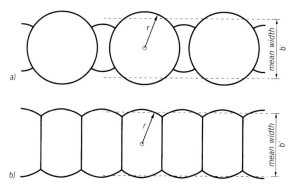

Fig. R 100-1. Schematic diagrams of plan views of cellular cofferdams
a) with circular cells, b) with diaphragm cells

(1) Circular cells, which are linked by small, connecting arcs, have the advantage that each cell can be individually constructed and filled, and is therefore independently stable. The connecting arcs required for sealing can be installed later. Junctions connect them to the stable circular cells. These junctions generally consist of specially shaped rolled sections or welded or bolted sections in which the connecting angle at the nodes can be varied between 30° and 38°. In order to avoid laminar tearing, only steel grades with the appropriate properties may be used for welded junctions piles (see R 67, section 8.1.6.1). In order to keep the unavoidable extra stresses at the junctions low, the clearance of the circular cells and the radius of the connecting arcs should be kept as small as possible. If necessary, bent piles may be used in the connecting arcs. Information on the calculation is to be found in [188].

(2) A succession of diaphragm cells with straight side walls must be used when the use of circular cells becomes unfeasible because the cell diameter is so large that the design value of the circumferential tensile forces is larger than the relevant design value of the tensile resistance of the straight web piling.

Since the individual diaphragm cells are not stable by themselves, a cofferdam of this type must be filled in stages unless other measures are taken to ensure stability during construction. Therefore special start buildings of circular cells must be constructed as stable units. In the design of long structures it is recommended to incorporate intermediate fixed points, especially if there is a danger of ship collisions, which can lead to destruction of large portions of the structure. All other conditions being equal, diaphragm cell cofferdams require more steel per linear meter than circular cell cofferdams.

(3) Mono-cell cofferdams are individual cells that can be used as foundations in open water. For example, the ends of moles, guide and fender structures at harbour entrances, or foundations for navigation aids (beacons etc).

8.3.1.3 Calculations for cellular cofferdams

(1) Verification of stability against failure of the cellular cofferdam for ULS 1B

The verification of stability against failure of cofferdams is carried out according to DIN 1054 for ULS 1B and is depicted in figs. R 100-2, R 100-3 and R 100-4. Equivalent width b' according to fig. R 100-1 is to be taken as the design width of the cellular cofferdam. It results from the conversion of the enclosed area into an equivalent rectangular area. If a cofferdam rests directly on rock (fig. R 100-2), a convex failure line occurs in a cross-sectional plane when there is failure between the toes of the cofferdam walls. The generating curve of this failure plane can be approximated initially by a logarithmic spiral for the characteristic value of the friction angle φ'_k, where the pole of the spiral represents the reference point for determining the acting and resisting cofferdam moments.

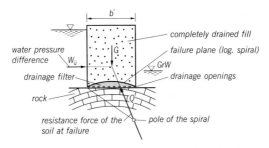

Fig. R 100-2. Cofferdam resting on rock, with drainage

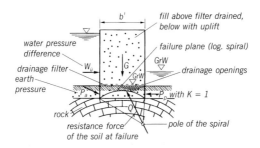

Fig. R 100-3. Cofferdam resting on rock overlain with other soil strata, with drainage

404

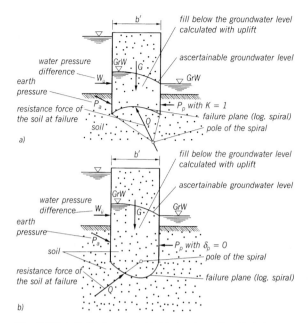

Fig. R 100-4. Cofferdam embedded in bearing soil, with drainage
a) In case of flat embedment
b) Additional investigation in case of deep embedment

All forces and resistances are determined with characteristic values; partial safety factors are entered into the calculation only for the design values of the cofferdam actions and resistances.

To determine the design value of the moment effect as a result of horizontal forces $W_{\ddot{u}}$, E_a and varying external forces, e.g. due to line pull force, the characteristic values of the individual moments are multiplied by the partial safety factors γ_G and γ_Q for forces and added together.

The design value of the resisting moment as a result of the vertical force G (dead load of cofferdam fill) is determined by dividing the characteristic moment M_{kG}^R by the partial safety factor γ_{Gl} of resistance against sliding.

$$M_{Ed} = M_{kG} \cdot \gamma_G + M_{w\ddot{u}} \cdot \gamma_G + M_{kQ} \cdot \gamma_Q \leq \frac{M_{kG}^R}{\gamma_{Gl}}$$

where:

M_{kG} = characteristic value of single moment due to earth pressure force

$M_{w\ddot{u}}$ = characteristic value of single moment due to water pressure force

405

M_{kQ} = characteristic value of single moment due to external variable force

M_{kG}^R = characteristic value of single moment due to the force of cofferdam fill

Use the partial safety factors for LS 1B according to section 0.2.2.4 for force and resistances depending on the respective loading case.

Verification of stability against failure is satisfied when the design value of the effected moment about the pole of the relevant failure plane is less than or equal to the design value of the resisting moment about the same pole. The relevant failure plane is the logarithmic spiral that produces the lowest design value for the resistance.

If the cofferdam rests on rock overlain by other soil strata (fig. R 100-3), or if the cofferdam is embedded in loadbearing loose rock (fig. R 100-4), the force have to be increased by the additional earth pressure of this soil layer and the resistance by the additional passive earth pressure. The passive earth pressure, considering the small deformations, is to be applied with a reduced magnitude – usually with $K_p = 1.0$ – and with a deeper embedment in the loose rock with K_p for $\delta_p' = 0$.

The principal force to be resisted by a cofferdam is generally the water pressure difference $W_{\ddot{u}}$. This results from the difference in the hydrostatic pressures $W_a - W_i$ acting on the outer and inner walls of the cofferdam respectively, and is applied down to the bottom of the outer, i.e. load-side, wall. The relevant water level W_i within the cofferdam enclosure need not always correspond with the level of the bottom.

When founded on loose rock, the resistance of the cofferdam required to prevent failure can be enhanced by

- widening the cofferdam
- choosing a filling material with a higher specific density and larger angle of friction
- draining the cells
- embedding the piles of the cofferdam deeper in the subsoil (After setting up all the sheet pile sections of the cell, the installation of the individual piles should be carried out step by step only, e.g. by a driving procedure that passes around the cell several times in which the penetration per round is low, see also section 8.3.1.5 (3).)

When the cross-sectional geometry of the cofferdam after deeper embedment and the in situ soil characteristics permit, verification against a failure must be carried out with a convex failure plane (fig. R 100-4a), as well as with a concave failure plane (fig. R 100-4b).

In the latter case the position of the spiral must be chosen so that its centre point lies below the line of action of E_p for $\delta_p = 0$ (fig. R 100-4). The above verification proves both safety against failure due to over-turning and also due to sliding.

(2) Verification of safety against failure of the sheet pile section for ULS 1B as a result of the circumferential tensile force

When analysing the cofferdam it can be assumed that the effects due to external forces such as water pressure and, if applicable, earth pressure can be accommodated by the monolithic block action of the cofferdam filling. To verify safety against failure of the straight web section, in this case it is adequate to investigate the cell cross-section at the level of the bottom of the excavation or watercourse because this is generally the position at which the relevant circumferential tensile force occurs.

However, in some circumstances it may be necessary to analyse the resistance to the circumferential tensile at several levels if the cofferdam is embedded in low-lying cohesive strata. Circumferential tensile force larger than at the level of the bottom can occur here due to the sudden leap in the value of the earth pressure or the smaller value of the passive earth pressure as well as any pore water pressure that may be present. Circumferential tensile force $(F_{ts,Ek})$ is determined to $Z = \Sigma\, p_i \cdot r$. The earth pressure 'p_i' inside the cell is calculated with $K_0 = 1 - \sin \varphi_k'$. Any water pressure difference effective within the cell generates an additional component for the circumferential tensile force. The design value of the circumferential tensile force $(F_{ts,Ed})$ in the individual wall elements may be determined using a simplified method according to ENV 1993-5, section 5.2.5 (20):

In the common wall:

$$F_{tc,Ed} = \sum p_{a,d} \cdot r_a \cdot \sin \varphi_a + \sum p_{m,d} \cdot r_m \cdot \sin \varphi_m$$

In the main wall:

$$F_{tm,Ed} = \sum p_{m,d} \cdot r_m$$

In the connecting arc:

$$F_{ta,Ed} = \sum p_{a,d} \cdot r_a$$

The design value of the circumferential tensile forces $(F_{tc,Ed}, F_{tm,Ed}, F_{ta,Ed})$ is calculated by multiplying effects due to hydrostatic and earth pressures and variable forces $(\Sigma\, p_a, \Sigma\, p_m)$ determined with the characteristic values of the forces by the associated partial safety factors for forces (according to section 0.2.2.4).

Information given in Section 8.3.1.3 may result in additional circumferential tensile forces. Verification of the sheet pile sections as well as the welded junctions is carried out according to DIN EN 1993-5, section 5.2.5.

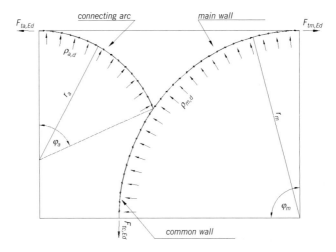

Fig. R 100-5. Circumferential tensile force(F_{ts}) in the individual wall elements of a cellular

Accordingly, this is verified for the sheet pile sections when the resisting tensile strength of the web and the interlock ($F_{ts,Rd}$) is equal to or greater than the design values of the hoop tension ($F_{tc,Ed}$, $F_{tm,Ed}$, $F_{ta,Ed}$).

$$F_{ts,Rd} \geq F_{tc,Ed} \quad \text{or} \quad F_{tm,Ed} \quad \text{or} \quad F_{ta,Ed}$$

where:

$$F_{ts,Rd} = \beta_R \cdot \frac{R_{k,S}}{\gamma_{MO}} \quad \text{(interlock)}$$

and

$$F_{ts,Rd} = t_w \cdot \frac{f_{yk}}{\gamma_{MO}} \quad \text{(web)}$$

where:

$R_{k,S}$ = characteristic value of interlock tensile strength resistance
f_{yk} = minimum yield point of steel to R 67, section 8.1.6.2
t_w = web thickness of straight web section
γ_{MO} = partial safety factor for sheet pile material
β_R = reduction factor for interlock tensile strength resistance (recommended value to EN 1993-5:2002 is $\beta_R = 0.8$)

Assuming that the junction is welded according to DIN EN 12 063, safety is verified when the resisting tensile strength of the junction ($\beta_T \cdot F_{ts,Rd}$

for interlock and web) is equal to or greater than the design value of circumferential tensile force ($F_{tm,Ed}$).

$$\beta_T \cdot F_{ts,Rd} \geq F_{tm,Ed} = \sum P_{m,d} \cdot r_m$$

The reduction factor β_T according to DIN EN 1993-5, section 5.2.5 (14), can be taken as

$$\beta_T = 0.9 \cdot \left(1.3 - 0.8\frac{r_a}{r_m}\right) \cdot \left(1 - 0.3 \tan \varphi_d\right)$$

For φ_d, select the design shear angle of the filling material to the circular cells.

(3) Verification of safety of the cellular cofferdam against hydraulic heave failure for LS 1B
For cofferdams not founded on rock, verification of bearing capacity of the subsoil is to be carried out in accordance with DIN 4017 in conjunction with DIN 1054, in which the equivalent width b' is taken as the cofferdam width. Please refer to section 8.3.1.3 (5).

(4) Verification of safety of the cellular cofferdam against ground failure for LS 1C
Verification of safety against ground failure according to DIN 4084 may also be required. This concerns, for example, backfilled cofferdams that form part of a waterfront structure. To carry out the analysis, position the failure plane with the load-side, ideal limit to the cofferdam width which coincides with the above value for the equivalent width b'.
Please refer to section 8.3.1.3 (5).

(5) Additional analyses in the case of water flows
- Any flow pressure present is to be taken into account in the analyses called for in points (1) to (4) above.
- Carry out an analysis to verify safety against failure of the subsoil as a result of hydraulic heave.
- Carry out an analysis to verify safety against failure of the subsoil as a result of subsurface erosion.
- Special sealing measures are required at the base of the sheet pile wall in the case of cofferdams founded on jointed rock or varying solid rock in order to rule out the aforementioned failure modes.

8.3.1.4 Constructional measures
Cellular cofferdams may be constructed on loadbearing subsoil only. Soft strata, especially if they occur near the bottom of the cofferdam, reduce the stability considerably due to the formation of mandatory failure planes. Replace such soils with granular fill inside the cofferdam or drain

them with vertical drains. If none of these measures are taken, circumferential tensile force increases after filling as a result of the pore water pressure difference that occurs, which has a detrimental effect when analysing the failure of the sheet pile section. Fine-grained soil to DIN 18 196 may not be used for the filling.

The filling should be particularly water-permeable in the case of enclosures to excavations in order to achieve the drawdown level for the dewatering.

Therefore, in order to minimise the dimensions of the cofferdam and to reach adequate stability, a soil with a high mass density γ or γ' and high internal angle of friction φ_k' should be used. Both these soil parameters can be improved by vibrating the filling.

(1) Cellular cofferdams as enclosures to excavations

In excavation enclosures founded on rock it must be possible to lower the water in the cofferdam at any time with a drainage system verified by an observation well to such an extent that it satisfies the stability analyses. Drainage openings at the bottom of the exposed wall, filters at the level of the excavation floor and good permeability of the entire fill are essential.

Experience has shown that the permeability of sheet piling interlocks under tension is low, meaning that no particular measures need to be taken here.

The part of the excavation enclosure subject to the load of the external hydrostatic pressure must be adequately watertight. It can be practical in some cases to plan additional sealing measures on the outboard side, e.g. underwater concrete.

(2) Cellular cofferdams as waterfront structures

In waterfront structures, especially in deep water, much of the cell fill is constantly submerged. Therefore, deep drainage systems cannot be used. When there are large and rapid fluctuations in the level of the water table, drainage of the cell fill and of the structure backfill can be of advantage by reducing the hydrostatic pressure difference (fig. R 100-6). In these cases the planned efficiency of the drainage measures is critical for the useful life of the waterfront structure.

The superstructure is to be designed and constructed so that
- the risk of local damage to the cofferdam cells due to ship impacts is prevented. This can be achieved, for example, through components to spread the load.
- The magnitude of the global effect of a ship impact is reduced by corresponding measures, e.g. fenders with a high absorption capacity, to such an extent that the stability of the cofferdam cells is not put at risk (fig. R 100-6).

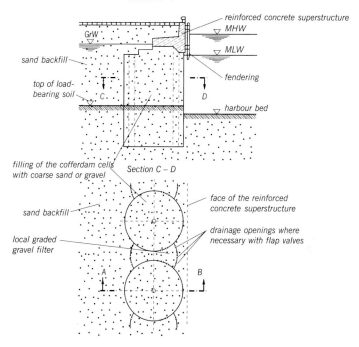

Section A – B

reinforced concrete superstructure
MHW
MLW

GrW

sand backfill

top of load-
bearing soil

fendering

harbour bed

Section C – D

filling of the cofferdam cells
with coarse sand or gravel

face of the reinforced
concrete superstructure

sand backfill

drainage openings where
necessary with flap valves

local graded
gravel filter

Fig. R 100-6. Schematic illustration of a waterfront structure constructed using circular cells and drainage

Components subject to significant vertical actions, e.g. due to crane operations, can be built on separate foundations, e.g. with an additional pile foundation, which can be positioned either adjacent to or even within the cofferdam. This avoids an increase in the circumferential tensile force and larger eccentricities for the resultants of the actions. Drive the loadbearing piles into the loadbearing strata beneath the base of the cofferdam so that their forces do not need to be considered when analysing the stability of the cellular cofferdam.

8.3.1.5 Building technology

(1) Use of cofferdams

Cofferdams can be constructed on land and in the water to suit their particular purpose. Generally, a cofferdam built to provide flood protection or to enclose an excavation is built in the dry. By contrast, the cofferdam designed exclusively as a waterfront structure is usually built from a floating or jack-up platform over the water.

411

- Construction on land

  The traditional case for constructing cofferdams on land is for weirs or power station sites on a river that is to be dammed. The cofferdam is built on dry ground in the flood plain during periods of low water in order to carry out excavations and building work within the protection of the cofferdam. The cofferdam elements (double-wall or cellular cofferdam; flat or profiled sheet pile elements) are set up on guides and, depending on the subsoil, driven several metres into the ground. The inside of the double-wall or cellular cofferdam is filled with non-cohesive soil, which is afterwards compacted to a high degree and, if possible, provided with drains.

  If the cofferdam is to be used as an enclosure for an excavation, it must be driven into the ground to the appropriate depth. Filling of the cells is not essential in this case – the in situ soil represents the filling. However, it may be necessary to improve the soil in the cofferdam through compacting it by means of vibroflotation, or by installing vertical drains to reduce the pore water pressure and hence increase the stability.

- Construction in water

  Basically, the difference between construction in water and construction on land is that when working in water all the operations have to be carried out from a floating or jack-up platform. When building a double-wall cofferdam, pile-driving trestles have to be set up first or otherwise jack-up platforms or pontoons will be required. Cellular cofferdams require working platforms to be built in the water. It may also be possible to pre-assemble a complete cell around a working platform erected on land. A floating crane is then used to lift the cell with the working platform and place it in position. This avoids the need to erect the individual pile elements on the open sea – often a tedious process.

Generally, the cofferdams serve as bases for retaining walls or wave chamber for incorporating the entire facilities for berthing ocean-going vessels.

The success or failure of the construction of waterfront structures comprising cellular cofferdams is very much influenced by the preparatory work. As individual cells without any filling are highly sensitive, non-loadbearing and easily damaged, we shall include at this point a number of remarks concerning the straight web section, the structure of the driving platform and driving methods.

(2) Diameter of circular cells

Regardless of the structural calculations, the actual diameter of a circular cell depends on the number of individual piles plus their rolling tolerances

412

and the play in the individual interlocks. Therefore, the diameter can vary between two values, which are important for the size of the guide ring, for which a minimum diameter must be specified.

(3) Installation of straight web sections
The proper installation of the piles requires at least two, on high cofferdams three, guide rings. These are placed around the driving platform, which is generally constructed as a space frame and suspended from several driven piles or pile trestles supported on a loadbearing seabed. All the piles are positioned around the driving platform and the rings tensioned accordingly. The piles are subsequently driven step by step, each about 50 cm per revolution, by means of vibrators or rapid-stroke hammers. It is also possible to sink the entire cell by using a multitude of vibrators, each one acting on several piles simultaneously. Prior to installation, the connections of all interlocks should be checked, if necessary with the help of divers.
On very tall cells it may be necessary to divide the piles into several lengths for easier handling. Basically, as only tensile forces from the internal pressure or outward wall friction forces due to the earth pressure have to be accommodated, a corresponding stagger can be provided without suffering any loss in load-carrying capacity.
One critical situation for the stability of a circular cell can occur after it has been set up and the driving platform has already been dismantled following installation of the piles. External actions, e.g. due to wave pressure, can cause a circular cell to collapse. Therefore, the interlocks are welded at the top of the cell and the inside filled to about one-third its height as quickly as possible. It is recommended to fill the entire circular cell cofferdam at the same time as removing the driving platform.

8.3.2 Double-wall cofferdams as excavation enclosures and waterfront structures (R 101)

8.3.2.1 General
In double-wall cofferdams the parallel steel sheet piling walls are driven into or otherwise placed on the bottom depending on the prevailing subsoil and hydraulic conditions, as well as on the structural requirements. The two walls must be connected by tie rods, and if the double-wall cofferdam rests on rock, at least two horizontal rows of tie rods must be provided.
Transverse walls and anchor cells as shown in fig. R 101-1 may also assist in the planning and executing of the construction work. They are also recommended for use in permanent structures of considerable length because they will confine any damage to the section in which the damage occurs. The length of the individual sections in which the cofferdam is

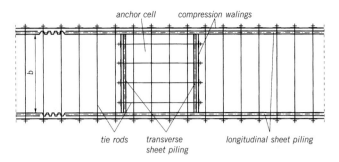

Fig. R 101-1. Plan of a double-wall cofferdam with anchor cells

constructed, including tie rods and fill, depends on the spacing of the transverse walls and anchor cells.

The remarks in R 100, section 8.3.1.4, regarding the fill apply here. When analysing a cofferdam with large external actions due to hydrostatic pressure, e.g. excavation enclosures in open water, the permanently effective drainage of its soil filling is crucial for minimising its dimensions. The fill is drained towards the excavation, for which weepholes to R 51, section 4.4, are sufficient.

Drainage can also be useful when using a cofferdam as a waterfront structure. Such cofferdams are drained towards the harbour side. But if there is a risk of clogging, drainage systems with flap valves to R 32, section 4.5.2, must always be used.

In the following, the external sheet pile wall to a cofferdam which is subject to the force of earth and hydrostatic pressures plus variable action is designated as the load-side wall, whereas the opposite sheet pile wall not subject to such loads is designated the air-, excavation-, or harbour-side wall depending on the particular use of the cofferdam.

8.3.2.2 Calculations

(1) Verification of stability for a double-wall cofferdam (LS 1B)
For calculation purposes, the width of the double-wall cofferdam is taken as the centre-to-centre distance b between the two sheet piling walls. For verification of the stability of double-wall cofferdams, essentially the same principles apply as for investigating the stability of cellular cofferdams – see R 100, section 8.3.1.3 (1). In contrast to figs. R 100-4a and 4b, the passive earth pressure E_p in front of the air-side sheet pile wall – owing to its greater deflection potential – is applied corresponding to a customary anchored sheet wall to R 4, section 8.2.4, with an angle of inclination $\delta_p < 0$. The following applies to the position of the failure planes to be investigated:

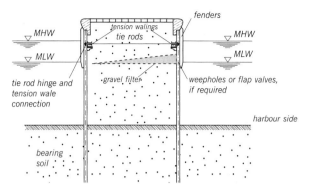

Fig. R 101-2. Schematic illustration of a mole structure built using a double-wall cofferdam

- Air-side sheet pile wall
  - For a sheet pile wall simply supported in the ground, the logarithmic spiral of the failure plane intersects the base of this wall.
  - For a sheet pile fixed in the ground, the logarithmic spiral of the failure plane intersects the point of zero shear force.
- Load-side sheet pile wall
  - The point of application of the logarithmic spiral in this case is generally at the same level as that of the air-side sheet pile wall.
  - If the load-side wall is shorter than the air-side wall, the failure plane on the load-side wall must be continued to the existing base.

In a double-wall cofferdam, a concave failure plane usually governs because of the deeper embedment of the sheet pile walls in the subsoil. The resistance to failure of the double-wall cofferdam can be increased through one or more of the following measures:

- Widening the cofferdam
- Selecting a fill material with better γ, γ' and φ'_k soil parameters
- Compacting the cofferdam fill, possibly including the subsoil as well
- Driving the cofferdam sheet pile deeper if this leads to a concave failure plane with which the limit state conditions against failure of the soil can be satisfied
- Additional anchor levels (But it should be investigated whether the more complex installation of another anchor level – e.g. underwater with the help of divers – is advisable.)

(2) Verification of stability for sheet piling and anchors (LS 1B)
In a filled cofferdam the load transfer into the subsoil relies on the cofferdam functioning as a compact block of soil. The moment load as a result of the horizontal actions (hydrostatic and earth pressure) with respect to

415

the point of rotation are transferred into the loadbearing subsoil by means of vertical stresses in the soil through the cofferdam filling acting as a monolithic block. The stresses in the soil change linearly over the width of the cofferdam and reach a maximum at the side opposite to the forces, i.e. the air- or harbour-side sheet pile wall. Due to the vertical surcharge stress, this sheet pile wall is subject to a higher active earth pressure. Experience has shown that the increase in active earth pressure can generally be taken into account with adequate accuracy by increasing the active earth pressure calculated with $\delta_a = +^2/_3 \, \varphi'_k$ by 25%. Another force acting on the air-side of the sheet pile wall is the water pressure difference, which results from a water level difference between the drawdown level within the fill and the air- or harbour-side water level. If the cofferdam fill is installed by using water-jetting and then compacted, the active earth pressure can rise to that of the hydrostatic pressure as a result of the filling effect.

If the in situ soil is used as the cofferdam filling, however, ancillary failure planes can develop within the fill after excavation work. Redistribution of the active earth pressure according to R 77, section 8.2.2.3, is then permissible. The air-side wall is calculated as an anchored sheet pile wall taking into account all the forces. If the sheet pile wall is embedded in loadbearing loose rock, the supporting passive earth pressure can be calculated with the angle of inclination as per R 4, section 8.2.4. The determination of the design values for the effects can be carried out for a sheet pile wall with single anchor as per R 77, section 8.2.2, and for a double anchor as per R 134, section 8.2.3.

The load-side sheet piling can be installed with a different section and can be shorter than the air- or harbour-side sheet piling if this is checked for the individual construction phases or if the requirements regarding watertightness and limiting underflow are satisfied.

Bearing stability verification for the load-sided wall shall take account of the following actions:

- transferring the anchor force of the air-side sheet pile wall (or the anchor forces with double anchors)
- external hydrostatic pressure
- external active and, where applicable, also passive earth pressure
- ship impacts, line pulls and other horizontal forces
- support provided by the soil filling

The distribution and magnitude of the earth support over the height of the wall must be applied such that the equilibrium condition $\Sigma H = 0$ is satisfied. If the load-side wall is fixed in the ground, the equivalent force C must be considered when checking the equilibrium.

$$\Sigma \text{ (characteristic forces on wall)} = \Sigma \text{ (earth support to wall)}$$

416

(3) Verification of stability for a double-wall cofferdam (LS 1B)
  on the lower failure plane

The stability of the anchorage at the lower failure plane is to be verified according to R 10, section 8.4.9. Here, the line of the lower failure plane can be approximated as follows for a single anchor:

- Load-side wall from the base of an equivalent anchor wall assumed to be simply supported (fig. R 101-3).

- Air-side wall
  – simply supported
    ... to the base of the sheet pile wall
  – fixed
    ... to the shear force zero point in the region of the fixity

The upper starting point for the lower failure plane can be placed deeper when the following aspects are verified:

- The design value of the soil support induced by the forces above the imaginary dividing line must be less than or equal to the design value of the partial resistance in the cofferdam above the dividing line.
- The design value of the wall effects in this area as a result of the aforementioned forces must be less than or equal to the design value of the section resistance.

The upper starting point for the lower failure plane can then be selected approximately as the shear force zero point for a fixed anchor wall when the following conditions apply:

- It must be possible to represent the design values of the soil support induced about by the forces and also the equivalent force C on the two opposite sides of the sheet pile wall as a passive earth pressure within the scope of the overall system. These values must be less than or equal to the design values for the passive earth pressures inside and outside the cofferdam on the two sides of the sheet pile wall.
- The design value of the wall effects in this area as a result of acting as a fixed anchor wall must be less than or equal to the design value of the section resistance.

If several anchors are used, an equivalent anchor wall can also be used in the calculations, but the imaginary dividing line below the bottommost anchor must be positioned in such a way that the lowest anchor cannot fail.

(4) Failure by hydraulic heave
See R 100, section 8.3.1.3 (3).

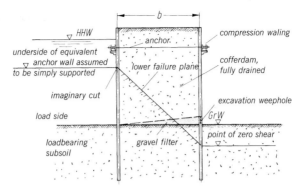

Fig. R 101-3. Verifying the stability of anchors for the deep failure plane to R 10, section 8.4.9

(5) Slope failure
See R 100, section 8.3.1.3 (4).

(6) Additional analyses in the case of water flow
See R 100, section 8.3.1.3 (5).

8.3.2.3 Constructional measures
See R 100, section 8.3.1.4.
Furthermore, for the individual structural members please refer to the relevant recommendations of the EAU. This applies, above all, to the tie rods and their proper installation.

(1) Excavation enclosures
See R 100, section 8.3.1.4 (1) (except the details regarding interlocks subject to tension).
The weepholes in the base area of the air- or harbour-side sheet pile wall are located in the web of the sheet pile section for expediency.
The walings for the transmission of the anchor forces are mounted on the outer side of the sheet piling as compression walings, if this does not create an objectionable hazard to ships. Waling bolts are not necessary with this solution and installation of the tie rods is simplified. However, the anchor sleeve must be made watertight on the load side.

(2) Waterfront structures, breakwaters and moles
The remarks in R 100, section 8.3.1.4 (2), are valid here accordingly (fig. R 101-2).

418

8.3.2.4 General advice for the construction of double-wall cofferdams

(1) Use of double-wall cofferdams
Similar to R 100, section 8.3.1.5 (1).

(2) Special aspects concerning the construction of double-wall cofferdams
- Special attention is to be given to the installation of a deep anchor. As these are usually underwater and can only be installed with the help of divers, simple but effective connections to the sheet pile wall are vital.
- Prior to filling of the cofferdam, the bottom surface inside the cofferdam should be cleaned in order to avoid increased earth pressure and hence higher loads on the low-level anchor.

8.3.3 Narrow partition moles in sheet piling (R 162)

8.3.3.1 General
Narrow partition moles in sheet piling construction are cofferdams at which the spacing of the sheet piling is only a few metres and thus considerably less than a customary cofferdam (R 101, section 8.3.2). These partition moles are subject to load chiefly from hydrostatic pressure difference, vessel impact, ice impact, line pull etc.

The sheet piling walls are tensioned against each other at or near the top and provided with compression stiffeners to ensure joint transmission of the external actions.

The space between the sheet piles is filled with sand or gravelly sand of at least medium degree of density.

8.3.3.2 Calculation assumptions for the partition mole as a non-anchored system (LS 1B)
For accommodating the external forces acting perpendicular to the partition mole centre line and transmitting them into the subsoil, the partition mole should be considered as a free-standing structure, fully fixed in the ground, comprising two parallel, interconnected sheet pile walls. The force of the soil fill between the two sheet piling walls is neglected when determining the stiffness of the system. It is assumed that both walls are interconnected by an essentially articulated connection at the top. A rigid connection may also be planned which, however, leads to large bending moments at the head of the sheet piling and thus to elaborate structural measures at the junction. In addition, it leads to axial forces in the sheet piling walls, which in some circumstances reduce the mobilisable passive earth pressure for the sheet pile wall in tension. The full magnitude of the mobilisable passive earth pressure cannot be used because this is partly required to resist the active earth pressure and, possibly, water pressure difference due to the filling of the mole.

419

This proportion must be determined in advance and deducted from the total mobilisable passive earth pressure.

As both sheet piling walls exhibit an essentially parallel bending behaviour due to the external actions, the bending moment in the total system can be distributed over both walls in the ratio of their flexural stiffnesses. The ultimate limit state conditions according to EC 3-5 are to be checked separately for each of the two sheet pile walls using the applicable design values of the bending moment proportions and the section resistances.

8.3.3.3 Calculation assumption for sheet piling walls anchored against each other

The individual sheet piling walls are loaded by active earth pressure from the fill and the surcharge on the partition mole, as well as by external forces. In addition, hydrostatic pressure differences have to be considered if the water table is higher inside the mole than in front of it. Generally, a negative surge load case which takes into account flooding of the mole with subsequent, brief lowering of the outer water level, should be taken as the water pressure difference in the calculation. In some circumstances an internal water level that drops at a much lower rate will cause a higher water pressure difference. The anchor connecting both walls is to be designed for the tension loads due to the aforementioned forces.

The relevant bending moment on which the design of the sheet piling walls is based generally corresponds to the total effect divided between the two sections for the fully fixed sheet piling structure as already determined in section 8.3.3.2.

8.3.3.4 Design

The walings, anchors and stiffeners must be calculated, designed and installed in accordance with the relevant recommendations.

Special significance should be given to the load transfer points for accommodating the external forces on the partition mole. An analysis of the forces acting parallel with the centre line of the mole is to be carried out according to R 132, section 8.2.12.

Transverse walls or anchor cells are to be provided as per R 101, section 8.3.2.

8.4 Anchors, stiffeners

All anchoring elements are to be so designed that their bearing capacity corresponds with the full internal bearing capacity of the anchors.

8.4.1 Design of steel walings for sheet piling (R 29)

8.4.1.1 Arrangement

The walings must transmit the anchor forces (support reactions) from the sheet piling into the anchors, and the resistance forces of the anchor walls into the anchors. Further more, they stiffen the sheet piling and facilitate alignment of the piling.

As a rule, these walings are installed as tension walings on the inboard side of the waterfront sheet piling. At anchor walls they are generally placed as compression walings behind the wall.

8.4.1.2 Design

Walings should be of heavy construction and ample design. Heavier walings of S 235 JR (formerly St 37-2) are preferable to lighter ones of S 355 JO (formerly St 52-3). Splices, stiffeners, bolts and connections must be properly made in accordance with the best steel construction and welding practice. Stressed welds must be made at least 2 mm thicker than is structurally required because of the danger of corrosion. The walings are conveniently constructed of two closely-spaced U-section steel channels, whose webs are at right angles to the plane of the sheet piling (see R 132, section 8.2.12.2, figs. R 132-1 to R 132-3). Where possible, the U-section steel channels are placed symmetrically about the point of application of the anchor rod, so that the anchor rod can rotate freely by the anticipated amount. Proper spacing of the two U-section steel channels is maintained by U-section steel channel stiffeners or by web plates. Additional stiffening of the U-section waling channels is necessary in the area of the anchor force transmission in heavy anchor systems, or if there is a direct connection between the anchor rod and the waling.

Splices of the waling should be placed where stress is at a minimum. A full cross-section splice is not required, but it must be adequate to carry the calculated stresses at the section.

8.4.1.3 Fixings

The walings are either supported on welded brackets or, especially with limited working space beneath the walings, suspended from the sheet piling. The design and attachment must be so that any vertical loads on the walings are satisfactorily transmitted into the sheet piling. Brackets facilitate the installation of the walings. Suspensions should not weaken the walings, and should therefore be welded to the walings or attached to the base plate of the waling bolts.

If the anchor force is transmitted directly (through hinges) into the tension waling at the inboard side, the waling must be attached to the piling with special care. The anchor force is transferred from the sheet piling into the walings through heavy bolts. They are placed in the centre between the two U-section waling channels and transmit their load through base plates, which are attached by tack welding to the walings. The waling bolts are made extra long, so that they can be used to align the sheet piling against the walings.

8.4.1.4 Inclined anchors

The connection of the inclined anchors must also allow for the transmission of vertical forces.

8.4.1.5 Extra waling

Sheet piling which has become severely misaligned due to improper driving is to be realigned by means of an extra waling, which remains in the structure.

8.4.2 Verification of bearing capacity of steel walings (R 30)

Walings and waling bolts should be designed for at least the force which corresponds to the bearing capacity of the selected anchor. Additionally, they must be so designed that all horizontal and vertical loads which would otherwise be applied are absorbed and transmitted into the anchors or into the sheet piling and anchor wall. The following loads are to be taken into consideration:

8.4.2.1 Horizontal loads

(1) The horizontal component of the anchor tensile forces, whose magnitude can be taken from the sheet piling calculation. Taking into consideration future deepening in front of the quay wall, it is recommended that the walings be of generous size, and capable of absorbing the permissible tensile force derived from the chosen anchor rod diameter.

(2) Design values of direct acting hawser pulls on mooring devices.

(3) Design value of vessel impact, depending on the size of the vessel, the berthing manoeuvre, current and wind conditions. Ice impact may be neglected.

(4) Compulsive forces which are introduced when aligning the sheet piling.

8.4.2.2 Vertical loads

(1) The dead load of the waling channels including stiffeners, bolts and base plates.
(2) That portion of the soil surcharge between the rear surface of the sheet piling to a vertical plane through the rear edge of the waling.
(3) The portion of the live load on the quay wall between the rear edge of the sheet piling capping and a vertical plane through the rear edge of the waling.
(4) The vertical component of the active earth pressure which acts on the vertical plane through the rear edge of the waling from the bottom edge of the waling to the surface of the ground. The active earth pressure is in this case calculated for plane sliding surfaces with $\vartheta_a = + \varphi'$.
(5) The vertical component of an inclined anchor tensile force with tension and compression walings, according to section 8.4.2.1 (1).

The loads stated under (1) to (5) are to be considered with their design values for LS 1B.

8.4.2.3 Loads for the calculation

In the structural calculation of the walings, the only horizontal loads generally included are the component of the anchor tensile force (section 8.4.2.1 (1)) and the hawser pull (section 8.4.2.1 (2)). On the other hand, the vertical loads (section 8.4.2.2) are all included. In order to make some allowance for the stresses from vessel impact and aligning the piling, it is recommended that the partial safety factors for the resistance be increased by 15%. When several walings lie over each other, the vertical loads are divided among the walings. In order to ensure the safe design of the attachment of the waling brackets, the loads are assumed to act at the rear edge of the waling.

8.4.2.4 Method of calculation

The loads included are resolved into component forces vertical and parallel to the sheet piling plane (main axes of the waling). It is to be assumed in the calculation that for the absorption of the forces, acting at right angles to the plane of the sheet piling, the walings are supported by the anchors, and for the parallel forces by brackets or suspensions. If the anchors are connected to the sheet piling, the pressure of the piling on the waling in the areas around the anchor connection has an adequate supporting effect so that it is sufficient here to suspend the waling on the inner side, as is normal for compression walings. The moment at support and the span moment resulting from the design value of the sheet piling support reaction force are generally calculated according to the equation $q \cdot l^2 / 10$.

8.4.2.5 Waling bolts

The waling bolts should be designed using the same principles as for the anchoring of the sheet piling, see R 20, section 8.2.6.1, but generously so in order to allow for corrosion and the stresses introduced when aligning the piling. With double anchoring, the bolts of the upper walings, structurally only slightly loaded, should be at least 32 mm (1¼″), preferably 38 mm (1½″) in diameter, to allow for vessel impact. The base plates of the waling bolts are to be designed in a manner that their bearing capacity corresponds with that of the waling bolts.

8.4.3 Walings of reinforced concrete for sheet piling with driven steel anchor piles (R 59)

8.4.3.1 General

In quay walls, it has been found that anchors consisting of steel anchor piles driven with a 1 : 1 batter are practicable and very economical.

This is especially valid when there are high lying layers of poor soil which make other anchors difficult or impossible, and circumstances where extensive earthwork would otherwise be necessary.

If the anchor piles are driven first and if the sheet piles cannot be placed with precision, the anchor piles are not always in the designed position relative to the sheet piling.

However, inaccuracies of this type are of no consequence when the sheet piling waling is constructed of reinforced concrete, and the local structural dimensions have already been considered in the reinforcement plans (fig. R 59-1).

If the reinforced concrete waling is constructed at a greater distance above the existing terrain, it would be advisable to equip the sheet piling wall with an auxiliary steel waling and to leave this in place until the piles are connected and the reinforced concrete waling has achieved its bearing capacity.

8.4.3.2 Construction of sheet piling walings

Reinforced concrete walings are anchored to the sheet piling by round or square steel bars which are welded to the sheet pile webs, evenly spaced except that additional bars are required at expansion joints (fig. R 59-1, pos. 4 and 5). The anchor force is transferred into the anchor piles in a similar manner (fig. R 59-1, pos. 1 to 3).

The steel connectors welded to the sheet piles and the anchor piles are generally made of S 235 JO (formerly St 37-3). They are forged flat at the connection points. Similarly, round steel bars of BSt 500 S may also be used.

Welding may be performed only by qualified welders under the supervision of a welding engineer. Only material whose welding properties

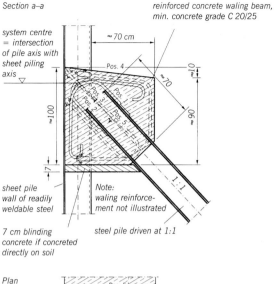

Section a–a

reinforced concrete waling beam, min. concrete grade C 20/25

system centre = intersection of pile axis with sheet piling axis

≈ 70 cm

Pos. 4

Pos. 3
Pos. 2
Pos. 1
Pos. 5

≈70

≈10

≈90

≈100

≈7

sheet pile wall of readily weldable steel

Note: waling reinforce- ment not illustrated

7 cm blinding concrete if concreted directly on soil

steel pile driven at 1:1

1:1

Plan

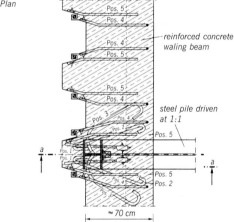

Pos. 5
Pos. 4

Pos. 4
Pos. 5

reinforced concrete waling beam

Pos. 5
Pos. 4

Pos. 3
Pos. 4

steel pile driven at 1:1

Pos. 2
Pos. 5

a

Pos. 1
Pos. 1

Pos. 5

a

Pos. 4
Pos. 3

Pos. 5
Pos. 2

≈ 70 cm

Fig. R 59-1. Reinforced concrete waling for steel sheet piling

are known, which are of uniformly good quality and are compatible with each other, may be used (see also R 99, section 8.1.18).

The concrete shall be at least of strength class C 20/25 with an aggregate grading in the favourable range between curves A and B. BSt 500 S is generally chosen for the reinforcing bars.

8.4.3.3 Construction of the connection between piles and waling

If there are layers of considerable thickness, composed of soils which are very susceptible to settlement, or if non-compacted backfill is to be placed to a great depth behind the wall, the pile connections should be constructed as hinges.

With more favourable soil conditions when considerable settlement or subsidence is not to be expected, the steel piles should be fixed into the reinforced concrete waling. Even with settlement-sensitive, thin-layered soils or with well compacted backfill of non-cohesive soil, such a connection can also be an economical solution. In order to allow for residual settlement or subsidence of the soil, and for the fixed-end influences, including those resulting from the deflection of the sheet piling, the fixed-end moment of the anchor pile along with the other horizontal and vertical forces acting on the waling must be taken into account in the design of the pile connection. In extremely yielding subsoil, this is to be determined for the yield strength $f_{y,k}$ taking account of the characteristic value of the axial force N_k acting in the pile. It is to be unfavourably introduced in the verification of the bearing capacity of the pile connection.

If the anchor piles are driven through relatively thin layers of soils susceptible to settlement or if the backfill is shallow, a correspondingly smaller additional connection moment may be applied.

The introduction of the internal forces in the anchor pile into the reinforced concrete waling at its point of connection with the waling is to be verified at this point. Thereby, the combined stressing of the pile top by axial force, shear force and bending moment is to be observed. If necessary, reinforcing plates may be welded laterally on the anchor pile, in order to improve the absorption of the forces. The reinforcing bars, which are otherwise formed as loops, may be welded to these plates. The voids which tend to form along the web of the anchor pile as a result of this construction must be carefully filled with concrete.

Only specially killed steels, resistant to brittle fracture, such as grade S 235 J2G3 (formerly St 37-3) or S 355 J2G3 (formerly St 52-3) may be used for piles and their connections in all quay walls with pile anchors exposed to comparatively large uncontrollable bending stresses.

8.4.3.4 Calculation

The waling loads are to be applied according to R 30, section 8.4.2. The horizontal component of the anchor force according to the sheet piling calculation is considered as a horizontal load action to be applied at the system centre (= intersection of the axis of the sheet piling with the axis of the pile). The waling, including its connections to the sheet pilling, is calculated as uniformly supported. Dead load, vertical surcharges, pile forces, bending moment and shear force of the anchor piles are considered to be actions and introduced with their design values.

The internal forces at the pile connection, which result from the soil surcharges on the pile by the backfill or the layers susceptible to settlement, are calculated on an assumed equivalent beam fixed in the waling and in the loadbearing foundation soil. If shielding of the sheet piling loading by the anchor piles has not been taken into account, the fixed-end moment and the shear force acting on the pile connection have to be considered only with reference to the connection between the waling and the sheet piling, and need not be calculated as acting on the sheet piling itself.

A reduction of the anchor pile cross-section at the point where the pile is fixed into the waling in order to reduce the connection moment and the accompanying shear force is not allowable because such a reduction can lead to pile failure especially if the workmanship has been faulty.

If instead of the fixed connection a flexible one is chosen, the additional internal forces at the pile connection arising from settlement or subsidence of the soil are also in this case to be checked and safely absorbed.

Verification of bearing capacity is to be provided for the design values of the internal forces E_d, which may be reduced depending on the loading case, as stated in section 8.2.0.2.

When the connection moment and the accompanying shear force in the anchor pile are based on the yield strength $f_{y,k}$, then the connection elements themselves may also be designed on the basis of the yield strength $f_{y,k}$.

For practical construction reasons, the dimensions of the concrete waling should not be smaller than those shown in fig. R 59-1. In order to allow for variations in the acting forces and in the pile anchors, the cross-section of the reinforcing steel should be increased by at least 20% beyond calculated requirements.

8.4.3.5 Expansion joints

Reinforced concrete walings can be cast with or without expansion joints. Their layout is based on R 72, sections 10.2.4 and 10.2.5. See section 10.2.3 for details of construction joints. If expansion joints are required, they should be arranged in such a way that they do not hinder the changes in length of the sections.

To provide mutual support of the building sections in the horizontal direction, the expansion joints have a joggled form, with dowels if necessary. The horizontal joggle is accommodated in the piled platform in the case of piled quay walls The joints are to be designed to prevent the backfilling from leaking out.

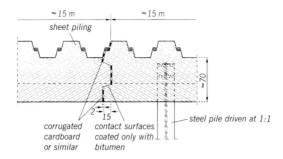

Fig. R 59-2. Joggle joint in a reinforced concrete waling

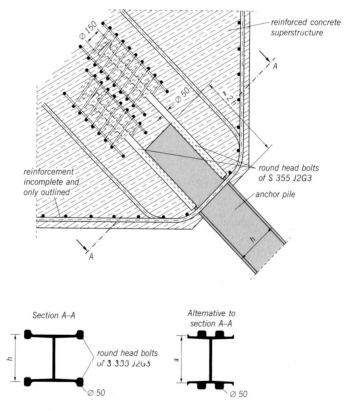

Fig. R 59-3. Example of an anchor pile connection to a reinforced concrete superstructure using so-called round head bolts

8.4.3.6 Detailing the heads of steel anchor piles to transmit forces into a reinforced concrete superstructure

The equipping of the top of steel anchor piles must be arranged, designed and dimensioned in such a way that the anchor forces of the connected structure can be absorbed in the framework of the allowable effects of actions. Additional stresses from deflection and shear of the anchor pile should be kept to a minimum in the connection zone. For this purpose, the pile should be embedded to about twice its height in the reinforced concrete (fig. R 59-3). It is then sufficient to dimension the connection steels and their welds in such a way that about the whole cross-sectional area of the anchor pile is connected.

With yielding subsoil under the anchor piles, stresses in the reinforced concrete superstructure are to be verified within the framework of the allowable effects of actions according to loading case 3. This is valid not only for the full anchor pile force but also for the loads from shear force and the bending moment at the anchor pile connection when the pile is stressed up to the yield strength.

Fig. R 59-3 shows a favourable connection solution with so-called "round head bolts" – as already used for bollard anchoring. Here one end of the round steel bar is upset to form a plate of up to three times the diameter of the round steel bar at the head. The end of the round steel bar to be welded to the tension pile is flattened to allow for good welding.

End anchoring in the concrete can also be achieved by welding cross bars or plates of a corresponding size to the round and square anchor bars.

8.4.4 Steel capping beams for waterfront structures (R 95)

8.4.4.1 General

Steel capping beams are designed with a view to meeting structural, operational and construction requirements. Furthermore, R 94, section 8.4.6.1 applies accordingly.

8.4.4.2 Structural and construction requirements

Steel capping beams serve as upper closure of sheet piling (fig. R 95-1). With adequate flexural rigidity (fig. R 95-2), capping beams can also be used for absorbing forces arising during the alignment of sheet piling tops, as well as for meeting certain operational requirements.

The sheet piling top can be aligned only if the sheet piling is sufficiently flexible and free during alignment to permit the necessary corrective deflections. With close spacing between capping beam and waling, the alignment of the sheet piling will be accomplished mostly with the waling. In service, the capping beam distributes non-uniform loads to the sheet pile tops and prevents non-uniform deflections of the sheet pile heads.

429

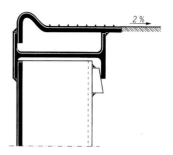

Fig. R 95-1. Rolled or pressed steel capping beam with bulb plate as nosing welded to steel sheet piling

Fig. R 95-2. Welded capping beam waling with high section modulus, otherwise as fig. R 95-1

Fig. R 95-1 shows a standard type of capping beam.

The greater the distance to the waling, the more important is an adequately high moment of inertia of the capping beam. Fig. R 95-2 shows a strengthened capping beam or a capping beam waling respectively.

Vessel impact is to be taken into account in dimensioning the capping beam. In order to avoid deflection or buckling, the capping beams shown in fig. R 95-1 will be provided with stiffeners welded to the capping beam and sheet piling, if the troughs in the sheet piling are so wide as to make this necessary.

If the capping beam also serves as waling, this capping beam waling is to be designed and dimensioned in accordance with R 29, section 8.4.1, and R 30, section 8.4.2.

8.4.4.3 Operational requirements

The top surface of the capping beam must be such that hawsers trailing over it will not be damaged or damage the top of the key. Furthermore, make sure that hawsers and lines (e.g. also thin heaving lines) cannot become caught in gaps etc. To protect the personnel working on the quay against slipping off, a portion of the capping beam should project slightly above the surface of the quay.

Horizontal capping beam surfaces should if possible be studded, chequered or similarly treated (figs. R 95-1 and R 95-2).

For heavy vehicle traffic, a guard rail is recommended.

If there is an outboard crane rail (R 74, section 6.3.4, fig. R 74-3), this is made part of the guide rail.

The berthing side of the capping beam must be smooth. Unavoidable edges are to be chamfered if possible. The design must be such that ships cannot catch under it and that the danger of capping beam sections being turned out by crane hooks is minimised (fig. R 95-3).

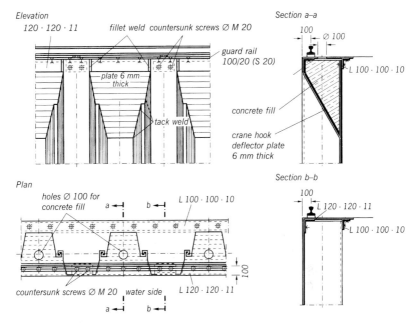

Fig. R 95-3. Special design of a steel sheet piling capping beam with crane hook deflector plates

8.4.4.4 Delivery and installation

The capping beam members are to be delivered unwarped and true to size. During manufacture in the workshop, the tolerances for the width and depth of the sheet pile sections and deviations during driving are to be taken into account. Where necessary, the capping beams are to be adjusted and aligned at the construction site. Capping beam joints are to be designed as full joints.

After installation of the capping beam, when it is located at an adequate height above HHW and is safe from wave action, sand is to be densely deposited in the area of the top of the sheet piling and replaced where necessary, in order to reduce settlement and protect the ground side of the sheet piling and the capping beam from corrosion.

If the capping beam can be flooded or inundated, lies in the wave action zone or is designed to lie below the water level and the water level can be lowered by passing ships, the risk exists that sandy backfill material will be washed out. The reason for this is that there is not a tight junction between the steel capping beam and the top of the sheet piling, as a rule. In order to prevent the washing out of sandy material, a tight connection is to be made in such cases between the steel capping beam and the

431

sheet piling, for example by backfilling the top of the sheet piling with concrete. In so doing, the vertical flange of the capping beam or the capping beam angle should embed adequately deep into the concrete and the concrete should be adequately secured in its position by welded-on claws or bolts. Furthermore, the backfill material in the area of the paving is to be covered with an adequately thick graded gravel filter of well-balanced composition.

8.4.5 Reinforced concrete capping beams for waterfront structures with steel sheet piling (R 129)

8.4.5.1 General
Structural, operational and construction aspects govern the design of reinforced concrete capping beams.

8.4.5.2 Structural requirements
In many cases, the capping beam serves not only to cover the sheet piling but also as stiffener and thus also to absorb horizontal and vertical loads. If in addition, it acts as a capping beam waling for transmitting anchor forces, it must be designed for ample strength, especially when it must carry a crane rail, resting directly on its upper surface.

Please refer to the applicable portions of R 30, section 8.4.2 concerning horizontal and vertical loads. In case bollards or other mooring facilities occur, the forces acting on these members must be added (R 153, section 5.11, R 12, section 5.12, R 102, section 5.13) insofar as line pull is not absorbed by special structural members. In addition, if a crane rail rests directly on a reinforced concrete capping beam (fig. R 129-2), the horizontal and vertical crane wheel loads are also to be absorbed (R 84, section 5.14).

In the structural calculations, it is advisable to treat the reinforced concrete capping beam as a flexible beam, elastically supported on the sheet piling both horizontally and vertically. In doing so, for heavy capping beams on quay walls for seagoing vessels, a modulus of horizontal subgrade reaction $k_{s,bh} = 25$ MN/m^3 can generally serve as criterion for the horizontal direction. The modulus for the vertical direction $k_{s,bv}$ depends largely on the section and length of the sheet piling, as well as the width of the capping beam. $k_{s,bv}$ must therefore be determined specially for each structure, but $k_{s,bv} = 250$ MN/m^3 can be used in preliminary calculations. Limit considerations are required for final dimensioning with the design rated on the least favourable case. Anchors connected to the sheet piling structure or to bollard foundations are to be taken into account separately. Special attention should be paid to the absorption of shrinkage and temperature stresses because expansion and/or contraction

432

of the capping beam can be greatly restricted by the connected sheet piling and by the soil backfill.

In order to allow for irregularities in the support by the sheet piling and possibly by anchors, the reinforcing steel should be increased by at least 20% over calculated requirements, in accordance with R 59, section 8.4.3. Please refer to R 72, section 10.2, regarding concrete grade, reinforcement and design.

The vertical loads to be absorbed in the plane of the sheet piling are generally introduced centrically into the sheet pile top. On this account, to prevent cleavage, sufficient lateral tension reinforcement is placed in the reinforced concrete capping beam directly above the sheet piling. For corrugated steel sheet piling, the concrete capping beam can be designed in a way that the vertical forces will be transmitted directly into the sheet piling. Hereby, the cross-sectional area of the piling is acting as a very narrow bearing. This so-called "edge bearing" must be designed in accordance with a construction supervisory permit issued by the IfBt. In case there are large concentrated loads, e.g. from a crane rail, a plate effect of the sheet piling should be ensured by properly welding the interlocks. Geometric conditions (see R 74, section 6.3) may make it necessary to support the craneway out of centre of the capping beam.

Verification is required of the safe transfer of all internal forces (e.g. bending moments, shear forces, etc.) in the transition range piling/capping beam.

8.4.5.3 Construction and operational requirements

The sheet pile top must be aligned as required before concrete is placed. For this, a steel waling included in the design or a temporary steel waling can be used. However, the sheet piling top can only be aligned by these means if the waling protrudes far enough out of a more or less yielding soil at the completion of driving. It is then possible to give the sheet pile structure good alignment at the top with the aid of the reinforced concrete capping beam. If it is necessary to ensure a concrete cover of adequate thickness over the water side of the sheet piling top, the width of the capping beam should be appropriately increased. Contingent on the design, the pile top should have a concrete cover of at least 15 cm on both water and land sides, and the depth of the capping beam should be at least 50 cm (figs. R 129-1 and R 129-2). Furthermore, the length of the sheet pile top embedded in the capping beam should be about 10 to 15 cm.

A reinforced concrete capping beam should be positioned adequately high above the water level so that the sheet piling wall immediately below the concrete capping beam is accessible for regular inspections and, if necessary, the renewal of any corrosion protection measures.

For waterfront structures on watercourses with an increased risk of corrosion (seawater, briny water) due to their location, it is advisable to position the reinforced concrete capping beam completely behind the sheet piling in order to prevent corrosion damage. In this case the sheet piling continues to the upper edge of the quay. This is an effective way of preventing increased corrosion at the transition from steel to concrete on the water side.

In order to prevent a ship's hull from catching under it, the capping beam may be designed as shown in fig. R 129-2, where it is provided with a wide universal bent plate at 2 : 1 or steeper at the water side. The lower edge of this steel plate is welded to the sheet piling.

If the concrete cover over the water side is omitted, a universal steel plate is generally installed on the water side of the sheeting (fig. R 129-1). It is welded to the flange of the sheet piling, as this method is more economical than a bolted connection. Anchor claws should be used over the sheet pile troughs in order to produce a solid connection between the steel plate and the concrete. The steel plate is to be chamfered on top (fig. R 129-1). Irregularities in the alignment of the sheet pile top up to about 3 cm can be corrected by the use of fillers.

The capping beam is provided with nosing and skid protection as per R 94, section 8.4.6, or DIN 19 703, at least at facilities serving sea-going traffic. The other provisions of the aforementioned section are also to be observed as applicable.

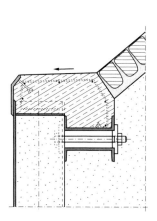

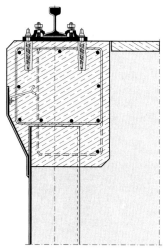

Fig. R 129-1. Reinforced concrete capping beam for corrugated sheet piling without water-side concrete cover on a partially sloped bank

Fig. R 129-2. Reinforced concrete capping beam for corrugated sheet piling with concrete cover on both sides and craneway resting directly on the surface

The stirrup reinforcement must be so designed that the portions of the capping beam which are separated by the embedded sheet piling are securely connected. To this end, the stirrups should either be welded to the webs of the sheet piles or inserted through holes or placed in slots, burnt in the sheeting. Such measures are not required with the certified "edge bearing" as per section 8.4.5.2. If transverse tension reinforcement to prevent cleavage is placed above the top of the sheet piling to carry off vertical loads, and if the stirrups associated with this reinforcement lie immediately above the upper surface of the sheeting, additional stirrups should be provided to ensure the structural integrity of the capping beam, including its lower portions.

Bulkheads composed of box steel sheet piling may be also built with a capping beam which requires no concrete cover over the water side of the pile tops (fig. R 129-3). In this case, the reinforcing steel must be placed in the sheet piling cells. To make this possible, the webs and flanges are cut away as necessary and holes are burnt in them for inserting the longitudinal rods.

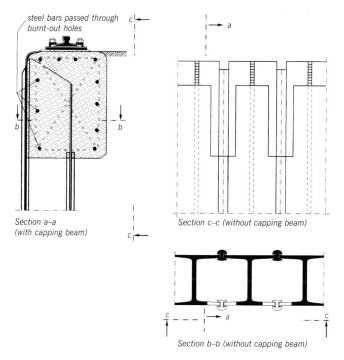

Fig. R 129-3. Reinforced concrete capping beam for box steel sheet piling without concrete cover on water side, with craneway resting directly on surface

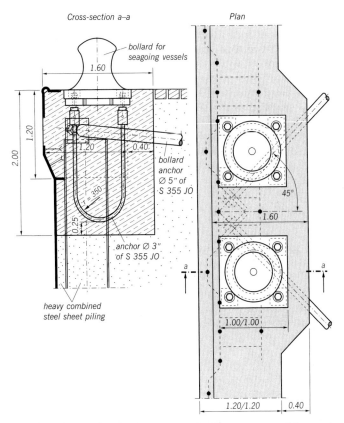

Fig. R 129-4. Heavy reinforced concrete capping beam to a quay wall for seagoing vessels, details of an anchored bollard foundation

The same procedure can be followed for combined steel sheet piling. Reinforced concrete capping beams can also be used as foundations for bollards, if the necessary local reinforcement is provided. Fig. R 129-4 shows an example of this for a heavy quay wall for seagoing vessels. In such cases, hawser pulls of large magnitude are best absorbed by heavy tie rods, in order to keep the elongation of the anchor connection and thus the bending moments in the capping beam as slight as possible. Prestressed steel cable anchors may prove to be a disadvantage if later excavation work behind the capping beam becomes necessary.

8.4.5.4 Expansion joints

Reinforced concrete capping beams can be constructed without joints when all the actions due to loads and restraint (shrinkage, creep, settlement, temperature) are taken into account (see R 72, section 10.2.4). In doing so, the theoretical crack widths must be limited taking into account the ambient conditions (see section 10.2.5). If expansion joints are planned, the lengths of the sections should be specified such that no significant restraint forces occur in the longitudinal direction of the sections. Otherwise, the restraint forces should be taken into consideration in relation to the substructure or subsoil.

The joints themselves must also be so constructed that the changes in length of the reinforced concrete capping beam at these points are not hampered by the sheet piling. To accomplish this for corrugated steel sheet piling, the following methods are suggested:

(1) The expansion joint is placed directly above a sheet pile web, which is coated with elastic material so that the required movements are possible.

(2) The expansion joint is placed above a sheet pile trough. The pile or piles of this trough should then be only slightly embedded in the reinforced concrete capping beam and must be covered with a substantial plastic coating, which at the same time ensures water-tightness in the vicinity of the joint.

An example for a joint in cases without the requirement of transferring shear forces is shown in fig. R 129-5.

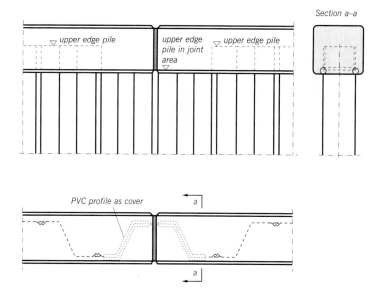

Fig. R 129-5. Expansion joint in a reinforced concrete capping beam

Expansion joints in heavy reinforced concrete capping beams are keyed for the transmission of horizontal forces. A certain key effect can be achieved in not so heavily stressed capping beams by the use of a steel dowel.

8.4.6 Top steel nosing for reinforced concrete walls and capping beams at waterfront structures (R 94)

8.4.6.1 General

For practical purposes, edges of reinforced concrete waterfront structures are provided with a carefully designed steel nosing. This is to protect both the edge and the lines running over it from damage caused by ship operations, and serves as a safety measure to prevent line handlers and other personnel working in this area from slipping off. The nosing must be so constructed that ships cannot catch under it. The same also applies to crane hooks (R 17, section 10.1.2).

If waterfront structures in inland harbours are frequently flooded so that there is a danger of ships grounding thereon, the nosing may not have any bulges or moulding.

8.4.6.2 Examples

Fig. R 94-1 shows the standard design for navigation locks which can also be used for waterfront structures in ports.

It is possible to build such waterfront structures, especially where cargo is handled, so that surface run-off will flow to the land-side, making weepholes as per fig. R 94-1 unnecessary.

The steel nosing shown in fig. R 94-1 can also be supplied with aperture angles $\neq 90°$, so that it can be fitted to a sloping upper surface or front face of the waterfront structure. It is supplied in lengths of about 2500 mm, which are welded before installation.

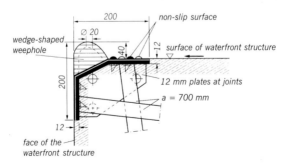

Fig. R 94-1. Nosing with weephole

438

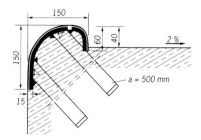

Fig. R 94-2. Special section of nosing frequently used in the Netherlands

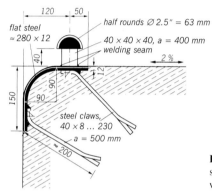

Fig. R 94-3. Nosing made of rounded sheet with foot railing in seaports and without in inland harbours

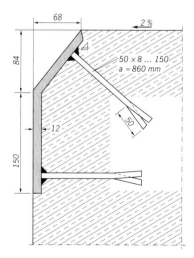

Fig. R 94-4. Nosing made of angular sheet without foot railing for non flood-free embankments in inland harbours

439

The design in fig. R 94-2 depicts a special section developed in the Netherlands and frequently used there with success. It is made of steel sheet of augmented thickness and has reinforced steel anchors, so that the upper hollow space created during concreting need not be grouted. The upper ventilation openings must be closed after concreting however, in order to minimise corrosion attacks on the inner surface.

The designs shown in figs. R 94-3 and R 94-4 have proven themselves at numerous German waterfront structures.

All types in figs. R 94-1 to R 94-4 must be carefully aligned, placed in the shuttering and securely fastened. The types in figs. R 94-3 and R 94-4 must be concreted flush against all contact surfaces in the course of concreting the quay wall. The inner surface of the nosing is to be cleaned of adhering rust with a wire brush before installation.

8.4.7 Auxiliary anchoring at the top of steel sheet piling structures (R 133)

8.4.7.1 General

For structural and economic reasons, the anchoring of a waterfront sheet piling wall is in general not connected at the top of the wall, but rather at some distance below the top. This applies especially to walls, when there is a large difference in elevation between harbour bottom and top of the wall. In this manner, the span in case of a single-anchored wall is decreased and thus also the moment in the span and the fixed end moment. Furthermore, increased redistribution of earth pressure takes place.

In such cases, the section above the anchor is frequently given auxiliary anchoring at the top, even if the customary sheet piling calculations (R 77, section 8.2.2) show that it sustains no load. Its function is to prevent the flexible upper sheet piling end from too large deflection during the final stage of construction and later at the occurrence of large, local unexpected operational loads. The auxiliary anchoring however is not taken into consideration in calculating the structural main system of the sheet piling structure.

8.4.7.2 Aspects affecting the installation

The height of the section above the anchor, for which the installation of an auxiliary anchor is useful, depends on various factors, such as the flexural rigidity of the sheet piling, the magnitude of the horizontal and vertical live loads, on operational requirements for the alignment of the top off the sheet piling wall and the like.

When a waterfront sheet piling wall is designed for crane loading, an auxiliary anchor should be added as near as possible to the top, unless circumstances make it advisable to place the main anchors quite near the top of the wall.

As a rule, loads on the section above the anchor by mooring hooks or posts also call for auxiliary anchoring. Although the anchoring for major line pull forces is likewise connected near the top of the wall, it is generally run to the main anchor wall and incorporated in the main anchoring system.

8.4.7.3 **Design, calculation and dimensioning of the auxiliary anchoring**
Tie rod bars with flexible connections at both ends are generally used for the auxiliary anchoring. For the calculation of the auxiliary anchoring an equivalent system is used as a basis in which the section above the anchor is considered to be fixed at the level of the main anchor. The load acts on this system similar to the load in the statics for the main system. It is to be observed that the load applied to the section above the anchor must be fully absorbed by both the auxiliary anchoring as well as by the main anchor.

In some cases, the auxiliary anchoring must also be calculated with the load assumptions according to R 5, section 5.5.5.

R 5, section 5.5, especially section 5.5.5 and R 20, section 8.2.6, as well as R 29, section 8.4.1 and R 30, section 8.4.2, also apply to the design, calculation and dimensioning of the auxiliary anchor waling. With regard to future need to straighten the top of the waling, and to absorb the force of moderate vessel impact, the auxiliary anchor waling should be dimensioned stronger than theoretically required. In fact, it is customary to make it identical with the main anchor waling.

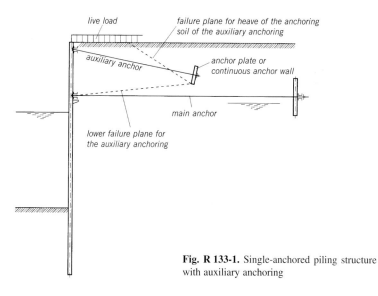

Fig. R 133-1. Single-anchored piling structure with auxiliary anchoring

441

If the sheet piling wall capping beam is also used for the connection of the auxiliary anchor, R 95, section 8.4.4 and/or R 129, section 8.4.5 are also to be observed.

The stability of the auxiliary anchoring is to be checked against both heave of the anchoring soil as well as for the lower failure plane, which extends to the connection point of the main anchor (fig. R 133-1). R 10, section 8.4.9, applies accordingly here.

8.4.7.4 Work on site

It is advisable to dredge the harbour bottom in front of the quay wall sheet piling after the auxiliary anchoring has been installed. If the reverse procedure is followed, the top of the sheet piling wall may move uncontrollably, so that later adjustment with only the auxiliary anchoring may not always be successful.

8.4.8 Threads for sheet piling anchors (R 184)

8.4.8.1 Types of thread

The following thread types are used:

(1) Cut thread (machined thread) (fig. R 184-1)
The outside thread diameter is equal to the diameter of the round bar steel or the upsetting.

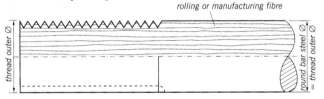

Fig. R 184-1. Cut thread

(2) Rolled thread (non-cutting thread produced in the cold state)
   (fig. R 184-2)
When using steel grades S 235 JRG2 (formerly St 37-2) and S 355 J2G3 (formerly St 52-3), the round bar steel or any upsetting must be machined or skimmed to the necessary extent before rolling the thread, in order to obtain a thread conforming to standards. For anchors with a rolled thread, the diameter of the round bar steel or upsetting may be somewhat smaller than for anchors with cut thread, without the loadbearing capacity declining.

With the methods shown, an outside thread diameter is produced, depending on the pre-machining, which is greater than the diameter of the starting material.

Drawn steels (up to ∅ 36 mm) do not need to be pre-machined.

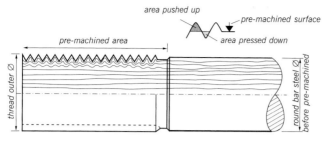

Fig. R 184-2. Rolled thread

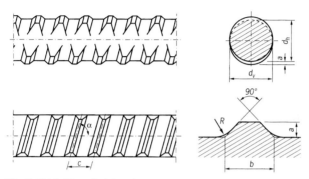

Fig. R 184-3. Hot-rolled thread

(3) Hot-rolled thread (non-cutting thread) (fig. R 184-3)
During hot rolling, the thread shaft receives two rows of thread flanks which lie opposite and supplement one another to form a continuous thread.

With the hot-rolled thread, the additional processing stage of rolling or cutting the thread is not required. The nominal diameter is applicable to the thread shaft. The actual cross-sectional dimensions differ slightly from this. Associated elements must be used for end anchorages and butt joint designs.

8.4.8.2   Margins of safety required
Please refer to R 20, section 8.2.6.3 in respect of the verification of bearing capacity and special production information.

8.4.8.3   Further information on thread types
- Rolled threads have a high profile accuracy.
- When the thread is rolled, cold forming occurs. This increases the strength and the yield point of the thread root and flanks, which has a favourable effect in respect of central loads.

443

- The thread root and flanks are particularly smooth in rolled threads and therefore have higher fatigue strength under dynamic loads.
- The production times for rolled threads are shorter than those for cut ones, but this advantage is more than offset by the necessary machining or skimming, provided drawn grades are not used.
- The course of the steel fibres is not interrupted in rolled or hot-rolled threads.
- Rolled threads with larger diameters are used primarily in centrally loaded anchors with dynamic stresses.
- Compared with the cut thread, the rolled thread achieves a weight saving of e.g. 14% in anchor rods with $\varnothing$ 2″ and 8% with $\varnothing$ 5″.
- In round bar steel anchors with rolled thread, no nuts, couplers or turnbuckles with rolled internal thread are necessary, especially since the stress on an internal thread is always smaller than on an external thread. When the internal thread is loaded, ring tensile forces are generated, and these provide support. Therefore, a combination of rolled external thread and cut internal thread can be selected without hesitation.

8.4.9 Verification of stability for anchoring at lower failure plane (R 10)

8.4.9.1 Stability for the lower failure plane for anchorages with anchor walls

Stability for the lower failure plane is verified using the method proposed by KRANZ [74], working on the basis of a section running behind the retaining wall, in the lower failure plane and behind the anchor wall. The lower failure plane always has a convex form and is approximated by a straight line. The analysis determines the anchor length required. Fig. R 10-1 shows the forces involved.

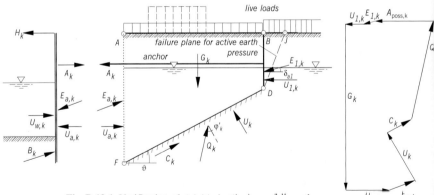

Fig. R 10-1. Verification of stability for the lower failure plane

444

Symbols used in fig. R 10-1 and forces acting on the failure body:

ϑ — inclination of lower failure plane

G_k — total characteristic weight of sliding body *FDBA*, plus live load if applicable

$E_{a,k}$ — characteristic active earth pressure (increased active earth pressure if applicable)

$U_{a,k}$ — characteristic hydrostatic pressure at section *AF* between soil and retaining wall

U_k — characteristic hydrostatic pressure at the lower failure plane *FD*

Q_k — characteristic resultant force at the lower failure plane due to axial force and maximum possible friction force (therefore inclined at angle φ'_k to the failure plane perpendicular)

C_k — characteristic shear force at the lower failure plane due to cohesion – its magnitude depends on the characteristic value of the cohesion and the length of the lower failure plane

$U_{1,k}$ — characteristic hydrostatic pressure on anchor wall *FD*

$E_{1,k}$ — characteristic active earth pressure plus live loads on the anchor wall *DB*

A_k — characteristic anchor force

In the analysis we must distinguish between the component $A_{G,k}$ due to permanent actions and the component $A_{Q,k}$ due to varying actions when considering the characteristic anchor force.

The analysis must be carried out for exclusively permanent loads and also for permanent plus varying loads. In the latter case, the components due to varying loads are to be applied in an adverse position. The force $A_{Q,k}$ resulting from these components is to be identified separately.

Stability at the lower failure plane is given when:

$$A_{G,k} \cdot \gamma_G \leq \frac{A_{poss,k}}{\gamma_{Ep}}$$

where $A_{poss,k}$ is determined from the polygon of forces with exclusively permanent loads, and

$$A_{G,k} \cdot \gamma_G + A_{Q,k} \cdot \gamma_Q \leq \frac{A_{poss,k}}{\gamma_{Ep}}$$

where $A_{poss,k}$ is determined from the polygon of forces with permanent plus varying loads.

Partial safety factors according to DIN 1054:

γ_G — partial safety factor for permanent actions

γ_Q — partial safety factor for varying actions

γ_{Ep} — partial safety factor for passive earth pressure

The analysis is based on the notion that a failure body ensues behind the retaining wall due to the transfer of the anchor force into the soil, and this body is bounded by the retaining wall, the anchor wall and the lower failure plane. In doing so, the maximum possible shear resistance in the lower failure plane is exploited, whereas the limiting value for the reaction at the base is not achieved. $A_{poss,k}$ is the characteristic anchor force that can be accommodated by the sliding body FDBA using the full shear strength of the soil. Therefore, defining the margin of safety by means of the anchor force actually represents the utilized shear strength of the soil. Like with all the other active earth pressure approaches, the equilibrium condition for the moment applied is not considered, i.e. the nature of the pressure distribution is not explained. It is sufficiently accurate to use the straight line DF as the relevant failure plane.

If the influence of the flowing groundwater on the stability in the lower failure plane is to be taken into account for a groundwater table descending towards the sheet piling, a flow net according to R 113, section 4.7, is required to determine the hydrostatic pressures on the retaining wall, the anchor wall and the lower failure plane.

8.4.9.2 Stability in unconsolidated, water-saturated, cohesive soils

The investigation is undertaken as in section 8.4.9.1. The active earth pressure is determined for the unconsolidated, water-saturated case according to R 130, section 2.6. The characteristic cohesion force $C_{u,k}$ is effective in the lower failure plane. The angle of internal friction is to be taken as $\varphi_u = 0$ for unconsolidated, water-saturated, virgin, cohesive soils.

8.4.9.3 Stability with differing soil strata

The calculation is carried out (fig. R 10-2) by first splitting up the soil mass between the sheet piling and anchor wall into as many sections as there are layers cut by the lower slip plane. This is done by passing imaginary vertical planes through the intersections of the lower failure plane with the boundaries of the layers. The method shown in fig. R 10-2 is now applied to all component sections in turn. If cohesion exists in individual layers, it is taken into consideration in the lower failure plane for the corresponding component sections. (Cohesion is not considered in the polygon of forces shown in fig. R 10-2.) The earth pressures in the vertical sections are assumed to act parallel to the surface.

Forces acting on the failure body (characteristic values) in fig. R 10-2:

$G_{1,k}$ total weight of sliding body F_1DBB_1 plus live load if applicable

$G_{2,k}$ total weight of sliding body FF_1B_1A

$E_{a,k}$ active earth pressure (across all soil strata)

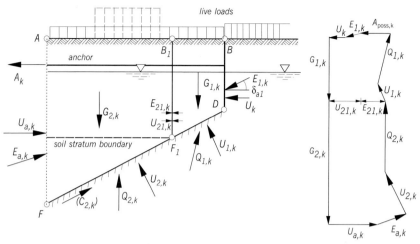

Fig. R 10-2. Verification of stability for the lower failure plane in stratified soil

$U_{a,k}$ hydrostatic pressure between soil and retaining wall AF
Ak   anchor force
$U_{1,k}$ hydrostatic pressure on lower slip plane in section F_1D
$U_{2,k}$ hydrostatic pressure on lower slip plane in section FF_1
$E_{1,k}$ active earth pressure with live load on anchor wall DB
U_k hydrostatic pressure on anchor wall DB
$U_{21.k}$ hydrostatic pressure on vertical separation plane F_1B_1
$E_{21,k}$ earth pressure in vertical separation plane F_1B_1

The stability at the lower failure plane is determined as specified in section 8.4.9.1.

8.4.9.4   Stability with fixed earth support of sheet piling
The method can be used with sufficient accuracy also in the case of fixed earth support, if the shear force zero point in the fixation area is taken as the theoretical point of the sheet piling to which the lower failure plane runs. This point coincides with the position of the greatest fixed end moment. Its position can therefore be taken from the sheet piling calculation. The active earth pressure in this case is determined only down to the theoretical base of the sheet piling, the anchor force present is taken from the sheet piling analysis for the fixed wall.

8.4.9.5   Stability with a fixed anchor wall
If the anchor wall is fixed in the ground, the lower failure plane should run to the theoretical point at the level of the shear force zero point in the fixed-end area of the anchor wall, according to section 8.4.9.4.

447

8.4.9.6 Stability with anchor plates

When analysing the lower failure plane, an equivalent anchor wall placed a distance of $\frac{1}{2} \cdot a$ in front of the anchor plates is assumed for individual anchor plates, where a designates the clear distance between the anchor plates.

8.4.9.7 Safety against failure of anchoring soil

In order to avoid failure of the anchoring soil mass, and the consequent upward displacement of the anchor plate or anchor wall, it must be proven that the design values of the resisting horizontal forces from the bottom edge of the anchor plate or anchor wall to the surface of the ground are greater than or equal to the sum of the horizontal component of the anchor force, the horizontal component of the active earth pressure on the anchor wall and possibly a hydrostatic pressure difference on the anchor wall.

The active and passive earth pressures acting on the anchor wall or on anchor plates positioned clear of the wall can be determined according to DIN 4085. A live load must be considered in the design in an adverse position, i.e. normally behind the anchor wall or anchor plate. Unfavourable, high groundwater levels, where they can occur, are also to be taken into consideration. In calculating the passive earth pressure on the anchor wall, the angle of inclination shall not be assumed to be greater than required to counteract the total of all acting vertical forces, including dead load and soil surcharge (condition $\Sigma V = 0$ at anchor wall).

In the case of simply supported anchor plates and walls, the anchor connection is generally located in the middle of the height of the plate or wall. For further details, see R 152, section 8.2.13, and R 50, section 8.4.10.

8.4.9.8 Stability of tension piles and grouted anchors – one anchor level

As an alternative to anchor walls or anchor plates, retaining walls can be secured with anchors that transfer the tensile forces into the soil by way of skin friction. Basically, there are three groups:

- anchor piles with and without grouted skin to section 9.2
- grouted, small-diameter piles to DIN EN 14 199 and DIN EN 1536
- grouted anchors to DIN EN 1537

Use fig. R 10-3 to calculate the necessary length of these anchors.
The similarity to anchorages with anchor walls is created by considering a vertical section through the soil in the centre of the force transfer length of the anchor as an equivalent anchor wall. In a grouted anchor the nominal length of grouting is regarded as the force transfer length. The active earth pressure $E_{1,k}$ at the equivalent anchor wall is always assumed to act parallel with the surface.

448

If the anchor spacing is greater than $\frac{1}{2} \cdot l_r$, the potential anchor force $A_{poss,k}$ must be reduced in the ratio of the anchor spacing a_A to this theoretical maximum value to produce the potential anchor force $A_{poss,k}{}^*$:

$$A_{poss,k}{}^* = A_{poss,k} \cdot \frac{\frac{1}{2} \cdot l_r}{a_A}$$

The analysis usually lies on the safe side when the remaining mobilisable pull-out resistance is not applied behind the toe of the equivalent anchor wall. A more accurate appraisal is necessary when the minimum anchor length l_r extends from the active sliding wedge to the end of the anchor ($l_w = l_r$) [22].

This analysis represents a considerable simplification. A more accurate, but also more complex, analysis is possible by varying the inclination of the failure plane to find the most unfavourable failure plane while taking into account the remaining mobilisable pull-out resistance behind the toe of the equivalent anchor wall (following DIN 4084, [227]). In doing so, use the characteristic value of the pull-out force that is transferred from the force transfer length behind the failure plane to the unaffected soil.

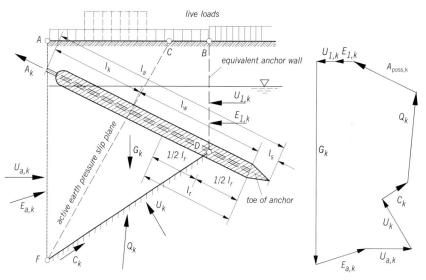

Fig. R 10-3. Verification of stability for the lower failure plane when using piles and grouted anchors (example: grouted pile)

Symbols used in fig. R 10-3 (the forces on the failure body are explained in fig. R 10-1):

l_a length of pile or anchor

l_s length of toe of pile (grouted piles only)

l_r required minimum load transfer length, or length of grouting in a grouted anchor, calculated from the design value of the anchor force A_d and the design value of the skin friction T_d of the pile ($l_r = A_d / T_d$)

T_d design value for skin friction, calculated from the design value of the pull-out resistance $R_{S,d}$ and the force transfer length l_0 in a tensile test ($T_d = R_{S,d} / l_0$)

l_k upper anchor pile length not effective structurally, which begins at the head of the anchor pile and ends upon reaching the active earth pressure slip plane or at the upper edge of the loadbearing soil where this is deeper

l_w the structurally effective anchor length, which extends from the active earth pressure slip plane or from the upper edge of the loadbearing soil to the end of the anchor (excluding toe of anchor); $l_w \geq l_r$ always applies, and for grouted skin anchors $l_w \geq 5.00$ m as well

8.4.9.9   Stability of tension piles and grouted anchors – several anchor levels

When using several anchor levels, each of the centre points of the force transfer paths must be intersected by one failure plane, unless a more accurate analysis is performed. In an arrangement with several anchor levels, if the lower failure plane intersects an anchor before or in the force transfer path, the force that can be transferred behind the failure plane in undisturbed soil may be considered (force $A^*_{2,k}$ in fig. R 10-4). The separated anchor force $A^*_{2,k}$ may be determined from a uniform distribution of the anchor force $A_{2,k}$ via the force transfer path l_r.

In the analysis we must distinguish between the permanent components ($A_{G,k}$) and the varying components ($A_{Q,k}$) of the anchor force.

The stability at the lower failure plane is assured when:

$$\sum (A_{G,k} - A^*_{G,k}) \cdot \gamma_G \leq \frac{A_{poss,k}}{\gamma_{Ep}}$$

where $A_{poss,k}$ is determined from the polygon of forces using exclusively permanent loads, and

$$\sum (A_{G,k} - A^*_{G,k}) \cdot \gamma_G + \sum (A_{Q,k} - A^*_{Q,k}) \cdot \gamma_Q \leq \frac{A_{poss,k}}{\gamma_{Ep}}$$

where $A_{poss,k}$ is determined from the polygon of forces using permanent and varying loads,

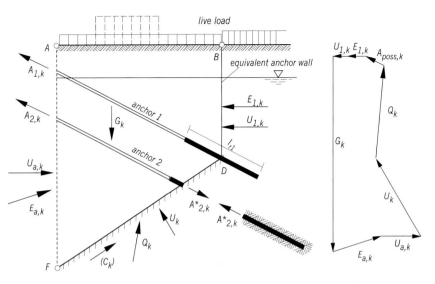

Fig. R 10-4. Verification of stability for the lower failure plane when using several anchors

where:

$\Sigma\,(A_{G,k} - A_{G,k}^{*})$ = total of all permanent components of characteristic anchor forces reduced by the forces transferred to the unaffected soil behind the failure plane

$\Sigma\,(A_{Q,k} - A_{Q,k}^{*})$ = total of all varying components of characteristic anchor forces reduced by the forces transferred to the unaffected soil behind the failure plane (safety factors γ_i as given in section 8.4.9.1)

8.4.9.10 Safety against slope failure

The verification of stability for the lower failure plane, and for anchor walls and plates the verification of adequate safety against the failure of the anchoring soil, replace the slope failure investigation to DIN 4084 normally required.

Irrespective of the stability at the lower failure plane, the safety against slope failure to DIN 4084 must still be verified if there are unfavourable soil strata (soft strata below the anchoring zone) or high loads behind the anchor wall or equivalent anchor wall, or particularly long anchors are used. Please refer to DIN 1054, sections 10.6.7 and 10.6.9, for further advice.

8.4.10 Sheet piling anchors in unconsolidated, soft cohesive soils (R 50)

8.4.10.1 General

If subsoil conditions exist which affect the design of sheet piling bulkheads as treated in R 43, section 8.2.16, special measures are also required in the anchors of these walls in order to avoid the harmful effects of differential settlement.

Even a sheet pile bulkhead, designed as floating, generally has its point in a layer of soil which is more firm than the upper layers. Therefore, in such cases, a movement of the soil in the region of the anchor relative to the sheet piling should be taken into consideration. This condition is aggravated, the more the soil settles and the less the sheet piling is displaced downwards. This can cause a considerable rotation of the anchor connection of the sheet piling. Inclinations of anchor rods of 1 : 3 have already been measured at quay walls of medium height, where the anchor rods were originally installed horizontally.

If the land end of the anchor rods is connected to a firmly founded structure, the situation is similar.

Differential settlement relative to the anchors is generally slight for anchor walls on floating foundation.

As observations of completed structures have shown, the anchor rod is taken downward by settlement of the soil, even by soft soil. It hardly indents into the downward pressing soil, so that it must bend considerably at its connection to a firmly founded structure.

Under the conditions described above, settlement of the subsoil in the entire anchoring area may vary greatly, so that greater or lesser settlement differences may occur even along the anchor length. Therefore, the anchor rod must be able to bend without being damaged. Tie rods with upset thread sections are to be recommended here, since they always have a greater elongation and a higher flexibility than anchors without upsetting.

8.4.10.2 Steel cable anchors

As an example, this requirement can be met with steel cable anchors. They are sufficiently flexible in all practical cases, without it being necessary to lower the allowable effects of actions. However because of corrosion danger, only patently keyed and sealed anchors, or prestressed anchors approved as permanent anchors should be used. It is important that the ends of the steel cable are perfectly insulated at the transition to the cable head or to a concrete structure.

Steel cable anchors must have provisions at one end for prestressing because of the great elongation which occurs in the anchor cable, and for restressing when anchoring is accomplished with floating anchor walls. The waterside end is usually used for this purpose. The steel cable anchor shaft ends there in a cable socket in which the cable wire ends

452

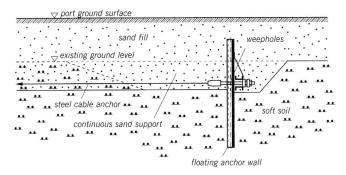

Fig. R 50-1. Floating anchor wall with eccentric anchor connection

are held by poured babbit metal. A short round steel bar with a continuous thread is attached to the cable head. It projects through the sheet piling, and is connected to the sheet piling on the water side by a hinged joint. The tensioning device is placed on the protruding threaded end. The surplus threaded end is burnt off after pre- or restressing has been completed.

For hinged anchor connection see R 20, section 8.2.6.3. In order to give the anchor end sufficient freedom of movement for the expected large rotation about the hinge, the two U-section waling channels must be spread wide apart. The required spacing however often exceeds that which is structurally allowable, so that the anchor must be attached below the waling. Care must then be taken that there is a satisfactory transmission of forces into the anchor by reinforcement on the sheet piling, or by supplementary attachments on the waling.

8.4.10.3 Floating anchor wall

In a floating anchor wall, the usual spacing of both U-section waling channels is generally sufficient; the anchor rod passes between them and is attached to the compression waling behind the anchor wall by means of a hinged joint (fig. R 50-1).

If the steel cable anchor terminates on the land side in the reinforced concrete waling of a floating anchor wall or the like, the end should be unravelled and broomed before it is embedded in the concrete. If the anchor cable is attached to a structure which has a rigid foundation, the connection must also be a flexible one because even a steel cable anchor must not be bent more than 5°.

8.4.10.4 Waling

Especially soft and weak zones in the subsoil must be expected in these cases, even when they are not apparent from the test borings. In order to bridge these intruding layers, the walings of the sheet piling and the anchor wall must be liberally overdesigned. In general, U 400 channels of grade S 235 JRG2 (formerly St 37-2) should be used for walings, in larger structures S 355 J2G3 (formerly St 52-3). Reinforced concrete walings must have at least the same bearing capacity. They are placed in sections of 6.00 to 8.00 m length, with joints keyed against horizontal movement (R 59, section 8.4.3).

8.4.10.5 Anchoring at pile bents or anchor walls

The design of the rear anchoring depends on whether or not considerable displacement of the quay wall coping can be accepted. If no displacement is permissible, anchoring must be by means of pile bents or the like. If some displacement is not objectionable, anchoring onto a floating anchor wall is possible. In this case, the horizontal pressures in front of the anchor wall must remain so slight that undesirable displacements will not occur. If local experience is not available, soil investigations and calculations, are required regarding the partial mobilisation of the passive earth pressure. If necessary, test loading is to be performed.

8.4.10.6 Design

If the harbour land area is filled to the top surface with sand, the design shown in fig. R 50-1 is recommended. The soft soil in front of the anchor wall is excavated to just below the anchor connection, and replaced by a compacted sand buffer of sufficient width. The anchors can be laid in trenches, which are either filled with sand or carefully tamped with suitable excavation soil. The anchor wall is then eccentrically connected, so that the allowable effects of horizontal actions are not exceeded in either the sand fill or in the area of the soft soil. In this case, it is sufficiently accurate if uniformly distributed pressures in both areas are used in design calculations. The magnitude of the pressure follows from the requirement that equilibrium must exist with reference to the anchor connection.

In this solution, irregularities in the anchor wall bedding are equalised. In order to prevent the build-up of hydrostatic pressure difference, weepholes must be placed in the anchor wall for the percolation of water.

8.4.10.7 Stability

The stability at the lower failure plane must be checked here with special care. The usual investigations according to R 10, section 8.4.9. for the final state are not sufficient in this case. The shear strength must be determined for the unconsolidated state, and used as a basis for the verification (initial strength). If the shear strain of the soft cohesive soils in the triaxial test as per DIN 18 137-2 exceeds 10%, the degree of shear strength utilisation is to be reduced according to DIN 4084.

8.4.11 Design and calculation of protruding corner structures with tie roding (R 31)

8.4.11.1 Impractical design

The quay wall sheet piling at corner structures should not be held by anchors running diagonally from quay wall to quay wall and thus are not connected at right angles to the wall axis as is customary. Otherwise damage can occur because the anchor forces create high, additional tensile forces in the walings whose highest stress occurs at the last diagonal anchor connection. Since corrugated steel sheet piling yields comparatively easily under horizontal loads in the plane of the wall, a considerable length of the quay wall is required for the transfer of tensile forces from the waling, through the sheet piles and into the foundation soil. The waling joints are especially endangered thereby. Special reference here is made to R 132, section 8.2.12.

8.4.11.2 Recommended crossing anchors

The tensile forces mentioned in section 8.4.11.1 do not occur in the waling if crossed anchoring is used as shown in fig. R 31-1. In order to prevent the anchors from interfering with each other, the rows of anchors and the walings must be offset in height. Clearance at the turnbuckles should be somewhat more than the diameter of the anchor rod.

Edge bollards should have independent, additional anchoring.

8.4.11.3 Walings

The walings at the sheet piling are steel tension walings, shaped to fit the quay wall. The walings at the anchor wall are pressure walings of steel or reinforced concrete. The transition from the walings of the corner section to the walings of the quay wall and anchor wall, as well as the intersection of the anchor wall walings, are so arranged that the walings can move independently of each other. The walings and copings have spliced joints with slotted holes where the anchor walls abut the quay wall.

8.4.11.4 Anchor walls

The position and construction of the anchor walls in the corner section depend on the design of the quay walls. The anchor walls at the corner are carried through to the quay wall (fig. R 31-1) but are driven staggered and down to the harbour bottom in the end stretch so that in case of damage, the loss of backfill at the especially critical corner area will stop at the anchor wall. This precaution is also recommended when individual anchor plates are used, such as reinforced concrete slabs, instead of anchor walls.

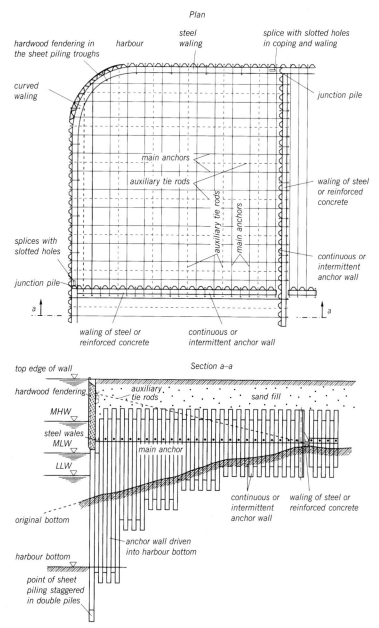

Fig. R 31-1. Anchoring of a protruding quay wall corner in a sheet piling structure at seaports

456

8.4.11.5 Timber fendering

Ships and the wall itself are protected if the sheet pile troughs at wall corners are fitted with suitable wharf timbers. This fendering should project about 5 cm beyond the outer face of the sheet piling (fig. R 31-1).

8.4.11.6 Rounded and reinforced concrete wall corners

Since protruding quay wall corners are especially exposed to damage by ship traffic, they should be rounded off if possible, and if need be, also strengthened by strong reinforced concrete wall.

8.4.11.7 Dolphin protection in front of the corner

Each quay wall corner should be protected by a resilient dolphin if ship traffic permits.

8.4.11.8 Verification of stability

Verification of stability is carried out individually for each quay wall, according to R 10, section 8.4.9. A special check for the corner section is not necessary if the anchors of the quay walls are carried through to the other wall as shown in fig. R 31-1.

Self-supporting pier heads are designed and calculated according to other principles.

8.4.12 Design and calculation of protruding quay wall corners with batter pile anchoring (R 146)

8.4.12.1 General

Protruding corners of quay walls are especially exposed to damage from passing ship traffic. In many cases, quay wall corners are at the end position of a berth for a large ship and hence must accommodate a heavily loaded bollard, as described in R 12, section 5.12.2. In seaports, they are also equipped with the required fendering, which must possess higher energy absorption capacity than that on the adjacent quay wall sections. Corners should be of sturdy construction and be as rigid as possible.

Batter pile anchoring can also be a beneficial design for such quay wall corners, so that the solution with pile anchoring elements is dealt with here as a supplement to the solution with tie rods according to R 31, section 8.4.11.

8.4.12.2 Design of the corner structure

The most appropriate design of quay wall corners with batter pile anchoring can differ greatly due to local conditions and future utilisation as a port facility. It depends largely on the structural design of the adjoining quay walls, on the difference in ground surface elevation and

457

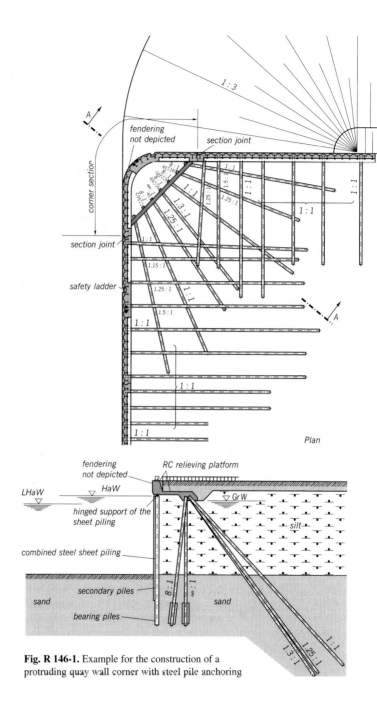

Fig. R 146-1. Example for the construction of a protruding quay wall corner with steel pile anchoring

on the angle enclosed by the wall sections forming the corner. As far as construction is concerned, the final design will be decisively influenced by the existing water depth and the nature of the subsoil.

In order to ensure proper positioning of the overlapping inclined piles occurring at the corners, definite requirements must be adhered to regarding the clearance between these piles at all points where they cross. Whereas the clearance between crossing piles which overlap above the existing bottom can be kept comparatively small (about 25 to 50 cm), the clearance between long piles below the bottom at all crossing points should be at least 1.0 m, or preferably 1.5 m, especially in compact soils where driving is difficult. In stony soils, where driving can still be executed, longer piles will probably drift out of line. Under these conditions, the clearance at greater depths should be at least 2.5 m. When calculating the clearance between piles, existing steel vanes are always to be taken into account.

In order to be able to meet these requirements, the spacing and inclination of the piles must be properly varied, but the inclination should be kept fairly uniform because of the varied bearing behaviour of individual piles of an interdependent pile group.

Should a deep foundation also be necessary at the quay wall corner for highly loaded bollards or other items of equipment such as anchoring constructions of conveyor belts and the like, the construction of a special reinforced concrete corner section with a deep-founded relieving platform is recommended in most cases. The latter has a hinged support on the sheet piling. This also applies in general to quay wall corners at which proper positioning of the piles cannot be achieved by altering pile inclinations and clearances. In such corner constructions, the tension piles required in the corner area are appropriately placed at the rear portion of the relieving platform. As a result, they lie within a plane which differs from that of the tension piles of the bordering quay wall sections, so that interference between the piles can be more readily avoided. Due to the additional compression piles required at the rear slab edge, as well as to the required relieving platform, such designs are much more costly. However, they ensure sound and proper construction. Fig. R 146-1 shows a typical example of this.

Sections 8.4.11.5 and 8.4.11.7 in R 31 also apply to quay wall corners with batter pile anchoring.

8.4.12.3 Use of scale models

In order to avoid future driving difficulties, a small but adequately exact scale model should be built and be available for checking during the project planning of difficult corner constructions. A larger model scaled at about 1 : 10 should be used later on the building site. In this model, each pile must be placed in the position in which it was actually installed, to allow for necessary corrections in the position or inclination of the remaining piles.

8.4.12.4 Verification of stability of the corner sections

In all designs of corner constructions with batter pile anchoring, verification is required of the stability of all piles in the entire corner area. Please refer to section 9. In so doing, each wall in the corner is to be considered individually. At corners with additional loads, from corner stations of a conveyor facility, from bollards, fenders and other items of equipment, it must be verified that the piles are also able to absorb these additional forces satisfactorily.

Should extensive changes take place in the piling during construction, their effects are to be verified in a supplementary calculation.

8.4.13 High prestressing of anchors of high-strength steels for waterfront structures (R 151)

8.4.13.1 Anchors for waterfront structures, particularly for sheet piling bulkheads and also for subsequent securing of other structures, such as walls on pile foundations and the like, are usually made of steel grades S 235 JRG2 (formerly St 37-2), S 235 J2G3 (formerly St 37-3) or S 355 J2G3 (formerly St 52-3). However, in certain cases high pre-stressing of the anchors can be beneficial, but this is only advisable when they consist of high-strength steels.

The high prestressing of anchors of high-strength steels can be practical or necessary for the following, among others:

- for limiting displacements, especially at structures with long anchors, making allowance for existing sensitive structures or when joining to subsequently driven front sheet piling and
- for achieving load transmission with pronounced active earth pressure redistribution (strongly reduced moment of span at increased anchor force), with the prerequisite that the structure be situated in at least medium densely deposited non-cohesive or stiff, cohesive soil.

For permanent anchors of high-strength steels, special significance is to be ascribed to technically sound corrosion protection in all cases. Any existing licences, for example, for grouted anchors according to DIN EN 1537, are to be observed.

8.4.13.2 Effects of high anchor prestressing on the active earth pressure

Anchor prestressing always lessens the displacement of the waterfront structure toward the water side, especially in its upper part. High prestressing can favour an increased redistribution of the active earth pressure toward the top. In this case, the resultant of the active earth pressure can move from the bottom third point of the wall height h above the harbour bottom to about 0.55 h upward, whereby the anchor force to be absorbed increases correspondingly. This active earth pressure redistribution is especially pronounced at quay walls extending above the anchor (cantilever).

In case an active earth pressure diagram deviating from the classic distribution according to COULOMB is to be achieved using fully-loaded, high-strength anchoring steels, the anchors must be locked off at about 80% of the anchor force determined for loading case 1 by prestressing.

8.4.13.3 Time for prestressing

The prestressing of the anchors may not begin until the respective prestressing forces can be absorbed without appreciable undesirable movements of the structure or its members. This requires corresponding backfill conditions and is to be considered in the planning of the construction stages and in the assumption regarding the absorption/transmission in the structure.

Experience shows that the anchors must be prestressed briefly beyond the planned value, since part of the prestressing force is lost again through yielding of the soil and of the structure when the adjacent anchors are stressed. This can be extensively avoided if the anchors are prestressed in several steps, which however can create difficulties in the execution of the work.

The prestressing forces are already to be subjected to spot checks during the construction period, in order to be able to make a correction in the scheduled prestressing, if necessary.

DIN EN 1537 is chiefly applicable to pressure-grouted anchors.

8.4.13.4 Further advice

For high anchor prestressing in limited waterfront sectors, the produced locally differing displacement possibility of the waterfront structure is to be taken into account. The prestressed zones act as rigid points, which are acted on by increased spatial active earth pressure and are to be adequately designed fore this. The anchor forces in such areas are always to be rechecked.

Where at all possible, one end of the prestressing anchor should be fabricated so as to always be accessible and be so designed that if necessary, the prestressing force or related displacement can also be checked subsequently and corrected. Otherwise, the anchor ends should have hinged connections.

Since bollards are loaded only from time to time, their anchors should not be made prestressed of high-strength steel but rather of strong round steel bars of S 235 JRG2 (formerly St 37-2), S 235 J2G3 (formerly St 37-3) or S 355 J2G3 (formerly St 52-3). The latter show only slight extension when subject to loads. Difficulties would crop up during earthwork behind bollard heads, if highly prestressed anchors of high-strength steels were used. However, these are frequently required for operational reasons, e.g. for laying lines of various types.

8.4.14 Hinged connection of driven steel anchor piles to steel sheet piling structures (R 145)

8.4.14.1 General

A hinged connection between driven steel anchor piles and sheet piling structure allows the desired and largely independent reciprocal rotation of the structural members to take place freely, thus creating an un-complicated structural condition which helps to keep the costs low.

Rotation in the proximity of the connection between the anchor pile and the sheet piling occurs inevitably as a result of deflection at the sheet pile wall. It can however also occur at the head of the steel anchor pile, especially when a downward movement of the active earth pressure sliding wedge is accompanied by severe settlement and/or subsidence of the natural or filled ground behind the sheet piling. In such cases, the hinged connection is preferable to the fixed connection described in R 59, section 8.4.3.

In accordance with the principles of steel construction, all parts of the connection must be designed for safety and effectiveness.

8.4.14.2 Advice for designing the hinge connection parts

The capability to rotate can be achieved by the use of single or double hinge pins or by the plastic deformation of a structural member (designed to be stressed to the yield point) suitable for this purpose. A combination of pins and flexible joint is also possible.

(1) Hinges, designed to be stressed to the yield point, should be located at a sufficient distance from butt and fillet welds, to prevent the yielding of welded connections as far as possible.
Fillet welds parallel to acting forces should lie in the plane of forces or in the plane of the tension transmitting member, in order to guard against the welds loosening or sealing off. If this is not possible, other means must be employed to accomplish this.

(2) Every weld transverse to the planned tensile force of the anchor pile can become effective as a metallurgical notch.

(3) Construction welds in difficult positions, which are not made true to stress and welding requirements, increase the probability of failure.

(4) In difficult connection designs, even with hinged connection, it is recommended that a calculation be made of the effect on the cross-sectional area probably forming a yielding hinge by the action of

462

the planned axial forces, in connection with possible supplemental stresses and the like (see R 59, section 8.4.3). When verifying hinges, designed to be stressed to the yield point, DIN 18 800 is to be taken into consideration.

(5) Notches caused by sudden discontinuity of stiffness, e.g. when there are oxy-burned notches in a pile and/or metallurgical notches from cross welds, as well as abrupt increases of steel cross-sections, e.g. due to welded-on, very thick straps, are to be avoided, especially in possible yield areas of tensile anchor piles because they can produce sudden failure without deformation (brittle fracture).

Several characteristic examples of hinged connections of steel anchor piles are shown in figs. R 145-1 to R 145-7.

8.4.14.3 Work on site

Contingent on local conditions and on the design, the steel anchor piles may be driven either before or after the sheet piling. If the location of the connection with respect to certain members of the sheet piling depends on the geometry, as would be the case if the connection must be made in the trough of a corrugated sheet pile, or on the bearing pile of combined sheet piling, it is important that the upper end of anchor piles be as close as possible to their designed position. This is best accomplished if the anchor piles are driven after the sheet piling. However, the construction of the connection must always be such that certain deviations and rotations can be compensated for and absorbed.

If the steel anchor pile is driven directly above the top of the sheet piling, or through a "window" in the sheet piling, the sheet piling can provide effective guidance for driving the anchor pile. If an anchor pile must be driven behind a double pile, a driving "window" can be provided by burning off the upper end of the double pile, hoisting it clear and later returning it to its original position and welding it in place.

Steel piles which are not embedded in the soil to the upper end leave a certain scope for aligning the head in the connection.

In determining the lengths of the anchor piles, an allowance of extra length should be made. This extra length depends on the type of the connection and is for the purpose of allowing the butt of the pile to be burned off in case it is damaged in driving or for the driving itself.

Slots for connection plates shall if possible be cut into the sheet piling and into the anchor piles only after the piles have been driven.

8.4.14.4 Detail design of the connection

The hinged connection to corrugated sheet piling is generally placed in the trough, especially if the interlock connection is on the centroidal axis, or, for combined sheet piling, on the web of the bearing piles.

463

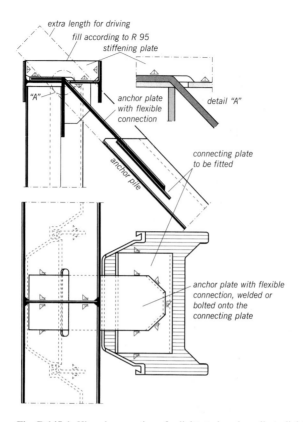

Fig. R 145-1. Hinged connection of a light steel anchor pile to light steel sheet piling, using anchor plate and flexible connection, designed to be partly stressed to the yield point

Where a wall is only lightly loaded, especially in an open canal section, the steel anchor pile may be connected to the capping beam, which is mounted on the top of the sheet piling (fig. R 145-1), or to a waling behind the sheet piling by means of a splice plate (anchor plate) and a flexible connection. Special attention must be paid to the danger of corrosion in such cases. Please refer to R 95, section 8.4.4 for waterfront structures with cargo handling and for berthing places.

Tension elements of round steel rods (fig. R 145-3), flat or universal flat steel straps (tension straps) are frequently installed between the connection in the trough or at the web and the upper pile end (figs. R 145-4 and R 145-5). Connections consisting of a threaded steel rod, nut, base plate and joint plate have the advantage that they may be tensioned.

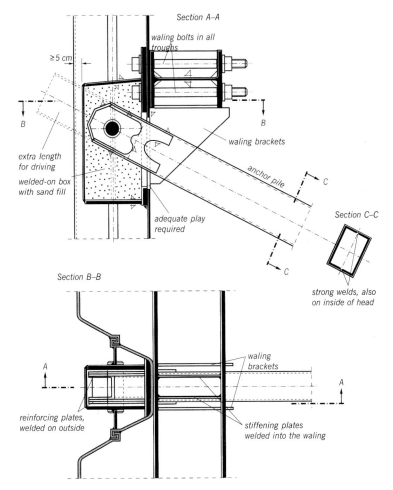

Section A–A

waling bolts in all troughs

≥5 cm

B

B

extra length
for driving

welded-on box
with sand fill

waling brackets

anchor pile

C

adequate play
required

Section C–C

Section B–B

C

strong welds, also
on inside of head

waling
brackets

A

A

reinforcing plates,
welded on outside

stiffening plates
welded into the waling

Fig. R 145-2. Hinged connection of a steel anchor pile to heavy steel sheet piling, using a hinge pin

In addition to the hinged connection in the sheet piling trough, in the capping beam or at the web of the bearing pile, an additional hinge may, in special cases, be installed near the top end of the anchor pile. This solution, depicted in fig. R 145-5 for the case with double bearing piles, can also be employed for single bearing piles, somewhat varied. The slots (burned openings) in the flanges of the bearing piles are to be taken deep enough under the connection plates to create sufficient freedom of movement for pile deformations and to rule out any compulsive forces

465

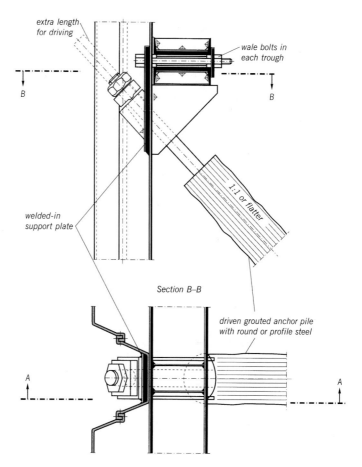

Fig. R 145-3. Hinged connection of a driven grouted anchor pile to heavy steel piling

which can arise from unwanted fixity, should contact between plates and pile occur. Care is also required to ensure that the intended hinge effect is not impaired by incrustations, sintering and corrosion at the connection elements. This is to be checked from the individual case and taken into account in the structural design.

The anchor pile can also be driven through an opening in a sheet piling trough and the hinge connected there by means of a supporting construction welded to the sheet piling (fig. R 145-2).

466

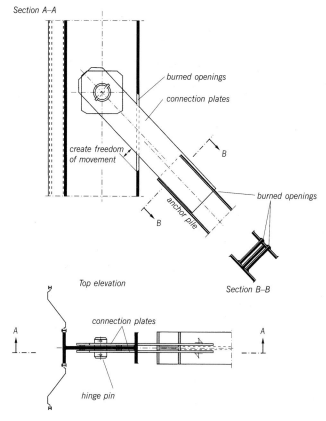

Section A–A

burned openings

connection plates

create freedom
of movement

B

anchor pile

B

burned openings

Top elevation

Section B–B

connection plates

A

A

hinge pin

Fig. R 145-4. Hinged connection of a steel anchor pile to a combined steel sheet piling with single bearing piles, using a hinge pin

If the connection is made at the water-side trough of a sheet piling wall, all construction members must terminate at least 5 cm behind the sheet piling alignment. Furthermore, the point of penetration should be carefully protected against soil running out and/or washing out (e.g. by means of a protective box as in fig. R 145-2).

Depending on the design selected, preference should be given to those connections which can be manufactured largely in the workshop and which provide adequate tolerances. Extensive fitting work at the site is very expensive and is therefore to be avoided if at all possible.

The connection shown in fig. R 145-6, for example, fulfils these conditions. All bearing seams on the bearing pile are welded in the workshop in the box (flat) position. However, this solution should only be used if

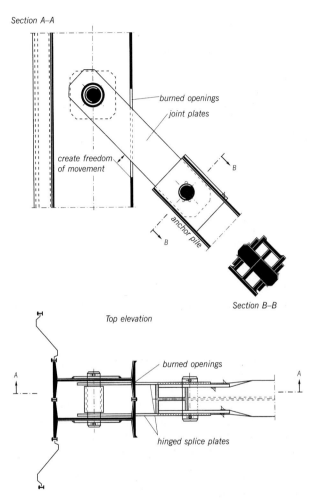

Fig. R 145-5. Hinged connection of a steel anchor pile to a combined steel sheet piling with double bearing piles, using hinged splice plates

the jaw bearing plates are installed after driving the anchor piles. In addition, the length of the bearing pile must be specified exactly; this connection is not possible when extending or shortening the pile.

In the solution shown in fig. R 145-7, the connection between anchor pile and sheet piling is created by loops which enclose a tubular waling welded into the sheet piling. Care is required here because the free rotation of the connection is prevented by friction between loops and tube. The

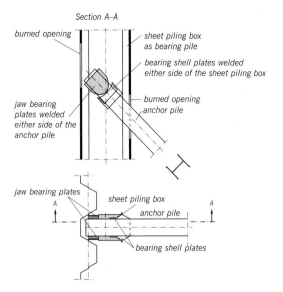

Fig. R 145-6. Hinged connection of a steel anchor pile to combined steel sheet piling using jaw bearings/bearing shells

loops are therefore to be dimensioned to allow for the adequate bearing reserves required for the resulting unequal load distribution. As a rule, deflections of the anchor piles and consequently displacement of the pile heads are to be expected, so that the connection has to be designed for transfer of shear forces besides the anchor tensile forces. This can be achieved with a bracket-type extension of the piles through to the tubular waling, or other suitable measures.

8.4.14.5 Verifying the bearing capacity of the connection
The anchor force determined by the sheet piling calculations is the primary factor in the design of the anchor connection. It is recommended however to design all anchor connection parts for the internal forces, which can be transmitted by the anchor system used. Loads from the water side, such as vessel impact, ice pressure or the subsidence of mines etc., can at times reduce the tensile force present in the steel pile or even change it into a compressive force. If necessary, checks should be made of the stresses in the connection and of the buckling stress on the free-standing portion of a partially embedded pile or on the pile connection. In some cases, ice impact must also be taken into account.

If possible, the connection should lie at the intersection of the sheet piling and the pile axis (figs. R 145-1, R 145-2, R 145-4 to 145-7). If

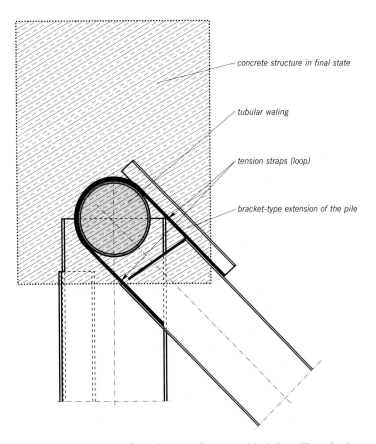

concrete structure in final state

tubular waling

tension straps (loop)

bracket-type extension of the pile

Fig. R 145-7. Connection of a steel anchor pile to a combined sheet piling using loops

there is a substantial deviation from this, additional moments in the sheet piling are to be taken into account.

The vertical and horizontal components of the anchor pile force are also to be taken into account in its connection to the sheet piling and, if every stressed wall element is not anchored, in the waling and its connections. If a vertical load due to soil effects must be expected, it must also be allowed for in the reaction forces and in the verification of the bearing capacity of connections. This is always the case when deflection of the anchor piles is to be expected. When the opening angle between anchor pile and sheet piling changes, the bearing capacity verification must also take account of the resulting changes in the tensile forces in the area of the connection structure, as well as resulting shear forces.

In case of connections in the trough, the horizontal force component is to be transmitted into the sheet piling web (fig. R 145-3) by a support plate of adequate width. The weakening of the sheet piling cross-section should be considered because reinforcing of the sheet piling may be necessary in this area.

A constant flow of tensile forces and safe transfer of the shear forces is to be observed in the connection elements, particularly in the area of the upper anchor pile end. If the flow of forces cannot be satisfactorily checked and confirmed at difficult, highly loaded connection structures, the calculated dimensions and stresses should be checked by at least two test loadings to failure on full sized working parts.

8.4.15 Armoured steel sheet piling (R 176)

8.4.15.1 Necessity

The increasing size of ships' vessels, the traffic of ship formations and the changed nature of locomotion have led to increased operational requirements being placed on waterfronts in inland harbours and waterways. In order to prevent damage to sheet piling structures also, these should have a surface which is as even as possible (R 158, section 6.6).

With dual piles of large widths, the susceptibility to damage increases as a result of diminution of the approach angle and the increased clearance between the backs of the piles. The requirement for a largely even surface is achieved with an armouring for the sheet pilling, in which steel plates are welded into or over the sheet piling troughs (fig. R 176-1).

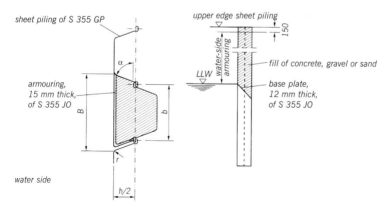

Fig. R 176-1. Armouring to a U-shaped sheet pile

8.4.15.2 Scope of application

However, owing to the technical and economic resources involved, armouring is recommended only for stretches of waterfront subjected to special loads.

In inland harbours, these are waterfronts with very heavy traffic, especially with push tows and large motor vessels, as well as waterfronts in areas of high risk, such as changes of direction in plan and guide walls in lock entrances.

8.4.15.3 Elevation

The sheet piling armouring is required for the elevation area of the waterfront in which contact between a ship and the sheet piling is possible from the lowest to the highest water level (fig. R 176-1).

8.4.15.4 Design

The design of the armouring and its dimensions depend primarily on the opening width B of the sheet piling trough. This is determined by the system dimension b of the sheet piling, the inclination α of the pile leg, the section height h of the piling wall, and the radius r between the leg and the back of the sheet piling (fig. R 176-1).

8.4.15.5 Pile shape and method of armouring

In designing the armouring, it must also be made clear whether the sheet piles are Z-shaped or U-shaped. It is also of significance whether the armouring is to be fitted by the workshop, and therefore before driving, or on site, and therefore after driving.

In the case of Z-shaped piles and armouring produced on site, the steel plates are supported over their full width as far as the interlocks. Because they project beyond the alignment, they also protect the interlocks (fig. R 176-2).

For Z-shaped piles, production in the workshop in accordance with fig. R 176-2 cannot be recommended because driving may then be impaired due to a lack of flexibility of the dual piles.

The subsequent installation of armouring is also possible with U-shaped piles, a completely even waterfront surface being achievable (fig. R 176-2).

During installation on site, pressure and adjustment measures will be unavoidable with either section type. This will produce a waterfront structure of uniform rigidity, but with less elasticity than a wall consisting of individual elements.

The best solution for installation and operation is offered by the piling wall of U-shaped piles with armouring produced in the workshop.

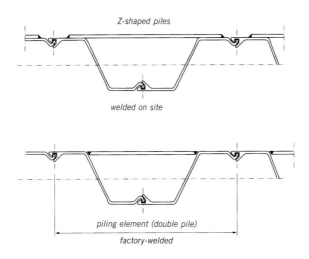

Z-shaped piles

welded on site

piling element (double pile)

factory-welded

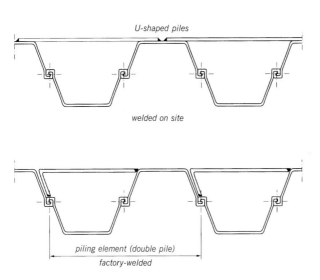

U-shaped piles

welded on site

piling element (double pile)

factory-welded

Fig. R 176-2. Different methods of armouring for Z- and U-shaped sheet piles

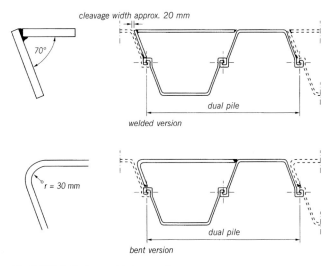

Fig. R 176-3. Factory-welded armouring

8.4.15.6 Details of on factory-applied armouring for U-shaped piles

In the case of installation by the workshop, the armouring is fixed to the interlock of the trough pile and the back of the peak pile. For technical reasons relating to driving, the interlock of the dual pile must also be welded, so that a rigid connection is created. Only then can the weld seams of the armouring withstand the driving process without damage. The pile back and the free leg of the peak pile are still available for elastic deformation (fig. R 176-2).

The armour plates can be welded together or bent (fig. R 176-3). The welded version produces the smaller width of the remaining cleavage because limits are set for the minimum radius for cold forming in the bent version. The cleavage width in the welded version is approx. 20 mm. With a system dimension for the dual pile of 1.0 m, an even piling wall is then achieved to approx. 98%.

8.4.15.7 Dimensioning

The armouring is generally not taken into account in the static calculations for the sheet piling. The thickness of the plate is derived from the support width corresponding to the opening width of the sheet piling troughs.

However, since the support width of the impact armouring is always greater than the width of the back of the peak pile, the armouring plates must be thicker than the backs of the piles. To avoid plate thicknesses ≥ 15 mm, it is recommended that the remaining space between armouring and the dual pile be filled up.

Elevation – dual pile with impact armour as ladder

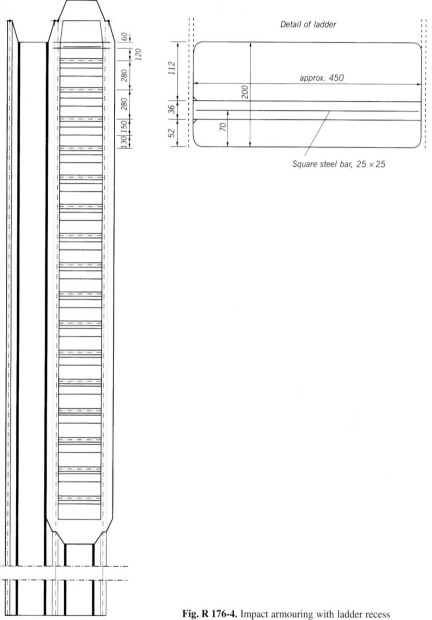

Detail of ladder

approx. 450

Square steel bar, 25 × 25

Fig. R 176-4. Impact armouring with ladder recess

475

8.4.15.8 Filling

A base plate is welded in to close off the bottom (fig. R 176-1) so that the space between armouring and the dual pile can be filled in. Sand, gravel or concrete can be used for filling purposes.

8.4.15.9 Access ladders and mooring facilities

The impact armouring is usually interrupted in sheet piling where there are access ladders and recess bollards. In order to keep the surface as smooth as possible here, armouring with step recesses (fig. R 176-4) can be used at the ladder pile, or a continuous step recess box (fig. R 176-5) can be used, leaving a space which can be filled according to section 8.4.15.8, insofar as this solution appears technically feasible and economically acceptable. Fig. R 176-6 shows the arrangement of recess bollards in an armoured wall. The requirements of R 14, section 6.11, and R 102, section 5.13, are observed.

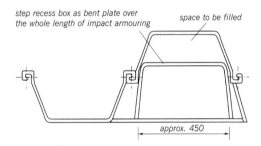

Fig. R 176-5. Impact armouring with ladder case

8.4.15.10 Costs

The extra costs for armoured sheet piling as compared to non-armoured solutions depend primarily on the ratio of the length of the armouring to the overall length of the sheet pile, and on the section. The delivery charges for the piling wall material are increased by 25 to 40%. In the case of U-section piles, it is cheaper and technically better to include armouring in the design from the outset and have it installed by the factory.

View of a dual pile with impact armouring and bollard recesses

upper edge dual pile

Section A–A

lock plate fitted and welded after driving

B

A — A

B

approx. 1500

Section B–B

area of bollard recess

min. 500

sloping concrete surface

vent hole

filled with concrete

Fig. R 176-6. Impact armouring with recess bollards

477

9 Anchor piles and anchors

9.1 General

Vertical constructions as waterfront structures, for example, quay walls, usually have to be provided with corresponding anchoring elements for overturning and sliding safety, for resisting the horizontal forces due to earth and water pressure, and for resisting forces from the superstructure such as line pulls and vessel impact. In the case of slight differences in terrain elevation, these forces can also be transferred via pile bents by means of corresponding design of the piled structure. Larger differences in terrain elevation, such as those regularly encountered in modern seaports and inland ports, require special anchoring elements.

The recommendations featured in this section for the structural calculations are based on the safety concept of verifying the theoretical bearing capacity for the limit state using partial safety factors (see section 0.1).

9.2 Anchoring elements

Essentially, the following types of anchor are available:

- displacement piles
- micropiles
- special piles

9.2.1 Displacement piles

9.2.1.1 **Steel piles** are rolled products which can be supplied as I-sections in long lengths. Their outstanding features include good adaptation to the prevailing structural, geotechnical and driving conditions. They can be installed by driving or vibrating as "bare" piles and are relatively insensitive to driving hindrances and heavy driving. The head of the pile can be lengthened if the local soil conditions call for this. Steel piles can be easily welded to other structural components of steel or reinforced concrete.

Steel grades S 235 JRG2 or S 355 JRG3 to DIN EN 10 025 should be used.

Hot-rolled wide-flange I-sections may also be used provided the grade of steel complies with Euronorm 53/edition 62 or DIN EN 10 025.

9.2.1.2 **Precast concrete piles** are driven into the ground as complete piles.

9.2.1.3 **Driven piles with grouted skin** consist of steel sections as described in section 9.2.1.1. However, these piles have a special toe construction to facilitate the installation of grout at the base of the pile. The toe of the pile in this case is a shoe made from sheet metal welded into a box form. The grouted skin pile is a "full displacement pile" which creates a prismatic void around the entire steel section of the driven pile into which a cement suspension is injected at the same time as driving. Alternatively, the top section of the shaft is simply filled with the cement suspension. The grout is conveyed via a grouting line to the toe of the pile where it escapes and fills the void left by the shoe.

The grout in the void sets (i.e. hydrated cement) and acts as a bonding medium between steel pile and subsoil to guarantee the skin friction resistance, which – depending on the subsoil – can achieve values of three to five times the value of the "bare" steel pile.

An additional effect of the soil displacement due to the driving is the prestressing of the subsoil in three dimensions, which leads to a further increase in the load-carrying capacity of the pile. Grouted skin piles are generally installed at angles of between 2 : 1 and 1 : 1.

9.2.1.4 **Driven grouted piles** are similar to grouted skin piles in some respects. To enlarge the toe of the I-section steel pile, approx. 20 mm thick steel flats are welded to all sides to increase the thickness of web and flanges. These displacement elements create a gap equal to their thickness around the vibrated pile and, like the grouted skin pile, a cement suspension is injected into this gap during the vibrating in order to increase the skin resistance.

Driven grouted piles can be vibrated into position at angles between vertical and 1 : 1. It is difficult to achieve a satisfactory degree of efficiency with vibrated raking piles. This can be defined as the "pile toe penetration per unit of time related to the vibration energy applied". Measures must be taken to minimise the amount of energy lost on its way from vibrator through the pile and into the subsoil. To do this, raking driven grouted piles are "prestressed" against the ground with an axial compressive force amounting to approx. 100 kN. This enables the pile to maintain permanent frictional contact with the subsoil, which allows the energy applied to be converted into movement of the pile more effectively.

9.2.2 Micropiles (diameter ≤ 300 mm)

9.2.2.1 **Self-boring micropiles** represent an alternative to grouted piles. A solid round steel bar is drilled into the ground. The advantage of this type of installation is less vibration and with less noise. The tip of the auger creates a ring gap in the soil which is immediately filled with a cement suspension under pressure. A defined length at the base of the pile is constructed as a grouted mass (hydrated cement) up to a diameter of 50 cm or even larger.

9.2.2.2   **Tubular grouted piles** represent yet another alternative to skin grouted piles. In this case a steel casing with an external thread is embedded to the required depth by drilling (with external jetting) section by section to match the lengths of the tie rod sections. Consequently, this method also has the advantage of involving less vibration and less noise. During construction, the steel acts as a casing to the drilled hole and in the final condition as a tension member. The external thread is used to connect the tie rod sections, to form the connection at the head of the pile and to guarantee a bond at the toe. The toe of the pile is formed with an underrreamed enlargement by extracting and re-inserting the tip of the auger several times combined with radial jetting to match the size of the toe. The ensuing void is subsequently filled with grout.
The thickness of the casing is 12–28 mm and the external diameter 73–114 mm.

9.2.2.3   **Grouted in situ concrete piles** in the form of non-prestressed micropiles have maximum shaft diameters of 150-300 mm. These are bored piles into which continuous longitudinal steel reinforcement is inserted before filling with concrete or grout. The actual grouting procedure for the structural bond with the subsoil is achieved by applying compressed air to the top of the in situ concrete. Subsequent grouting is also possible provided appropriate technical precautions, such as re-grouting tubes and sleeve pipe, are taken.

9.2.2.4   **Composite piles** in the form of non-prestressed micropiles have shaft diameters between 100 and 300 mm. They are installed as bored piles but can also be driven ("driven grouted pile" with small cross-section), have a continuous prefabricated steel loadbearing element and are filled with a cement suspension. The grouting procedure is achieved by applying a higher fluid pressure to the grout. Subsequent grouting is also possible provided appropriate technical precautions, such as re-grouting tubes and sleeve pipe, are taken (DIN EN 1536 or DIN EN 14 199).

9.2.3   Special piles/anchors

9.2.3.1   **Retractable raking piles** are fully prefabricated steel tension members consisting of a steel pile section and an anchor plate. When used as anchor piles they are attached to the steel pile wall with a structural connection above the water level. The anchor plates are installed on the bottom of the watercourse and filled. They are installed at angles between 0° and 45°. In contrast to a raking pile, however, the pile's resistance is not formed by the skin friction, but rather by adding the vectors of vertical soil weight and horizontal passive earth pressure on and in front of the low-lying anchor plate.

The prefabrication includes forming the head of the steel pile in such a way that a later connection between anchor pile and quay wall permits rotation. The adjoining bearing piles of the sheet pile wall to be anchored are fitted with a compatible steelwork connection to accommodate this pile head and in which the pile can rotate freely. At the toe of the pile the anchor plate is welded on at right-angles to the axis of the pile via a connecting construction. This consists, preferably, of corrugated sheet pile sections. The pile toe is lifted for installation. The connection between head of pile and sheet piling permits rotation but also transfers the forces. Afterwards, the toe of the pile – still suspended from the crane – is lowered into position, allowing the shaft of the pile to rotate about the pivot and the toe of the pile to be placed in position on the bottom of the watercourse in the fixed anchor plate. Following this "hinging procedure", the pile is in its final position. The geotechnical conditions can be further improved if the anchor plate is vibrated with an underwater vibrator into the bottom of the watercourse for a depth equal to half its height. Afterwards, a hydraulic backfill is introduced, but must be placed in the following way: the first charge of filling material must be placed directly in front of and on the anchor plate. This enables part of the pile resistance to be mobilised.

9.2.3.2 **Bored piles with jet-grouted underreaming.** In terms of their construction these involve drilling followed by jetting. A double-wall casing with a detachable boring shoe is sunk to the depth required. A rolled HEM steel section is inserted into the casing to act as a tension member. After detaching the shoe, the casing is withdrawn. As the double-wall casing is extracted, a radial high-pressure jet at the position of the underside of the casing creates a relatively large injected grout mass. Pile resistances of up to 4-5 MN are possible with this type of pile.

9.2.3.3 **Tie rods (laid in the ground)**

If at the time of construction there is neither any buildings nor extensive buried cables or pipes behind the quay wall, tie rods with anchor plates can be laid in the ground. The combination of upset T-heads and threaded ends, temporary anchor supports, sleeves and turnbuckles plus anchor plates made from corrugated sheet piles or precast (reinforced) concrete components render possible a solution that can be varied to suit the most diverse problems. Such tie rods are generally limited to angles of approx. 8° to 10° owing to the extent of the earthworks required. Horizontal tie rods are also suitable for special structures such as mole heads, quay wall projections and returns, where they can be anchored across the opposite quay walls. In such cases a raking pile solution would cause considerable geometrical problems in trying to arrange and accommodate the piles.

9.2.3.4 Grouted anchors to DIN EN 1537

Prestressed grouted anchors can also be used (solid bar anchor grade St 1080/1230 or St 950/1050, or stranded anchor grade St 1570/1770). Perfect corrosion protection is essential here because these are high-quality prestressing steels.

9.3 Safety factors for anchors (R 26)

The safety of anchorages with anchor elements that transfer tension forces into the soil via skin friction must be verified according to DIN 1054. The stability at the failure slip plane is to be verified according to R 10, section 8.4.9. DIN EN 1537 applies with respect to installation, testing and monitoring of grouted anchors.

9.4 Pull-out resistance of piles (R 27)

9.4.1   The characteristic pull-out resistance is taken to be the load that starts to pull out the pile. If this is not clearly revealed by the load–lifting line, the characteristic pull-out resistance is taken to be the load that does not yet endanger the presence and use of the structure upon lifting the pile (in the pile axis). In the case of waterfront structures, the remaining lift can generally amount to around 2 cm. The characteristic pull-out resistance can be determined as the "critical displacement load" via the creep dimension k_s according to DIN EN 1537. The creep dimension should not exceed 2 mm.

9.4.2   The pull-out resistance for grouted anchors – both for temporary and permanent anchors – is stipulated according to DIN EN 1537. The pull-out resistance for driven piles and grouted piles is stipulated as per DIN 1054.

9.4.3   In preliminary designs the pull-out resistance can also be determined approximately by static penetrometer tests, if both the cone resistance and the local skin friction are measured with suitable penetrometers. In so doing, the results of boreholes sunk in the vicinity must also be taken into account in order to ascertain to which type of soil each of the test results refer. Guide values for characteristic pile resistances are stated in DIN 1054 appendix B and C.
In preliminary designs the stated skin friction tensions of grouted skin piles may be increased by factors varying between 2 and 4.

9.4.4   When determining the pull-out resistance of an individual pile within a group of piles, the group effect must also be taken into consideration.

9.4.5   If the pull-out resistance is not achieved during test loading, the maximum tensile force reached during the test is taken as the characteristic pull-out resistance.

9.4.6   Under certain circumstances, the so-called "hyperbola method" ([5, 6]) can be used to estimate the failure loads – as asymptotic lines of the load–displacement curve. It should be noted that this method can also lead to grossly incorrect estimations, particularly in densely compacted, non-cohesive soils and in cohesive soils with a semi-firm to firm consistency [189].

9.5   Design and installation of driven steel piles (R 16)

9.5.1   Design

With suitable subsoil, steel piles can be equipped with welded-on steel vanes, thus improving their bearing capacity. Vane piles should only be used in soils without any hindrances, and preferably in non-cohesive soils, and must reach far enough into the loadbearing subsoil. In the case of cohesive layers, the vanes should be positioned underneath these layers, and open driving channels should be closed, e,g, by grouting. Such vanes must be designed and arranged so as not to impede the driving process excessively and without damage to the vanes themselves. The design of the vanes and the elevation at which they are positioned must be carefully adapted to suit the respective soil conditions. Care must be taken that saturated, cohesive soils are displaced but not compacted during driving. In non-cohesive soils, a highly compacted solid packing of soil can develop primarily around the vane area due to driving vibrations, which correspondingly increases the bearing capacity, but at the same time makes the driving process difficult. Therefore, the subsoil must be thoroughly investigated with boreholes followed by very careful exploration with penetrometer tests and soil mechanics laboratory tests before using vane piles in the execution of works. Pile cross-sections suitable for driving and resistant to bending should be used in heavy soil or with long piles. In non-cohesive soil the vanes of vane piles should be at least 2.00 m long so that the required soil arching (plug formation) is achieved in the cells. The clear intervals between cell walls should be less than 30 to 40 cm to guarantee plug formation (see also [7]).

The vanes are placed symmetrically around the pile axis and begin generally just above the end of the pile toe, so that at least an 8 mm weld can be placed between vanes and pile toe. The upper end of the vanes must also be given a correspondingly strong transverse weld. The welds are subsequently extended for a length of 500 mm on both sides of the vanes in the longitudinal direction of the pile. Intermittent weld beads are sufficient for the rest (interrupted weld seam).

The contact surface of the vanes shall be sufficiently wide in consideration of the restraint forces (in general at least 100 mm). Cross-sections and positions of the vanes should promote cell formation. Depending on the soil conditions, the vanes can also be mounted higher up the pile shaft.

9.5.2 Driving

Secure guidance is required when driving piles at a shallow angle. Slow-stroke hammers are basically preferable to rapid-stroke hammers because of their longer force action but also for environmental aspects (noise, vibration). However, rapid-stroke hammers can result in an increase in the bearing capacity in non-cohesive soils because of their "vibration" effect. Energy losses resulting from the inclined position must be taken into account when rating the hammer weight. The free length of the pile shaft projecting beyond the driving guide should be such that the permissible bending stresses in the pile will not be exceeded during installation. The consequences of any jetting assistance must be taken into account. For further details see [78].

9.5.3 Embedment length

The pull-out resistance of shallow raking anchor piles (critical tensile force), or the required embedment length of the piles in the loadbearing subsoil, can be estimated using the guide values for limit skin friction stated in DIN 1054 appendix C. The skin friction is to be referred to the outer, developed pile surface area. These values must be checked by adequate subsoil investigations, e.g. cone penetration tests, with the prerequisite that the piles are not subjected to any notable vibrations. The final pull-out resistance must in any case be specified by a sufficient number of test loads.

Settlement of the ground creates shear forces on the side or skin surfaces of the piles resulting from negative skin friction; in raking piles, the additional deformations and loads in the piles are to be taken into account, together with their effect on the pile connection details.

9.6 Design and loading of driven piles with grouted skin (R 66)

9.6.1 General

A skin grouted pile is a driven steel pile that is driven into the ground under simultaneous grouting with mortar and is suitable for accommodating large tensile and compression forces. Skin grouted piles differ regarding construction and bearing capacity in many ways from anchors in accordance with DIN EN 1537, but have much in common with piles as per DIN EN 14 199,

484

Use and type presume accurate knowledge of the soil conditions and soil properties, above all in the loadbearing foundation area. Non-cohesive soils with comparatively large pore volumes are particularly suitable. Suitability of the subsoil is to be verified carefully, particularly for accommodating a tensile load via skin friction and movement of the piles under permanent load, particularly in cohesive soils. Skin grouted piles can be constructed above and below the groundwater table. Skin grouted piles are especially recommended for anchoring waterfront walls because the inner interlocking of the hardened grout with the subsoil produces good utilisation of the high inner loadbearing capacity of the steel pile. Long-term environmental compatibility of the grout with groundwater/soil must be guaranteed. This is generally the case with cement compounds. Any risk to the grout from aggressive substances in the groundwater or soil must be checked with examinations as per DIN 4030.

9.6.2 Calculation of piles

The bearing capacity of skin grouted piles depends essentially on the following factors: structurally effective grouting length, circumference of pile shoe, type of soil and depth of cover.

The circumference of the pile shoe is to be taken as the circumference of the grouting body. The skin friction surface is thus defined as a product of the pile shoe circumference and structurally effective grouting length. The driving grouting length and anchor angle should be selected so that the structurally effective anchoring length l_w lies in a uniform, bearing stratum as far as possible. This makes it easier to determine the values to be included in the design (in strata with differing bearing capacities, the values of the stratum with the lower bearing capacity have to be taken to avoid the so-called zipper effect).

For skin friction, the values in DIN 1054 appendix C can be used for determining the theoretical critical load at the preliminary design stage. These values will still have to be checked by loading tests. The critical tension force as per R 27, section 9.4, and permissible load as per R 26, section 9.3, taken as basic criteria for a project, must be checked in any case by adequate loading tests. A minimum of two loading tests is specified in DIN 1054 for grouted piles, but for at least 3% of the piles. If the grouting length is the same as the pile length, i.e. grouting takes place along the length l_k, the characteristic pull-out resistance for use in the stability verification is:

$$Q_k' = Q_k \cdot \frac{l_w}{l_k + l_w}$$

where Q_K is the critical tension load determined in the test (see also R 10, fig. R 10-3).

9.6.3 Construction

The cross-section of the pile shoe is adapted to the shape of the pile shaft. In general, the dimensions for the cross-section vary between 450 and 2000 cm^2, for the circumference between 0.80 and 1.60 m.

The spacing of the piles can be selected according to requirements. It must be ensured, however, that this produces a technically sound installation. Therefore, the centre-to-centre spacing should be at least 1.60 m. When anchoring sheet piling in which the anchor piles have been placed directly in the trough, the spacing is many times the double sheet pile width. The space and time sequence of the pile construction is to be coordinated in such a way that the setting of grouting mortar in neighbouring piles is not disturbed.

The bearing capacity of the skin grouted piles is chiefly contingent on proper and expert workmanship. Installation may therefore only be awarded to contractors with experience and who provide a guarantee for diligent work. Particular significance is attached to the grouting. The capacity of the mixing and grouting plant is to be adjusted to the capacity of the driving plant. At the lower end of the shaft, the pile has a projecting, closed, wedge-shaped point of length l_s. Depending on the type of pile, the pile shoe is either permanently welded to the shaft or constructed as a detachable unit (to be left in the soil). It produces a ring-gap in the ground while being driven which is constantly filled with a grout compound under pressure. A steel pipe or plastic hose is fastened to the pile shaft for conveying the grout compound to the pile shoe. Interruptions may not take place during driving so that the grout compound does not set before the pile has been completely constructed.

The grout compound consists of cement, fine sand, water, trass and usually a swelling agent. The composition of the grout compound depends on the respective soil type and on the degree of density of the foundation soil. Grout compounds consisting of hydrated cement only result in greater adhesion and friction forces between steel sections and grout mass, and between grout mass and subsoil; however, their use should be restricted to scarcely permeable strata.

The consumption of grout compound per metre of pile depends on the following factors: theoretical ring-gap cross-area, degree of density and volume of pores in the subsoil, as well as grouting pressure. The ratio of the consumption of grout compound to theoretical ring-gap volume is designated the consumption factor. It can be generally expected to amount to at least 1.2, but can be substantially higher; values of up to 2.0 are not unusual. If the consumption factor is less than 1.2, the piles should be checked.

Loading tests as per DIN EN 1997-1 are to be carried out as per DIN 1054. The loading tests may not be carried out until the grout compound has hardened.

9.7 Construction and testing (R 207)

The construction of the various types of pile must be carried out in accordance with the relevant standards:

- large bored piles: DIN EN 1536
- displacement piles: DIN EN 12 699
- micropiles: DIN 4128 or DIN EN 14 199

If the above standards do not apply, building authority approvals, either general or specific to a project, are normally required.

9.8 Anchoring with piles of small diameter (R 208)

As a rule, the bearing capacity (characteristic value) depends on the outer bearing capacity, i.e. according to the in situ subsoil. Depending on diameter, this lies between 300 and 900 kN; however, higher load-bearing capacities are also possible. Loading tests are usually required in order to specify the outer loadbearing capacity.

9.9 Connecting anchor piles to reinforced concrete and steel structures

See R 59, section 8.4.3.

9.10 Transmission of horizontal loads via pile bents, diaphragm walls, frames and large bored piles (R 209)

9.10.1 Preliminary remarks

The complete solution for anchoring horizontal loads to land-side pile bents is contained in section 11, which can also be used for all intermediate stages through to single pile bents, both with and also without upper screening plate, and with intermediate tension element, e.g. tie rod.

In special cases, however, simpler solutions are possible, as explained in section 9.10.2.

In addition, under specific conditions, anchoring to vertical structures can also be advisable and economical. As examples of this, diaphragm walls and large bored piles are described in section 9.10.3.

9.10.2 Pile bent anchorages for special cases

The relatively large compression pile loads are frequently transmitted by means of in situ driven piles for economic reasons. If these are typical end-bearing piles, there will be no resulting action on the waterfront structure if the embedment depth reaches to a straight line rising at an angle of 1 : 2 from the shear force zero point of the waterfront structure.

The influence on the waterfront structure in this case is negligible. This principle applies both to single pile bents as per fig. R 209-1 and to several rows of compression piles as per fig. R 209-2.

If the pile loads are carried by skin friction and end bearing, the pile forces create active earth pressures, i.e. actions on the waterfront structure which are then to be multiplied by the partial safety factors as usual. The magnitude of the actions particularly depends on the properties of the subsoil and the pile angle. It is to be determined according to DIN 4085.

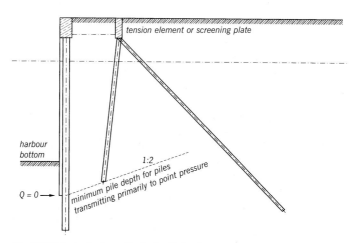

Fig. R 209-1. Anchoring with single pile bents

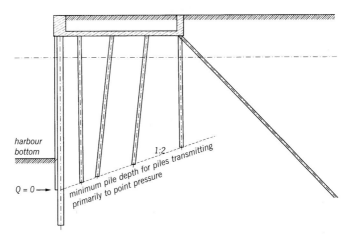

Fig. R 209-2. Pile bent anchoring with several pressure piles

If necessary, the effects of the peak pressure on active earth pressure as per DIN 4085 are to be taken into account.

As regards designing compression piles, it should be noted that – depending on construction method and point in time of constructing the pressure piles with respect to erecting the waterfront structure – deflection of the waterfront structure creates moments in the compression piles depending on their distance from the waterfront structure. On the other hand, an improvement in the soil friction angle behind the waterfront structure resulting from the compaction effect of driving the compression piles can also have a positive effect on the whole waterfront structure.

9.10.3 Special anchorages

In general, the most favourable way of accommodating anchor forces in a waterfront structure will be to transfer the anchor forces to the anchoring elements by the shortest possible route. With favourable soil conditions, this is by way of horizontal anchoring to anchor plates, on pile bents in lower-lying bearing soil layers, or direct anchoring to the waterfront structure by means of raking piles.

But when the soil conditions for driving the waterfront structure and pile bents are unfavourable, in the form of hindrances or rocky formations in the subsoil, special structures may be necessary.

The diaphragm wall technique is representative for such structures.

In principle, transverse diaphragm wall plates can be used to stabilise a diaphragm wall waterfront structure without additional anchoring.

Examples of such structures are known. Attention is drawn to the fact that there are not yet any generally accepted calculation procedures for such solutions. In some cases, the land-side diaphragm wall plate has been replaced by large bored piles.

10 Waterfront structures, quays and superstructures of concrete

10.1 Design principles for waterfront structures, quays and superstructures (R 17)

10.1.1 General principles

The aspects of durability and robustness are particularly important when designing waterfront structures of concrete, reinforced concrete or prestressed concrete. The intended service life must be taken into account, which for harbour engineering projects can be shorter than for general structures or waterways. Waterfront structures are exposed to attack by changing water levels, water and soil aggressive to concrete, ice loads, ship impacts (berthing loads or accidents), chemical influences due to the cargo stored and handled at the port, etc. It is therefore not sufficient to design the concrete components of waterfront structures simply according to the structural requirements.

Also relevant are requirements for simple construction without complicated formwork, and for the simple integration of sheet pile walls, pile foundations and similar components. Such aspects lead to structural and constructional measures that may exceed the minimum requirements of DIN 1045. They must be agreed between the designer, the checking structural engineer and the building authority responsible.

The recommendations given here take account of the loads common in harbour engineering and their more severe consequences for the construction and maintenance of components in contact with water.

10.1.2 Nosing, edge protection

Concrete walls have a 5×5 cm chamfer at the upper edge of the wall, or are correspondingly rounded off, or are reinforced on the water side with steel angles when cargo is handled above them; R 94, section 8.4.6, must be observed where applicable. A specially installed nosing for the protection of the wall and to safeguard line handlers from slipping, must be constructed in such a way that the water can easily drain away. At quay walls with a sheet pile front and reinforced concrete superstructure, the reinforced concrete cross-section extends about 15 cm beyond the front face of the sheet piling. The water-side lower edge of the reinforced concrete superstructure should, however, be positioned at least 1 m above mean tidal high water or the mean water level to avoid any severe corrosion at this point. The bevel underneath is constructed at a slop ofabout 2 : 1 so that vessels and cargo-handling equipment cannot become caught underneath. The bevel is fitted with a folded steel plate protection which is connected flush to the face of both the sheet piling and the concrete.

10.1.3 Facing

Facing to the concrete can be omitted if good-quality concrete is used. If a facing is required as protection against unusual mechanical or chemical wear, or for appearance purposes, the use of basalt, granite or hard-burned brick is recommended. Ashlar masonry stones or slabs along the front upper edge of a wall must be secured against displacement and lifting.

10.1.4 Precast elements

These are useful for solving construction problems underwater or in the splashing zone or for avoiding reductions in quality in these areas. The unavoidable construction joints in this case, especially at points in the structure where the stresses are high, must therefore be carefully designed, and constant supervision during construction is essential.

10.2 Design and construction of reinforced concrete waterfront structures (R 72)

10.2.1 Preliminary Remarks

The recommendations regarding structural calculations given in this section are based on the safety concept according to DIN 1055-100.

To guarantee sufficient reliability, numerical analyses of the limit states of bearing capacity and serviceability must be carried out and the structure designed taking into account constructional recommendations and the parameters for ensuring durability.

In general, DIN 1045-1, DIN 1045-2 and DIN 1045-3 apply, together with the regulations listed in those standards.

10.2.2 Concrete

The strength and deformation parameters are given in DIN 1045-1, section 9.1. The concrete properties must be specified according to DIN EN 206-1 while taking DIN 1045-2 into account.

When selecting the type of concrete, the appropriate exposure classes (i.e. degree of exposure) are relevant, taking into account the environmental conditions (DIN 1045-1, section 6.2). The minimum requirements regarding concrete grade, concrete cover and the analyses to limit crack widths are regulated by means of the exposure classes.

The exposure class for chemical attack due to natural soils and groundwater should be assessed according to DIN EN 206-1 table 2 in conjunction with DIN 4030.

A low-shrinkage concrete mix with low heat of hydration is important where significant expansion of the concrete components (see section 10.2.4) is expected or large cross-sections are in use.

Important factors are a concrete as dense as possible, intensive curing and adequate concrete cover to the steel reinforcement.

The concrete cover should be larger than that given in DIN 1045, i.e. at least $c_{min} = 50$ mm and with a nominal dimension of $c_{min} = 60$ mm. See section 10.2.5 with regard to limiting the crack widths under serviceability loads.

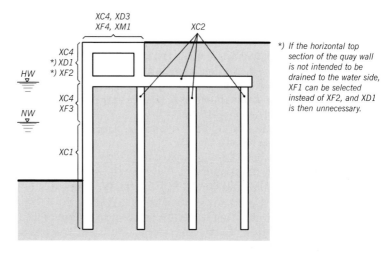

Fig. 72-1. Example of exposure classes for a quay wall in tidal freshwater

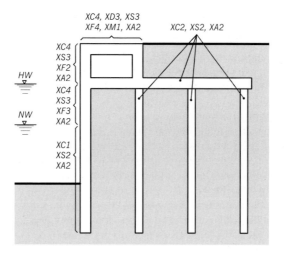

Fig. 72-2. Example of exposure classes for a quay wall in seawater

The strength class, minimum cement content, maximum water/cement (w/c) ratio and other requirements selected on the basis of the degree of exposure are given in ZTV-W, LB 215 [118] and in the case of large solid components (smallest dimension ≥ 80 cm) according to the DAfStb directive covering such components.

Components not fully covered by the loads given in DIN 1045 under the respective exposure class require special regulations to be included in the specification.

Figs. R 72-1 and R 72-2 show the exposure classes for typical waterfront structure components in both seawater and freshwater environments.

10.2.3 Construction joints

Construction joints should be avoided wherever possible. Unavoidable construction joints are to be specified prior to starting the concreting works and must be constructed in such a way that all the stresses and strains can be accommodated.

Construction joints may not impair the durability of a component. This means that careful planning and preparation of construction joints (see also DIN 1045-3, section 8.2) and, if necessary, subsequent treatment (e.g. grouting) are essential. The local conditions must be suitable for such work.

10.2.4 Long structures

Long linear structures can be built with or without movement joints. The decision regarding positions and number of movement joints is to be taken with the aim of optimising the economic efficiency, durability and robustness of the structure. In doing so, the influences of the subsoil, the substructure and the constructional details are to be taken into account. If waterfront structures are built without joints, then in addition to the actions due to the loads, additional actions due to restraint (shrinkage, creep, settlement, temperature) must be included in the calculations. Full restraint is to be presumed unless more detailed calculations show otherwise. The analysis of crack widths due to loads and restraint is then particularly important in order to avoid damage caused by excessive cracking (see also section 10.2.5).

See section 10.2.3 for information on construction joints.

The joints are to be designed in such a way that the changes in length of the individual units of the structure are not hindered.

Joggle joints help to provide mutual support for the sections of the structure in the horizontal direction. Dowels can also be advantageous in certain circumstances. The horizontal joggle is positioned in the piled platform in the case of piled quay walls. Joints must be designed to prevent any backfill from leaking out.

10.2.5   Limiting the width of cracks

Owing to the danger of more severe corrosion, reinforced concrete waterfront structures must be constructed in such a way as to prevent the occurrence of any cracks that might have a detrimental effect on the durability of the structure. If concrete technology measures are not sufficient on their own, an analysis of the crack widths is necessary, taking into account the local conditions and influences (DIN 1045-1, section 11.2).

The crack width should be chosen so that a self-healing effect can take place. As a rule, this can be assumed for theoretical crack widths with $w_k \leq 0.25$ mm.

Where there is an increased risk of corrosion, e.g. in the tropics and with prestressing steel, higher demands are to be placed on limiting crack widths. All cracks which do not close of their own accord must be permanently sealed by injecting a suitable compound according to ZTV-ING [216].

The restraint stresses can be reduced by choosing a sensible concreting sequence in which the inherent deformations of subsequent concrete pours as a result of dissipating heat of hydration and shrinkage are not excessively hindered by concrete components constructed earlier and which have already cooled significantly. This can be achieved with larger components, e.g. a pier structure, if beams and slabs are constructed monolithically in one pour.

Limiting the crack widths in components with large cross-sectional dimensions can be achieved by including minimum reinforcement according to the Supplementary Technical Contractual Conditions – Hydraulic Engineering (ZTV-W) – (works section 215), part 1, section 11.2 [118]. Rostasy [230] contains further principles.

10.3   Formwork in marine environments (R 169)

10.3.1   Principles for the design of the formwork

(1) In the zone influenced by the tide and/or wave action, formwork should be avoided wherever possible, e.g. by using prefabricated elements (fig. R 169-1, water side), by positioning the concrete construction at a high level, or similar measures.

(2) Concreting work in the zone influenced by the tide and/or waves should take place – as far as possible – in calm weather periods.

(3) Areas difficult to reach, such as the underside of pier slabs, should, if possible, be placed in forms with the use of permanent formwork, such as concrete slabs, corrugated steel sheets, or similar.

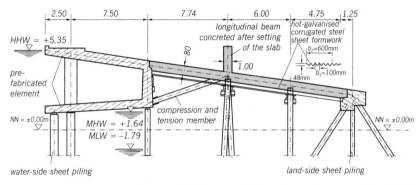

Fig. R 169-1. Execution example of a quay wall in marine environment, with prefabricated reinforced concrete element and rear corrugated steel sheet formwork

10.3.2 Construction of the formwork

(1) Formwork that is to be used more than once should be sturdy, easy to repair and readily transportable.

Prefabricated timber formwork or large steel formwork elements have proven worthwhile in practice; these can be installed and moved rapidly and in large units, so that their use in an endangered area is limited to comparatively brief periods. Reference is also made here to portable formwork.

(2) Formwork in marine environments should largely be able to react elastically if struck by wave action, which is the case, for example, with properly designed corrugated steel sheet bottom formwork at suitable elevation (fig. R 169-1).

The corrugated steel sheet panels must be secured against lifting and equipped for the connection to the concrete, as permanent formwork, e.g. with galvanised wires embedded in the concrete, or with other anchorages.

Owing to the joints between the individual panels, corrugated steel sheet formwork cannot be classed as part of the corrosion protection for the slab reinforcement.

10.4 Design of reinforced concrete roadway slabs on piers (R 76)

10.4.1 Refer to R 5, section 5.5 for the actions on piers used as roadways. If it is anticipated that cargo will also be stacked on piers, the designer is recommended to assume a uniformly distributed live load of at least 20 kN/m^2 at the most adverse position.

10.4.2 Contrary to DIN 1045-1, section 13.3.1, a minimum thickness of 20 cm is recommended for pier slabs.

495

10.5 Box caissons as waterfront structures in seaports (R 79)

10.5.1 General

Box caissons often offer economical solutions for retaining heavily loaded, high vertical banks in areas with good loadbearing soil, and especially for construction rejecting into the open harbour water.

Box caissons consist of rows of floating reinforced concrete sections that are designed for floating stability, usually with additional ballast. After being launched and positioned over a properly prepared bearing area, they are then filled and backfilled with sand, stones or other suitable material. To reduce the edge pressures, the water-side cells are often left unfilled. When they are in place, the top is barely above the lowest working water level. A reinforced concrete superstructure is placed on top of the caisson to give additional rigidity to the structure and to form the coping to the front wall. Any irregular settlement and horizontal displacement which occur when the caissons are set in place and during backfilling can be compensated for by suitable shaping of the reinforced concrete top.

The front wall of the caissons must be resistant to mechanical and chemical wear.

10.5.2 Calculation

Apart from the verification of stability in the final condition, structural conditions such as floating stability of the caissons are to be investigated for the conditions during construction, during launching, when they are set in place and during backfilling. In addition, safety against bottom erosion is to be verified for the final condition.

In contrast to DIN 1054, the bottom joint may not gape under any combination of characteristic loads.

Stability of a box caisson must also be verified for the longitudinal direction. This must take account of centre support as well as support at the ends only. In these extreme case investigations, the partial safety factor γ, which applies to reinforced concrete in the normal loading class, may be divided by 1.3.

10.5.3 Safety against sliding

An especially careful investigation is required to determine if silt will be deposited on the foundation surface in the period between the completion of the foundation bed and setting the box caissons in place. If this is possible, it must be verified that a sufficient safety factor remains against sliding of the caissons on the in situ foundation bottom. The need for this precaution is similarly valid for the surface between the existing subsoil and the fill placed over a dredged area.

Safety against sliding can be improved at reasonable cost by a roughening the concrete bottom. The degree of roughness must be matched by the

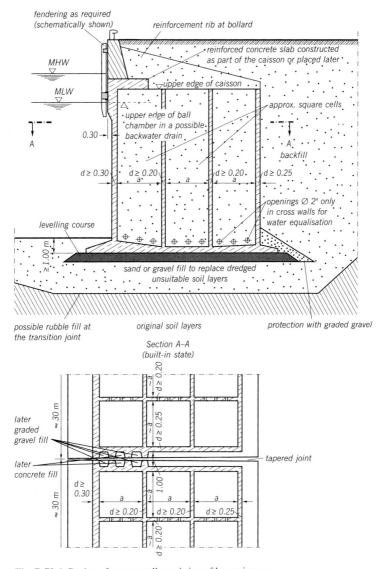

Fig. R 79-1. Design of a quay wall consisting of box caissons

mean grain size of the foundation joint material. If the bottom surface of the concrete is correspondingly roughened, the angle of friction between the concrete and the foundation surface is to be assumed to be equal to the angle of internal friction φ' of the foundation material, but for a smooth bottom surface, only $^2/_3$ φ' of that of the foundation material. Safety against sliding can also be increased founding at a greater depth. The least favourable combination of water pressures is to be presumed at the bottom and sides of the caissons when analysing sliding. These can be caused by hydraulic backfilling or by tidal changes, precipitation, etc. Hawser pull must also be taken into consideration.

10.5.4 Construction

The joint between two adjoining box caissons must be so designed that any unexpected uneven settlement of the caissons when being set in place, during filling and backfilling, will be absorbed without damage. On the other hand, in the final condition it must be sufficiently tight to guarantee no washout of the backfill.

A design with a tongue and groove joint extending from top to bottom may be used only when the movements of adjoining caissons is expected to be minimal, even when there is a satisfactory solution to the sealing. A solution according to fig. R 79-1 has proven practical. Here, four vertical reinforced concrete ridges are arranged on each of the side walls of the caisson such that they face each other on both sides of the joint and form three chambers when the caissons are in place. As soon as the adjoining caisson is installed, the two outer chambers are filled with suitably graded gravel to form a seal. The centre chamber is flushed clean after backfilling of the caissons when the majority of the settlement has taken place, and is carefully filled with underwater concrete or concrete in sacks.

The danger of washout under the foundation slab exists when there is a great difference in water level between the front and rear of the caissons. In such cases the layers of the bedding must exhibit filter stability with respect to each other and to the subsoil. To reduce high water pressure differences, backwater drains can be successfully used as per R 32, section 4.5.

When there is a risk of scour from currents and wave action, adequate scour protection is to be provided as per R 83, section 7.6.

10.5.5 Construction

The box caissons must be founded on a well-levelled foundation of stones, gravel or sand. When weak soil layers exist in the foundation area, these must be removed and replaced with sand or gravel before installing the caisson (R 109, section 7.9).

ISCHEBECK ®
TITAN

TITAN tension pile
for permanent applications
(tie back unstressed) to EN 14199

- full grout cover
- not prestressed
- pile head cast into old wall

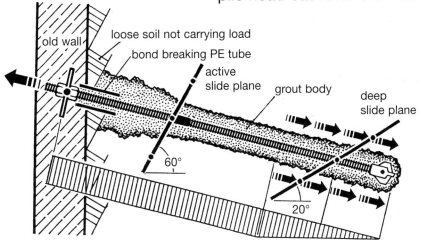

old wall
loose soil not carrying load
bond breaking PE tube
active slide plane
grout body
deep slide plane
60°
20°

FRIEDR. ISCHEBECK GMBH
P.O. Box 1341 · D-58242 ENNEPETAL · TEL. +49 2333 8305 0 · FAX +49 2333 8305 55
E-MAIL: export@ischebeck.de · INTERNET: http://www.ischebeck.de

Stahlbau – Kalender *Hrsg.: U. Kuhlmann*

Jährliche Schwerpunkte:

Schlanke Tragwerke

DASt-Richtlinie 019; Schweißen
im Stahlbau; Schlanke Stab-
tragwerke; Träger mit profilierten
Stegen; Maste und Türme;
Gerüstbau; Radioteleskope;
Membrantragwerke

Stahlbau-Kalender 2004

2004. IX, 802 S. 589 Abb.
167 Tab. Geb.
€ 129,-* / sFr 204,-
ISBN 3-433-01703-4

Verbindungen

Verbundbau-Kommentar DIN
18800; Mechanische
Verbundmittel; Betondübel;
Steifenlose Anschlüsse;
Klebeverbindungen; Zugstangen

Stahlbau-Kalender 2005

2005. 990 S. 885 Abb.
146 Tab. Geb.
€ 129,-* / sFr 204,-
Fortsetzungspreis:
€ 109,-*/ sFr 172,-
ISBN 3-433-01721-2

Dauerhaftigkeit von Stahlbauten

Stahlbaunormen, Anwendung von
DIN 18800-7, DASt-Richtlinie 009,
Ermüdung, Bewertung bestehen-
der Brücken, Behälter, Korrosions-
schutz, Zerstörungsfreie Prüfung,
Stahlwasserbau

Stahlbau-Kalender 2006

2006. Ca. 700 S,
ca. 450 Abb, Geb.
Ca. € 129,-* / sFr 204,-
Fortsetzungspreis:
Ca. € 109,-*/ sFr 172,-
ISBN 3-433-01821-9

Erscheint April 2006

Ernst & Sohn
Verlag für Architektur und
technische Wissenschaften GmbH & Co. KG

www.ernst-und-sohn.de

Für Bestellungen und Kundenservice:
Verlag Wiley-VCH
Boschstraße 12
69469 Weinheim
Telefon: +49(0) 6201 / 606-400
Telefax: +49(0) 6201 / 606-184
E-Mail: service@wiley-vch.de

10.6 Pneumatic caissons as waterfront structures in seaports (R 87)

10.6.1 General

Pneumatic caissons may offer an advantageous solution for retaining high banks when their installation can be accomplished from the shore. The pneumatic caissons for the quay wall are then first sunk making use of accessibility from the existing terrain; dredging work in the harbour basin is carried out subsequently.

Pneumatic caissons can also be constructed as box caissons when an adequate bearing foundation does not exist or cannot be provided in the sinking area, or when the levelling of the foundation bed creates special difficulties, as with a rocky bottom. The construction fundamentals given in R 79, section 10.5.1, are then equally applicable to pneumatic caissons.

10.6.2 Analysis

R 79, section 10.5.2, applies here. Also important are the normal bending and shear force analyses for the caisson sinking conditions in the vertical direction as a result of uneven surcharge on the caisson cutting edges, and analyses for the horizontal bending and shear forces due to unequal active earth pressure.

In view of the location and preparation for the foundation and the firm keying of the caisson edges and the concrete plug in the working chamber with the foundation layer, the caisson may be assumed to have a normal level foundation. Hence, in contrast to R 79, section 10.5.2, paragraph 2, the bottom joint may gape, but the minimum distance of the resultant from the front face of the caisson due to the actions of characteristic loads shall not be less than $b/4$.

The danger of washout in front of and under the foundation bed is to be investigated for large water pressure differences. If necessary, special safeguards are to be provided to prevent washout, such as soil stabilisation from the working chamber or similar. However, it may be preferable – for economic reasons – to deepen or widen the foundation bottom.

In the final condition, a special verification of stresses resulting from uneven bearing for the longitudinal direction is not necessary for pneumatic caissons.

With especially large dimensions, however, it is recommended that a verification of the stresses be provided for a bottom pressure distribution according to BOUSSINESQ.

10.6.3 Safety against sliding

R 79, section 10.5.3, is valid here.

10.6.4 Design

R 79, section 10.5.4, paragraphs 1 to 3, apply here. Fig. R 87-1 shows a method which has been used successfully in practice for closing the joint between sections of pneumatic caissons. Sheet pile interlocks are embedded in the side walls during construction of the caissons. After the caissons are in their final position, with the customary 40 to 50 cm space between them, flexible closure piles are driven into the two pairs of interlocks, thus closing off the space between the caissons. The enclosure formed by the two closure piles is cleaned out and filled with underwater concrete if the subsoil is firm. In poor subsoil the space is filled with crushed rocks and can be grouted later. The front closure pile may lie flush with the face of the caisson. However, it may also be set back in order to form a shallow recess for installing an access ladder or the like.

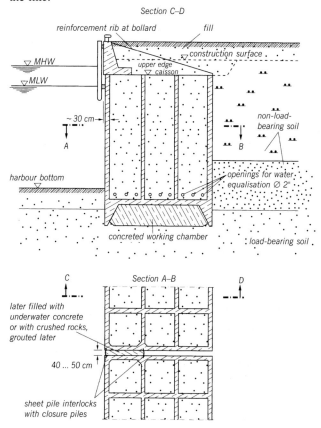

Fig. R 87-1. Design of a pneumatic caisson wall with subsequent harbour dredging

If large fluctuations in water level occur, the elevation of the bottom joint must be such that an adequate safety factor against underscouring is ensured, or the differences in water pressure can be compensated for by using suitable backfill and drainage devices.

10.6.5 Construction

Pneumatic caissons installed from the shore are sunk from the ground on which they were previously constructed. As a rule, the soil in the working chamber is excavated almost exclusively under compressed air, or jetted and pumped out. If the soil at the predicted foundation depth proves to be too weak, the caisson must be sunk to a corresponding deeper level. When the required foundation depth has been reached, the bed is adequately levelled and the working chamber concreted under compressed air.

Pneumatic caissons floated into position must first be set down on the existing or deepened bed. In general, a rough levelling of the bed suffices, as the narrow caisson cutting edges easily penetrate into the soil so that smaller irregularities in the foundation surface are of no importance. Subsequently, the caissons are sunk and concreted in the manner already described.

10.6.6 Frictional resistance during sinking

The frictional resistance depends on the characteristics of the subsoil, on the position of the caisson with respect to the groundwater and on the construction of the caisson. It is influenced by:

(1) Type of soil, density and strength of existing strata
    (non-cohesive and cohesive soils)
(2) Groundwater level
(3) Embedded depth of caisson
(4) Plan form and size of caisson
(5) Geometry of cutting edge and outer wall surfaces

The determination of the necessary "sinking overweight" for any given sinking condition is less a matter of exact calculation than of experience. It is sufficient in general if the "overweight" (total of all vertical forces without taking friction into account) is adequate to overcome a skin friction of 20 kN/m^2 on the embedded caisson surface. At a lower overweight (modern reinforced concrete caissons), the consideration of additional measures to reduce friction, like the use of lubricants, e.g. bentonite, is recommended.

10.7 Design and dimensioning of quay walls in block construction (R 123)

10.7.1 Basic remarks on design and construction

Waterfront structures in block construction can be built successfully only if loadbearing soil is present below the base of the foundation. If necessary, the bearing capacity of the in situ soil should be improved (e.g. by compaction) or the inadequate soil replaced.

The dimensions and the weight of the individual blocks must be determined in consideration of the construction materials available, the fabrication and transportation options, the capacity of the plant for placing the blocks, and the conditions to be expected with regard to the site location, including wind, weather and wave action during construction and under operating conditions. In economic terms it is better to use fewer, larger blocks because the time taken by individual activities such as setting up forms, striking, transporting and placing does not depend on the size of the elements. The buoyancy can be used to relieve the means of transport during transport to the building site, provided the blocks are transported immersed.

The buoyancy during placing of the blocks is often used to reduce the effective dead load and enable a corresponding longer reach of the crane used to place the blocks. However, in this case the blocks must be so large, or rather, heavy that they can withstand wave action. The plant required is chosen according to this.

Blocks of 600 to 800 kN effective dead load are frequently chosen, especially when using a floating crane.

The blocks have to be shaped and placed in such a way that they will not be damaged during installation. If the blocks are simply stacked, which is recommended in settlement-sensitive subsoils, large joint widths can be avoided only at great expense. These joint widths can, however, be accepted if suitable backfill material is used. Generally, the optimum economic solution should be sought regarding allowable joint width and backfill material. The blocks can be connected with tongue and groove joints or with I-shaped blocks. If continuous vertical joints are to be avoided, this is achieved, for example, by placing the blocks at an inclination of 10–20° from the vertical. The support can consist of, for example, blocks placed horizontally, a sunk box caisson or the like. Wedge-shaped blocks form the transition. The latter can also be utilised if it becomes necessary to correct the inclination. The sloping position of the blocks facilitates a minimum joint width between the individual blocks. However, this does increase the number of block types. All blocks for this type of construction are made with tongue and groove interlocks to key their adjacent surfaces together. The tongue projection lies on the exposed side of the blocks already in place, and the succeeding blocks are guided interposition so that their groove engages with this tongue, along which they then slide downward into place.

A bed of rubble and hard ballast at least 1.0 m thick is placed between the loadbearing subsoil and the block wall (Fig. R 123-1). The surface must be carefully graded and levelled – usually with special plant and with the aid of divers. In sediment-laden waters, this bed must be thoroughly cleaned before the blocks are placed, so that the foundation joint does not become a failure plane. This is especially important where the blocks are simply stacked.

In order to prevent the bed from sinking into the subsoil, especially in fine-grained, non-cohesive soil, its pore volume must be filled with suitably graded gravel. In addition, a graded gravel filter can be placed between the foundation bed and the subsoil. If the foundation soil is very fine-grained but not cohesive, a separating layer of non-woven material should be placed as a stabilisation measure between the foundation soil and the graded gravel filter.

Depending on the plant used, block construction may be particularly successful in areas with severe wave action and in countries in which there is a scarcity of skilled workers. Besides the use of suitable heavy plant, however, this method chiefly requires extensive diving work in order to ensure and check the required careful execution of both the foundation bed and the block placing and backfill work. For further information regarding construction please refer to [255].

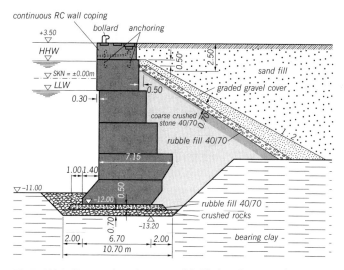

Fig. R 123-1. Cross-section of a quay wall in blockwork construction

503

10.7.2   Nature of forces applied

10.7.2.1   Active and passive earth pressures

Active earth pressure may be applied because we can presume the wall movements necessary to mobilise it. With the generally minimal foundation depth of block walls, the passive earth pressure cannot be taken into account.

10.7.2.2   Water pressure difference

If the joints between the individual blocks have good permeability and if rapid water table equalisation is ensured by the proper choice of backfill material (fig. R 123-1), the water pressure difference on the quay wall can be assumed as only half the height of the highest waves to be expected in the harbour basin, at the most unfavourable water level as per R 19, section 4.2. Otherwise, the water pressure difference as per R 19 is to be added to half the wave height. In cases of doubt, reliably functioning backwater drainage systems can be installed, even if wave action is present. On the other hand, experience has demonstrated that there is no method for sealing the block joints properly. A permanently effective filter layer, which will certainly prevent washout, is to be installed behind the quay wall or between a backfill of coarse material and a subsequent fill of sand and the like (fig. R 123-1).

10.7.2.3   Stresses due to waves

When waterfront structures of block construction must be built in areas in which high waves can occur, special stability investigations are required. In doubtful cases a special investigation must be carried out by means of model tests to determine whether breaking waves can occur. If this is the case, the risks that must be anticipated with regard to the stability and service life of a block wall are so great that this method of construction can no longer be recommended. The relation between the water depth d at the wall to the wave height H can be used as a criterion for evaluating whether breaking or only reflected waves occur. Where $d \geq 1.5 \cdot H$, it can generally be assumed that only reflected waves will occur; see also R 135, section 5.7.2 and R 136, section 5.6.

Wave pressures act not only on the face of the wall but also propagate in the joints between the individual blocks. The water pressure in the joints can momentarily reduce the effective block weight to a greater extent than the buoyancy and consequently reduce the friction between blocks to the point where the stability of the wall is endangered. At the moment the wave recedes, the pressure drop in the narrow joints, which is also influenced by the groundwater, takes place more slowly than along the outer surface of the quay wall, so that higher water pressure occurs in the joints than that attributable to the water level in front of the wall. At the same time, however, the active earth pressure and the water pressure difference from behind remain fully effective. This condition, too, can be relevant to stability.

10.7.2.4 Hawser pull, vessel impact and crane loads
The relevant recommendations such as R 12, section 5.12, R 38, section 5.2, R 84, section 5.14, and R 128, section 13.3, are applicable here.

10.7.3 Analyses, design and configuration

10.7.3.1 Base of wall, soil pressure, stability
The block wall cross-section must be designed such that the soil pressure at the foundation joint due to dead load is distributed as uniformly as possible. This can normally be achieved without any difficulty, by a suitable design of the base with a spur on the water side projecting beyond the face of the wall, and by placing a protruding "knapsack" on the land side (fig. R 123-1).

In order to prevent the formation of hollow space under blocks which project beyond the back of the wall, such blocks must be trimmed at an angle steeper than the angle of internal friction of the backfill material (fig. R 123-1).

The soil pressures are also to be checked for all important stages of construction. The quay wall must be backfilled at about the same time as the blocks are placed to counteract landward overturning movements or excessive pressures at the heel of the foundation bed (fig. R 123-2). Besides the allowable soil pressures, the safety against sliding, foundation failure and slope failure must be verified.

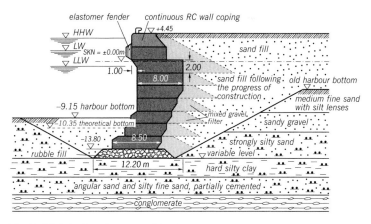

Fig. R 123-2. Design of a quay wall in blockwork construction in an earthquake region

Please refer to R 79, section 10.5.3, with respect to and sliding.

Possible changes in the harbour bottom due to scouring, but above all due to foreseeable deepening, are to be taken into account. In the course of later port operations, the condition of the harbour bottom in front of the wall must be checked at regular intervals; if necessary, suitable protective measures are to be undertaken immediately.

In anticipation of a known tendency of the wall to overturn towards the water when in use, the quay wall must be initially designed with a slight batter. The crane track gauge can change later due to unavoidable wall movements and should always be designed to the adjustable.

10.7.3.2 Horizontal joints in the block wall

Safety against sliding and the position of the resultant of the forces acting must also be checked in the horizontal joints of the block wall for all essential construction stages and for the final condition. In contrast to the foundation joint, theoretical tension in the joints may be allowed here up to the centroidal axis, if all adverse forces are applied simultaneously.

10.7.3.3 Reinforced concrete capping beam

The reinforced concrete beam, which is to be cast in situ and installed on the top of every block wall, serves to compensate for inaccuracies in block placement, to distribute concentrated horizontal and vertical loads and to compensate for local variations in soil pressures and support conditions at the base, as well as construction inaccuracies. Since unequal settlement of the block wall will occur, concrete should not be placed before settlement has essentially ceased. In order to accelerate the settling process, a temporary surcharge on the wall, e.g. additional layers of concrete blocks, is practicable. In doing so, the settlement behaviour must be measured constantly.

Normally, relief due to excavations in cohesive soil does not need to be considered in the structural analyses because it is replaced by the growing load of the wall.

In calculating the internal forces in the capping beam due to vessel impact, hawser pull and side thrust of crane wheels, it can be generally assumed that the capping beam is rigid in comparison to the block wall supporting it. This assumption lies generally on the safe side.

In the calculation for the vertical forces on the capping beam, primarily for the crane wheel loads, the bedding module method is normally used. Where major non-uniform settlement or subsidence of the block wall is to be expected, the internal forces in the capping beam, however, are to be determined by means of comparison investigations with various support conditions, "riding" on the supported middle section or on the end sections. In this case the dead load of the capping beam is also to be taken into account. An analysis of crack widths to R 72, section 10.2.5, is required. If necessary, joints between blocks are to be specified.

The capping beams are joggled at the block joints only for transmitting horizontal forces. Because the manner in which the block wall will eventually settle is not clearly predictable, keying for the vertical forces is not recommended.

Protection against abrupt changes in alignment due to settlement must be provided for rail supports at block joints. For this purpose, short bridges are inserted across the block joints, permitting the crane rail to be laid across the bridge without a joint in the rail.

Mutual, effective keying is required for the transmission of horizontal forces between capping beam and block wall. An anchorage can be provided instead of keying.

10.8 Construction and design of quay walls using the open caisson method (R 147)

10.8.1 General

Open caissons are used in seaports for waterfront structures and berthing heads, but also as foundation elements for other types of structure, although this is very rare. Like the pneumatic caissons as per R 87, section 10.6, they can be constructed on terrain lying on the site of the sinking area over the water table, or in a driven or floating spindle frame, or can be floated into position and then sunk as finished boxes using a jack-up platform or floating elements. Open sinking requires lower labour and installation costs than pneumatic caissons, and can be used in greater depths of water. However, the same degree of positioning accuracy cannot be achieved. In addition, this solution does not produce equally reliable loadbearing conditions in the foundation bottom. Any hindrances encountered during the sinking process can only be bypassed or removed with considerable difficulties. Founding on sloping rock surfaces always requires additional measures.

The construction principles for waterfront structures outlined for box caissons in R 79, section 10.5.1, also apply to open caissons similarly. For other aspects, please refer to section 3.5 in [7].

10.8.2 Analysis

R 79, sections 10.5.2 and 10.5.3, and R 87, section 10.6.2, are to be taken into account.

10.8.3 Design

Open caissons may be either rectangular or circular on plan, depending on operational and constructional considerations.

Because of the inherent funnel shaped excavations, open caissons with a rectangular layout will not assume a position on their cutting edges which is as uniform as can be expected when circular caissons are used. This results in an increased risk of deviation from the design position. If a rectangular shape is necessary, it should therefore be a compact one.

Cross section C–D

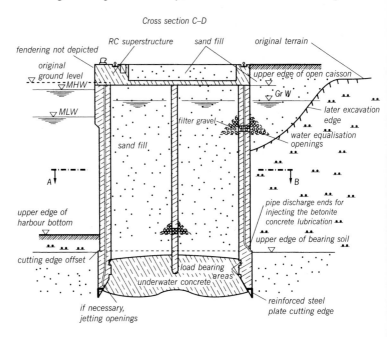

Cross-section A–B

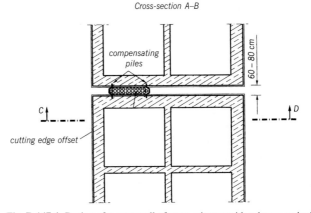

Fig. R 147-1. Design of a quay wall of open caissons with subsequent harbour dredging

Since the excavation and sinking procedures are difficult to check, and because an open caisson can be only lightly ballasted, the caisson should have thick walls, so as to make the dead load of the caisson definitely greater than the expected wall friction force, taking uplift into account. A rigid steel cutting edge is fitted to the base of the outer walls. Alternatively, a cutting edge of high-strength concrete (at least grade C 80/95) or steel fibre-reinforced concrete is conceivable. At the bottom of the cutting edge, jetting pipes discharging towards the caisson interior can be of help in loosening non-cohesive excavation soil (fig. R 147-1, section C-D, showing jets on the water side).

The bottom edges of partition walls must terminate at least 0.5 m above the bottom surface of the caisson cutting edges in order to avoid loads being transmitted into the subsoil.

The outer walls and partition walls should have reliable contact surfaces which can be easily cleaned after sinking for the reliable transmission of loads into the underwater concrete bottom.

When constructing foundations for open caissons, the unavoidable loosening of the soil in the foundation bed and adjacent to the caisson surface encourages significant settlement and an incline in the completed structure. This must be allowed for in the dimensioning and structural design, as well as during the course of construction.

The 2nd and 3rd paragraphs of R 79, section 10.5.4, apply here in their entirety. The solution for the joints shown in R 87, section 10.6, fig. R 87-1, is recommended. However, filter gravel is preferable to inflexible material for sealing the space between adjacent caissons because it can accept the movement caused by settlement without being damaged.

A space of 40 to 50 cm between adjacent caissons, as given for pneumatic caissons in R 87, section 10.6.4, is adequate – considering the excavation method – for open caissons only when the actual embedment is not deep and hindrances are not expected – including strata of embedded, solid, cohesive soil. If sinking is difficult, the clearance should be increased to 60–80 cm. Compensating piles of adequate width, or loop-like arrangements of more deformable pile-chains may be used as closures.

10.8.4   Advice on construction

If construction takes place in the dry, the loadbearing capacity of the soil in the erection area must be carefully checked and observed because soil under the cutting edges that yields excessively or too unevenly may lead to problems. The latter condition may cause the cutting edge to break. The soil within the caisson is removed by grab buckets or pumps, while the water level inside the caisson must always be kept no lower than that of the water on the outside in order to prevent ground heave.

When a row of caissons is being sunk, a sequence of 1, 3, 5 ... 2, 4, 6 may be advisable because then the same active earth pressure will act on both faces of each caisson.

The sinking of a caisson can be greatly facilitated by lubricating its outer surface, above the offset near the base, with a thixotropic liquid such as a bentonite suspension. In order to make sure that the lubricant actually covers the entire surface, it should not be poured in from above, but instead injected by means of pipes cast in the caisson walls. These discharge directly above the offset near the base. A deflecting steel plate may be added to make sure that the liquid is distributed properly. (A pipe of this kind is shown on the land side in fig. R 147-1). The injection of liquid must proceed with such care that it will not break through into the excavation cavity and flow away. The vertical distance up to the offset on the outer wall should be generously designed. Special care is called for if the surface of the excavation floor shows extensive subsidence after underlying loosened sand has been compacted.

After the designed embedment depth has been reached, the bottom must first be carefully cleaned. Only then may the floor slab of underwater or Colcrete concrete be installed.

10.8.5   Frictional resistance during sinking

The suggestions regarding pneumatic caissons given in R 87, section 10.6.6, also apply to open caissons. However, since open caissons cannot be ballasted to the same extent as pneumatic caissons, special importance is attached to thixotropic lubrication of the outer surface at greater embedment depths. Experience shows that it reduces the mean skin friction to less than 10 kN/m^2.

10.8.6   Preparing the subsoil

Non-cohesive soil liable to liquefaction is to be compacted or replaced in and around the area of the foundation. Owing to the loosening associated with the excavations within the open caisson, subsequent compaction of the soil below the bottom of the excavation is necessary when employing the open caisson method.

10.9   Design and dimensioning of large, solid waterfront structures (e.g. block construction, box or pneumatic caissons) in earthquake areas (R 126)

10.9.1   General

In addition to the general conditions contained in R 123, section 10.7, recommendation R 124, section 2.13, must also be taken into consideration.

In determining the horizontal mass forces of the waterfront structure, it should be noted that these are derived from the mass of the respective components and the earth wedge in the backfill. In this connection, the mass of the pore water in the soil is also to be taken into account.

10.9.2 Active and passive earth pressures, water pressure difference, live loads
The information in sections 2.13.3, 2.13.4 and 2.13.5 of recommendation R 124 is also applicable here.

10.9.3 Safety aspects
Please refer, above all, to R 124, section 2.13.6.
In block construction, even when considering the seismic forces, the eccentricity of the resultant in the horizontal joints between the individual blocks may only be so large that no theoretical gaping occurs over and beyond the centroidal axis under characteristic loads.

10.9.4 Base of the wall
No gaping may extend beyond the axis of the centre of gravity at the joint between base of wall and subsoil (where no gaping is allowed without an earthquake) when subjected to the characteristic loads.

10.10 Application and design of bored pile walls (R 86)

10.10.1 General
With proper construction, structural design and dimensioning, bored pile walls can also be used for waterfront structures. Besides economic and technical reasons for their use, they also comply with the demands for safe, essentially vibration-free and/or quieter construction methods.

10.10.2 Construction
Bored pile walls enables the construction of straight or curved walls to suit the particular plan layout required.
The following types of bored pile walls can be constructed, depending on pile spacing:

(1) Overlapping bored pile wall (fig. R 86-1)
The centre-to-centre spacing of the bored piles is smaller than the pile diameter. At first, the primary piles (1, 3, 5, ...) of plain concrete are installed. They are overlapped during the construction of the intermediate, reinforced, secondary piles (2, 4, 6, ...). In special cases, three plain piles can also be arranged adjacent to each other. The overlap is normally 10–15% of the diameter, but at least 10 cm. Such walls are generally then as good as watertight. Structural interaction of the piles can be

presumed for loading normal to the pile wall and in the plane of the wall at least occasionally, but is generally not considered in the design of the loadbearing secondary piles. Besides being transmitted by load-distributing capping beams, vertical loads can also be distributed between the adjacent piles given adequate roughness and accuracy of the overlapping surfaces, and to a certain extent also by shear forces. However, in walls with a short length on plan, outward deviation of the pile toes from the plane of the wall must be prevented by fixing the pile toes firmly in an especially resistant subsoil.

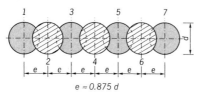

$e \approx 0.875\,d$

Fig. R 86-1. Overlapping bored pile wall

(2) Tangential bored pile walls (fig. R 86-2)
For practical construction requirements, the centre-to-centre spacing of the bored pile is at least 5 cm greater than the pile diameter. As a rule, each pile is reinforced. Watertightness of this wall can only be achieved by additional measures, such as overlapping columns using jetting or other injection methods. The wall cannot be expected to accommodate longitudinal loads through plate action.

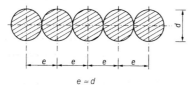

$e \approx d$

Fig. R 86-2. Tangential bored pile wall

(3) Open bored pile wall (fig. R 86-3)
The centre-to-centre spacing of the bored piles can amount to several times the pile diameter. The spaces are closed off with structures of timber, sprayed concrete or steel plates.

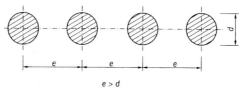

$e > d$

Fig. R 86-3. Open bored pile wall

10.10.3 Construction of bored pile walls

The construction of closed bored pile walls as shown in figs. R 86-1 and R 86-2 presumes a high drilling accuracy, which generally requires good guidance of the casing.

If possible, construction operations for bored pile walls should take place on dry ground, either on a natural ground or on a filled island, or on a jack-up platform.

Soil excavation is accomplished by means of cable grabs, Kelly grabs, rotary drills, or suction or airlift methods, inside of and from behind the advancing end of a casing. In rigid soil the casing may be omitted. Instead, water pressure or support slurries are used. Obstacles are cut away with chisels or by core drilling. Attention must be paid to an adequate overpressure from the water filling or support slurries in the bored hole or in the casing, which as a rule should be kept at least 1.5 m above the groundwater level.

With appropriate guidance, cased holes can also be drilled at an angle to the vertical.

Overlapping contiguous piling is mostly built with the aid of a machine that forces the casing into the ground or with cutter ring, with a rotary and/or compressive motion. Here, the bottom edge of the casing forms the drill bit. The primary piles of plain concrete are suitably concreted with cement CEM III B or a mixture of Portland cement and fly ash (at least 25–40%), and the sequence of operations is defined so that – depending on the capacity of the drilling machine – when the secondary piles are placed, the concrete strength will normally not exceed 3 to 10 MN/m^2. The strength difference of the primary piles to be placed must be kept as small as possible in order to prevent directional deviations of the secondary piles. Unreinforced primary piles in overlapping bored pile walls may use concrete with a lower strength than grade C 30/37.

When building this type of wall in open water, sacrificial casing pipes are required above the river bottom, unless precast piles are placed in the borings and held in place by grouting or in-situ concrete. DIN 1045 and DIN EN 1536 are applicable with respect to cleaning the bottom joint, placing concrete, concrete cover and design of reinforcement.

When support slurries are used, stability-reducing influences from the soil and/or groundwater (e.g. elevated salinity or organic soil components) etc. must be counteracted by selecting suitable clays and/or additives.

10.10.4 Advice on construction

In cased drilling, inadvertent turning of the reinforcing cage cannot be ruled out when extracting the casing. For this reason, a radially symmetric arrangement of the reinforcement should always be used unless workmanship and inspection are very careful. Accidental moving of the reinforcing cage can be prevented by placing some freshly mixed concrete on a plate installed for this purpose at the bottom of the cage, and/or by coordinating the maximum grain size of the concrete aggregate and the space between the reinforcing cage and the casing and by placing the concrete at an appropriate rate.

The pile reinforcement must be stiffened to an adequate extent in order to maintain the required concrete cover and to rule out deformation of the reinforcing cage.

Welded stiffeners to ZTV-ING part 2 [216] have proved worthwhile. The minimum measures indicated are adequate only for piles of small diameter to about 1 m. For large-diameter piles (d = approx. 1.30 m), e.g. at 1.60 m spacing, stiffener rings of 2 $\varnothing$ 28 mm BSt 420 S bars with 8 spacers $\varnothing$ 22 mm, l = 400 mm, are recommended, which are welded to each other and to the longitudinal pile reinforcement.

The piles are designed on the basis of DIN 1054 and DIN EN 1997-1. Crack widths are to be checked according to R 72, section 10.2.5. Minimum reinforcement equal to 0.8% of the concrete cross-section of the pile for a shaft diameter $D < 50$ cm or $\varnothing$ 20 mm, e $\leq$ 20 cm, and helical or hoop reinforcement $\varnothing$ 10 mm at a spacing of 24 cm for a shaft diameter $D \geq 50$ cm is required (see [118], section 11.2).

As a rule, unless the piles are held by an adequately stiff superstructure with only a small distance to the anchoring level, the wall requires walings to distribute the anchor loads. Walings may be omitted in anchored overlapping or tangential bored pile walls constructed in granular soil of medium density or in semi-rigid cohesive soil, provided that at least every other pile, or every other diaphragm between piles, is anchored. At the same time, however, the pile wall at both ends must be of adequate length with a tension-resistant waling.

Connections to adjacent structural members should, if possible, be made only by means of the reinforcement at the pile head and in the rest of the wall, only in special cases and then only by means of recesses or special built-in connections.

10.11   Application and design of diaphragm walls (R 144)

10.11.1   General

The remarks in R 86, section 10.10.1 also apply to diaphragm walls.

In situ concrete walls, which are built in sections by the slurry trench method, are called diaphragm walls. A special grab excavator or cutter digs a trench between guide walls which is filled continuously with supporting slurry. After the slurry has been cleaned of impurities and homogenised, the reinforcement is installed and concrete is poured using the tremie method, forcing the slurry to the surface, where it is pumped away.

514

Diaphragm walls are described in detail in DIN EN 1538 (DIN 4126, DIN 4127), which contains detailed data on:

- supplying the slurry ingredients; preparation, mixing, swelling, storage, placing, homogenising and reprocessing of the supporting slurry,
- reinforcement, concreting, and
- stability of the slurry-supported trench.

The remarks in DIN EN 1538 (DIN 4126) are also to be carefully observed for all diaphragm walls in waterfront structures. For other details, please refer to [82] to [88], DIN EN 1538, DIN 4126 and [7], section 3.5.

Diaphragm walls are generally erected as continuous structures and are practically watertight, with thicknesses of 60, 80 and 100 cm, or thicknesses of 120 and 150 cm in the case of quay walls with large ground surface elevation. Where high loads prevail, a diaphragm wall consisting of a succession of T-shaped elements can also be used instead of a plain wall. Since the corners between the wall and the stem of the T-section tend to break, particularly in the upper zone (low hydrostatic pressure and also low flow pressure), this must be taken into account when designing the guide walls (adequate depth) and in designing the soil masses and concrete masses. Construction with such T-shaped elements in very loosely stratified or soft soils is only recommended with additional measures, such as soil improvement beforehand.

A wall curved on plan is replaced by a chord line. The length of an individual element (segment) is limited by the stability of the slurry-supported trench. Where high groundwater levels, a lack of cohesion in the soil, neighbouring heavily loaded foundations, vulnerable buried services or the like are encountered, the normal maximum length of 10 m is reduced to a minimum of about 2.80 or 3.40 m, corresponding to the usual opening width of a diaphragm wall grab.

In suitable soil and with sound construction, diaphragm walls can transmit large horizontal and vertical loads into the subsoil. Junctions with other vertical or horizontal structural members are possible with special junction elements which are cast in or subsequently dowelled in position, if need be with recesses. Fair-face concrete surfaces can be achieved with suspended precast elements, whose use is, however, limited to a depth of 12–15 m on account of the high loads involved.

10.11.2 Verification of stability of the open trench

In order to assess the stability of the open trench, the equilibrium is investigated at a sliding wedge. The dead load of the soil and any surcharges from neighbouring structures, construction plant or other live loads and the external water pressure act as loads. The pressure of the slurry, the full friction on the slip plane which leads to active earth pressure, and the friction on the side surfaces of the sliding wedge as well as any cohesion, act as resisting forces. In addition, the resisting force of the stiffened guide wall may be taken into account. This force is especially significant for high-level slip planes because the shear stress is hardly effective here in non-cohesive soil. With deep slip planes, the influence of the guide wall is so small as to be negligible.

As regards the stability of the open trench and safeguarding the excavated walls against failure, please also refer to DIN EN 1538, DIN 4126 and [7], section 3.5.

Verification of failure of a sliding body must be provided for all depths, insofar as loads from structures exist. The highest groundwater levels expected during the construction work must be taken into consideration. In tidal areas the critical external water level must be stipulated or determined on the basis of the intended supporting slurry level. If it is anticipated that the allowable external water level will be exceeded, e.g. following storm flooding, an open trench must be filled in good time.

10.11.3 Composition of the slurry

A thixotropic clay or a bentonite suspension is used as the slurry for supporting the trench walls. With respect to its composition, suitability tests, processing with mixing and swelling times, desanding, etc., please refer to DIN EN 1538, DIN 4126 and DIN 4127.

It must be particularly borne in mind that the ionic equilibrium of the clay suspension changes unfavourably with the influx of salts, which may occur at structures in seawater or in strongly saline groundwater. Flocculation occurs, which can cause a decrease in the bracing capacity of the slurry. Salt water-resistant bentonite suspensions must therefore be used when building diaphragm walls in such areas. The following mixes have proven effective in actual practice:

(1) Slurry mixed with freshwater (tap water), 30 to 50 kg/m^3 Na-bentonite and 5 kg/m^3 CMC (carboxy-methyl-cellulose), a protective colloid.

(2) Slurry mixed with seawater and at least 100 kg/m^3 clay and salt-resistant minerals, e.g. attapulgit or sepiolith. A polymer (1–5 kg/m^3) may be added to the slurry to check filtrate water elution.

There is a large variety of mixes. In any case, suitability tests must be carried out before using any mix for construction purposes. These tests must take into account the salinity of the water, the soil conditions and any other special features (e.g. where the water passes through coral). The contamination of a slurry under salt water conditions is best shown by the rise in filtrate water elution (see DIN 4127). Special care is also necessary regarding soil contamination, soil particles of peat or lignite

516

and the like. Unfavourable effects, however, can be at least partially neutralised by adequate additives. Suitability tests of the supporting slurries are strongly recommended in such cases.

10.11.4   Details for construction of a diaphragm wall

In general, excavation of a diaphragm wall segment starts from the ground surface between the guide-walls, which as a rule are 1.0 to 1.5 m high and consist of lightly reinforced concrete. Depending on the soil conditions and loads from the excavation and lifting plant for the stop ends, they are designed as continuous wall sections, reciprocally supported outside the excavation area, or as retaining walls. Existing structural members are suitable as guide walls if they reach to an adequate depth and can absorb the pressure of the supporting slurry and other loads, and allow the operation of excavation and lifting plant.

The slurry is enriched with more and more ultra-fine particles as the excavation work progresses and must therefore be constantly checked, and replaced if the suspension density required is no longer achieved. Generally, the support slurry can be used several times. Routine checks are to be made of the density, filtrate water elution, sand level and flow limit of the supporting slurry at the construction site.

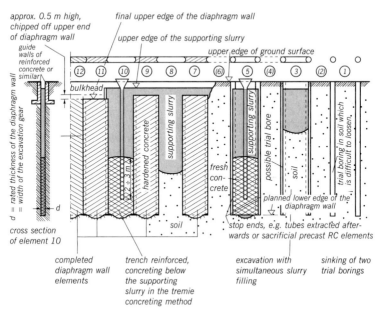

Fig. R 144-1. Example of construction of a diaphragm wall

517

According to DIN EN 1538 and DIN 4126, placing of the reinforcement and the concreting work must proceed immediately after uninterrupted excavation of the soil, particularly to avoid disturbing the ground by relaxing the tension. If reinforcement and concreting work is not possible immediately, the trench base must be cleared of any deposited soil particles before reinforcement work takes place.

Fig. R 144-1 shows details of the construction of a diaphragm wall.

In some cases, however, stepwise production of the segments in the sequence 1, 2, 3 etc., is preferable. The stop ends should be as narrow as possible to minimise the zone without reinforcement.

10.11.5 Concrete and reinforcement

Please refer, above all, to the detailed remarks in DIN EN 1538 and DIN 4126.

To achieve a good concrete flow, unfavourable reinforcement concentrations or recesses in the steel must be avoided when designing the reinforcement. Profiled reinforcing bars are preferable because of their better bonding characteristics. In order to obtain the required concrete cover of at least 5 to 10 cm, depending on the use as an excavation enclosure or as a permanent structure, an ample number of generously sized bar spacers should be used.

The following are recommended as minimum reinforcement:

- vertical, each side:
  $\varnothing$ 20 mm, $e \leq 20$ cm, ribbed reinforcing bars grade BSt 500 S

- horizontal, each side:
  $\varnothing$ 14 mm, $e \leq 20$ cm, with adequate shear reinforcement, ribbed reinforcing bars grade BSt 500 S

Attention must be paid to adequate installation stiffness of the cages, particularly in the case of minimum reinforcement.

10.11.6 Advice on the calculation and design of diaphragm walls

Due to their stiffness against bending and slight deformations, diaphragm walls must normally be designed for increased active earth pressure. The assumption of the active earth pressure is justified only when the required displacement is available for complete activation of the shear stresses in the slip planes through adequate yielding of the base of the wall, the supports and adequate yielding of the anchorages and horizontal deflection. For large differences in elevation with head displacements in the centimetre range, e.g. for quay walls for sea-going vessels with a difference in elevation ≥ 20 m, active earth pressure redistribution as per R 77, section 8.2.2, is possible. The potential deformation behaviour is to be taken into account in every single case.

Full fixity of the foot of the wall in the ground is generally not attainable with upper anchoring or support because of the stiffness of the wall. It is therefore advisable in a wall calculation according to BLUM to assume only partial fixity, or to calculate with elastic foot fixity according to the coefficient of subgrade reaction method or the coefficient of compressibility method. If on the other hand the finite element method is used, correct material characteristics are of paramount importance. The angle of inclination in the active and passive zone is essentially dependent on the type of soil, progress of the work and standing time of the open trench. Coarse-grained soils result in an extremely rough excavation wall, whereas fine-grained soils lead to comparatively smooth trench wall surfaces. Slower progress of work and longer standing times encourage deposits from the slurry (formation of filter cake). Owing to their dependency on the aforementioned factors, the angles of inclination are usually assumed to lie between the following limits:

$$0 \leq \delta_{a,k} \leq \frac{1}{2} \cdot \varphi'_k \quad \text{or} \quad -\frac{1}{2} \cdot \varphi'_k \leq \delta_{p,k} \leq 0$$

Walings for props or anchors can be formed by placing additional transverse reinforcement in the elements. If the wall elements are not too wide, one central support or anchor is sufficient; for wider elements, two or more are required, which are equally spaced symmetrically about the centre line. If necessary, check for punching.

Design is carried out according to DIN 1045 and DIN 19 702. Analysis of the crack widths is carried out according to R 72, section 10.2.5.

In view of the unfavourable effect of a probable residual film of slurry on the steel, or of fine sand deposits, the bond stresses of horizontal reinforcing bars must comply with the bond conditions of DIN 1045, section 12.4. Good bonding conditions can usually be assumed for vertical bars, but it is recommended to increase the anchorage lengths at the top and bottom of the diaphragm wall.

10.12 Application and construction of impermeable diaphragm walls and impermeable thin walls (R 156)

10.12.1 General

Impermeable diaphragm and impermeable thin walls consist of a material of minimal permeability which is frequently made from a mixture of clay, cement, water, fillers and additives. This material is placed in the required thickness in a flowable consistency, according to different methods in natural soil or in fills of higher permeability, and hardens to form a body of low permeability.

The use of these impermeable diaphragm walls limits the ingress of groundwater or other (often harmful) liquids to a very low extent.

Impermeable thin walls cannot be used if larger pressure gradients are to be expected during the construction and setting period. The following have proven themselves as fields of application for impermeable thin walls in connection with a diverse range of port facilities and waterfront structures:

- Enclosures for environmentally harmful spoil from maintenance dredging in harbours, harbour approaches, stretches of rivers in industrial areas and the like.
- Sealing of waterfront embankments on the shore and along inland waterways against high outer water or dammed river water levels.
- Protection of water-loaded embankments against erosion, suffosion, groundwater ingress and discharge etc.
- Hydraulic separation of industrial facilities, tank farm areas, etc. from surrounding harbours and groundwater as protection against spreading of harmful liquids.
- Reducing land-side water pressure differences on waterfront structures.
- Closing of larger excavations in port areas, so that the water table can be lowered without endangering adjacent waterfront structures and other facilities.

When selecting the method of construction, besides economic considerations, attention has to be paid to the limit of application of the method available. The essential points hereby are:

- Depth of the impermeable diaphragm wall.
- Thickness of the impermeable diaphragm wall, contingent on its resistance to erosion. The hydraulic gradient i, the coefficients of permeability k for the cured impermeable wall compound and the bordering ground, the unconfined compressive strength and, last but not least, the duration of loading are decisive here.
- Suitability of the intended construction method for erection of a completely watertight wall (with overlapping of the individual wall elements) under the prevailing soil conditions and obstacles to be expected.
- Safe embedment in slightly permeable subsoil.

10.12.2 Method of construction

(1) Impermeable diaphragm wall [7, 89, 90, 155, 190–196]
The impermeable wall is built according to the diaphragm wall method (DIN V 4126-100, DIN 4127 and R 144, sections 10.11.1 to 10.11.4), but the thicknesses are as a rule 60 cm or more. Excavation depths of more than 100 m are known. The thickness of the impermeable diaphragm wall is contingent on the effective hydraulic gradient and depends on the movements of the wall which are to be expected under loads from water pressure difference in the area of changing strata.

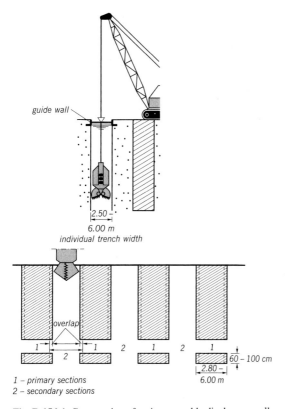

Fig. R 156-1. Construction of an impermeable diaphragm wall

Impermeable diaphragm walls may also be employed where there are inclusion of rock or similar, as these obstacles can be broken up in the trench and excavated. By checking the excavation spoil, the reliable connection to the stratum of low permeability can be checked. In rock, the embedding of the wall is attained by chiselling or cutting.

The wall is generally constructed in sections without stop ends between adjacent sections, whereby the work proceeds with an overlap of 25 to 50 cm at the edges of each section (fig. R 156-1).

Two methods have proven effective for the construction of impermeable diaphragm walls:

(2) Single-phase method

The trench is sunk under protection of a supporting slurry mixed with cement, which remains in the trench after the excavation (phase 1) and hardens there. The construction of each section in one phase favours the necessary rapid progress of construction, so that there is no impairment to setting of the sealing compound, which starts after about six hours. If the construction work is slowed down, work can progress in the trench for up to 12 hours if special compounds are used. Higher slurry consumption rates occur when passing through highly permeable soil layers, or if the grab is operated too quickly, thus preventing the formation of a filter cake.

For long excavation times and when passing through fine-grained soils, the suspension can become so enriched with solids that the grab "floats" and cannot dig deep enough. This also occurs with high filtrate water elution in deep trenches.

A special form of the single-phase method is the dry diaphragm wall, used for earth dams [90], also called a trench wall. Here, no supporting slurry is used because the wall section is always only 1–2 m deep and is therefore stable and can be filled with a soil–aggregate mix.

(3) Two-phase method

The trench is prepared in the customary manner with using a supporting slurry. After reaching the final depth, the sealing compound is applied using the tremie method in a second step (phase 2) and the supporting slurry withdrawn. The difference in density between the impermeable wall compound and the support slurry containing soil should be 0.5–0.6 t/m^3 as otherwise the compound replacement cannot be assured.

This method is more expensive than the single-phase method. The impermeable wall compound, however, is generally more homogeneous than in the single-phase method. The two-phase method is therefore used, for instance, when work is progressing slowly because of obstacles, or for large depths (> 30 m).

(4) Impermeable thin wall

To build impermeable thin walls, a steel section of 500 to 1000 mm web height (thin wall pile) is introduced into the soil strata to be sealed, and compacted, preferably with vibration hammers. The trench thus created is grouted with an impermeable wall material upon withdrawal of the steel section. A continuous thin wall is created through a continuous, overlapping repetition of this procedure. Instead of I-section steel profiles, other shapes can also be used (e.g. depth vibrators with extra straps).

A mixture with the greatest possible density – cement, clay or bentonite, rock flour or fly ash and water – is used as the impermeable wall material. Impermeable thin walls are preferably used for temporary works.

The impermeable thin wall compound is placed under pressure and in so doing penetrates into the voids of loosely deposited soils.

The minimum thickness of the diaphragm wall is vital for evaluating its effectiveness; this is determined by the section thickness of the pile toe or other precast parts used for making the trench. The individual work

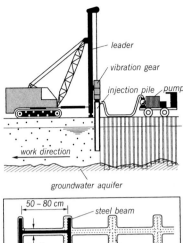

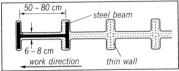

Fig R 156-2. Construction of a thin wall

steps are executed with overlapping [91]. Adequate overlapping of neighbouring sections must be maintained (fig. R 156-2).

The advantage of the impermeable thin wall over the impermeable diaphragm wall lies in its lower cost and in the shorter time needed for its construction.

Its use is limited, however,

- when encountering soil with obstacles which is unsuitable for driving and vibrating,
- when the depth of the wall is more than 15 to 30 m, depending on the nature of the subsoil, if customary plant is used,
- when a comparatively high water pressure difference acts on a thinner diaphragm wall,
- when a soil stratum that tends to flow lies above a coarse-grained permeable layer of soil, no sealing compound may flow downward to compensate for the suspension loss of the lower layer. (This effect can be increased through driving and vibration influences, but then there is the risk that no impermeable wall compound penetrates into the transition zone.),
- when the natural soil exhibits strong flow and/or settlement tendencies,
- when the transition to the groundwater aquifer is recognised only with difficulty, especially with a varying granulometric composition in the transition area. (In addition to observing the soil brought to the surface by the driving or vibrating plant, suitable measurements and further observations are then required.)

523

10.12.3 Raw materials for impermeable wall compounds

(1) Clay

In addition to suitable natural clays and clay powder, bentonite is preferably used.

Commercial bentonite has varying properties. Furthermore, the flow properties and the water-binding capacity of bentonite suspensions are evidently changed by cement additive. Attention must be paid to this when preparing the impermeable wall compound.

(2) Cement

Commercial cements have proven effective. Blastfurnace cements with a high foundry sand component are particularly advantageous, or Portland cement/fly ash mixtures with a high ash content.

(3) Fillers

Basically, all neutral sands, dusts, powders and granulates may be used whose maximum aggregate size remains suspended in the clay-cement suspension. In respect of the settling tendency, there are advantages in fillers of low density. On the other hand, however, the self-compacting flow is favoured by a high density. When impermeable diaphragm walls are erected in the single-phase method, the prevailing soil determines the enrichment with fines. When the two-phase method is used, the grain size of the filler in the impermeable wall material (contingent on the diaphragm wall thickness and the compound processing plant) may be in the magnitude of up to 30 mm. No special attention need be paid to the granulometric composition in case of fine fillers. But it must be uniform in case of coarse grain mixtures.

(4) Impermeable wall compounds resistant to harmful substances

Silicate binders are used for impermeable wall compounds which are resistant to harmful substances. These impermeable wall compounds have a high density and must be installed using the two-phase method.

(5) Water

The mixing water must be neutral. Acid water may lead to flocculation of the bentonite and, just like acid-reacting filler, reduce the liquid limit of the impermeable wall material. Light alkalisation of the mixing water by adding a little soda or caustic soda has proven effective.

(6) Additives

Where difficult groundwater conditions prevail, protective colloids are recommended to stabilise the clay suspension. These additives reduce the filtering-out of the water and at the same time increase the liquid limit of the supporting slurry.

10.12.4 Requirements for impermeable wall compounds

The material properties and behaviour of the impermeable wall compound during the placing and in the final condition are to be coordinated with the type of impermeable wall and its purpose and are to be verified by tests. Ready-mixed impermeable wall mixtures have proven successful. Cooperation with an institute experienced in material tests on impermeable wall compounds is recommended.

(1) State during installation

The flow behaviour of the impermeable wall compound is determined by the liquid limit τ_F. It must be so high that the granular components contained in the impermeable wall compound remain securely in suspension at least until the onset of setting.

In the single-phase method, the upper limit of the liquid limit τ_F of the impermeable wall compound is to be defined so that excavation is not impeded and not too much impermeable wall compound is removed.

For the two-phase method, the upper limit of τ_F results from the demand for self-compacting flow in the supporting slurry. The complete displacement of the supporting suspension also presupposes that the impermeable wall compound has a considerably greater density than the supporting slurry. It is therefore necessary to replace the supporting slurry with a cleaned or fresh suspension before placing the impermeable wall compound.

The impermeable wall compound must be stable. It must absorb or discharge as little water as possible and may settle to only a limited extent during the setting phase. Partial sedimentation of the cement and increased filtration of water in the suspension is only desirable under load from hydrostatic water pressure, for partial increase of the cement portion in the more heavily loaded lower area of the impermeable diaphragm wall. The setting behaviour of the impermeable wall compound must ensure that the setting process is not disturbed by excavation work in the single-phase method.

(2) State after curing of the compound

A permeability coefficient k of the cured impermeable wall compound should be verified by suitability tests after 28 days up to a hydraulic gradient of $i = 20$. The permeability requirements for the impermeable wall compound and those for the cured wall differ. The permeability of the wall compound produced in the laboratory from the raw materials (see section 10.12.5) should be smaller by 10 than the permeabilities ascertained from the wall samples. Since the volume of percolating water infiltrating the diaphragm wall per unit of time is also of significance, the desired ratio of k to wall thickness d (conductivity coefficient in s^{-1}) should be indicated in the design for the maximum hydraulic gradient occurring $i = \Delta h/d$ (Δh = water level difference between interior and exterior).

Usually, an unconfined compressive strength in accordance with DIN 18 136 of 0.2 to 0.3 MN/m^2 is required, and 0.4 to 0.7 MN/m^2 for thin walls. This strength is to be tested after 28 days using samples taken from the mixer. A substantial increase in strength is not desirable because this would decrease the deformability of the impermeable wall compound, and deformations in the soil could lead to local damage with greater erodibility. Where possible, the deformability of the impermeable wall compound should be of the same magnitude as that of the soil.

In view of the erosion resistance of the impermeable wall compound, the hydraulic gradient must remain limited to $i = 20$ for permanent loads on a single-phase impermeable diaphragm wall with a strength of 0.2–0.3 MN/m^2.

In impermeable thin walls, whose impermeable wall compound as a rule contains more solid matter than with impermeable diaphragm walls, a hydraulic gradient of $i = 30$ should not be exceeded for permanent loads with the confirmed minimum thickness.

Influences that could effect the behaviour of the walls in working state, e.g. consolidation, shrinkage, etc., are to be taken into account properly in the material composition of the walls.

10.12.5 Testing methods for impermeable wall compounds

DIN V 4126-100 and DIN 4127 are relevant for testing.

The flow behaviour of the impermeable wall compound can be ascertained by measuring the liquid limit with the ball harp apparatus or the outlet times according to the Marsh funnel test.

The stability of the impermeable wall compound at the placing stage may be seen in the degree of settlement (percentage settlement), which as a rule should not exceed 3%.

The workability limit up to which the impermeable wall compound may be moved without damage due to excavation work, is determined through the strength of test cylinders made from impermeable wall compound, moved for appropriate periods.

The permeability is determined in the laboratory in accordance with DIN 18 130. If need be, permeability with respect to defined test liquids is also to be determined in respect of durability. Normally, 28- or 56-day-old samples are examined. In practical applications the generally higher permeability of the diaphragm wall in the completed structure compared with the laboratory tests can be taken into account.

The unconfined compressive strength is determined according to DIN 18 136, whereby the dimensions of the samples are to be coordinated with the material being investigated. It is a reliable value for evaluating the bearing capacity, where this is of significance. As a rule, the age of the samples is 28 or 56 days.

Samples from the completely set wall should be taken for investigation only in special cases, and then in cooperation with an experienced institute which fixes the extent and type of sampling.

10.12.6 On-site testing of the impermeable wall compound

The properties of the impermeable wall compound before being placed are compared with the nominal values determined from the suitability test.

For impermeable diaphragm walls according to the single-phase method, samples taken from various depths in the trench with grab or special sampling tins are to be investigated.

The on-site check is carried out in laboratory by measuring the following properties of the impermeable wall compound:

1. Mass density
2. Support characteristics (ball harp apparatus or pendulum apparatus)
3. Flow characteristics (Marsh funnel)
4. Stability (settling test, filter press)
5. Sand components at various depths
6. pH value

It is useful to determine the water permeability of the finished wall as a function of time, starting after approx. 28 days, on reserved samples in the laboratory in accordance with DIN 18 130, and/or by sinking tests in boreholes [144] which should form the basis for acceptance. These measurements are generally taken into account for the sealing effect.

10.13 Inventory before repairing concrete components in hydraulic engineering (R 194)

10.13.1 General

Construction operations for repairs to concrete components are likely to be successful only if they take accurate account of the causes of the defects or damage. Since several causes are usually involved, a systematic investigation of the actual situation should first be carried out by a qualified engineer.

Since a correct assessment of the causes of defects or damage is an important prerequisite for a lasting repair, some recommendations for determining the actual situation and for troubleshooting are given below (see also section 1.2 of DIN 31 051).

The following recommendations also apply in principle to the testing of concrete components within the scope of surveys to R 193, section 15.1. The individual investigations listed may exceed the scope of normal testing and are therefore listed separately here.

(1) Description of item
- Year of construction
- Stresses arising from use, operation, environment
- Existing stability verifications
- Soil investigations
- Construction drawings
- Special features of the erection of the structure

(2) Inventory of components affected by the damage
- Nature, position and dimensions of components
- Building materials used (type and quality)
- Description of damage (nature and extent of damage with dimensions of damaged areas)
- Documentation (photographs and sketches)

(3) Necessary investigations
The nature and extent of the investigations necessary for troubleshooting are laid down on the basis of the conclusions arising from sections 10.13.1 (1) and 10.13.1 (2).

10.13.2 Investigations of the structure
More detailed information on the actual condition of the structure can be obtained from the following investigations:

(1) Concrete
- Discolouration, moisture penetration, organic growths, efflorescence/honeycombing, concrete spalling, flaws
- Surface roughness
- Adhesive strength
- Watertightness, cavities
- Depth of carbonation
- Chloride content (quantitative)
- Crack propagation, widths, depths, lengths
- Crack movements
- Condition of joints

(2) Reinforcement
- Concrete cover
- Corrosion attack, degree of rusting
- Reduction in cross-section

(3) Prestressing elements
- Concrete covering,
- Condition of grouting
  (ultrasound, radiography, endoscopy if necessary)
- Condition of tensile steel

- Existing degree of pretension
- Grouting mortar (SO_3 content)

(4) Components
- Deformation
- Forces
- Vibration behaviour

(5) Sampling on structure
- Efflorescence/honeycombing material
- Concrete components
- Drilling cores
- Drilling dust
- Reinforcing components

10.13.3 Investigations in the laboratory

(1) Concrete
- Mass density
- Porosity/capillary action
- Water absorption
- Water penetration depth (waterproof concrete)
- Wear resistance (as per DIN 52 108)
- Micro air pore content
- Chloride content (quantitative, in different depth zones)
- Sulphate content
- Compressive strength (as per DIN 1048)
- Modulus of elasticity (as per DIN 1048)
- Mix proportions (as per DIN 52 170)
- Granulometric composition
- Disc strength (as per DIN 1048)
- Depth of carbonation
- Surface tensile strength (as per DIN 1048) at various depth horizons

(2) Steel
- Tensile test
- Fatigue test

10.13.4 Theoretical investigations
Structural calculations of the bearing stability and deformation behaviour of the structure or individual components before and after repair.
Estimation of carbonation progress and/or preliminary chloride enrichment with or without repair.

10.14   Repair of concrete components in hydraulic engineering (R 195)

10.14.1 General

Hydraulic structures are subject to particularly unfavourable environmental stresses arising from physical, chemical and biological factors. In addition, in harbour facilities and other areas which must be kept accessible for operational reasons, for example, the effects of de-icing salts and other harmful impurities must be anticipated, along with occasional cargo items which are harmful to concrete.

Apart from live loads and impact and friction forces from vessels, the physical effects result primarily from the repeated drying out and wetting of the concrete, the constant temperature fluctuations with severe frost effects on water-saturated concrete and the action forms of ice. Apart from the effects of de-icing salts and cargo goods resulting from requirements for use in some cases, chemical stresses are caused primarily by the salinity of the seawater. Chlorides that have penetrated into the concrete can destroy the passive layer of the reinforcement. In areas with a prevailing adequate presence of oxygen and moisture in the concrete, e.g. above the splashing zone, the reinforcement can be corroded. Biological stresses occur primarily through the growth of algae and the resulting metabolic products.

Such factors can lead to cracks and surface damage to the concrete and corrosion damage to the reinforcement. Particularly at risk are components in the spray and splashing areas, especially components in contact with seawater and those in the direct vicinity of coasts in strongly salt-laden air. Fig. R 195-1 shows a diagram of seawater attack on reinforced concrete.

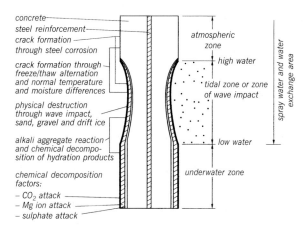

Fig. 195-1. Diagram to show the attack of sea water on reinforced concrete as per [133]

If the damage discovered indicates that repair work is necessary on concrete components, an engineering expert according to the instructions for protection and repair of concrete components [134] must always be consulted to determine the actual condition, assess the damage and plan the repair measures. The lasting success of the work depends significantly on the expertise of the personnel involved, the quality and suitability of materials used, and the care exercised in both execution and monitoring. Protection and repair measures on components that can be protected from contact with water at least while the work is being carried out (work above water) should be carried out according to ZTV-W LB 219 [138]. ZTV-ING [216] should be used as the standard work when filling cracks and hollows in such parts.

For protection and repair measures that have to be carried out underwater, trial repairs should verify that the intended work can lead to the desired result under the boundary conditions of the particular case concerned. Verification of the quality of the construction materials should be provided in the form of suitability and quality tests, coordinated with the prevailing component conditions.

Measures in the context of cathodic corrosion protection should be planned on the basis of [179].

Repair work should only be carried out by contractors with adequate expertise and experience in this field, complying with the requirements regarding material and personnel according to [134, 138] or [216].

10.14.2 Assessing the actual situation

The effects of the defects and damage on the stability, serviceability and durability of the structure must be assessed on the basis of a careful inventory in accordance with R 194, section 10.13.

The actions and loads affecting the structure requiring repairs should be researched as accurately as possible because this results in the requirements to be fulfilled by the materials or construction method to be used.

10.14.3 Planning the repair work

10.14.3.1 General

The repair requirements result from a comparison of the actual situation as per section 10.14.2 with the intended required state at the end of the repair work. The repair targets for the repair work should be defined as precisely as possible. According to [134] and [138], a distinction should be made between measures for the protection and repair of the concrete and measures for the restoration or preservation of the corrosion protection in the reinforcement. When it comes to filling cracks, it should be clarified whether the cracks should be simply closed or sealed, or whether the edges of the crack should be bonded together to transfer forces or to allow expansion.

The possible effects of protection and repair work on the durability and bearing capacity of the component or the whole structure must be investigated. Special attention should be given to any unfavourable changes in the performance of the structure and changes in the loadbearing behaviour (increase in dead load, redistribution of loads, etc.).

When elaborating the repair concept, special consideration should be given to the fundamentally different boundary conditions for reinforcement corrosion in structures above and below water, particularly the different exposure to the oxygen required for the corrosion process.

Cracks must be investigated to see which causes are behind the crack formation and which loads or deformations have to be anticipated here in future.

10.14.3.2 Repair plan

An engineering expert should draw up a repair plan for all repair work. The plan should describe all relevant details for the execution of repairs, including preparation of the sub-base, type and quality of materials being used, selected construction method and subsequent treatment, and quality assurance.

The repair plan should be drawn up where possible on the basis of [138] or [134].

The repair plan should contain details referring to the following points (among others):

(1) Repair principles/fundamental solutions as per [134] and [138].

(2) Requirements for contractors/staff, e.g.
- Suitability certificate for the nozzle operator for sprayed concrete work
- Suitability certificate for handling plastics in concrete construction work

(3) Sub-base preparation
- Aim of the sub-base pretreatment and type of pretreatment method
- Extent to which concrete is to be removed or the reinforcement exposed
- Degree to which rust has to be removed from the reinforcement

(4) Concrete replacement
- Type and quality of materials and methods to be used, e.g.
  - concrete
  - sprayed concrete
  - sprayed concrete/mortar with plastic additive (SPCC)
  - cement mortar/concrete with plastic additive (PCC)
- Formwork
- Layer thicknesses

- Additional reinforcement
- Construction joints

(5) Cracks
- Filling material
- Filling methods

(6) Joints
- Preliminary work
- Type of joint sealing material
- Execution

(7) Surface protection systems
- Type of system
- Layer thicknesses

(8) Curing
- Type
- Duration

(9) Quality assurance
- Basic tests
- Suitability tests
- Quality monitoring

10.14.4 Performance of repair work

10.14.4.1 General

The performance of repair work above water is described in detail in ZTV-W LB 219 [138] and ZTV-ING [216].

Before applying cement-bonded concrete replacement material (concrete, sprayed concrete, SPCC, PCC), the concrete sub-base should be adequately pre-wetted (for the first time 24 hours in advance). However, before placing the concrete replacement, the contact surfaces should have dried to such an extent that they appear matt-damp.

Adequate curing is vital for the success of repair work. Cement-bonded concrete replacement should be cured with water in the first few days after placing. This applies particularly to thinner layers of concrete replacement using PCC.

In view of the different raw materials involved, differences in colour between the existing concrete and the concrete replacement must always be expected for local repairs using cement-bonded concrete replacement. The following details generally do not apply to measures with cathodic corrosion protection.

10.14.4.2 Sub-base preparation

(1) General

The plan should not state the type of sub-base pretreatment but the aim to be fulfilled with the sub-base pretreatment.

At the end of the sub-base pretreatment work, a check should be carried out to establish whether the concrete sub-base possesses the friction strengths stipulated for the intended repair work.

(2) Work above water level

To create a good bond, the concrete sub-base must be uniformly sound and free of separating, inherent or foreign substances. Loose and brittle concrete together with all foreign substances, such as growths of algae or barnacles, oil or paint residues, are to be removed. The extent to which concrete then has to be removed and the reinforcement exposed in order to achieve the repair aim depends on the basic method selected according to [134] or [138] and should be stated in the repair plan.

Before placing a cement-bonded concrete replacement, on conclusion of sub-base preparation work, as a rule cones of aggregate grains should be exposed near to the surface measuring ≥ 4 mm in diameter.

On conclusion of sub-base preparation work, loose corrosion products on exposed reinforcement and exposed cast-in parts should be removed. For corrosion protection by restoring the alkali milieu as per [134] or [138], the degree to which rust is removed must comply at least with normal purity degree Sa 2, but Sa 2½ where corrosion protection is provided by coating the reinforcement. Rust removal from reinforcement in the case of chloride-induced reinforcement corrosion may only be carried out by high-pressure water jetting (≥ 600 bar).

The following methods can be used for sub-base preparation, depending on the purpose:

- mortising
- cutting
- grinding
- blasting with
  - solid blasting agents
  - water/sand mixtures
  - high-pressure water

Material removed during sub-base pretreatment and mixtures resulting from the process are to be disposed of properly according to the waste management regulations.

(3) Work under water

The information according to section 10.14.4.2 (2) applies accordingly. Depending on the purpose, sub-base pretreatment can be carried out using the following methods:

- hydraulically driven cleaning equipment
- underwater blasting with
  - solid blasting agents
  - high-pressure water

10.14.4.3 Repairs with concrete

(1) General
Repairs with concrete are to be given preference, particularly for repairs to large areas with thicker layers, from both a technical and economic point of view.

(2) Work above water level
Repairs with concrete should be carried out on the basis of ZTV-W LB 219 [138]. This stipulates specific requirements for the composition and properties of the concrete, contingent on the loads affecting the component.

(3) Work under water
This work is to be carried out according to section 10.14.4.3 (1). Proper placement and compaction of the concrete without segregation is to be ensured by adding a suitable stabiliser certified by the DIBt (German Institute for Construction Technology, Berlin) or according to the rules for underwater concrete as per DIN 1045, section 6.5.7.8.

10.14.4.4 Repairs with sprayed concrete

(1) General
The sprayed concrete method has proven effective for the repair of concrete structures in hydraulic engineering and is probably the most frequently used repair method.

(2) Work above water level
See section 10.14.4.3 (2). Specific requirements for the composition of the concrete mixture and the properties of the finished sprayed concrete as a function of the actions on the component are stipulated in [138].
ZTV-W B 219 makes a basic distinction between sprayed concrete in layers of up to approx. 5 cm thick, which can be placed without reinforcement, and sprayed concrete in layers > 5 cm thick, which must also be reinforced and connected to the structure by means of anchoring elements. The surface of the sprayed concrete is to be left in its rough sprayed state. If a smooth or specially textured surface is required, mortar or sprayed mortar is to be applied after the sprayed concrete has hardened, and processed accordingly.

(3) Work under water
Not applicable

10.14.4.5 Repairs with plastic-modified sprayed concrete (SPCC)

(1) General
The use of SPCC can be particularly advantageous for thinner layers because the plastic additives improve certain concrete properties, such as water retention value, adhesion strength or watertightness. In addition, the plastic additives also make it possible to achieve a deformation behaviour that can be compared with the existing concrete. Only plastic additives insensitive to moisture may be used.

(2) Work above water level
See section 10.14.4.3 (2). The SPCC layer thickness should be between 1 and 5 cm when spread onto the surface. Additional reinforcement is not required for these thicknesses.
The surface of the SPCC should be left in its rough sprayed state. If a smooth or specially textured surface is required,
- in the case of a single layer, hardening of the SPCC is to be followed by application of a mortar that is compatible with SPCC and which can then be processed accordingly,
- in the case of several layers, the last sprayed layer is processed accordingly.

(3) Work under water
Not applicable.

10.14.4.6 Repairs with cement mortar/concrete with plastic additives (PCC)

(1) General
PCC is particularly suitable for the repair of small eruption areas. PCC is applied to the contact surface by hand or machine. However, in contrast to sprayed concrete or SPCC, compaction is by hand in both cases. Only plastic additives insensitive to moisture may be used.

(2) Work above water level
See section 10.14.4.3 (2). The PCC layer can be up to approx. 10 cm thick for local repairs.

(3) Work under water
Special products are available for this application.

10.14.4.7 Repairs with reaction resin mortar/reaction resin concrete (PC)

(1) General
PC products are practically impervious to water vapour. This is just one of the main reasons why the use of PC products is restricted to local repairs or work under water.

(2) Work above water level
PC products should only be used in exceptional cases and only for local repairs [216].

(3) Work under water
Special products are available for this application.

10.14.4.8 Sheathing of components

(1) General
The damaged component is sheathed with a watertight covering which is sufficiently resistant to the anticipated mechanical, chemical and biological attacks. The protective covering may be placed around the component to be protected, either with or without an adhesive bond. The aim of the procedure is to prevent the ingress of water, oxygen or other substances between the covering and the component. The procedure can be used both above and below water.

(2) Cleaning and preparing the sub-base
Carry out the work in accordance with section 10.14.4.2 (2) or 10.14.4.2 (3).

(3) Sheathing of concrete with a prefabricated concrete shell attachment
Requirements relating to the prefabricated part:
- watertight concrete free from capillary pores (water/cement ratio ≤ 0.4) with high frost resistance
- layered reinforcement

The space between the shell attachment and the pretreated concrete is filled by injecting a low-shrinkage cement-lime mortar with high frost resistance.

(4) Sheathing of concrete with a prefabricated fibre-reinforced concrete shell attachment
Suitable fibres:
- steel fibres
- alkali-resistant glass fibres

Requirements and execution are in accordance with section 10.14.4.8 (3).

(5) Sheathing of concrete with in situ fibre-reinforced concrete
As for section 10.14.4.8 (4).

(6) Sheathing of concrete with a plastic shell, can be used for columns
Requirements relating to the plastic shell:
- resistant to UV radiation (above water level only)
- resistant to surrounding water
- watertight and adequately diffusion-resistant

- if necessary, adequate mechanical resistance to anticipated influences, e.g. ice load, bed load and contact with vessels

The space between the shell and the existing concrete is filled as in section 10.14.4.8 (3).

(7) Winding flexible foil around the component, can be used for columns

Clean and prepare the sub-base in accordance with section 10.14.4.2 (2) or 10.14.4.2 (3).

Treat the reinforcement with corrosion protection and fill damaged areas above water as per section 10.14.4.3 to 10.14.4.7. Wind flexible foil around the supports.

Requirements relating to the system:
- resistance to UV radiation
- resistance to the surrounding water
- imperviousness to water and gas
- adequate mechanical resistance to anticipated external effects, such as ice load,
- watertight interlocks between foil edges and tight upper and lower connections with the support so that neither fluid nor gaseous substances can penetrate between foil and sub-base

10.14.4.9 Coating the component

(1) General

As an additional measure against penetration of harmful substances into the concrete, particularly chlorides and carbon dioxide in the case of steel and prestressed concrete components (when no sheathing is provided in accordance with 10.14.4.8), a coating on the cleaned, prepared component which may have been repaired with concrete can prove effective.

(2) Work above water level
See section 10.14.4.3 (2).
Coating may be used only when there is no risk of moisture penetrating from the rear.

(3) Work under water
Special products are available for this application.

10.14.4.10 Filling cracks

The work should be carried out on the basis of ZTV-ING [216] as far as possible. Cracks in hydraulic engineering components with their frequently high water saturation levels can best be filled with cement lime/ cement suspensions for positive connections and polyurethane for expansion-compatible connections.

10.14.4.11 Formation and sealing of joints

- Clean joints and extend existing joint gap if necessary.
- Repair damaged edges with epoxy resin mortar.
- Install joint sealing in accordance with the applicable specifications and directives.

The closure and filling of joints can be carried out accordingly as per DIN 18 540. [137] is to be observed for joints in traffiked areas.

11 Piled structures

11.1 General

The piled structures dealt with in the following sections always require the horizontal head deformations to be calculated, and they must be checked for serviceability. In the case of large differences in ground levels, it is worth considering whether raking anchors as per section 9 should be used in addition to pile bents. In doing so, check the compatibility of the deformations.

Piles can be subjected to changing loads (tension/compression) from hawsers pulling on bollards, lateral crane impact and tidal influences. Suitability of the chosen pile system must therefore be checked for alternating loads.

The recommendations for structural calculations featured in this section are based on the concept of partial safety factors (see section 0.2).

11.2 Determining the active earth pressure shielding on a wall below a relieving platform under average ground surcharges (R 172)

A wall can be more or less shielded from the active earth pressure by means of a relieving platform, depending, above all, on the position and width of the platform, but also on the shear strength and compressibility of the soil behind the wall and below the structure. This can have a favourable influence on the earth pressure distribution which is relevant to ascertaining the internal forces. With uniform, non-cohesive soils and average ground surcharges (usually 20 to 30 kN/m^2 as a uniformly

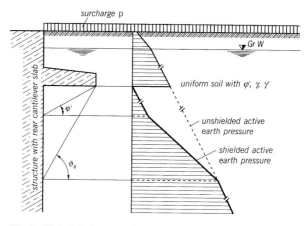

Fig. R 172-1. Solution according to LOHMEYER with uniform soil

distributed load), the active earth pressure shielding can be determined according to LOHMEYER [93] (fig. R 172-1). As can be easily confirmed by CULMANN investigations, the use of the LOHMEYER method is highly applicable under the foregoing prerequisites.

With stratified, non-cohesive soil, the assumptions according to fig. R 172-2 or R 172-3 offer approximate solutions, whereby the calculation as per R 172-3 can be carried out with a PC even for multiple changes of strata.

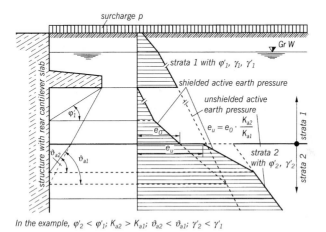

Fig. R 172-2. Solution according to LOHMEYER, expanded for stratified soil (solution possibility 1)

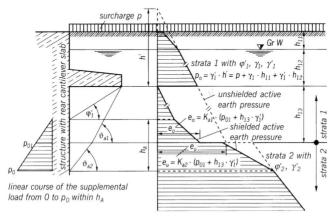

Fig. R 172-3. Solution according to LOHMEYER, expanded for stratified soil (solution possibility 2)

In cases of doubt involving multiple changes of strata or the different ground surcharges, the active earth pressure can be ascertained with the help of a method according to R 171, section 2.4.3, or DIN 4085.

If the soil also has a cohesion c', the shielded active earth pressure can be taken approximately in such a manner that at first the shielded active earth pressure distribution is determined without taking c' into account, and is subsequently superimposed on the cohesion influence:

$$\Delta e_{ac} = c' \cdot K_{ac}$$

(K_{ac}: coefficient for active earth pressure when taking account of cohesion, see DIN 4085). This procedure, however, is only permissible when the degree of cohesion is small in comparison with the total active earth pressure. A more accurate determination is also possible here with use of the expanded CULMANN method as per R 171, section 2.4.

The same applies to calculating the effect of earthquakes, taking into account R 124, section 2.13.

The calculations according to figs. R 172-1 to R 172-3 may not be used in cases where several relieving platforms have been installed one above the other. Furthermore, irrespective of the shielding, the total stability of the structure is to be verified for the corresponding limit states to DIN 1054, whereby full active earth pressure is to be applied in the relevant reference planes.

11.3 Active earth pressure on sheet piling in front of piled structures (R 45)

11.3.1 General

More and more, piled structures with rear sheet piling must be strengthened by driving an additional front sheet piling wall in order to allow for greater water depths. This new front sheet piling will be stressed by the earth support pressure of the existing front sheet piling, and often by soil stresses from the bearing piles, which may begin to take effect at a level immediately below the new harbour bottom.

Also in cases of new piled structures, it can happen that a new front sheet pile wall lies in the area influenced by pile forces.

The loads acting on the sliding wedges and on the new sheet piling can only be approximated. The following statements apply in the first instance to non-cohesive soil for ascertaining the internal forces.

It is presumed that verification of overall stability has been provided according to DIN 4084 for LS 1C, and that the embedment depth of the new front quay wall was thus defined.

In this context, fig. R 45-1a shows an example of a possible failure mechanism. Further failure mechanisms, in which it is primarily the

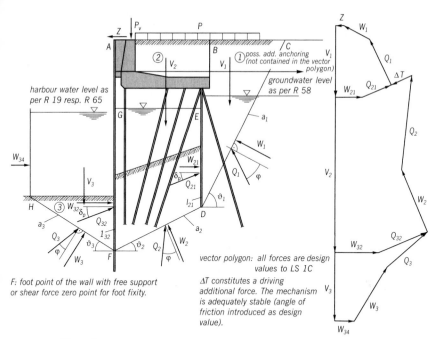

Fig. R 45-1a. Overall stability of an embankment with superstructure. Example for a failure mechanism.

incline of the outer failure lines of sliding wedges 1 and 3 that varies, must show whether the vector polygon with the design values for actions and resistances can always be closed without supporting additional force ΔT.

In fig. R 45-1a, the symbols have the following meanings:

V_1, V_2, V_3	Dead load forces of sliding wedges 1, 2, 3, including their share of the changing and permanent live loads and vertical water loads
W_1, W_2, W_3	Water pressure forces on the outer failure planes a_i of sliding wedges 1, 2, 3, normal to the failure planes a_i
Q_1, Q_2, Q_3	Failure plane forces of the outer failure planes a_i of sliding wedges 1, 2, 3 inclined at the angle of friction φ to the vertical on the slip plane
Q_{21}, Q_{32}	Failure plane forces of the inner failure planes i_{21}, i_{32}, inclined at δ_p for sheet piling, at φ in the soil
W_{21}, W_{32}	Water pressure forces, acting on the vertical inner failure plane i_{21} (DEB), i_{32} (FG) from left to right

543

W_{34}	Water pressure force on the perpendicular to H, acting to the right
P_v	Vertical load
Z	Horizontal load
ΔT	Additional force for fulfilling equilibrium
a_1, a_2, a_3	Outer failure planes of sliding wedges 1, 2, 3
i_{21}, i_{32}	Inner failure planes between sliding wedges 2 and 1, 3 and 2
φ	Angle of friction
δ_p	Angle of passive earth pressure
$\vartheta_1, \vartheta_2, \vartheta_3$	Angle of the failure planes

11.3.2 Load influences

The active earth pressure acting on the new front sheet piling is influenced by:

(1) The active earth pressure from the soil behind the quay wall. It is generally referred to the plane of any existing rear sheet piling or to a vertical plane through the rear edge of the superstructure (rear reference plane). It is calculated with straight slip planes for the existing ground level and the existing surcharge. The presumed angle of earth pressure has to comply with DIN 4085.

(2) The toe support reaction of the any existing rear sheet piling (Q_1 in figs. R 45-1b and d, inclined at δ_p to the horizontal).

(3) The flow pressure in the sections of earth behind the existing front sheet piling caused by the difference between groundwater and harbour water levels.

(4) The dead load of the earth mass lying between the existing front sheet piling and the rear plane of reference, acting together with the active earth pressure according to (1). In the case of existing rear sheet piling this is the toe support reaction required which is transferred to the soil between the two sheet piling walls.

(5) The pile forces that result from the vertical and horizontal super-structure loads. In calculating the pile forces, the reaction at the upper support of the front sheet piling must be included when an additional anchoring system independent of the piled structures is not used (fig. R 45-1a).

(6) The resistance Q of the soil between the front sheet piling and the rear plane of reference defined in (1), for forward movement of the structure (fig. R 45-1b to d).

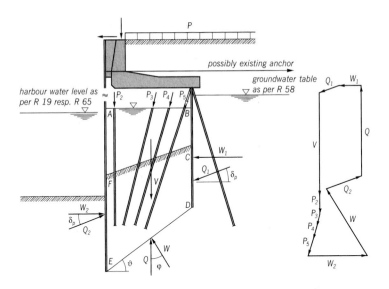

Fig. R 45-1b. Investigation of various failure mechanisms to ascertain the actions on a new front sheet piling wall for an embankment below an elevated structure; 1st example of a failure mechanism

In fig. R 45-1b, the symbols have the following meanings:

V	Dead load of the soil $FCDE$ including water load $ABDE$
W_1, W_2	Water pressure forces from right, left
W	Water pressure force on failure plane ED
Q	Failure plane force of failure plane ED
Q_1	Characteristic value of required toe support force of rear sheet piling
Q_2	Supporting force required for equilibrium
P_2–P_5	Axial forces resisted by sliding wedge $CDEF$
φ	Angle of friction
δ_p	Angle of passive earth pressure
ϑ	Failure plane angle

545

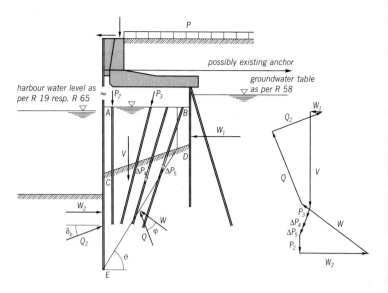

Fig. R 45-1c. Investigation of various failure mechanisms to ascertain the actions on a new front sheet piling wall for an embankment below an elevated structure; 2nd example of a failure mechanism

In fig. R 45-1c, the symbols have the following meanings:

V	Dead load of the soil *CDE* including water load *ABDE*
W_1, W_2	Water pressure forces from right, left
W	Water pressure force on failure plane *ED*
Q	Failure plane force of failure plane *ED*
Q_2	Supporting force required for equilibrium
P_2–P_3	Axial forces resisted by sliding wedge *CDE*
$\Delta P_4, \Delta P_5$	Axial forces from skin friction, resisted by sliding wedge *CDE*
φ	Angle of friction
δ_p	Angle of passive earth pressure
ϑ	Failure plane angle

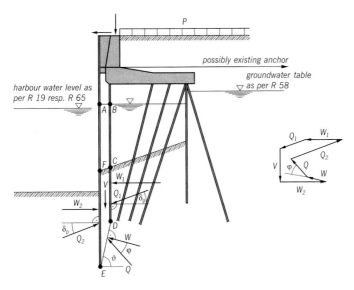

Fig. R 45-1d. Investigation of various failure mechanisms to ascertain the actions on a new front sheet piling wall for an embankment below an elevated structure; 3rd example of a failure mechanism

In fig. R 45-1d, the symbols have the following meanings:

V	Dead load of the soil $FCDE$ including water load $ABDE$
W_1, W_2	Water pressure forces from right, left on the section surfaces DB, EA
W	Water pressure force on failure plane ED
Q	Failure plane force of failure plane ED
Q_1	Characteristic value of required toe support force of existing front sheet piling wall
Q_2	Supporting force required for equilibrium
φ	Angle of friction
δ_p	Angle of passive earth pressure
ϑ	Failure plane angle

11.3.3 Assumed loads for determining the active earth pressure on the new front sheet piling

The assumed loads for new sheet piling driven in front of an existing quay wall are shown in figs. R 45-1b to d. According to DIN 1054 loading case LS 1B should be used to verify the stability of the toe support of the new front sheet piling. The procedure is given in section 0.2.3: first determine the actions and resistances using characteristic values. The

547

relevant characteristic value for the toe support force Q_2 is derived by investigating several failure mechanisms, examples of which are shown in figs. R 45-1b to d. Varying actions should be increased by the ratio γ_Q/γ_G in these analyses so that verification of the limit state only needs to be carried out with the partial safety factor for permanent actions γ_G. The value of the toe support force Q_1 of any existing rear sheet piling (section 11.3.2 (2)) is taken from the characteristic values of the support reaction and internal forces for this wall. It is taken as an action on the soil body between this wall and the new front sheet piling. An inner section is considered on the active side of the new front sheet piling which cuts the sheet piling walls P_1 and P_2 together with the piles P_3 to P_5 underneath the piled structure with relieving platform.

Fig. R 45-1b shows the structural system and one of the failure mechanisms to be examined together with the corresponding vector polygon. The mechanism (see figs. R 45-1c and d) which requires the maximum toe support force Q_2 governs. This force is then to be taken as the characteristic action on the wedge of soil in front of the new sheet piling. The design value of the toe support force Q_2 is to be compared with the design value of the passive earth pressure for LS 1B in the limit state equation.

To calculate the design values of the internal forces in the front wall, Q_2 is multiplied by the partial safety factor for permanent actions. In selecting a direction for Q_2, the boundary conditions for active earth pressure on the new front wall may not be infringed. The distribution of active earth pressure on the new front wall depends on the wall's scope for movement. The distribution may be selected according to section 8.2, or as per DIN 4085. The pile forces are to be determined from a stability verification of the superstructure using the characteristic values of the actions. The forces carried here via skin friction and end bearing are transferred to the earth body between the two walls (ΔP forces in the figs. R 45-1b to d).

The effect of a flow force resulting from differences in level between groundwater and harbour water is to be taken into account by using the water pressure forces in all failure lines or element limits examined in conjunction with the mass density of the saturated soil. Figs. R 45-1c and d show examples of further mechanisms to be examined which are characterised by steeper failure lines through to a failure line between the toe or shear force zero point for a fixed new front wall and the toe of the existing front wall (fig. R 45-1d). Force Q_1 is now the support force of the existing front sheet piling and is found in a similar manner to Q_1 for the rear sheet piling.

11.3.4 Calculation for cohesive soils

A similar procedure can be used here. In soils with an effective cohesion c', the cohesion force $C' = c' \cdot l$ is also taken into consideration along the corresponding failure plane examined. In the case of water-saturated normally consolidated soils, the value c' is replaced by the value c_u, where $\varphi' = 0$.

11.3.5 Loads from water pressure difference

The water pressure difference acting on the front sheet piling depends, among other factors, on the soil conditions, the depth of the soil behind the wall and the availability of a drainage system. In new constructions with front sheet piling only and soil extending up to the underside of the relieving platform, the water pressure difference is considered as acting directly on the sheet piling as described in R 19, section 4.2. If in cases with rear sheet piling, the earth surface is below the outer water table, and if the new wall has a sufficiently large number of water drainage openings, R 19 applies to the assumed water pressure difference acting on the old sheet piling and the assumed flow pressure is according to section 11.3.2 (3). In such cases a water pressure difference equal to half the height of the waves anticipated in the harbour is assumed to act directly on the front sheet piling as a precautionary measure. As a rule, it is usually adequate to presume a difference in water table of 0.5 m as a characteristic value.

11.4 Calculation of planar piled structures (R 78)

There are basically two versions of planar piled structures available to compensate for differences in terrain elevation (see also drawings in R 200, section 6.8.2):

- Piled structure raised clear of an underwater embankment with sheet piling on the land side which compensates for the remaining difference in ground elevation at the top of the embankment.

- Piled structure with sheet piling on the water side:
  - strengthening structure in front of and over an existing piled structure to increase the design depth, whereby an existing embankment is usually preserved. The sheet piling should extend down to a certain depth – at least to the foot of the embankment – in an open design to prevent any build-up of differential water pressure;
  - construction of a new piled structure with fully backfilled sheet piling. The superstructure platform supported on the piles effectively screens the earth pressure on the sheet piling from surcharges (fig. R 78-1).

The loadbearing cross-section of the aforementioned waterfront structures, which are one-dimensional linear structures in the longitudinal

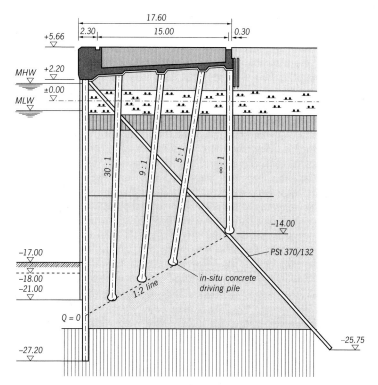

Fig. R 78-1. Executed example of a pile-founded structure with sheet piling on the water side

direction, is a planar piled structure, whose action and resistance is stated per continuous meter or per system grid dimension.

Bollard and fender actions can also be distributed in proportion to the calculation cross-section using the plate effect of the relieving platform within a section of the structure. However, if high concentrated tension loads arise due to the type of construction, and these loads are to be accommodated by a group of anchor piles, then the loadbearing capacity of the pile group is to be verified by suitable earth static calculations.

The internal forces for piles and superstructure can be ascertained in the plane case in an elementary manner:

The structural problem of the platform plate with the piles arranged in its plane can be represented correctly by means of an elastically supported continuous beam. The piles are represented by springs in the direction of the pile axis, the corresponding spring stiffness values depending on the pile characteristics. Here, the designer must ensure that the piles do not influence each other while carrying the loads, e.g. by ensuring an adequate spacing, and that any interaction between the piled platform and the piles remains negligible.

If these piles are relatively stiff, e.g. driven in situ concrete piles, it is possible to assume a rigid support in the calculations.

A plane frame bearing structure consisting of several pile "posts" and a "beam" representing the superstructure is equivalent to the continuous beam on elastic supports.

Fixed or pinned supports for the piles in the soil and on the superstructure platform, lateral bedding (and thus pile bending) as well as axial bedding can be handled by standard 2D software.

Accordingly, the analysis of waterfront structures with a "rigid" super-structure will only be applicable in special cases, e.g. when existing old quay structures with solid pilecaps are to be recalculated (fig. R 157-1). When analysing the anchorage to a sheet pile wall driven at a later date and considering the load deformation behaviour of the pile system, take into account the fact that the anchorage of the existing sheet pile wall is already under load and must be considered when calculating the new earth pressure, hydrostatic pressure and line pull loads. The designer must consider how the total loads are distributed among the old and new anchor piles. In order to minimise the deformations and avoid overloading the old anchors, the anchors to a new sheet pile wall can be prestressed by allowing for displacement of the anchor connection point to suit the local conditions.

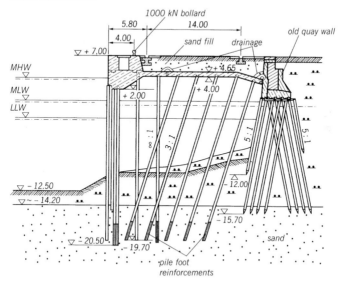

Fig. R 157-1. Executed example of a forward extension of a quay wall with elastic RCC relieving platform on steel piles

552

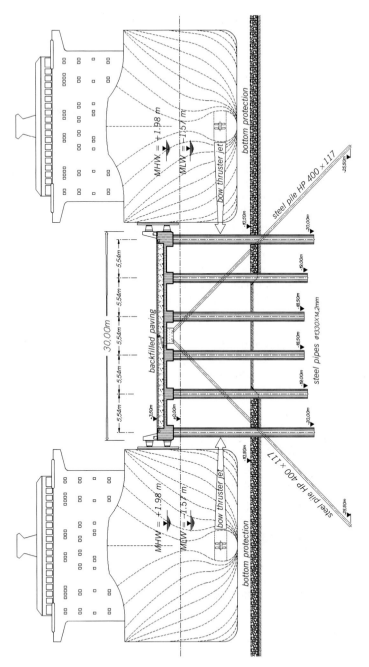

Fig. R 157-2. Example of a harbour pier

11.5 Design and calculation of general piled structures (R 157)

A general piled structure can be regarded as a three-dimensional structure of linear members on which a relieving platform is built as a superstructure. The loads acting are distributed over all piles through the combined loadbearing action of the superstructure as platform and plate, for effective carrying of the loads.

11.5.1 Special structures in the course of linear structures

Special structures are required when the piled structure walls described in R 78, section 11.4, have non-constant elements in some areas, such as cranks and corners, moles and pier heads, or if there are offset sections and differences in elevation at the front edge of the quay, e.g. like with Ro-Ro ramps. The foundation piles required here must be arranged so as to allow for the geometry of such special structures and to suit the space available under the relieving platform.

A plane arrangement of the piles is then normally no longer possible; on the contrary, highly complicated configurations are produced with a large number of intersection points and piles crossing in three dimensions, whose connection points to the superstructure may be positioned at several different elevations.

11.5.2 Free-standing piled structures

Free-standing, tall piled structures with elevated platforms are used mainly in the following cases:

- Soil with adequate bearing capacity is only present at a greater depth
- An essentially free passage for wave action is to be created
- Reduction of wave energy is by way of an arrangement of embankments or pile-supported structures instead of quay walls or other harbour structures with vertical walls
- Economic issues take priority

The relieving platforms for quay structures are mainly constructed from reinforced concrete, with steel sections usually being used for the foundation piles.

However, in hot climates both reinforced concrete and prestressed concrete piles are frequently used because here the specific requirements due to the effects of ice action are superfluous.

The way in which the superstructure platform rests on the foundation piles can be selected according to functional requirements or on the basis of the specific dimensions of the superstructure. The whole structural range is conceivable here: from statically determinate through to support with very few degrees of freedom, whereby extensive fixing of the possible degrees of freedom can be advantageous with regard to minimum system displacements.

Fixed degrees of freedom for the supports are expressed in the system calculations only in the larger number of equations to be solved simultaneously. However, this it not relevant when suitable computer programs are used because this is merely a numerical problem and not a structural engineering one.

Dimensioning the piles and determining the driving depths must take account of the fact that no permanent pile settlements occur even when assuming the least favourable combination of actions.

Part of the foundation must be arranged as pile bents so that the relieving platform cannot displace significantly in a horizontal direction. At the same time, this therefore reduces the bending moment load in all piles, provided there are no other significant horizontal actions, e.g.:

- lateral pressure on piles from flowing soil masses,
- strong currents, ice pressure, ice impact and the like.

If the above-mentioned assumptions are applicable, the piles can be calculated with sufficient accuracy assuming pinned connections at the relieving platform and at the pile toes, even if they are not designed as such structurally. Constructional measures at the pile head are to be arranged to deal with any unwanted fixity. This may be necessary in the case of large changes in length of the superstructure platform caused by temperature fluctuations.

Displacements of the pile heads caused by shrinkage during the construction period can be catered for by casting a long superstructure platform in sections, leaving wide strips between. These strips are then sealed with in situ concrete once the majority of the shrinkage has ceased. This procedure can usually keep the influence of shrinkage on the construction within harmless limits.

Relieving platforms are usually designed to be only 50 to 75 cm thick, so that they are to be seen as susceptible to bending, related to length changes in the piles. This can help to prevent the transfer of larger bending moments when, at the same time, the pile arrangement is sensibly adjusted to possible load positions. Actions occurring at each particular load position are thus carried essentially by the immediately adjacent piles, so that piles further away experience only a minimum action.

The platform thickness is selected on the basis of the bending moments and also with regard to ensuring a flawless transfer of the shear forces in structural and constructional terms from the concrete platform to the pile heads as a pile load; with a fixed head, this also applies to the subsequent pile bending moments.

There should be no direct traffic on the relieving platforms, a sand fill on the platform offers further operational and constructional advantages:

services, ducts, etc. can be accommodated within the sand layer, and this also makes it possible to reduce the loads on the relieving platform and the piles from locally influences, thanks to this compensating layer. In particular, sand fills ≥ 1.0 m thick are generally extremely favourable because this thickness makes it possible to accommodate all necessary services within the sand fill, and no vibration actions from vehicle operations need to be taken into account for the substructure.

If the actions from road and rail traffic are assumed to be uniformly distributed loads, the layer thickness must be selected according to R 5, section 5.5.

11.5.3 Structural system and calculations

An appropriate representation of the relieving platforms with 3D pile structures for structural calculations results in a model consisting of the elastic platform on elastic supports, i.e. from a structural point of view and in terms of carrying the load, the piled relieving platform acts as a flat slab supported on the piles with or without column heads. As in R 78, section 11.4, the piles are idealised as elastic springs acting in the axial direction, and the pile length between the hinges is taken to be the elastic length.

If it is not possible to restrict the procedure to approximate solutions, such as elastic continuous beam calculations with load carried in two orthogonal directions as per R 78, more demanding computing methods have to be used which can only be carried out using computers because of the numerical workload involved.

These computing methods are available in the form of software packages for platforms and folding structures, and work, for example, using the deformation method or finite element method.

These tools can be used to solve tasks with any degree of complexity for an elastic relieving platform on elastic piles, with adequate accuracy and without any fundamental difficulties.

In addition, computer programs for loadbearing platforms on elastic supports provide a possibility for calculating the loadbearing structure with good approximation. Relieving platforms with strengthening beams over the pile heads are taken into account realistically as loadbearing platforms with beams of different stiffness.

An adequate representation of the 3D pile structure serving as a support for the platform or loadbearing platform must be elaborated individually with the elements and tools provided in the software.

The loads acting on the whole relieving platform and piled structure are then most favourable when the piles are arranged in such a way that the negative bending moments on all supports in one axis and the maximum pile loads are all approximately equal.

However, this ideal optimum cannot always be achieved because of the frequent presence of marginal disturbances, such as influences from crane operations, hawser pull, vessel impact and the like, particularly in view of the fact that the pile positioning is also affected by structural and constructional irregularities.

11.5.4 Advice on construction (see also [96])

In order to arrive at the most economical overall solution possible, the following aspects are to be observed, among others:

- If necessary, the berthing forces of larger vessels should be absorbed by fendering in front of the platform with heavy-duty fender panels (fig. R 157-2). If this is not entirely possible, the fender must transfer that part of the impact action into the piled structure, which can not be absorbed by the fender structure.
- A heavy mooring bollard may also be installed at the top of the wall and combined with the fender construction.
- Local, limited horizontal forces, e.g. from line pull and vessel impact, will be spread over all piles in a block through the superstructure platform which is very stiff in its plane.
- In the case of small ships, fender piles are to be installed as protection for the structure piles and the ships themselves.
- Craneway beams are integrated into the reinforced concrete relieving platform as a structural component.
- If necessary, vertical loads from crane operations are accommodated by additional piles on the craneway axis.
- In order to disrupt the bending moment progression as little as possible, folds in the relieving platform may only be located over a row of piles (fig. R 157-1).
- In tidal areas it is practical to plan the lower edge of the relieving platform above MHW to remain independent of the normal tide when producing the relieving platform (figs. R 157-1 and R 157-2).
- Wave action is to be taken into account when designing the formwork for the relieving platform.
- The rows of vertical and raking piles are offset with respect to each other (fig. R 157-1).
- Horizontal keying is to be arranged as a rule between the blocks of relieving platforms.
- In the case of a long block of a free-standing pier platform, the horizontal longitudinal forces are absorbed by means of pile bents arranged in the middle of the block with the piles having the shallowest inclination possible.
- In the case of a long and also wide block, the horizontal forces in the transverse direction are absorbed by additional pile bents at both block ends (approximately in the longitudinal axis of the structure), again with the piles having the shallowest inclination possible.

- As a result of the absorption of the horizontal forces in the manner described above, the stresses in the relieving platform and piles are minimised.
- The relieving platform of a large pier can be concreted on sliding or travelling formwork which is supported chiefly on vertical piles.
- Special attention is to be paid to possible corrosion problems at the point of fixity of steel piles. This applies principally to zones with high corrosion risks (seawater, briny water).

A concrete piled platform to a large pier can also be cast on sliding or travelling formwork which is preferably supported on vertical piles. Components which deviate from the standard cross-section and thus disturb construction progress, such as the nodes of the raking pile bents named above, require additional measures:
As far as possible, the raking piles are driven in from the relieving platform after concreting, through driving recesses left in the super-structure, and the pile bent is connected by means of local reinforced concrete plugs in a subsequent concrete pour.
The driving recesses can consist of the shrinkage seals named in section 11.5.2, which may have to be widened locally for this purpose. The designer must check to what extent the stability of the blocks separated by joints is guaranteed, and whether temporary stiffening over these recesses is necessary.

11.6   Wave pressure on piled structures (R 159)

See section 5.10.

11.7   Verification of overall stability of structures on elevated piled structures (R 170)

See section 3.4.

11.8   Design and dimensioning of piled structures in earthquake zones (R 127)

11.8.1   General

Please refer to R 124, section 2.13, regarding the general effects of earthquakes on piled structures, permissible stresses and the required safety factors. However, the risk of resonance phenomena should be checked for tall and slender structures.
When designing piled structures in earthquake areas, it must be taken into account that as a result of the seismic effects, the superstructure with its full live loads and other masses result in additional horizontal

mass forces on the structure and its foundations. The cross-section must therefore be designed to reach an optimum solution between the advantage of shielding the active earth pressure through the relieving platform and the disadvantage of the extra forces to be absorbed from seismic acceleration.

11.8.2   Active and passive earth pressures, water pressure difference, live loads

The remarks in sections 2.13.3, 2.13.4 and 2.13.5 in R 124 apply accordingly. It must be observed that in the case of an earthquake, the influence of the live load prevailing behind the relieving platform including the additional soil dead load – resulting from the additional horizontal seismic actions – acts at a shallower angle than normal so that the shielding is less effective.

11.8.3   Absorption of the horizontal mass forces of the superstructure

The horizontal mass forces arising from an earthquake can act in any direction. At right-angles to the quay wall they can generally be absorbed without any difficulties by means of raking piles.

There may be problems involved in accommodating pile bents in the longitudinal direction of the structure. If the soil backfill to front sheet piling extends up to the underside of the relieving platform, the longitudinal horizontal loads can also be transferred by pile bending by embedding the piles in the ground. However, it must be verified that the resulting displacement is not too great. Limit values of about 3 cm apply here.

A superstructure built over an embankment results in a considerable reduction of the total active earth pressure loads acting on the structure. In this case the relieving platform should be designed to be as light as possible, to achieve minimum horizontal mass forces.

11.9   Stiffening the tops of steel pipe driven piles (R 192)

When driving steel pipes, there is a risk of the tops bulging in the upper section of the pipe, particularly in the case of pipes with relatively small wall thicknesses. This can mean that the pipes cannot be driven to the planned depth. In order to prevent bulging, the pile head must be stiffened in such cases. Various methods are available for this; the following measures have proven effective:

(1) Weld several steel brackets approx. 0.80 m long to the external wall of the pipe vertically (fig. R 192-1). This method is relatively simple and economical, as welding takes place on the outside only.
(2) Weld steel plates approx. 0.80 m long inside the head of the pipe to form a cross (fig. R 192-2). However, this method is more labour-intensive than the method described under (1).

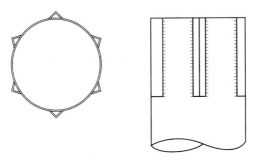

Fig. R 192-1. Bracing with brackets welded on externally

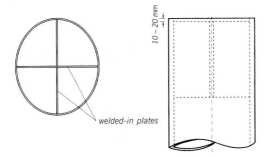

10 – 20 mm

welded-in plates

Fig. R 192-2. Bracing with welded-in plates

12   Embankments

12.1   Slope protection (R 211)

12.1.1   General

Embankments in loose rock are permanently stable against unsteady hydraulic loads only at a very shallow angle. On steeper banks slope protection is necessary to guarantee the stability against hydraulic loads and adequate overall stability of the embankment.

In selecting the slope of the embankment, any technical advantages of a shallow slope must be weighed against the disadvantages of the increase in the area of the embankment which needs to be stabilised, as well as the inefficient land use of this increased area. In other words, a slope should be chosen which results in a proper balance between construction and maintenance costs on the one hand, and the value and possible productive and ecological use of the land on the other.

The following conditions are valid for the design:
* the angle of the protected area should be as steep as possible while still ensuring stability, and
* the construction works should be carried out by mechanical plant wherever possible.

The information given in this section frequently relates to inland waterways but also applies in principle to any shoreline.

12.1.2   Loads on inland waterways

Hydraulic loads induced by navigation may be subdivided into propeller jet, drawdown on both sides of the ship (limited by the bow and stern waves), backflow, follow-up current and secondary wave systems. These components form individual loads on the waterfront and its revetment. The backflow acts predominantly along the embankment below the still water table while, for example, the transverse stern wave and the secondary wave system act on and in the vicinity of the still water table. For details of wave loads due to water inflow/outflow and vessel movements see R 185, section 5.8, and R 186, section 5.9.

The bank revetments must be designed to resist the hydraulic shear forces and flow pressures. The question as to which loading component is a governing factor for the design always depends on the anticipated extent of navigation as well as on vessel drive power and the cross-section of the navigation channel.

The natural water level differences plus those that occur due to passing vessels and the tides also result in loads on the embankment. In this case a distinction must be made between the upward water pressure below the revetment of the bank, which strengthens as the permeability of the revetment decreases, and the hydraulic gradient in the filter layer and in the subsoil.

Slope protection must be designed in such a way that hydrostatic or flow pressures acting upwards do not cause the stones to slide or lift.

On permeable revetments there is a constant exchange between the water in the watercourse and the groundwater. In a restricted navigation cross-section, the drop in the water level as a vessel passes takes place faster than the corresponding drop in pressure in the pore water, depending on the permeability of the subsoil. The result is an excess pore water pressure in the soil whose decrease towards the surface can be approximated with good accuracy using an exponential function [223]. An excess pore water pressure in the soil reduces the soil's shear strength and hence leads to a loss in stability which must be taken into account when designing the embankment and the revetment. Soils with low permeability which, however, exhibit no cohesion, e.g. silty fine sands, are particularly at risk.

12.1.3 Structure of slope protection

Slope protection can consist of the following elements, although not all of these elements are inevitable in every situation:

A revetment consists of (from top – outside – to bottom – inside):
- surface layer
- cushioning layer
- filter/separating layer
- impermeable lining
- levelling layer

Other elements in a revetment are the toe protection and, if required, connections to other components.

12.1.3.1 Surface layer

The surface layer is the uppermost, erosion-resistant layer of slope protection, which may be permeable or impermeable. The design is essentially governed by hydraulic aspects: currents and waves may not lead to a shifting of the elements in the slope protection. Impacts must be absorbed without damage. At the same time, energy has to be dissipated. The advice regarding scour protection in R 83, section 7.6.3, also applies in principle to the surface layers of embankments.

The surface layer is frequently dumped material. In recent years very good formulas for the design of such surface layers have been developed through experimentation and fundamental research (see, for example, [100, 100a]). The stone material of the surface layer must be solid, hard and exhibit a high weight density, be frost- and weather-resistant as well as light-fast. Concrete columns can be used instead of natural stone.

In order to increase the resistance to unsteady hydraulic loads, dumped materials can be partly or fully grouted with a cement bonded mortar. Partial grouting must be carried out in such a way that the surface layer retains sufficient flexibility and permeability. Ideally, partial grouting will lead to conglomerates interlocked with each other that possess the adaptability and erosion resistance of larger single elements. A similar effect is achieved with concrete blocks laid to form a sort of paving, provided the surface layer is sufficiently permeable and the stones are joined together. However, the weight per unit area is limited with such blocks. Impermeable (fully grouted) surface layers form both an impermeable lining (see also section 12.1.3.5) and a protective layer with increased resistance to erosion and other mechanical damage. They are less thick than a permeable revetment but are rigid and, from the ecological viewpoint, disputed.

12.1.3.2 Cushioning layer

Cushioning layers are called for in special situations to protect against very high loads (e.g. especially large armourstones on a geotextile filter). They must satisfy the filter criteria with respect to the adjoining layers (see section 12.1.3.3).

12.1.3.3 Filter/Separating layer

Basically, slope protection must be built to prevent material transport out of the subsoil. For this purpose, filter layers with appropriately coordinated properties may be required between the subsoil and the surface layer. Filter stability is essential for all the layers.

The filter should be designed according to geohydraulic aspects (i.e. governed by pore water flows and their interaction with the particle structure). Both graded gravel and geotextile filters are suitable (see R 189, section 12.5). In both cases water permeability and filter reliability (filter function and separating function) are the characteristic design parameters.

As a result of the turbulent inflow and the alternating through-flow, graded gravel and geotextile filters in slope protection and bottom protection – in contrast to their use for drainage applications with a one-way flow only – are exposed to high unsteady hydraulic loads. Therefore, the properties of the filter must be especially carefully matched to the soil, the surface layer and the loads.

12.1.3.4 Levelling layer

Levelling layers are used to create a level base where this cannot be achieved through excavation/dredging directly. Furthermore, a levelling layer can also be used in a sense of a filter layer over very inhomogeneous soil, which necessitates a filter with very different local variations. In such cases the particle distribution of the levelling layer roughly matches the in situ, coarser soil material and acts as a filter for the finer material. A levelling layer may actually amount to a soil replacement measure, e.g. if an embankment cannot be given a clean profile because the in situ soil may fluidize.

12.1.3.5 Impermeable lining

Impermeable linings are required when the loss of water (e.g. from a waterway) becomes excessive or when interaction with the groundwater is to be prevented. Basically, we distinguish between two types of linings: surface linings which combine sealing and protecting functions (a covering or a riprap layer fully grouted with asphalt or cement bonded mortar), and separate linings (clay, soil mixtures, geosynthetic clay liner). For advice on mineral lining, see R 204, section 7.15.

An impermeable revetment – in contrast to the fully permeable solution – leads to limited excess water pressure which must be considered in the design. The magnitude of the excess water pressure depends on the magnitude and rate of the fluctuations in the water level on the embankment and the groundwater levels behind the revetment at the same time. Such excess water pressure reduce the potential friction force between the lining and the underlying material.

Asphalt exhibits a viscous behaviour. This may, on the one hand, lead to roots and rhizomes penetrating such surface layers and, on the other, to undesirable creep phenomena. Asphalt armour layers in the form of coverings or fully grouted layers should be regarded as rigid solutions when a not inconsiderable ground deformation below the revetment due to erosion, suffusion or subsidence takes place faster than the creep process in the asphalt (which is always very slow). As the viscous behaviour of such a revetment has not yet been thoroughly researched, when using asphalt revetments the friction resistance may never be less than the dead load component of the revetment in the direction of the slope in order to provide a margin of safety against creep (creep criterion). All loading cases require that the dead load component of the revetment perpendicular to the slope is always greater than the maximum hydrostatic pressure directly underneath so that the surface layer can never be lifted (uplift criterion).

12.1.3.6 Toe protection

The forces acting down the slope may be only partly accommodated through friction between the revetment and the subsoil, the remainder must be transferred into the subsoil via a toe support. If all the forces had to be transferred via friction, this would frequently result in uneconomically thick revetments or excessively shallow embankments. On embankments that extend to the base of the watercourse, the revetment therefore generally continues into the subsoil at the same angle, unless it joins to bottom protection. The embedment depth should be not less than 1.5 metres in the case of soils at risk of erosion (see, for example, [124] and [129]).

The alternative is to provide sheet piling at the foot of the embankment in all soils suitable for driving. This corresponds to the design of a partial embankment behind a sheet pile wall (see also R 106, section 6.4, and R 119, section 6.5).

12.1.3.7 Transitions

The transitions to structures and covering materials or the subsoil require special attention. Any changes in the elevation of the construction or the loads must be taken into account. Many cases of damage at such transitions – attributable to errors during design and/or construction – are known in practice. When joining a revetment to a sheet pile wall or another part of the construction, care must be taken to ensure a good transfer of forces, filter stability and erosion protection at the junction. Likewise, the points at which filter layers and seals connect to structures must be very carefully planned and built.

12.2 Embankments in seaports and tidal inland harbours (R 107)

12.2.1 General

In port areas where bulk cargo is handled, at offshore berths and in the vicinity of harbour entrances and turning basins, the shores can be sloped and permanently stabilised even if the tidal range and other water level fluctuations are quite large, provided that no heavy or long-lasting mud deposits are expected. However, certain fundamental construction aspects must be observed if extensive and constant maintenance work is to be avoided.

As large seagoing vessels may not, as a rule, proceed under their own power in harbours, it is principally the large tugboats and inland vessels, as well as occasionally the small seagoing vessels and motor coasters, which are capable of eroding the shore by the action of the propeller jet or bow and stern waves, and this damage may extend to about 4 m below the water level (see R 83, section 7.6). The ongoing development of drives (bow and stern thrusters, azimuth drives) can cause particular actions which may need special solutions in individual cases (e.g. ferry terminals).

12.2.2 Design examples with permeable revetment

Fig. R 107-1 shows a solution built in Bremen.

The transition from the protected to the unprotected section consists of a horizontal berm, 3.00 m wide, covered with rip-rap. Above this berm,

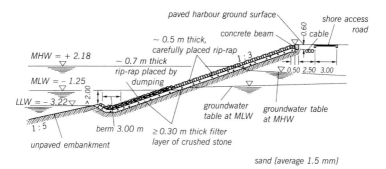

Fig. R 107-1. Standard construction of a harbour embankment in Bremen, with permeable revetment, example

the stone revetment is placed on a 1 : 3 slope. A concrete beam, 0.50 m × 0.60 m, grade B 25, lies flush with the port ground surface, forming the upper boundary to the revetment. The lower part of the revetment consists of heavy rip-rap which was dumped in a layer about 0.70 m thick to an elevation just above MLW. Extending upward from this point is a cover layer, about 0.50 m thick, consisting of rip-rap which has been carefully placed to form an interlocking protective rock cover which cannot be damaged by wave energy.

A 3.00 m wide shore access road, constructed for heavy vehicles (fig. R 107-1) was built 2.50 m inboard of the aforementioned concrete beam, for maintenance work on the stone revetment and as a walkway to the ship berths. Where required, the power supply cables for the port facilities and the shore beacons, as well as for the telephone lines, etc., were buried in the strip between the concrete beam and the shore access road.

Fig. R 107-2 shows a typical embankment cross-section in the Hamburg. In this solution an abutment consisting of crushed bricks, approx. 3.5 m³/m, was placed at the revetment toe on a geogrid. This was covered by the revetment in two layers for most of the embankment height.

The embankment revetment was taken down generally to only 0.70 m below MLW, taking account of prevailing soil conditions. Any scouring below can be repaired by simply adding more crushed bricks. Although ice action can tear pieces of crushed brick away, the costs involved in repairing such damage is comparatively negligible in Bremen.

In order to take account of the idea of a natural design of waterfront embankment with this kind of crushed stone revetment, an embankment with a vegetation pocket has been developed in Hamburg. The normal cross-section as per fig. R 107-2 is expanded in the area of MSL +0.40 m by approx. 8.00 to 12.00 m, depending on the space available. The

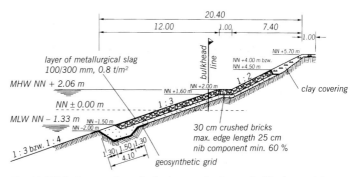

Fig. R 107-2. Construction of a harbour embankment in Hamburg with permeable revetment, example

resulting pocket is filled with approx. 0.40 to 0.50 m clay; here the substructure of crushed bricks is reinforced to a thickness of 0.50 m. Between clay and crushed bricks, a 0.15 m thick layer of Elbe sand is inserted as an aerating zone.

The vegetation pocket is planted, depending on the habitat horizons, with bulrushes (*Schoeno-Plectus Tabernaemontani),* sedge (*Carex Gracilis)* and reeds (*Phragmitis Australis*). Above the berm at MSL +2.0m, this also concerns the normal stone revetment, willow cuttings are inserted in the revetment. Such planting activities should be performed in April/May because of the better growth conditions.

The special design with vegetation pocket, however, is only possible in areas with sufficient space or moderate wave run up from ship traffic and wave action.

Fig. R 107-3 shows a corresponding cross-section.

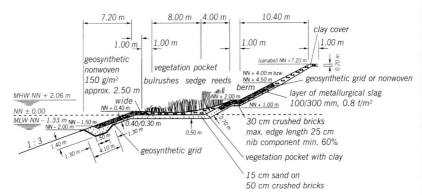

Fig. R 107-3. Construction of a harbour embankment with permeable revetment and vegetation pocket in Hamburg, example

566

Fig. R 107-4 shows a construction with permeable revetment selected for the Port of Rotterdam. Apart from the revetment itself, it is generally identical to the Rotterdam solution with an impermeable revetment, so the remarks made in section 12.2.3 are also applicable here with regard to further details.

Fig. R 107-5 shows another solution for a permeable harbour embankment.

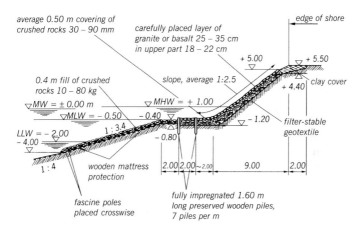

Fig. R 107-4. Construction of a harbour embankment in Rotterdam, with permeable revetment

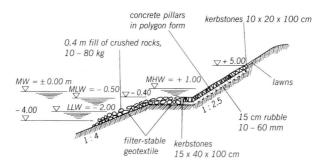

Fig. R 107-5. Construction of a harbour embankment near Rotterdam, with permeable revetment

12.2.3 Design examples with impermeable revetment

Fig. R 107-6 shows an impermeable revetment as developed and tested in Rotterdam, with a so-called "open toe" for reducing the excess water pressure. This "toe" consists of coarse gravel fill ≥ 30 mm dia., confined between two rows of tightly spaced wooden piles, which have been fully impregnated with an environmentally friendly preservative and are 2.00 m long and approx. 0.2 m thick. Subsequently, the coarse gravel layer in the "open toe" at the lower end of the asphalt-filled rubble fill is covered with a permeable layer of 25 to 35 cm granite or basalt paving stones.

There is a geotextile filter under the coarse gravel layer, which also extends below a substantial part of the impermeable cover layer.

The submerged part of the revetment, which in sand has a slope of 1 : 4, begins with the 2.00 m wide berm adjacent to the "open toe". The embankment is covered with a wooden mattress to a point about 3.50 m below MLW. On top of this, tightly wrapped fascines are placed horizontally, parallel to the shore. The top layer consists of crushed stone to a thickness of between 0.30 and 0.50 m. The surcharge of the cover layer should be about 3 to 5 kN/m^2, depending on the strength of the wave action.

The asphalt-grouted crushed stone revetment extends from the land-side row of wooden piles to about 3.7 m above MLW, and has a mean slope of 1 : 2.5. In individual cases its thickness may need to be adapted to the magnitude of the water pressures acting on its underside. In the normal case, the thickness tapers from 0.5 m at the bottom to 0.3 m at the top. The stones have an individual weight of 10 to 80 kg.

The upper edge of the revetment, for a vertical height of about 1.3 m, has a slope of 1 : 1.5. The lower portion of this zone is paved with

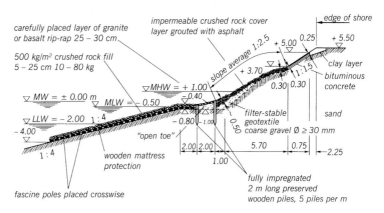

Fig. R 107-6. Construction of a harbour embankment in Rotterdam, with impermeable revetment

bituminous concrete to a thickness of 0.30 to 0.25 m, while the remainder, up to the elevation of the shoreline, is at the same angle and covered with clay only to a height of 0.50 m. This unpaved strip will facilitate subsequent installation of pipes and other services.

12.3 Embankments below quay superstructures behind closed sheet piling (R 68)

12.3.1 Loading on the embankment

Besides the static soil loads, water currents in the longitudinal direction of the quay wall and groundwater flowing transverse to the structure may affect the embankments. The latter is particularly detrimental when the groundwater table in the embankment is higher than the outer water level, so that the seepage flow exits from a slope (R 65, section 4.3). The inclination of the embankment and its protection must therefore be adapted to the position of the relevant water table, the magnitude and frequency of the water level fluctuations, the lateral groundwater flow, the subsoil and the structure as a whole, so that the stability and erosion resistance of the embankment are ensured.

12.3.2 Risk of silting up behind the sheet piling

There is a risk of silting up behind sheet piling in tidal areas if the ingress of outer water is not reliably prevented by appropriate measures. This can cause considerable additional costs and consequences for structure. Generally, the ingress of outer water is accepted and the accumulation of silt is prevented by weepholes in the sheet piling just above the toe of the embankment. The spacings and the cross-sections of the weepholes must be selected in accordance with the prevailing conditions. The hydraulic influence of the weepholes must be planned with particular care for the inward and outward tidal flow conditions. In some cases, facilities for extracting off the silt must be provided.

12.4 Partially sloped embankment in inland harbours with large water level fluctuations (R 119)

See R 119, section 6.5. The information given in R 211, section 12.1, applies to the design of the slope protection.

12.5 Use of geotextile filters in slope and bottom protection (R 189)

12.5.1 General

Geotextiles in the form of woven, non-woven and composite fabrics are used for slope and bottom protection.

Hitherto, plastics such as polyacrylic, polyamide, polyester, polyvinyl alcohol, polyethylene and polypropylene have proven suitable as rot-resistant materials for geotextile filters. Information on their properties is given in [100].

Geosynthetics that are to be used for slope and bottom protection must comply with the properties listed in DIN EN 13 253. Limiting values for these properties result from the individual applications. Examples of limiting values can be found in [100a] and [131].

The advantage of geotextiles lies in their mechanical prefabrication, as a result of which very uniform material properties can be achieved. Geotextiles are also suitable for installation underwater, provided certain installation rules and product requirements are observed.

Geotextile filters exhibit a very low self-weight. In some circumstances this can have an effect on the stability and a thicker armour layer may be required. In non-cohesive fine-grained soils there is the risk of soil liquefaction and erosion below the geotextile due to the action of waves in the splashing zone and below. This can lead to deformation of the revetment as a result of redistribution of the soil. To prevent this, a sufficiently heavy revetment must be selected and the mechanical filter stability guaranteed [128].

12.5.2 Design principles

With regard to mechanical and hydraulic filter efficiency, installation stresses such as tensile and punching forces, and durability with respect to friction stresses with unbonded cover layers, geotextile filters for slope and bottom protection can be designed in accordance with the rules stated in [100], [128] and [165]. [128] and [165] contain design rules for dynamic loads based on experience with unsteady hydraulic loads. [128] contains design rules based on turbulent flow tests ("soil type method"), aimed at unsteady hydraulic loads. Both sets of recommendations are essentially based on German experience. International experience and design principles are to be found in [145] to [147] and [159], for example.

When selecting geotextile filters, in addition to their mechanical and hydraulic filter efficiency, adequate resistance to the stresses and strains during construction must be guaranteed. When being installed underwater among waterborne traffic, relatively thick ($d \geq 4.5$ mm) or heavy ($g \geq 650$ g/m^2) geotextiles have proven efficient [128].

12.5.3 Requirements

When laying underwater, the tensile strength at failure must be min. 1200 N/10 cm in the longitudinal and transverse directions.

With armour layers of riprap, the punching resistance must be verified [130]. This is not usually a problem with stone weights up to 30 kg.

If abrasive movements of the armourstones can occur under wave or flow loads, the abrasion resistance of the geotextile must be verified [130].

12.5.4   Geocomposits

If necessary, the properties of the geotextile can be improved by using special measures. A number of examples are given below:

Geocomposite with a coarser bottom fabric – designed properly to the soil grain size – can achieve a certain interlocking effect with the subsoil and hence help to stabilise the boundary layer. However, as such supplementary fabric can also have negative effects, it is better to increase the surcharge on the geotextile.

Fascine grids on top of geotextiles have long since been a workable solution when building fascine mattresses (willow mats). They enable the geotextile to be laid without folds and creases, and increase the positional stability of the riprap on top of the geotextile. A combination of woven and nonwoven fabric can be used to increase the friction between geotextile and subsoil.

A factory-applied mineral inlay of sand or some other granular material ("sand mattress") increases the weight per unit area of the geotextile, which lends the geotextile a certain positional stability, even when exposed to lighter wave and flow loads (up to 1 m/s for a filling weight of approx. 8 to 9 kg/m^2), before the loose material is placed. It also reduces the risk of folds and creases.

When a bentonite inlay is used, this enables an impermeable lining to be constructed, provided the overlaps are suitable. The sealing function of this is equivalent to other sealing systems (e.g. mineral linings to R 204, section 7.15, or impermeable armourlayers as in R 211, section 12.1.2).

12.5.5   General construction advice

Before installation, delivery in accordance with the contract must in all cases be checked following the appropriate delivery conditions, e.g. listed in [130] and [131]. The geotextiles supplied must be carefully stored and protected from UV radiation, the effects of the weather and other damaging influences.

In order to rule out operational defects, correct positioning of the top and bottom sides must be ensured when placing multilayer geotextile filters (composite fabrics) with filter layers graded by pores.

When laying, take care to avoid folds and creases because these can create channels for water and the possibility of soil transport.

Nailing to the subsoil at the top edge of the embankment is allowed only if the geotextile will not be subjected to stresses and strains during the further course of construction. An improvement on rigid fixing is to build the geotextile into a trench along the top edge of the embankment.

This permits controlled movement of the geotextile under the high loads of the subsequent construction phases.

As geotextiles can float or rise when laid underwater, they must be fixed in position by a surface layer or a cushioning layer immediately after laying. Geotextiles should not be installed at temperatures below +5 °C. Particularly important regarding the soil-retaining capacity of a geotextile filter is the careful joining of the individual sheets, which is possible by sewing or by overlapping. In the case of sewing, the strength of the seam must comply with the required minimum strength of the geotextile. Overlaps must be at least 0.50 m when laying in the dry with an embankment inclination of 1 : 3 or less, but 1.00 m when laying in underwater and when the slope is steeper. In the case of soft subsoil, the designer should check whether, in certain cases, larger overlaps should be used. Transverse shortening of the geotextile sheets when placing stone layers must not result in uncovered areas.

On-site seams and overlaps are permitted only in the direction of the slope. If overlaps across the slope are unavoidable in exceptional circumstances, the strip lower down the slope must overlap the upper strip.

When laying above the water level, prevent the relatively lightweight geotextile being blown by the wind.

When installing geotextile filters underwater, even with operating waterborne traffic, the following aspects should also be observed to attain a crease-free geotextile filter placement, a level placement without distortion and a complete cover with sufficient overlapping:

- The construction site must be marked in such a way that it may be passed by all vessels only at slow speed.
- The subgrade must be carefully prepared and must be free of stones.
- The placing equipment must be positioned in such a way that currents and drawdown from passing vessels cannot impair the placement procedure and the geotextile is not subjected to any unacceptable forces (equipment on stilts is advantageous).
- The risk of the geotextile strips floating must be prevented by suitable placement techniques. It is advantageous to press the geotextile onto the soil during placement. The time lag between placing the geotextile and placing the stone layer must be kept to a minimum, and the stone layer should follow closely behind the geotextile.
- Fixing of the geotextile strips on the placing equipment must be released when placing the stone layer.
- Riprap layers are to be placed on embankments with geotextiles from bottom to top.
- Underwater placement should be allowed only if the contractor has proved that he can fulfil the requirements.
- Inspection by divers is indispensable.

13   Dolphins

13.1   Design of resilient multi-pile and single-pile dolphins (R 69)

13.1.1   Design principles and methods

Dolphins are either berthing dolphins designed for the action of ship forces, or mooring dolphins designed for the forces of line pull, wind pressure and flow pressure. The elastic–elastic method is used for their design.

Upon ship impact the impact force $F_{S,k}$ and the cross-sectional values for number of sections and dimensions are varied to such an extent that the required work capacity A is reached (exist $A_k = \frac{1}{2} \cdot F_{impact,k} \cdot f$). At the relevant impact force the stresses caused by the design values of the effects in the dolphin section must be less than or equal to the yield stress, and the deflection f at the point of application of the impact action must be acceptable for the port operations.

Under the force of line pull $F_{ten,k}$ or wind pressure $F_{wind,k}$ and/or flow pressure $F_{Q,k}$ the stresses caused by the design values of the effects on the dolphin section must be less than or equal to the maximum permissible stresses.

If the dolphin is to be designed for both the above functions, the design task cannot be clearly solved due to the excessive number of boundary conditions. In such cases the only answer is to find a suitable compromise between the requirements and the section to be selected which achieves satisfactory results from the engineering, operational and economic viewpoints.

The following assumptions apply when determining the passive earth pressure:

(1) Dolphin width
Resilient dolphins and multi-pile dolphins can be calculated by taking account of the total dolphin width B measured orthogonal to the direction of the force. In case of resilient multi-pile dolphins this is the distance between the outer edges of the edge piles.

(2) Specific density
The effective specific density of the soil strata involved is taken to be the specific weight density $\gamma'_{k,i}$ of submerged soil.

(3) Inclination of passive earth pressure
Like sheet pile walls, dolphins can be calculated with the maximum possible angle of inclination of the passive earth pressure. When using curved slip planes in the calculations, this means assuming a maximum of $\delta_p = \varphi'_k$, when using straight slip planes (permissible up to max. $\varphi'_k \leq 35°$) assuming a maximum of $\delta_p = -\frac{2}{3} \varphi'_k$ provided the condition $\Sigma V_k = 0$ is satisfied (fig. R 69-1). Otherwise, take a shallower angle for

the passive earth pressure resultant. An unfavourable upward vertical component of the line pull force is to be taken into consideration for mooring dolphins.

The vertical downward forces, in addition to the weight of the dolphin and the mass of submerged soil within the perimeter of the dolphin, can be taken to be the vertical characteristic skin friction on the side surfaces $\alpha \cdot t$ parallel to the direction of movement of the dolphin, and for $\delta_p > 0$ the vertical component $C_{v,k}$ of the equivalent force C_k as well.

Verification of bearing capacity at ultimate limit state ULS 1B

(A) The following partial safety factors are to be used for predominantly static forces:

(1) Berthing dolphin
This type of loading is classed as an extreme loading case according to section 5.4.4, which means that all partial safety factors for effects and resistances are equal to 1.00:

- actions:
    - impact force $F_{\text{impact,k}}$ $\qquad$ $\gamma_Q = 1.00$

- resistances:
    - passive earth pressure $E_{p,k}$ $\qquad$ $\gamma_{Ep} = 1.00$
    - yield point of steel $f_{y,k}$ $\qquad$ $\gamma_M = 1.00$

(2) Mooring dolphin
This type of loading is classed as load case 2 according to section 5.4.2, but the partial safety factor for variable forces is taken as 1.20. Like with sheet pile walls, the partial safety factor for passive earth pressure is selected according to section 8.2.0:

- actions:
    - line pull, line pull due to wind pressure, and
    - flow pressure $F_{P,k}$, $F_{W,k}$, $F_{S,k}$ $\qquad$ $\gamma_Q = 1.20$

- resistances:
    - passive earth pressure $E_{p,k}$ $\qquad$ $\gamma_{Ep} = 1.15$
    - yield stress of steel $f_{y,k}$ $\qquad$ $\gamma_M = 1.10$

(B) For not predominantly static forces take into account the fact that the permissible stresses for fatigue strength are lower than those for static strength.

The information in R 20, section 8.2.6.1 (2) apply to the capacity of the parent metal and the circumferential or butt welds to withstand stresses, i.e. verification of fatigue strength at working stress level must be carried out according to DIN 19 704 bearing in mind DIN 18 800 part 1, El. 741.

If the utilisation of the capacity to withstand stresses due to not pre-dominantly static forces is higher than allowed in DS 804 for steel grade S 355 J2G3 as per DIN EN 10 025 (formerly designated grade St 52-3 N), then verification is required of the permissible stress amplitudes of the fatigue strength in the parent metal and at the welds.

The fatigue strength depends greatly on the quality of the steel surface; it can decrease by up to 50% when subject to corrosion, which must be taken into account particularly for facilities being built in tropical marine conditions.

As the fatigue strength of welded connections is practically independent of the steel grade, the use of heat-treated fine-grained structural steels should be avoided if possible in areas where not predominantly static forces occur if there are welds transverse to the direction of the principal stresses.

13.1.2   Dolphins in non-cohesive and cohesive soils

The characteristic passive earth pressure force is calculated as a three-dimensional passive earth pressure $E_{ph,k}^r$ to E DIN 4085, section 6.5.2, eq. (88): $E_{ph,k}^r = E_{pgh,k}^r + E_{pch,k}^r + E_{pph,k}^r$. Here, $E_{ph,k}^r$ is the sum of the characteristic passive earth pressure force components for dead load, cohesion and, if applicable, bottom surcharge p_0 from the bottom to the depth h under consideration. The three-dimensional stress condition is taking into account by means of the theoretical equivalent dolphin widths to DIN 4085, section 6.5.2, eq. (84) to (87). These depend on the type of force and the respective depth h below the calculation bottom in relation to the dolphin width B.

The three-dimensional passive earth pressure distribution required for programs for linear member structures results from differentiation of eq. (88). It should be noted here that the areas $B \geq 0.3\,h$ and $B < 0.3\,h$ must be considered separately in each case when calculating the equivalent dolphin widths. Therefore, the progression of the passive earth pressure is evaluated for the position "near the surface" and starting from this boundary down to the theoretical base of the dolphin ("low position").

In cohesive strata the shear parameters from the undrained tests must be assumed owing to the "fast" load application by the impact force F_{impact}, i.e. angle of friction $\varphi_u' = 0$ and shear strength c_u'.

The characteristic equivalent force $C_{h,k}$ can be calculated according to the following equation by neglecting the active earth pressure from the condition $\Sigma H = 0$ (as is usual for dolphin calculations) (see fig. R 69-1):

$$C_{h,k} = E_{ph,mob.} - \Sigma F_{h,k,i}$$

($C_{h,k}$ is to be used for verifying $\Sigma V_k = 0$)

$\Sigma F_{h,k,I}$ = sum of characteristic actions

$E_{ph,mob.}^{r}$ = mobilisable three-dimensional passive earth pressure due to characteristic actions

$= E_{ph,k}^{r} / (\gamma_Q \cdot \gamma_{Ep})$

$E_{ph,k}^{r}$ = characteristic three-dimensional passive earth pressure

γ_Q = partial safety factor for forces

γ_{Ep} = partial safety factor for passive earth pressure

$C_{h,d} = C_{h,k} \cdot \gamma_Q$ = equivalent force C_h for calculating the additional embedment depth Δt

The additional embedment depth Δt required (fig. R 69-1) to accommodate the equivalent force C_h can be calculated with reference to R 56, section 8.2.9, and the terms used there:

$$\Delta t = \frac{C_{h,d}}{2 \cdot \dfrac{e_{ph,k}^{r'}}{\gamma_{Ep}}} = \frac{1}{2} \cdot C_{h,k} \cdot \gamma_Q \cdot \frac{\gamma_{Ep}}{e_{ph,k}^{r'}}$$

$e_{ph,k}^{r'}$ = ordinate of the characteristic three-dimensional passive earth pressure at the theoretical base on the equivalent force side

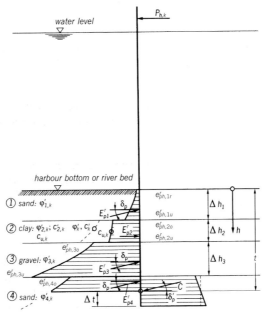

Fig. R 69-1. Dolphin calculation in stratified soil, three-dimensional passive earth pressure E_p^r and passive earth pressure e_{ph}^r ordinates according to DIN 4085

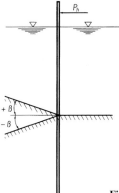

Fig. R 69-2. Dolphin at a sloping harbour bottom or river bed

The direction of the BLUM equivalent force C can be inclined at an angle of up to $\delta_p' = +^2/_3\,\varphi'$ to the perpendicular to the dolphin axis within the scope of the condition $\Sigma\,V = 0$.

For a sloping harbour bottom or river bed, the ground slope angle β is to be taken into account for the passive earth pressure coefficient K_{ph} in front of the dolphin and for K_{ph}' at the theoretical base on the equivalent force side (fig. R 69-2). The ground slope should lie within the sector between $\beta = +^1/_3\,\varphi_k'$ and $\beta = -^2/_3\,\varphi_k'$, where in these cases negative angles of inclination for the passive earth pressure may be selected only up to a value of $\delta_p = -^1/_3\,\varphi_k'$.

13.2 Spring constant for the calculation and dimensioning of heavy-duty fenders and berthing dolphins (R 111)

13.2.1 General

The spring constant c [kN/m] is defined as the ratio between the impact force P [kN] and the elastic deformation f [m] occurring in its line of force:

$$c = \frac{P}{f}$$

It is of special significance for the calculation and dimensioning of heavy-duty fendering and elastic berthing dolphins at berths for large ships. It determines the maximum impact force and deflection for the working capacity required for the energy absorption of the approaching ship according to the demands of port operations. The maximum allowable impact force for the type of ship concerned may not be exceeded.

Information concerning the allowable impact force that a particular ship can absorb on its hull or inner structural framing should be obtained from the shipping line concerned, or from Germanic Lloyd. In general, concentrated forces should be avoided. Structural elements to distribute the pressure (fender panels) are therefore required for heavy impact forces.

Not only the basic static and impact values are to be considered in selecting the spring constant c, but other aspects as well, especially with regard to navigational and structural requirements. Even under clearly defined conditions, only a narrow choice of values exists for the spring constant, but this limited latitude should be exploited in the selection of the correct spring constant. In high-grade fender systems and shock absorbers, provisions are made for accidental excessive impact by incorporating failure elements, break-off bolts and the like. By these means, both ship and structure are protected against damage. For the allowable effects of actions in the various loading cases, as well as the suitable steel grades and qualities, see R 112, section 13.4.

Since the spring constant corresponds to the stiffness of the fenders or the dolphin, a high spring constant produces hard berthing of a ship, a low spring constant a soft, and thus a less risky, one.

13.2.2 Determining factors for the selection of the spring constant

13.2.2.1 The magnitude of the required energy absorption capacity A [kNm] is determined by the vessel sizes encountered and their berthing speeds (see R 128, section 13.3, in conjunction with R 40, section 5.3).

Therefore, the smallest actual value of the spring constant min c will be determined by the formula

$$\min c = \frac{2\,A}{\max f^2}$$

insofar as the horizontal deflection f on reaching the yield point is stipulated or limited by max f for nautical, port operations or structural reasons.

13.2.2.2 A further criterion for the minimum stiffness of a berthing dolphin with or without simultaneous mooring tasks results from the structural load-carrying capacity of the structure by P_{stat} when classified in loading case 2. If a safety factor of 1.5 is chosen because of the inability to make an accurate determination of the maximum acting static load, the spring constant is:

$$\min c = \frac{1.5 \cdot P_{stat}}{\max f}$$

13.2.2.3 The upper limiting value of the spring constant c is determined by the maximum permissible impact force P_{impact} between ship's hull and fender or dolphin during berthing manoeuvres, provided max f has not already been determined for the maximum permissible impact force P_{impact}:

$$\max c = \frac{P_{impact}^2}{2\,A}$$

Permanent dynamic influences, primarily from strong waves, may also be of importance when selecting the spring constant, under certain circumstances.

13.2.3 Special conditions

The equations in section 13.2.2 fix only the limits within which the spring constant is to be selected. The final determination must take the following aspects into account:

13.2.3.1 Except for special circumstances, such as the desire for a larger energy absorption capacity at berths for large vessels, or the remarks in section 13.2.3.2, max f in general should not exceed about 1.5 m because the impact between ship and dolphin during the berthing manoeuvre will otherwise be so gentle that the helmsman will not be able to judge the movement and position of the ship properly in relation to the dolphin.

13.2.3.2 When calculating with the static force P_{stat} of the dolphin, the mutual dependency in the fender–ship–hawser system must be considered. This is especially important at berths exposed to high winds and/or long swells. In such cases, just as for berths on the open sea, model tests should always be carried out.

As shown by past experience, the following should be observed:

(1) Stiff hawsers, i.e. short lines or steel hawsers, require stiff fenders.
(2) Soft hawsers, i.e. long lines or manila, nylon, propylene and polyamide lines, require soft fenders.

In case (2) the result is always in both hawsers and dolphins.

13.2.3.3 The maximum permissible impact force P_{impact} between ship and berthing dolphin is determined on the one hand by the type of ship being berthed, and on the other by the design and construction of the dolphin, especially if it is equipped with berthing aprons and the like. Another requirement at berths for large ships is that the berthing pressure between ship and dolphin shall not exceed 200 kN/m² or in some cases even 100 kN/m².

A higher berthing pressure can be allowed if it is proven that the hull and the inner structural framing of the ships using the berth are able to absorb this. Please refer to section 13.2.1, 3rd paragraph, for information on stressing of the hull.

13.2.3.4 If a berth is such that a ship must moor simultaneously at rigid structures and resilient berthing dolphins, select the maximum spring constant value for the dolphin. If the fender structure thus becomes too stiff for the maximum permissible impact force during berthing of the ship, full separation between fender and mooring structure is required. In any case, thorough investigations are necessary. The same is true for the degree of softness of exposed berthing dolphins and protective dolphins in front of piers, jetties, guide walls and lock entrances.

13.2.3.5 If a berth is to be equipped with dolphins with different energy absorption capacities, the dolphins should be designed, if at all feasible, so that the yield stress of the material in every dolphin will be reached at the same horizontal deflection. Uniform stressing of all dolphins is ensured for forces acting centrally on the ship, especially by wind and waves. Furthermore, a uniform type of dolphin pile can the be used as a rule for the entire berth.

When differing water levels occur due to tide, wind effects etc., the dolphins should be equipped with a berthing apron, to ensure a fairly constant elevation at which the ship's berthing pressure is to be absorbed. Uniform aprons for heavy dolphins of different capacities at a berth should be used only when no appreciable dynamic stresses occur due to wind or swell, which could damage the lighter dolphins.

13.2.3.6 Starting with the dimensioning of the heavy dolphin with energy absorption capacity A_h and spring constant c_h, the following then applies to the lighter one with energy absorption capacity A_l and spring constant c_l:

$$c_l = c_h \cdot \frac{A_l}{A_h}$$

If the stiffness of the lighter dolphins determined by this equation is too small, they should be designed to meet the given requirements. In a case

$$c_h = c_l \cdot \frac{A_h}{A_l}$$

is valid for heavy dolphins.

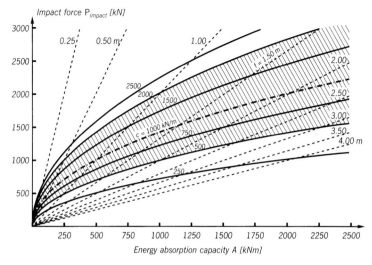

Fig. R 111-1. Magnitude of spring constant c and deflection f at berthing dolphins, contingent on energy absorption capacity A and impact force P_{impact}

13.2.3.7 Fig. R 111-1 shows the magnitude of the spring constant c, as well as that of the deflection f contingent on energy absorption capacity A and impact force P_{impact} for berthing dolphins. In the normal case, a spring constant c should be selected that lies between the curves for $c = 500$ and 2000 kN/m, as close as possible to the curve for $c = 1000$ kN/m.

13.3 Impact forces and required energy absorption capacity of fenders and dolphins in seaports (R 128)

13.3.1 Determination of impact forces
According to R 111, section 13.2.1, the maximum permissible impact force is equal to the product of the spring constant and the maximum permissible deflection of the resilient berthing dolphins, fenders, shock absorbers or the like at the point of contact with the ship ($P_{impact} = c \cdot f$ [kN]). For nautical reasons, the deflection f is generally limited to max. 1.50 m at berths for large ships (see also R 111, section 13.2.3.1).

13.3.2 Determining the energy absorption capacity required

13.3.2.1 General
When berthing, the movement of a ship generally consists of transverse and/or longitudinal motion and rotation around its centre of gravity, so that at first generally only one dolphin or fender is struck (fig. R 128-1).

581

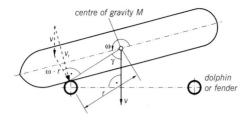

centre of gravity M

dolphin
or fender

Fig. R 128-1. Depiction
of a berthing manoeuvre

Governing the approach energy thereby is the impact velocity of the
ship v_r at the fender, whose magnitude and direction results from the
vector addition of the velocity components v and $\omega \cdot r$. With full frictional
contact between ship and fender, the impact velocity of the ship, which
is then identical with the deformation velocity of the fender, is reduced
to $v_r = 0$ in the course of the impact. The centre of gravity of the ship,
however, will generally remain in motion, albeit with a different
magnitude and direction of rotation.

The ship therefore retains a portion of its original energy of motion,
even at the time of maximum fender deformation. Under certain conditions, this can lead to the situation that after contact with the first fender,
the ship turns to the second one, which can produce a still stronger
berthing impact.

13.3.2.2 Numerical determination of the energy absorption capacity required [102]

The portion of the motion energy of the ship to be absorbed by one
dolphin in the course of the berthing impact represents the energy
absorption capacity the dolphin must possess in order to avoid damage
to ship and/or dolphin. In the general case depicted in fig. R 128-1, this
energy absorption capacity amounts to:

$$A = \frac{G \cdot C_m \cdot C_s}{2 \cdot (k^2 + r^2)} \cdot [v^2 \cdot (k^2 + r^2 \cdot \cos^2 \gamma) + 2 \cdot v \cdot \omega \cdot r \cdot k^2 \cdot \sin \gamma + \omega^2 \cdot k^2 \cdot r^2]$$

For case $\gamma = 90°$, i.e. a berthing angle of $0°$, this assumption can be
simplified to:

$$A = \frac{1}{2} \cdot G \cdot C_m \cdot C_s \cdot \frac{k^2}{k^2 + r^2} \cdot (v + \omega \cdot r^2)$$

$$= \frac{1}{2} \cdot G \cdot C_m \cdot C_s \cdot \frac{k^2}{k^2 + r^2} \cdot v_r^2$$

$$= \frac{1}{2} \cdot G \cdot C_m \cdot C_s \cdot C_e \cdot v_r^2$$

where:

A = energy absorption capacity [kNm]
G = displacement of berthing ship as per R 39, section 5.1 [t]
k = mass radius of gyration of the ship [m]
  (It can generally be taken to be $0.25 \cdot l$ for large ships with a high block coefficient.)
l = length of ship between perpendiculars [m]
r = distance of ship's centre of gravity from point of contact with dolphin [m]
υ = translation velocity of motion of the ship's centre of gravity at time of first contact with dolphin [m/s]
ω = turning speed of ship at time of first contact with dolphin [angle in radians per second = 1/s]
γ = angle between velocity vector v and distance r [°], $\gamma = 90°$ berthing angle to R 60, section 6.14.4
v_r = resulting impact velocity of ship on dolphin [m/s]
C_m = mass factor (see below) [1]
C_s = stiffness factor (see below) [1]
C_e = eccentricity factor = $k^2 / (k^2 + r^2)$ [1]

The mass factor C_m includes the influence of hydrodynamic pressure, suction and water friction which the turbulent water (hydrodynamic mass) exerts on the ship as it begins to stop.
The evaluation and comparison of diverse approaches for C_m in the specialist literature on this subject results in average values between 1.45 and 2.18.

The designer is recommended [232] to use the following values:
• for a large clearance under the keel (0.5 t) $C_m = 1.5$
• for a small clearance under the keel (0.1 t) $C_m = 1.8$
where t is the draught of the ship [m]

The values for keel clearances between 0.1 t and 0.5 t can be obtained through linear interpolation. The stiffness factor C_s takes into account a reduction of the impact energy through deformations of the ship's hull, depending on the properties of ship and dolphin in a mutual interaction. A value of $C_s = 0.90$ to 1.0 may be assumed. The upper limit value applies to soft dolphins and smaller ships with stiffer sides, the lower one for hard dolphins and larger ships with relatively soft sides. Please also refer to R 111, section 13.2.

13.3.2.3 Advice
If a ship is towed to a berth with tug assistance, it may be presumed that it is hardly still moving under its own power in the direction of its longitudinal axis, and that during berthing the ship's side lies practically

parallel with the line of the dolphins. For the inner dolphins of a row of dolphins, there is inevitably a greater distance between the ship's centre of gravity and dolphins it has struck. In this case the velocity vector v can therefore be assumed to be at right-angles to the distance r ($\gamma = 90°$) and the simplified formula for determining A may be used for designing the inner dolphins. Please refer to figs. R 40-1 and R 40-2 in section 5.3. On the other hand, in the calculation of the outer dolphins of a row of dolphins, the simplified calculation assumption may not be used because the ship's centre of gravity here in the direction of the dolphin line can also approach close to the dolphin point.

Besides, it must be considered in all cases that the ships cannot always be brought to their berths centrally. A distance between the ships' centre of gravity and the centre of the berth of $e = 0.1 \cdot l \leq 15$ m (parallel to fender line) should always be used as a basis for the calculations of berthing dolphins, e.g. at a tanker berth.

13.3.3 Practical application

If the energy theoretical absorption capacity required A has been determined, there must be an agreement between the absorption capacity required A, the permissible impact force P_{impact} and the resulting smallest spring constant min c desired. This can be achieved with the data given in R 111, sections 13.2.2.2 and 13.2.2.3, as well as by considering practical aspects. R 111, section 13.2.3.3, gives the maximum permissible impact force P_{impact} expressed in berthing force per m^2 hull plating for larger ships. Special attention must be given to the relationship between the three factors A, P_{impact} and f as shown in fig. R 111-1.

As has been pointed out in R 111, section 13.2.3.5, 2nd paragraph, the most efficient utilisation of the calculated capacity for energy absorption requires that the ship's berthing force be transmitted to the dolphin or other fendering system at a point whose elevation remains constant. In the case of resilient dolphins, the centre of gravity of the berthing force can, if necessary, be prevented from drifting downward by fitting the dolphin with a suitable fender apron which provides a fixed point at which the impact force will be transmitted.

13.4 Use of weldable fine-grained structural steels for resilient berthing and mooring dolphins in marine construction (R 112)

13.4.1 General

If dolphins of high energy absorption capacity are required, it is advisable to fabricate them from higher-strength, weldable, fine-grained structural steels. Material details are given in R 67, section 8.1.6.

13.4.2 Loading assumptions

In the design and calculation of dolphins, a distinction must be made as to whether they are stressed:

(1) predominantly by static loads (vessel impact, line pull), or
(2) by not predominantly static loads.

The following criterion can be used to determine the loads:

Dolphins are stressed predominantly statically when the proportion of the alternating stresses from waves or swells is low in comparison to the stresses from vessel impact and line pull, and when in examining the occurrence of stress reversal, the following stresses are not exceeded when using the design wave height H_{des} (= characteristic wave height $H_{1/3}$) over a design period of 25 years:

- 30% of the respective minimum yield point $f_{y,k}$ of the parent metal, insofar as no butt welds run transverse to the principal direction of stress,
- 100 MN/m^2 for the parent metal if butt welds run transverse and continuous fillet welds run longitudinally to the principal direction of stress,
- 50 MN/m^2 for the parent metal, where fillet welds end or run transverse to the principal direction of stress.

In areas where heavy swells occur, the loads on the dolphins are not predominantly static unless they are protected from the effects of swells by other means, such as prestressing against another structure, for example.

In the case of not predominantly static loads, this loading case is generally applicable for design purposes.

Attention is drawn to the general effect of the stress and stability problems on piles with larger cross-sectional shape and thin walls, and especially in loading cases where stresses up to yield are allowed. A structural check should be carried out for this when using large-diameter pipes (see DIN 18 800-4).

13.4.3 Structural design

13.4.3.1 The type of action determines the basic requirements for:

(1) choice of steel grade, allowable effects of actions and cross-sectional shape of the individual piles,
(2) manufacturing and material thickness, and
(3) structural design and welding techniques.

13.4.3.2 For practical reasons, the upper part of a dolphin, e.g. the uppermost pipe section, is composed of weldable, fine-grained structural steel of lower strength. This simplifies welded connections to tie beams and other structural parts.

A wall thickness is to be chosen which will permit all necessary welding work to be carried out on site without any preheating if possible. This is especially important in tidal harbours and where there is heavy wave action.

13.4.3.3 Welds between the individual parts (e.g. pipe sections) shall be made in the factory and, as far as possible, in parts of the dolphin with low stresses.

13.4.3.4 In the case of not predominantly static actions, special significance is attached to welded joints transverse to the flexural tensile stress. Therefore, the following is to be observed for circumferential welds and butt joints transverse to the direction of the applied force:

(1) In case of different wall thicknesses at the welded joint, the transition from the thicker plate to the thinner one is to be milled flat at a taper of 4 : 1, or shallower if possible. At least rough analytical verification is to be provided for the force transition in large-diameter pipes.

(2) Surface layers are to be formed without notches. The height of the weld reinforcement shall not exceed the material thickness by more than 5%, if possible.

(3) At piles without access, a full root weld is required. The transition between seam and plate is to be kept flat, without damaging splatter notches. If backing rings are used, they must be of ceramic material.

(4) At piles with access, the joints are to be welded on both sides. Root layers are to be veed out.

(5) At dolphin piles of high-strength, weldable, fine-grained structural steels, a filler metal is to be used whose quality corresponds to that of the parent metal, as recommended by the manufacturer.

(6) Welding material and workmanship must be such that the values for heat penetration specified by the manufacturer will be observed.

(7) Follow the recommendations of [103].

(8) Once welding has been completed, limited local stress relieving with appropriate temperature control is possible. Otherwise, section 13.4.3.3 is to be observed.

13.4.3.5 R 99, section 8.1.18, applies to all welding work. Circumferential welds and butt welds must be checked with non-destructive testing methods, and X-rayed if possible.

13.4.3.6 Dolphin piles of fine-grained structural steels generally have long delivery times and should therefore must be ordered well in advance. It is recommended that a certain number of spare piles be purchased to allow for any possible damage during construction or when in service.

14 Experience with waterfront structures

14.1 Average service life of waterfront structures (R 46)

As a result of changes in port operations or harbour traffic, large seaports must frequently be deepened and their waterfront structures strengthened or replaced long before their stability is at risk. Their service life is therefore often much shorter than their structural life, so that sometimes only about 25 years can be expected for such installations. Compared with that, an average service life of 50 to 60 years can be expected for sheet pile structures on inland waterways or in small fishing, ferry or dyke harbours on the North Sea and Baltic Sea coasts.

The most accurate estimation of the respective average service life is useful, for example, for preparing costs–benefits analyses, for the design, for the internal operations calculation and for fiscal assessments. However, the data is not intended as a basis for fixing the valuation in damage cases since in such cases it is not a matter of a statistically calculated value, but rather of the value actually applicable at the time of the damage.

Considering the anticipated average service life, the types of construction preferred are those that can later be strengthened with acceptable costs and minimum operational disturbances, and in such cases can be adapted to a deepened harbour bottom.

Waterfront structures for which an especially short service life is expected should be constructed in a manner that they may easily be demolished and rebuilt.

The aforementioned comments on the average service life do not apply to damage or failure due to accidents, and generally do not apply to dolphins of any kind.

In the case of accidents, the remaining service life of the damaged dolphin, apart from the structural design, depends particularly on the type and the intensity of use, so that general recommendations for determining the remaining service life cannot be given.

14.2 Operational damage to steel sheet piling (R 155)

14.2.1 Causes of damage

Besides the predominantly static loads used as a basis for design assumptions, steel sheet piling in the waterfront area of ports is mostly also subjected to high dynamic stresses deriving from shipping and cargo handling operations. This danger grows with increasing opening width of the sheet piling troughs.

The box-like construction of modern push lighters, motor cargo vessels and motor tankers can easily lead to damage to the sheet piling as a

587

result of impacts on corners and edges of the ship's hull during berthing manoeuvres, especially on inland waterways. Another cause of damage can be the catching of pointed objects, such as ship's anchors, in the sheet piling troughs, or crane hooks under the capping beams. Such types of damage are rather rare in seaports because sheet pile walls are increasingly set back from the front edge of the quay.

Especially extreme impairments of the waterfront structure occur when there is a head-on impact of the vessel in accidents, which can take place when the vessel deviates from its course or when the coupling lines of a push tow rupture. When handling aggressive materials such as salts, these may come into contact with the sheet piling and initiate corrosion. This also applies to aggressive groundwater.

Sheet pile walls without fender systems or fender piles are constantly exposed to mechanical loads and hence an increased risk of corrosion because the protective surface layer of corrosion products is being constantly worn away by ships. The same is true for the landing pontoons frequently used in the Baltic Sea. The chafing and scuffing caused by mechanical means and the ensuing premature corrosion (see also R 35, section 8.1.8.2 (2)) can also be caused by the chains and suspended weights often used to secure fender systems to sheet pile walls.

14.2.2   Types of damage

Contact with the ship's hull can lead to bulges with compressive and upset zones. Cracks, ruptures, holes and interlock failure, the latter mainly in Z-section sheet piles, can also be found, especially in older sheet piling bulkheads which do not possess today's steel grades. Furthermore, capping beam plates can be torn off. Overloading of the waterfront structure or corrosion can lead to anchor connection constructions being torn off and subsequent displacement of the sheet piling. Eventually, increased anchor forces may also tear sheet piling bulkheads in the area of the anchor connections.

The influence of aggressive materials leads to a reduction of the thickness of steel plates, especially when they have acted in the tidal zone for a long period.

14.2.3   Rectifying damage

If damage to a sheet piling bulkhead has been proved, its extent should at first be ascertained accurately to verify if individual structural members or the stability of the entire waterfront structure have become endangered. The damage can be repaired by welding on plates and angles or by attaching steel plate elements, which bridge a single sheet piling trough or more if necessary, provided that the sheet piling steel is weldable. If a sheet piling section has been weakened in the trough area, a solution may be to install reinforcing sections, which do not protrude beyond the line of the wall, and connect them to the existing sheet piling. In case of major damage, it may prove necessary to extract individual sheet piles or entire sections of the wall, and to replace them with new material. Where the damage is minor, it often suffices to cut out sheet piles and to increase the height of the piles in question with new material. Special care must be taken if an anchored sheet pile is concerned and the anchor must be loosened. The engineer must check whether auxiliary anchoring is required. In the case of very severe damage, especially if the sheet piling has deviated from its alignment to a great extent, it may be necessary to secure a section of the waterfront by driving a new sheet piling bulkhead in front of the old one. In any case, investigations must be carried out to determine whether the sheet piling damage has resulted in the outflow of soil and whether voids could have developed behind the sheeting.

14.2.4 Preventative measures for avoiding or reducing damage

When dimensioning a sheet piling structure, it must be investigated to what extent additions must be made to take account of port operations beyond the structural/constructural requirements. These may be, for instance, that the next stronger pile section is used for either all sheet piles or only the water-side ones. Reinforcing plates can be welded to the backs of sheet piles. A suitable example is the armoured sheet piling dealt with in R 176, section 8.4.16. Early recognition of irregularities in the sheet piling structure enable measures to be initiated to prevent or reduce more serious damage. It is therefore recommended to check the thickness of the wall and carry out underwater investigations with divers – possibly with an underwater camera – and record any damage (corrosion, cracks, etc.) systematically (R 193, section 15). Depending on the importance of the structure, later measurements should be taken into account during planning and building (e.g. by integrating inclined measuring sleeves).

14.3 Steel sheet piling waterfront structures under fire loads (R 181)

14.3.1 General

Fire loads can be initiated by burning substances floating on the water, e.g. in the case of burning products escaping as a result of an accident, or by a fire on land. A burning ship moored at the waterfront can also be the cause of temperature loads on the waterfront structure.

The increase in temperature can be derived from the standard temperature curve in accordance with DIN 4102, which applies to an enclosed burning area. According to this, the maximum temperature of approx. 1100 °C is reached after 180 minutes. However, in the case of waterfront structures, much more favourable local circumstances are to be found.

This is a fire in the open air, in which the heat generated can escape without restriction. In the case of a fire on the water surface, the heat is additionally absorbed by the water.

In the case of fires in the open air, temperatures of 800 °C are not exceeded in the flame zone [119]. It can generally be assumed that the reduced maximum value of 800 °C will definitely not be reached in waterfront structures located in the open air, provided the fire is extinguished within three hours.

Therefore, fire-fighting measures should aim at reducing the temperature as quickly as possible and preventing contact with further combustible substances. In addition, the steel sheet piling should be relieved, if possible, e.g. by removing live loads and stored items.

Otherwise, fire protection measures must comply with general legislation and statutory instruments. The fire protection requirements should be determined and stipulated on the basis of local circumstances. For this reason, reference is made in particular to the "Directive for requirements relating to facilities for handling hazardous fluids in waterways" [120].

14.3.2 Effects of fire loads on a waterfront structure of steel sheet piling

If steel piling is subjected to a high temperature, its mechanical properties will change. In this case the critical steel temperature (crit T) is of particular importance. This is the temperature at which the yield strength of the steel reduces to the steel stress existing in the sheet piling. In sheet piling steels in accordance with R 67, section 8.1.6, crit T is 500 °C.

If in the stressed area the permissible stress is not exploited, crit T will increase up to 650 °C if the existing stress is only 1/3 of the permissible stress.

If the temperatures remain below 500 °C, there is no hazard to the stability of a steel sheet piling structure in any circumstances. After cooling, the initial values assumed in the structural calculation can again be used as the basis for the mechanical properties.

In fire loads, only the air-side surface of the sheet piling is affected, provided it is above the water level. All remaining surfaces are not subject to loads.

If it is surrounded by air and/or water, the sheet piling surface opposite to the side exposed to the fire contributes to cooling the exposed surface. However, this also applies to soil material, particularly with groundwater in permeable soil.

Generally, the zone with the greatest heat increase does not lie in the area in which the steel sheet piling is fully loaded. With single-anchor sheet piling with a small section of wall above the anchor, the area of maximum stress is generally located below the water level and is thus not subject to the fire load.

The circumstances are likewise favourable in the case of unanchored sheet piling fixed in the soil and standing in water. Conversely, it should be noted in the case of sheet piling fixed at the top that the restraining moment may lie in the area of the fire load above water level.

With anchored sheet piling restrained in the soil or at the top, flow articulations may form in the area subject to fire loads. On the one hand, these may increase deflection, but on the other hand they may increase the safety against the failure state. Sheet piling of this type with statically indeterminate support should therefore be assessed more favourably than statically determinate sheet piling simply supported in the earth and with a single anchor.

The greater the wall thickness for the same surface area, the more slowly the component will heat up and the greater the duration of fire resistance will be, although this has little effect on the relatively thin sheet piles. Owing to the plane and therefore smaller surface with an increased steel mass and owing to the insulating effect of the gap between it and the load-bearing sheet piling, armoured sheet piling (R 176, section 8.4.15) is more resistant to the effects of heat than non-armoured sheet piling, provided the armouring is not structural. The connection construction of anchored sheet piling has a relatively small surface with a great piling wall thickness, so the construction heats up more slowly. The anchoring element located in the ground behind the sheet piling is not subject to fire load and can dissipate a limited proportion of the heat. However, in particularly endangered areas of waterfront structures, it is recommended that special protection for the anchor connections should always be provided.

14.3.3 Investigations of fire loads on a sheet piling structure

A 2.70 m high sheet piling surface with a reinforced concrete capping beam with nosing and a paved embankment in accordance with fig. R 129-1, section 8.4.5, was exposed to a burning oil slick floating on the surface of the water for up to one hour. After extinguishing the fire, the following was discovered:

In the area of the fire, the steel sheet piling exhibited pronounced colour changes. However, the mechanical and technological properties determined from steel samples taken from the section of the sheet piling subject to the most intense loading still complied with the technical supply conditions. Accordingly, the steel had suffered no loss of quality as a result of the fire.

Measurements up to 7 m below the upper edge of the sheet piling revealed that no unusual deformation had occurred under the fire load. According to the structural analysis, the sheet piling was stressed at the elevation of the water level with 1/3 its yield stress.

There was visible damage to the surface of the reinforced concrete capping beam in the form of local spalling 1 to 2 cm thick. The steel nosing had become detached from the concrete capping beam almost all the way around the beam and had cracked at the welded butt joints. In the area of the ladder recess, the reinforced concrete capping beam was subject to such intense loading that the reinforcement was exposed in places and cracks could be observed. The access ladders were partly distorted, apparently as a result of exposure to the fire on all sides.

Damage had also occurred in the paved embankment and in the revetments. The sandstone paving with mortar joints had bulged in places, so that the mortar joints were cracked.

On a free-standing dolphin and designed as an enclosed steel pile, the fire exposure on all sides had led to a bulge approx. 4 m above the water level, or 6 m below the top edge of the dolphin. After the fire, the mechanical and technological properties of the steel grade corresponded to sheet piling special steel S 355 GP. The individual samples taken showed the same results, although they were taken from different elevations which had also been subject to different fire loads.

The damage described leads to the conclusion that the duration and intensity of the fire loads had not been minor. Nonetheless, the steel sheet piling exhibited high fire resistance, apparently since the boundary conditions (soil backfilling) had greatly diminished the effects of the fire load. Steel sheet piling structures backfilled with soil need not be classified in accordance with DIN 4102 part 2 [126].

15 Monitoring and inspection of waterfront structures in seaports (R 193)

15.1 General

Regular inspections of waterfront structures are necessary, taking account of individual cases, in order to detect in good time any changes to their stability, functions and structural conditions. This is the true basis on which the safety of the structure can be guaranteed.

Careful and regular inspections of structures facilitates preventive maintenance work so that major repair costs or premature renewal of the structure can be avoided. Taking each individual case into account, the inspection of the structure consists of:

- monitoring the structure by way of straightforward visual inspections without any large equipment, and
- checking the structure by way of investigations by an experienced engineer with the appropriate equipment.

When planning and performing inspections of structures, the relevant provisions of statutory instruments and other regulations may need to be considered, e.g.:

- DIN 1076
- DIN 19 702
- VV-WSV 2101 [139]
- VV-WSV 2301 [163]
- ETAB [45]
- "Life cycle management of port structures – general principles" [136]
- "Inspection, maintenance and repair of maritime structures exposed to material degradation caused by a salt water environment" [258]

15.2 Records and reports

The results of observations and inspections together with measures implemented and their costs are to be recorded in a suitable form as an important aid for evaluating necessary maintenance measures. Therefore, the findings of all inspections are to be recorded.

The basis for observations and inspections are the as-built drawings and, where applicable, the logbook for the structure, which contains major data such as cross-sections, presumed loads, water levels and characteristic soil values, structural calculations, modifications completed and maintenance work, together with the results of earlier inspections.

15.3 Performing inspections of the structure

15.3.1 Monitoring the structure

The nature and extent of monitoring measures must be stipulated for every structure depending on the local boundary conditions. For example, the following aspects are among those that should be ascertained and recorded during a site visit:

- surface damage or changes
- settlement, subsidence, displacement
- changes in joints and structural connections
- absence of or damage to fixtures and fittings
- improper use
- function of the drainage system
- damage to sheet piling
- scouring or silting in front of the sheet pile wall (see R 80, section 7.1, and R 83, section 7.6)

Even parts of the structure not readily accessible, e.g. gangways, jetties and shafts, should be inspected if at all possible.

15.3.2 Checking the structure

Inspection of the structure should always be carried out by an engineer who is well acquainted with the structure, who can assess the structural, constructional and hydromechanical conditions of the structure, bearing in mind the requirements concerning its use, and also provide guidance for any diving activities deemed necessary. The nature and extent of inspection measures must be stipulated for every structure depending on the local boundary conditions. For example, the following aspects are among those that should be measured and/or examined depending on importance and/or boundary conditions:

- position and extent of damage to the quay wall, also underwater using divers
- condition and function of drainage systems
- condition of earlier repairs
- condition of anti-corrosion coatings
- condition of a cathodic corrosion protection system
- settlement and subsidence behind the quay wall
- soundings of the bottom of the watercourse in front of the quay wall
- seals in joints and at junctions in the structure
- movement at joints, bearings and supports
- damage to concrete (including reinforcing steel)
- measurements of horizontal movement (also head deformation) and vertical movement (settlement, uplift)

594

- measurements of residual wall thicknessess and/or average/maximum rates of corrosion (e.g. by ultrasound, see R 35, section 8.1.8.2)

Further measurements may be necessary in individual cases. If specific measurements (anchor forces, inclinometer, potential field, etc.) are carried out, a specialist should be called in to perform such work and assess the results.

The findings of the inspection must be assessed with respect to the stability, safety and long-term functioning of the structure.

Any defects found are to be assessed with respect to any causes suspected or actually established. Any further procedures required based on this information must be recorded in the documents/reports.

15.4   Inspection intervals

The frequency and intensity of inspections of waterfront structures depend on the age of the structure, its general condition, the construction materials used, the subsoil conditions, environmental influences and operational requirements and loads. To guarantee proper monitoring and inspection based on the aspects given year, it is recommended to specify the inspection intervals in the structure documentation. Advice on inspection intervals can be found in the documents listed in section 15.1, but must be adjusted to suit the individual needs of each particular case.

The first inspection of the structure should be carried out before handover to the client, and another one prior to the expiry of any guarantee(s). These should include levels and alignments, also a zeroing to establish a reference point for position and elevation. Other follow-up measurements are to be stipulated as part of the checking of the structure.

After any extraordinary loads which may have caused damage to the structure, e.g. accidents, extreme loads due to floods or fire, the possible effects on the loadbearing capacity of the structure have to be determined by special investigations.

Annex I Bibliography

I.1   Annual technical reports

The basis of the collected publication is the Annual Technical Report of the Committee for Waterfront Structures, published in the journals "DIE BAUTECHNIK" (since 1984 "BAUTECHNIK") and "HANSA", in issues:

HANSA   87 (1950) No. 46/47, p. 1524

DIE BAUTECHNIK   28 (1951), No. 11, p. 279 – 29 (1952), No. 12, p. 345
                  30 (1953), No. 12, p. 369 – 31 (1954), No. 12, p. 406
                  32 (1955), No. 12, p. 416 – 33 (1956), No. 12, p. 429
                  34 (1957), No. 12, p. 471 – 35 (1958), No. 12, p. 482
                  36 (1959), No. 12, p. 468 – 37 (1960), No. 12, p. 472
                  38 (1961), No. 12, p. 416 – 39 (1962), No. 12, p. 426
                  40 (1963), No. 12, p. 431 – 41 (1964), No. 12, p. 426
                  42 (1965), No. 12, p. 431 – 43 (1966), No. 12, p. 425
                  44 (1967), No. 12, p. 429 – 45 (1968), No. 12, p. 416
                  46 (1969), No. 12, p. 418 – 47 (1970), No. 12, p. 403
                  48 (1971), No. 12, p. 409 – 49 (1972), No. 12, p. 405
                  50 (1973), No. 12, p. 397 – 51 (1974), No. 12, p. 420
                  52 (1975), No. 12, p. 410 – 53 (1976), No. 12, p. 397
                  54 (1977), No. 12, p. 397 – 55 (1978), No. 12, p. 406
                  56 (1979), No. 12, p. 397 – 57 (1980), No. 12, p. 397
                  58 (1981), No. 12, p. 397 – 59 (1982), No. 12, p. 397
                  60 (1983), No. 12, p. 405 – 61 (1984), No. 12, p. 402
                  62 (1985), No. 12, p. 397 – 63 (1986), No. 12, p. 397
                  64 (1987), No. 12, p. 397 – 65 (1988), No. 12, p. 397
                  66 (1989), No. 12, p. 401 – 67 (1990), No. 12, p. 397
                  68 (1991), No. 12, p. 398 – 69 (1992), No. 12, p. 710
                  70 (1993), No. 12, p. 755 – 71 (1994), No. 12, p. 763
                  72 (1995), No. 12, p. 817 – 73 (1996), No. 12, p. 844
                  75 (1998), No. 12, p. 992 – 76 (1999), No. 12, p. 1062
                  77 (2000), No. 12, p. 909 – 78 (2001), No. 12, p. 872
                  79 (2002), No. 12, p. 850 – 80 (2003), No. 12, p. 903
                  81 (2004), No. 12, p. 980 – 82 (2005), No. 12

I.2   Books and papers

[1]   Report of the Sub-Committee on the Penetration Test for Use in Europe, 1977. (Copies of this report are available from: The Secretary General, ISSMFE, Department of Civil Engineering, King's College London Strand, WC 2 R 2 LS, U. K.)

[2]   SANGLERAT: The penetrometer and soil exploration. Amsterdam, London, New York – Elsevier Publishing Company 1972.

[3]   LANGEJAN, A.: Some aspects of the safety factor in soil mechanics, considered as a problem of probability. Proc. 6. Int. Conf. Soil MecH. Found. Eng. Montreal 1965, Vol. 2, p. 500.

[4]   ZLATAREW, K.: Determination of the necessary minimum number of soil samples. Proc. 6. Int. Conf. Soil Mech. Found. Eng. Montreal 1965, Vol. 1, p. 130.

[5]   ROLLBERG, D.: Bestimmung des Verhaltens von Pfählen aus Sondier- und Rammergebnissen, Forschungsberichte aus Bodenmechanik und Grundbau FBG 4, Techn. Hochschule Aachen, 1976.

[6]   ROLLBERG, D.: Bestimmung der Tragfähigkeit und des Rammwiderstands von Pfählen und Sondierungen, Veröffentlichungen des Instituts für Grundbau, Bodenmechanik, Felsmechanik und Verkehrswasserbau der Techn. Hochschule Aachen, 1977, No. 3, pp. 43–224.

[7]   Grundbau-Taschenbuch, 6th edition, Part 1-2001, Part 2-2001 and Part 3-2001, Ernst & Sohn Verlag, Berlin.

[8]   KAST, K.: Ermittlung von Erddrucklasten geschichteter Böden mit ebener Gleitfläche nach CULMANN. Bautechnik 62 (1985), No. 9, p. 292.

[9]   MINNICH, H. und STÖHR, G.: Erddruck auf eine Stützwand mit Böschung und unterschiedlichen Bodenschichten. Die Bautechnik 60 (1983), No. 9, p. 314.

[10]   KREY, H.: Erddruck, Erdwiderstand und Tragfähigkeit des Baugrundes. 5th edition, Berlin: Ernst & Sohn 1936 (out of print); see in [7].

[10a] JUMIKIS: Active and passive earth pressure coefficient tables. Rutgers, The State University. New Brunswick/New Jersey: Engineering Research Publication (1962) No. 43. CAQUOT, A., KÉRISEL, J. und ABSI, E.: Tables de butée et de poussée. Paris: Gauthier-Villars 1973.

[11]   BRINCH HANSEN, J. und LUNDGREN, H.: Hauptprobleme der Bodenmechanik. Berlin: Springer 1960.

[12]   BRINCH HANSEN, J: Earth Pressure Calculations. The Danish Technical Press, Copenhagen 1953.

[13]   HORN, A.: Sohlreibung und räumlicher Erdwiderstand bei massiven Gründungen in nichtbindigen Böden. Straßenbau und Straßenverkehrstechnik 1970, No. 110, Bundesminister für Verkehr, Bonn.

[14]   HORN, A.: Resistance and movement of laterally loaded abutments. Proc. 5. Europ. Conf. Soil Mech. Found. Eng. Madrid, Vol. 1 (1972), p. 143.

[15]   WEISSENBACH, A.: Der Erdwiderstand vor schmalen Druckflächen. Mitt. Franzius-Institut TH Hannover 1961, No. 19, p. 220.

[16]   PIANC-Report „Seismic Design Guidelines for Port Structures". MarCom Report of WG 34, Balkema Publishers 2001.

[17]   TERZAGHI, K. VON und PECK, R. B.: Die Bodenmechanik in der Baupraxis. Berlin/ Göttingen/Heidelberg: Springer 1961.

[18]   DAVIDENKOFF, R.: Zur Berechnung des hydraulischen Grundbruches. Die Wasserwirtschaft 46 (1956), No. 9, p. 230.

[19] KASTNER, H.: Über die Standsicherheit von Spundwänden im strömenden Grundwasser. Die Bautechnik 21 (1943), No. 8 and 9, p. 66.

[20] SAINFLOU, M.: Essai sur les digues maritimes verticales. Annales des Ponts et Chaussées, tome 98 II (1928), übersetzt: Treatise on vertical breakwaters, US Corps of Engineers (1928).

[21] SPM: Shore Protection Manual. US Army Corps of Engineers, Coastal Engineering Research Center, Vicksburg, USA, 1984.

[22] HEIBAUM, M.: Kleinbohrpfähle als Zugverankerung – Überlegungen zur System-standsicherheit und zur Ermittlung der erforderlichen Länge. In: Institut für Boden-mechanik, Felsmechanik und Grundbau der Technischen Universität Graz (organizer): Bohrpfähle und Kleinpfähle – Neue Entwicklungen (6. Christian Veder Kolloquium, Graz, 1991). Graz: Institut für Bodenmechanik der Technischen Universität.

[23] WALDEN, H. und SCHÄFER, P. J.: Die winderzeugten Meereswellen, Part II, Flach-wasserwellen, No. 1 und 2. Einzelveröffentlichungen des Deutschen Wetterdienstes, Seewetteramt Hamburg 1969.

[24] SCHÜTTRUMPF, R.: Über die Bestimmung von Bemessungswellen für den Seebau am Beispiel der südlichen Nordsee. Mitteilungen des Franzius-Instituts für Wasserbau und Küsteningenieurwesen der Technischen Universität Hannover, 1973, No. 39.

[25] PARTENSCKY, H.-W.: Auswirkungen der Naturvorgänge im Meer auf die Küsten – Seebauprobleme und Seebautechniken. Interocean 1970, Vol. 1.

[26] LONGUET-HIGGINS, M. S.: On the Statistical Distribution of the Heights of Sea Waves. Journal of Marine Research, Vol. XI, No. 3 (1952).

[27] KOHLHASE, S.: Ozeanographisch-seebauliche Grundlagen der Hafenplanung, Mit-teilungen des Franzius-Instituts für Wasserbau und Küsteningenieurwesen der Universität Hannover (1983), No. 57.

[28] WIEGEL, R. L.: Oceanographical Engineering. Prentice Hall Series in Fluid Mechanics, 1964.

[29] SILVESTER, R.: Coastal Engineering. Amsterdam/London/New York. Elsevier Scientific Publishing Company, 1974.

[30] HAGER, M.: Untersuchungen über Mach-Reflexion an senkrechter Wand. Mitteilungen des Franzius-Instituts für Wasserbau und Küsteningenieurwesen der Technischen Universität Hannover, (1975), No. 42.

[31] BERGER, U.: Mach-Reflexion als Diffraktionsproblem. Mitteilungen des Franzius-Instituts für Wasserbau und Küsteningenieurwesen der Technischen Universität Hannover (1976), No. 44.

[32] BÜSCHING, F.: Über Orbitalgeschwindigkeiten irregulärer Brandungswellen. Mit-teilungen des Leichtweiß-Instituts für Wasserbau der Technischen Universität Braun-schweig, (1974), No. 41.

[33] SIEFERT, W.: Über den Seegang in Flachwassergebieten. Mitteilungen des Leicht-weißInstituts für Wasserbau der Technischen Universität Braunschweig, (1974), No. 40.

[34] BATTJES, J. A.: Surf Similarity. Proc. of the 14th International Conference on Coastal Engineering. Copenhagen 1974, Vol. I, 1975.

[35] GALVIN, C. H. Ir.: Wave Breaking in Shallow Water, in Waves on Beaches, New York: Ed. R. E. MEYER, Academic Press. 1972.

[36] FÜHRBÖTER, A.: Einige Ergebnisse aus Naturuntersuchungen in Brandungszonen. Mitteilungen des Leichtweiß-Instituts für Wasserbau der Technischen Universität Braunschweig (1974), No. 40.

[37] FÜHRBÖTER, A.: Äußere Belastungen von Seedeichen und Deckwerken. Hamburg: Vereinigung der Naßbaggerunternehmungen e. V., 1976.

[38] MORISON, J. R., O'BRIEN, M. P., JOHNSON, J. W. und SCHAAF, S. A.: The Force Exerted by Surface Waves on Piles. Petroleum Transaction, Amer. Inst. Mining Eng. 189 (1950).

[39] MACCAMY, R. C. und FUCHS, R. A.: Wave Forces on Piles: A Diffraction Theory. Techn. Memorandum 69, U. S. Army, Corps of Engineers, Beach Erosion Board, Washington, D. C. Dec. 1954.

[40] Reports of the International Waves Commission, PIANC-Bulletin No. 15 (1973) and No. 25 (1976), Brussels.

[41] HAFNER, E.: Bemessungsdiagramme zur Bestimmung von Wellenkräften auf vertikale Kreiszylinder. Wasserwirtschaft 68 (1978), No. 7/8, p. 227.

[42] HAFNER, E.: Kraftwirkung der Wellen auf Pfähle. Wasserwirtschaft 67 (1977), No. 12, p. 385.

[43] STREETER, V. L.: Handbook of Fluid Dynamics. New York 1961.

[44] KOKKINOWRACHOS, K., in: „Handbuch der Werften", Vol. 15, Hamburg 1980.

[45] Bundesverband öffentlicher Binnenhäfen, Empfehlungen des Technischen Ausschusses Binnenhäfen, Neuss.

[46] EAK 2002 – Empfehlungen für Küstenschutzbauwerke: Ausschuss für Küstenschutzwerke der DGGT und der HTG, „Die Küste", No. 65-2002, Westholsteinische Verlagsanstalt Boyens &Co., Heide i. Holst.

[47] BURKHARDT, O.: Über den Wellendruck auf senkrechte Kreiszylinder. Mitt. Franzius-Institut Hannover, No. 29, 1967.

[48] DET NORSKE VERITAS: Rules for Design, Construction and Inspection of Fixed Offshore Structures 1977.

[49] DIETZE, W.: Seegangskräfte nichtbrechender Wellen auf senkrechte Pfähle. Bauingenieur 39 (1964), No. 9, p. 354.

[50] DANTZIG, D. VON: Economic Decision Problems for Flood Prevention. „Econometrica" Vol. 24, No. 3, p. 276. New Haven 1956.

[51] Report of the Delta Committee. Vol. 3 Contribution II. 2, p. 57. The Economic Decision Problems Concerning the Security of the Netherlands against Storm Surges (Dutch Language, Summary in English). Den Haag 1960, Staatsdrukkerij en uitgeversbedrijf.

[52] Richtlinien für Regelquerschnitte von Schiffahrtskanälen, Bundesverkehrsministerium, Abt. Binnenschiffahrt und Wasserstraßen, 1994.

[53] Beziehung zwischen Kranbahn und Kransystem, Ausschuss für Hafenumschlagtechnik der Hafenbautechnischen Gesellschaft e. V., HANSA 122 (1985), No. 21, p. 2215 und 22, p. 2319.

[54] KRANZ, E.: Die Verwendung von Kunststoffmörtel bei der Lagerung von Kranschienen auf Beton. Bauingenieur 46 (1971), No. 7, p. 251.

[55] Construction and Survey Accuracies for the execution of dredging and stone dumping works. Rotterdam Public Works Engineering Department. Port of Rotterdam. The Netherlands Association of Dredging, Shore and Bank Protection Contractors (VBKO). International Association of Dredging Companies (IADC), March 2001.

[56] KOPPEJAN, A. W.: A Formular combining the TERZAGHI Load-compression relationship and the BUISMAN secular time effect. Proceedings 2nd Int. Conf. on Soil Mech. and Found. Eng. 1948.

599

[57] HELLWEG, V.: Ein Vorschlag zur Abschätzung des Setzungs- und Sackungsverhaltens nichtbindiger Böden bei Durchnässung. Mitt. Institut für Grundbau und Bodenmechanik, Universität Hannover 1981, No. 17.

[58] KWALITEITSEISEN VOR HOUT (K. V. H. 1980).

[59] WIRSBITZKI, B.: Kathodischer Korrosionsschutz im Wasserbau. Hafenbautechnische Gesellschaft e. V., Hamburg 1981.

[60] WOLLIN, G.: Korrosion im Grund- und Wasserbau. Die Bautechnik 40 (1963), No. 2, p. 37.

[61] BLUM, H.: Einspannungsverhältnisse bei Bohlwerken, Ernst & Sohn, Berlin 1931.

[62] ROWE, P. W.: Anchored Sheet-Pile Walls. Proc. Inst. Civ. Eng. London 1952, Paper 5788.

[63] ROWE, P. W.: Sheet-Pile Walls at Failure. Proc. Inst. Civ. Eng. London 1956, Paper 6107 and relevant discussion 1957.

[64] ZWECK, H. und DIETRICH, Th.: Die Berechnung verankerter Spundwände in nichtbindigen Böden nach ROWE [62], Mitteilungsblatt der Bundesanstalt für Wasserbau, Karlsruhe 1959, No. 13.

[65] BRISKE, R.: Anwendung von Erddruckumlagerungen bei Spundwandbauwerken. Die Bautechnik 34 (1957), No. 7, p. 264, and No. 10, p. 376.

[66] BRINCH HANSEN, J.: Spundwandberechnungen nach dem Traglastverfahren. Internationaler Baugrundkursus 1961. Mitteilungen aus dem Institut für Verkehrswasserbau, Grundbau und Bodenmechanik der Technischen Hochschule Aachen, Aachen (1962), No. 25, p. 171.

[67] LAUMANS, Q.: Verhalten einer ebenen, in Sand eingespannten Wand bei nichtlinearen Stoffeigenschaften des Bodens. Baugrundinstitut Stuttgart, Mitteilung 7 (1977).

[68] OS, P. J. VAN: Damwandberekening: computermodel of BLUM. Polytechnisch Tijdschrift, Editie B, 31 (1976), No. 6, pp. 367–378.

[69] FAGES, R. und BOUYAT, C.: Calcul de rideaux de parois moulées et de palplanches (Modèle mathématique intégrant le comportement irréversible du sol en état élastoplastique. Exemple d'application, Etude de l'influence des paramètres). Travaux (1971), No. 439, pp. 49–51 and (1971), No. 441, pp. 38–46.

[70] FAGES, R. und GALLET, M.: Calculations for Sheet Piled or Cast in Situ Diaphragm Walls (Determination of Equilibrium Assuming the Ground to be in an Irreversible Elasto-Plastic State). Civil Engineering and Public Works Review (1973), Dec.

[71] SHERIF, G.: Elastisch eingespannte Bauwerke, Tafeln zur Berechnung nach dem Bettungsmodulverfahren mit variablen Bettungsmoduli, Ernst & Sohn, Berlin/Munich/Düsseldorf 1974.

[72] RANKE, A. und OSTERMAYER, H.: Beitrag zur Stabilitätsuntersuchung mehrfach verankerter Baugrubenumschließungen. Die Bautechnik 45 (1968), No. 10, pp. 341–350.

[73] LACKNER, E.: Berechnung mehrfach gestützter Spundwände, 3rd edition, Berlin, Ernst & Sohn, 1950. Siehe auch in [7].

[74] KRANZ, E.: Über die Verankerung von Spundwänden, 2nd edition, Ernst & Sohn, Berlin 1953.

[75] WIEGMANN, D.: Messungen an fertigen Spundwandbauwerken. Vortr. Baugrundtag. Dt. Ges. für Erd- und Grundbau, May 1953, Hamburg 1953, pp. 39–52.

[76] BRISKE, R.: Erddruckverlagerung bei Spundwandbauwerken, 2nd edition, Berlin, Ernst & Sohn, 1957.

[77] BEGEMANN, H. K. S. Ph.: The Dutch Static Penetration Test with the Adhesion Jacket Cone (Tension Piles, Positive and Negative Friction, the Electrical Adhesion Jacket Cone) LGM-Mededelingen (1969), Vol. 13, No. 1, 4 und 13.

600

[78] SCHENK, W.: Verfahren beim Rammen besonders langer, flachgeneigter Schrägpfähle. Bauingenieur 43 (1968), No. 5.

[79] LEONHARDT, F.: Vorlesungen über Massivbau, Part 4, 2nd edition, Springer, Berlin/ Heidelberg/New York 1978.

[80] Dynamit Nobel: Sprengtechnisches Handbuch, Dynamit Nobel Aktiengesellschaft (Ed.), Troisdorf, 1993.

[81] PRIEBE, H.: Bemessungstafeln für Großbohrpfähle. Die Bautechnik 59 (1982), No. 8, p. 276.

[82] WEISS, F.: Die Standfestigkeit flüssigkeitsgestützter Erdwände. Bauingenieur-Praxis Berlin/Munich/Düsseldorf: Ernst & Sohn, 1967, No. 70.

[83] MÜLLER-KIRCHENBAUER, H., WALZ, B. und KILCHERT, M.: Vergleichende Untersuchung der Berechnungsverfahren zum Nachweis der Sicherheit gegen Gleitflächenbildung bei suspensionsgestützten Erdwänden. Veröffentlichungen des Grundbauinstituts der TU Berlin, No. 5, 1979.

[84] FEILE, W.: Konstruktion und Bau der Schleuse Regensburg mit Hilfe von Schlitzwänden. Bauingenieur 50 (1975), No. 5, p. 168.

[85] LOERS, G. und PAUSE, H.: Die Schlitzwandbauweise – große und tiefe Baugruben in Städten. Bauingenieur 51 (1976), No. 2, p. 41.

[86] VEDER, Ch.: Beispiele neuzeitlicher Tiefgründungen. Bauingenieur 51 (1976), No. 3, p. 89.

[87] VEDER, Ch.: Die Schlitzwandbauweise – Entwicklung, Gegenwart und Zukunft, Österreichischer Ing. Z. 18 (1975), No. 8, p. 247.

[88] VEDER, Ch.: Einige Ursachen von Mißerfolgen bei der Herstellung von Schlitzwänden und Vorschläge zu ihrer Vermeidung. Bauingenieur 56 (1981), No. 8, p. 299.

[89] CARL, L. und STROBL, Th.: Dichtungswände aus einer Zement-Bentonit-Suspension. Wasserwirtschaft 66 (1976), No. 9, p. 246.

[90] LORENZ, W.: Plastische Dichtungswände bei Staudämmen. Vorträge Baugrundtagung 1976 in Nürnberg, Deutsche Gesellschaft für Erd- und Grundbau e. V., p. 389.

[91] KIRSCH, K. und RÜGER, M.: Die Rüttelschmalwand – Ein Verfahren zur Untergrundabdichtung. Vorträge Baugrundtagung 1976 in Nürnberg, Deutsche Gesellschaft für Erd- und Grundbau e. V., p. 439.

[92] KAESBOHRER, H.-P.: Fortschritte an der Donau im Dichtungsverfahren für Stauräume. Die Bautechnik 49 (1972), No. 10, p. 329.

[93] BRENNECKE/LOHMEYER: Der Grundbau, 4th edition, Vol. II, Ernst & Sohn, Berlin 1930.

[94] NÖKKENTVED, C.: Berechnung von Pfahlrosten, Ernst & Sohn, Berlin 1928.

[95] SCHIEL, F.: Statik der Pfahlgründungen. Berlin: Springer, 1960.

[96] AGATZ, A. und LACKNER, E.: Erfahrungen mit Grundbauwerken, Springer, Berlin 1977.

[97] Technische Lieferbedingungen für Wasserbausteine – edition 2003 (TLW) – des Bundesministers für Verkehr, Bau- und Wohnungswesen, Verkehrsblatt (2004), No. 11.

[98] Uferschutzwerke aus Beton, Schriftenreihe der Zementindustrie, Verein deutscher Zementwerke e. V., Düsseldorf (1971), No. 38.

[99] FINKE, G.: Geböschte Ufer in Binnenhäfen, Zeitschrift für Binnenschiffahrt und Wasserstraßen (1978), No. 1, p. 3.

[100] Report of the PIANC Working Group I-4, Guidelines for the design and construction of flexible revements incorporating geotextiles for inland waterways, Supplement to PIANC-Bulletin No 57, Brussels 1987.

[100a] Report of the PIANC Working Group II-21 „Guidelines for the design and construction of flexible revetments incorporating geotextiles in marine environment, Supplement to PIANC-Bulletin No 78/79, Brussels 1992.

[101] BLUM, H.: Wirtschaftliche Dalbenformen und deren Berechnung. Die Bautechnik 9 (1932), No. 5, p. 50.

[102] COSTA, F. V.: The Berthing Ship. The Dock and Harbour Authority. Vol. XLV, (1964), Nos. 523 to 525.

[103] Stahl-Eisen-Werkstoffblatt 088. Schweißbare Feinkornbaustähle, Richtlinien für die Verarbeitung, Düsseldorf: Verlag Stahleisen.

[104] Zulassungsbescheid für hochfeste, schweißgeeignete Feinkornbaustähle StE 460 und StE 690, Institut für Bautechnik, Kolonnenstraße 301, 10829 Berlin.

[105] BINDER, G.: Probleme der Bauwerkserhaltung – eine Wirtschaftlichkeitsberechnung, BAW-Brief No. 1, Karlsruhe 2001.

[106] LUNNE, T., ROBERTSON, P. K., POWELL, J. J. M.: Cone Penetration Testing in Geotechnical Practice, Spon Press, London 1997.

[107] BYDIN, F. I.: Development of certain questions in area of river's winter regime, III. Hydrologic Congress, Leningrad 1959.

[108] SCHWARZ, J., HIRAYAMA, K., WU, H. C.: Effect of Ice Thickness on Ice Forces, Proceedings Sixth Annual Offshore Technology Conference, Houston, Texas, USA 1974.

[109] KORZHAVIN, K. N.: Action of ice on engineering structures, English translation, U. S. Cold Region Research and Engineering Laboratory, Trans. T. L. 260.

[110] Germanischer Lloyd: Vorschriften für Konstruktion und Prüfung von Meerestechnischen Einrichtungen, Band I – Meerestechnische Einheiten – (Seebauwerke). Hamburg: Eigenverlag des Germanischen Lloyd, July 1976.

[111] Ice Engineering Guide for Design and Construction of Small Craft HarborS. University of Wisconsin, Advisory Report SG-78-417.

[112] HORN, A.: Bodenmechanische und grundbauliche Einflüsse bei der Planung, Konstruktion und Bauausführung von Kaianlagen. Mitt. d. Inst. f. Bodenmechanik und Grundbau, HSBw München, No. 4, und Mitt. des Franzius-Instituts für Wasserbau, (1981), No. 54, p.110.

[113] HORN, A.: Determination of properties for weak soils by test embankments. International Symposium „Soil and Rock Investigations by in-situ Testing"; Paris, (1983), Vol. 2, p. 61.

[114] HORN, A.: Vorbelastung als Mittel zur schnelleren Konsolidierung weicher Böden. Geotechnik (1984), No. 3, p. 152.

[115] SCHMEDEL, U.: Seitendruck auf Pfähle. Bauingenieur 59 (1984), p. 61.

[116] FRANKE, E. und SCHUPPENER, B.: Horizontalbelastung von Pfählen infolge seitlicher Erdauflasten. Geotechnik (1982), p. 189.

[117] DBV-Merkblatt. Begrenzung der Rißbildung im Stahlbeton- und Spannbetonbau. Deutscher Beton-Verein e. V.

[118] Zusätzliche Technische Vertragsbedingungen – Wasserbau (ZTV-W) für Wasserbauwerke aus Beton und Stahlbeton (Leistungsbereich 215).

[119] RÜPING, F.: Beitrag und neue Erkenntnisse über die Errichtung und Sicherung von großen Mineralöl-Lagertanks für brennbare Flüssigkeiten der Gefahrenkasse A I. Dissertation, Hannover 1965.

[120] Richtlinien für Anforderungen an Anlagen zum Umschlag gefährdender flüssiger Stoffe im Bereich der Wasserstrassen. Erlaß des Bundesministers für Verkehr vom 24.7.1975, Verkehrsblatt 1975, p. 485.

[121] MAYER, B. K., KREUTZ, B., SCHULZ, H.: Setting sheet piles with driving aids, Proc. 11th Int. Conf. Soil MecH. Found. Eng. San Francisco, 1985.

[122] ARBED, S. A., Luxembourg: Europäisches Patent 09. 04. 86, Patenterteilung am 9.4.86, Patentblatt 86/15.

[123] PARTENSCKY, H.-W.: Binnenverkehrswasserbau, Schleusenanlagen. Berlin, Heidelberg, New York, Tokyo: Springer-Verlag, 1986.

[124] Merkblatt „Anwendung von Regelbauweisen für Böschungs- und Sohlensicherungen an Wasserstraßen (MAR)", edition 1993. Bundesanstalt für Wasserbau, Karlsruhe.

[125] Kanal- und Schiffahrtsversuche 1967. Schiff und Hafen, 20 (1968), No. 4–9. See also 27. Mitteilungsblatt der Bundesanstalt für Wasserbau, Karlsruhe, Sept. 1968.

[126] TUNNEL-SONDERAUSGABE APRIL 1987. Internationale Fachzeitschrift für unterirdisches Bauen. Gütersloh: Bertelsmann.

[127] DEUTSCH, V., und VOGT, M.: Die zerstörungsfreie Prüfung von Schweißverbindungen – Verfahren und Anwendungsmöglichkeiten. Schweißen und Schneiden 39 (1987), No. 3.

[128] Merkblatt „Anwendung von geotextilen Filtern an Wasserstrassen (MAG)", edition 1993, Bundesanstalt für Wasserbau, Karlsruhe.

[129] „Grundlagen zur Bemessung von Böschungs- und Sohlensicherungen an Binnen- wasserstraßen". Bundesanstalt für Wasserbau: Mitteilungsheft 87, Karlsruhe 2004.

[130] Richtlinien für die Prüfung von geotextilen Filtern im Verkehrswasserbau (RPG), Bundesanstalt für Wasserbau, Karlsruhe 1994.

[131] Technische Lieferbedingungen für Geotextilien und geotextilverwandte Produkte an Wasserstraßen (TLG) – edition 2003 – des Bundesministeriums für Verkehr, Bau- und Wohnungswesen, Verkehrsblatt 2003, No. 18.

[132] Zusätzliche Technische Vertragsbedingungen – Wasserbau (ZTV-W) für Böschungs- und Sohlensicherungen (Leistungsbereich 210).

[133] Concrete International. Detroit 1982, pp. 45–51.

[134] Richtlinie für Schutz und Instandsetzung von Betonbauteilen, Part 1 to 4; Deutscher Ausschuss für Stahlbeton DAfStB 1990, 1991, 1992.

[135] DIERSSEN, G., GUDEHUS, G.: Vibrationsrammungen in trockenem Sand. Geotechnik 3/1992, p. 131.

[136] PIANC-Report of Working group II-31 „life cycle management of port structures – General principles" in PIANC-Bulletin No. 99, Brussels 1998.

[137] FGSV-820, Merkblatt für die Fugenfüllung in Verkehrsflächen aus Beton. Forschungs- gesellschaft für Straßen- und Verkehrswesen e. V., Fassung 1982.

[138] Zusätzliche Technische Vertragsbedingungen – Wasserbau (ZTV-W) für Schutz und Instandsetzung der Betonbauteile von Wasserbauwerken (Leistungsbereich 219).

[139] VV-WSV 2101 BAUWERKSINSPEKTION, herausgegeben vom Bundesminister für Verkehr, Bonn, 1984, erhältlich bei der Drucksachenstelle der Wasser- und Schiffahrtsdirektion Mitte, Hannover.

[140] Report of the PIANC Working Group II-9, Development of modern Marine Terminals Supplement to PIANC-Bulletin No. 56, Brussels 1987.

[141] BJERRUM, L.: General Report, 8, ICSMFE, (1973) Moskau, Vol. 3, p. 124.

[142] WROTH, C. P.: „The interpretation of in situ soil tests", 1984, Géotechnique 34, No. 4, pp. 449–489.

[143] SELIG, E. T. und MCKEE, K. E.:„Static and dynamic behavior of small footings", Am. Soc. Civ. Eng., Journ. Soil Mech. Found. Div., Vol. 87 (1961), No. SM 6, Part I, pp. 29–47). (Vgl. Horn, A. – Bauing. (1963) 38, No. 10, p. 404).

[144] HORN, A.: „Insitu-Prüfung der Wasserdurchlässigkeit von Dichtwänden" (1986), Geotechnik 1, p. 37.

[145] RANKILOR, P. R.: „Membranes in ground engineering" (1981), Wiley, New York.

[146] VELDHUYZEN VAN ZANTEN, R.: „Geotextiles and Geomembranes in Civil Engineering" (1994), Balkema, Rotterdam/Boston.

[147] KOERNER, R. M.: „Design with Geosynthetics", Prentice-Hall (1997), Englewood Cliffs, N. Y.

[148] HAGER, M.: Eisdruck, Kap. 1.14 Grundbau-Taschenbuch, 5th edition, Part 1, Ernst & Sohn, Verlag für Architektur und technische Wissenschaften, 1996.

[149] Merkblatt Anwendung von Kornfiltern an Wasserstrassen (MAK), edition 1989, Bundesanstalt für Wasserbau, Karlsruhe.

[150] Kunststoffmodifizierter Spritzbeton. Merkblatt des Deutschen Betonvereins e. V. Wiesbaden.

[151] HEIN, W.: Zur Korrosion von Stahlspundwänden in Wasser, Mitteilungsblatt der BAW No. 67, Karlsruhe 1990.

[152] HEIN, W.: Korrosion von Stahlspundwänden im Wasser, HANSA, 126. annual set 1989, No. 3/4 Schiffahrtsverlag „Hansa", C. Schroedter & Co., Hamburg.

[153] Richtlinie für die Prüfung von Beschichtungssystemen für den Korrosionsschutz im Stahlwasserbau (RPB), edition 2001, Bundesanstalt für Wasserbau, Karlsruhe.

[154] FEDERATION EUROPÉENNE DE LA MANUTENTION, Section I, Rules for the design of hoisting appliances, Booklet 2: Classification and loading on structures and mechanisms F. E. M. 1.001. 3rd edition, 1987. Deutsches National-Komitee Frankfurt/Main.

[155] Deutscher Verband für Wasserwirtschaft und Kulturbau e. V. (DVWK): Dichtungselemente im Wasserbau DK 626/627 Wasserbau; DK 69.034.93 Abdichtung Hamburg, Berlin, Verl. Paul Parey 1990.

[156] HENNE, J.: Versuchsgerät zur Ermittlung der Biegezugfestigkeit von bindigen Böden Geotechnik 1989, No. 2, pp. 96 cont.

[157] SCHULZ, H.: Mineralische Dichtungen für Wasserstraßen, Fachseminar „Dichtungswände und Dichtsohlen", June 1987 in Braunschweig, Mitteilungen des Instituts für Grundbau und Bodenmechanik, Techn. Universität Braunschweig, No. 23, 1987.

[158] SCHULZ, H.: Conditions for day sealings at joints, Proc. of the IX. Europ. Conf. on Soil MecH. and Found. Eng., Dublin, 1987.

[159] HOLTZ, R., CHRISTOPHER, R., BERG, R.: Geosynthetic Engineering. Richmond (Canada), BiTech, 1997.

[160] HTG Kaimauer-Workshop SMM '92 Conference: Kaimauerbau, Erfahrungen und Entwicklungen, Beiträge, HANSA, 129. annual set 1992, No. 7, pp. 693 cont. and No. 8, pp. 792 cont.

[161] DÜCKER, H. P. und OESER, F. W., Der Bau von Umschlag- und Werft-Kaimauern, erläutert am Beispiel von 4 Neubauprojekten, der Bauingenieur 59 (1984), pp. 15 cont.

[162] SPARBOOM, U.: Über die Seegangsbelastung lotrechter zylindrischer Pfähle im Flachwasserbereich; Mitteilungen des Leichtweiß-Instituts der TU Braunschweig, No. 93, Braunschweig 1986.

[163] VV WSV 2301 Damminspektion, herausgegeben vom Bundesminister für Verkehr, Bonn, erhältlich bei der Drucksachenstelle der WSD Mitte, Hannover.

[164] Zusätzliche Technische Vertragsbedingungen – Wasserbau (ZTV-W) für Technische Bearbeitung (LB 202).

[165] Deutscher Verband für Wasserwirtschaft und Kulturbau e. V. (DYWK): Anwendung von Geotextilien im Wasserbau DVWK Merkblatt 221/1992.

[166] DAVIDENKOFF, R. und FRANKE, O. L.: Untersuchungen der räumlichen Sickerströmung in einer umspundeten Baugrube im Grundwasser, Bautechnik 42, No. 9, Berlin 1965.

[167] DAVIDENKOFF, R.: Deiche und Erddämme, Sickerwasser-Standsicherheit, Werner-Verlag Düsseldorf, 1964.

[168] DAVIDENKOFF, R.: Unterläufigkeit von Bauwerken, Werner-Verlag Düsseldorf, 1970.

[169] ALEXY, M., FÜHRER, M., KÜHNE, E., Verbesserung der Schiffahrtsverhältnisse auf der Elbe bei Torgau: Vorbereitung, Ausführung und Erfolgskontrolle, Jahrbuch der Hafenbautechnischen Gesellschaft e. V., Hamburg, 50th Volume 1995, pp. 71 cont.

[170] RÖMISCH, K.: Propellerstrahlinduzierte Erosionserscheinungen in Häfen, HANSA, 130. annual set 1993, Nr. 8 und – Spezielle Probleme – HANSA, 131. annual set 1994, No. 9.

[171] PIANC Report of the 3rd International Wave Commission, Supplement of the Bulletin No. 36, Brussels 1980.

[172] Report of the PIANC Working Group II-12: Analysis of Rubble Mound Breakwaters, Supplement to PIANC-Bulletin No. 78/79, Brussels 1992.

[173] CIRIA/CUR: Manual on the use of rock in coastal and shoreline engineering. CIRIA Special Publication 83, CUR Report 154, Rotterdam, A. A. Balkema 1991.

[174] BRUNN, P.: Port Engineering, London 1980.

[175] STÜCKRATH, T.: Über die Probleme des Unternehmers beim Hafenbau. Mitteilungen d. Franziusinstituts für Wasserbau und Küsteningenieurwesen der TU Hannover, No. 54, Hannover 1983.

[176] ALBERTS, D. und HEELING, A.: Wanddickenmessungen an korrodierten Stahlspundwänden; statistische Datenauswertung, Mitteilungsblatt der BAW No. 75, Karlsruhe 1996.

[177] Zusätzliche Technische Vertragsbedingungen – Wasserbau (ZTV-W) für Korrosionsschutz im Stahlwasserbau (Leistungsbereich 218).

[178] Zusätzliche Technische Vertragsbedingungen – Wasserbau (ZTV-W) für kathodischen Korrosionsschutz im Stahlwasserbau (Leistungsbereich 220).

[179] Kathodischer Korrosionsschutz für Stahlbeton, Hafenbautechnische Gesellschaft e. V. (HTG) Hamburg 1994.

[180] Allgemeine Verwaltungsvorschrift zum Schutz gegen Baulärm – Geräuschemissionen, sowie – Emissionsmeßverfahren, Carl Heymanns Verlag KG, Cologne 1971.

[181] Richtlinie 79/113/EWG vom 19.12.1978 zur Angleichung der Rechtsvorschriften der Mitgliedsstaaten betreffend die Ermittlung des Geräuschemissionspegels von Baumaschinen und Baugeräten (Amtsbl. EG 1979 Nr. L 33 p. 15).

[182] 15. Verordnung zur Durchführung des BImSchG vom 10.11.1986 (Baumaschinen-LärmVO).

[183] VDI-Richtlinie 2714 (01/88) Schallausbreitung im Freien – Berechnungsverfahren.

[184] VDI-Richtlinie 3576: Schienen für Krananlagen, Schienenverbindungen, Schienenbefestigungen, Toleranzen.

[185] Empfehlungen und Berichte des Ausschusses für Hafenumschlagtechnik (AHU) der Hafenbautechnischen Gesellschaft e.V, Hamburg.

[186] EAB-100: Empfehlungen des Arbeitskreises „Baugruben" (EAB) auf der Grundlage des Teilsicherheitskonzeptes EAB-100; Ed. Deutsche Gesellschaft für Geotechnik (DGGT), Essen, Ernst & Sohn, Berlin 1996.

[187] RADOMSKI, H.: Untersuchungen über den Einfluß der Querschnittsform wellenförmiger Spundwände auf die statischen und rammtechnischen Eigenschaften, Mitt. Institut für Wasserwirtschaft, Grundbau und Wasserbau der Universität Stuttgart, No. 10, Stuttgart 1968.

[188] CLASMEIER, H.-D.: Ein Beitrag zur erdstatischen Berechnung von Kreiszellenfange-dämmen, Mitt. Institut für Grundbau und Bodenmechanik, Universität Hannover, No. 44/1996.

[189] FEDDERSEN, I.: Das Hyperbelverfahren zur Ermittlung der Bruchlasten von Pfählen, eine kritische Betrachtung. Bautechnik 1982, No. 1, pp. 27 cont.

[190] BAUMANN, V.: Das Soilcrete-Verfahren in der Baupraxis, Vorträge der Baugrundtagung 1984 der DGEG in Düsseldorf, pp. 43 cont.

[191] BAUER, K.: Einsatz der Bauer-Schlitzwandfräse beim Bau der Dichtwand am Brombachspeicher und an der Sperre Kleine Roth, Tiefbau-BG 10/1985, pp. 630 cont.

[192] STROBL, Th. und WEBER, R.: Neuartige Abdichtungsverfahren im Sandsteingebirge, Vorträge der Baugrundtagung 1986 der DGEG.

[193] STROBL, Th.: Ein Beitrag zur Erosionssicherheit von Einphasen-Dichtungswänden, Wasserwirtschaft 7/8 (1982), pp. 269 cont. und Erfahrungen über Untergrundabdich-tungen von Talsperren. Wasserwirtschaft 79 (1989), No. 7/8.

[194] GEIL, M.: Entwicklung und Eigenschaften von Dichtwandmassen und deren Über-wachung in der Praxis, s+t 35, 9/1981, pp. 6 cont.

[195] KARSTEDT, J. und RUPPERT, F.-R.: Standsicherheitsprobleme bei der Schlitzwandbau-weise, Baumaschinen und Bautechnik 5/1980, pp. 327 cont.

[196] MESECK, H., RUPPERT, F.-R., Simons, H.: Herstellung von Dichtungsschlitzwänden im Einphasenverfahren, Tiefbau, Ingenieurbau, Straßenbau 8/79, pp. 601 cont.

[197] ROM Recomendaciones para Obras Maritimas (Englische Fassung) Maritime Works Recommendations (MWR): Actions in the design of maritime and Harbor Works (ROM 0.2-90), Ministerio de Obras Publicas y Transportes, Madrid 1990.

[198] Japan Society of Civil Engineering: The 1995 Hyogoken-Nanbu Earthquake – Investigation into Damage to Civil Engineering Structures – Committee of Earthquake Engineering, Tokyo 1996.

[199] Report of the joint Working Group PIANC and IAPH, in cooperation with IMPA and IALA, PTC II-30: Approach Channels – A Guide for Design; Supplement to PIANC-Bulletin No. 95, Brussels 1997.

[200] Report of the PIANC Working Group PTC I-16: Standardisation of Ships and Inland Waterways for River/Sea Navigation; Supplement to PIANC-Bulletin No. 90, Brussels 1996.

[201] Verordnung über die Schiffs- und Schiffsbehältervermessung (Schiffsvermessungs-verordnung SchVmV) dated 5. July 1982 (BGBl. I S. 916) changed by die Erste Verordnung zur Änderung der Schiffsvermessungsverordnung dated 3. September 1990 (BGBl. I S. 1993).

[202] HUDSON, R. Y.: Design of quarry stone cover layers for rubble mound breakwaters. Waterway Experiment Station, Research Report No. 2-2, Vicksburg, USA 1958.

[203] HUDSON, R. Y.: Laboratory investigations of rubble mound breakwaters. Waterway Experiment Station Report, Vicksburg, USA 1959.

[204] MEER, J. W. VAN DER: Rock slopes and gravel beaches under wave attack. Delft Hydraulics Publication No. 396, Delft 1988.

[205] MEER, J. W. VAN DER: Conceptual design of rubble mound breakwaters Delft Hydraulics Publication No. 483, Delft 1993.

[206] OWEN, M. W.: Design of seawalls allowing for wave overtopping. Hydraulics Research, Wallingford, Report No. Ex 294, 1980.

[207] MEER, J. W. VAN DER; JANSSEN, J. P. F. M.: Wave run-up and waver overtopping at dikes and revetments. Delft Hydraulics Publication No. 485, Delft 1994.

[208] ABROMEIT, H.-U.: Ermittlung technisch gleichwertiger Deckwerke an Wasserstraßen und im Küstenbereich in Abhängigkeit von der Trockenrohdichte der verwendeten Wasserbausteine. Mitteilungsblatt der Bundesanstalt für Wasserbau, No. 75, Karlsruhe 1997.

[209] KÖHLER, H.-J.: Messungen von Porenwasserüberdrücken im Untergrund. In: Mitteilungsblatt der Bundesanstalt für Wasserbau, Karlsruhe 1989, No. 66, pp. 155–174.

[210] KÖHLER, H.-J.,HAARER, R.:Development of excess pore water pressure in over-consolidated clay, induced by hydraulic head changes and its effect on shett pile wall stability of a navigable lock. In: Proc. Of the 4th Intern. Symp. On Field Measurements in Geomechanics (FMGM 95), Bergamo 1995, SGEditorial Padua, pp. 519–526.

[211] KÖHLER, H.-J.: Porenwasserdruckausbreitung im Boden, Messverfahren und Berechnungsansätze. In: Mitteilungen des Instituts für Grundbau und Bodenmechanik, IGB-TUBS, Braunschweig 1996, No. 50, pp. 247–258d.

[212] KÖHLER, H.-J.: Grundwasserabsenkung – Einfluss auf die Baugrubensicherheit. In: Tagungsband Seminar Grundwasserabsenkungsanlagen, Landesgewerbeanstalt Bayern (LGA), Nürnberg 1997, pp. 1–21.

[213] ANDREWS, J. D.; MOSS, T. R.: Reliability and Risk Assessment, Verlag Longman Scientific & Technical, Burnt Mill (UK), 1993.

[214] RICHWIEN, W.; LESNY, K.: Risikobewertung als Schlüssel des Sicherheitskonzepts – Ein probabilistisches Nachweiskonzept für die Gründung von Offshore-Windenergieanlagen. In: Erneuerbare Energien 13 (2003), No. 2, pp. 30–35.

[215] SCHUELLER, G. I.: Einführung in die Sicherheit und Zuverlässigkeit von Tragwerken, Verlag Ernst und Sohn, Berlin 1981.

[216] Zusätzliche Technische Vertragsbedingungen und Richtlinien für Ingenieurbauten (ZTV-ING). Bundesanstalt für Straßenwesen (Ed.), Verkehrsblatt-Sammlung No. S 1056, Verkehrsblatt-Verlag, Dortmund 2003.

[217] HERDT, W.; ARNDTS, E.: Theorie und Praxis der Grundwasserabsenkung. Ernst & Sohn, Berlin 1973.

[218] Hochwasserschutz in Häfen – Neue Bemessungsansätze – Tagungsband zum HTG-Sprechtag October 1996, Hafenbautechnische Gesellschaft (HTG) e. V., Hamburg.

[219] OUMERACI, H., KORTENHAUS, A.: Anforderungen an ein Bemessungskonzept, HANSA, 134. annual set, 1997, pp. 71 cont.

[220] KORTENHAUS, A., OUMERACI, H.: Lastansätze für Wellendruck, HANSA, 134. annual set, 1997, No. 5, pp. 77 cont.

[221] HEIL, H., KRUPPE, J., MÖLLER, B.: Berechnungsansätze für HWS-Wände und Uferbauwerke, HANSA, 134. annual set, 1997, No. 5, pp. 77 cont.

[222] Richtlinie „Berechnungsgrundsätze für private Hochwasserschutzwände und Uferbauwerke im Bereich der Freien und Hansestadt Hamburg", May 1997, Amtlicher Anzeiger, Part II of the Hamburgisches Gesetz- und Verordnungsblatt, No. 33, 1998.

[223] KÖHLER, H.-J.; SCHULZ, H.: Bemessung von Deckwerken unter Berücksichtigung von Geotextilien. 3. Internationale Konferenz über Geotextilien in Wien 1986, Rotterdam, A. A. Balkema, 1986.

[224] TAKAHASHI, S.: Design of Breakwaters. Port and Harbour Research Institute, Yokosuka, Japan 1996.

[225] CEM: Costal Engineering Manual Part VI. Design of Coastal Projects Elements. US Army Corps of Engineers, Washington, D. C. 2001.

[226] CAMFIELD, F. E.: Wave Forces on a Wall. J. Waterway, Port, Coastal and Ocean Engineering, ASCE, New York (1991), Vol. 117 No. 1, pp. 76–79.

607

[227] HEIBAUM, M.: Zur Frage der Standsicherheit verankerter Stützwände auf der tiefen Gleitfuge. Technische Hochschule Darmstadt, Fachbereich Konstruktiver Ingenieurbau, Diss., 1987. Erschienen in: Franke, E. (Ed.): Mitteilungen des Instituts für Grundbau, Boden- und Felsmechanik der Technischen Hochschule Darmstadt, No. 27, 1987.

[228] IAHR/PIANC: Intern. Ass. For Hydr. Research/Permanent Intern. Ass. Of Navigation Congresses. List of Sea State Parameters. Supplement to Bulletin No. 52, Brussels 1986.

[229] OUMERACI, H.: Küsteningenieurwesen. In: Lecher, K. et al.: Taschenbuch der Wasserwirtschaft, 8. völlig neu bearbeitete Auflage, Berlin 2001, Paul Parey Verlag, Chapter 12, pp. 657–743.

[230] ROSTASY, F. S., ONKEN P.; Wirksame Betonzugfestigkeit bei früh einsetzendem Temperaturzwang; Deutscher Ausschuss für Stahlbeton – No. 449, Beuth Verlag, Berlin, 1995.

[231] HAGER, M.: Ice loading actions, Geotechnical Engineering Handbook, Vol. 1: Fundamentals, Chap. 1.14, Ernst & Sohn, Berlin 2002.

[232] PIANC-Report „Guidelines for the Design of Fender Systems: 2002". MarCom Report of WG 33, 2002.

[233] DREWES, U., RÖMISCH, K., SCHMIDT E.: Propellerstrahlbedingte Erosionen im Hafenbau und Möglichkeiten zum Schutz für den Ausbau des Burchardkais im Hafen Hamburg, Mitteilungen des Leichtweiß-Instituts für Wasserbau der Technischen Universität Braunschweig 1995, No. 134.

[234] Report of the PIANC Working Group 22 „Guidelines for the Design of Armoured Slopes under open Piled Quay Walls", Supplement to PIANC-Bulletin No. 96 (1997) of the Permanent International Navigation Congress.

[235] RÖMISCH, K.: Strömungsstabilität vergossener Steinschüttungen. Wasserwirtschaft 90, 2000, No. 7–8, pp. 356–361.

[236] RÖMISCH, K.: Scouring in Front of Quay Walls Caused by Bow Thruster and New Measures for its Reduction. V. International Seminar on Renovation and Improvements to Existing Quay Structures, TU Gdansk (Poland), May 28–30, 2001.

[237] EAO – Empfehlungen zur anwendung von Oberflächendichtungen an Sohle und Böschung von Wasserstrassen. Mitteilungsblatt der Bundesanstalt für Wasserbau, Karlsruhe 2002, No. 85.

[238] RAITHEL, M.: Zum Trag- und Verformungsverhalten von geokunststoffummantelten Sandsäulen, Schriftenreihe Geotechnik, Universität Kassel, No. 6, 1999.

[239] Deutsche Gesellschaft für Geotechnik DGGT: Empfehlungen für Bewehrungen aus Geokunststoffen – EBGEO, Verlag Ernst & Sohn, Berlin 1997.

[240] KEMPFERT, H.-G.: Embankment foundation on geotextile-coated sand columns in soft ground. Proceedings of the first european geosynthetics conference EurGeo 1/ Maastricht/Netherlands, October 1996.

[241] ZAESKE, D.: Zur Wirkungsweise von unbewehrten und bewehrten mineralischen Tragschichten über pfahlartigen Gründungselementen, Schriftenreihe Geotechnik, Universität Kassel, No. 10, 2001.

[242] KEMPFERT, H.-G., STADEL, M: Berechnung von geokunststoffbewehrten Tragschichten über Pfahlelementen, Bautechnik Jahrgang 75, 1997, No. 12, pp. 818–825.

[243] MÖBIUS, W., WALLIS, P., RAITHEL, M., KEMPFERT, H.-G., GEDUHN, M.: Deichgründung auf Geokunststoffummantelten Sandsäulen. HANSA, 139. annual set, 2002, No. 12, pp. 49–53.

[244] KEMPFERT, H. G., RAITHEL, M., GEDUHN, M.: Practical Aspects of the Design of Deep Geotextile Coated Sand Columns for the Foundation of a Dike on very soft soils. International Symposium Earth Reinforcement IS Kyushu, Fukuoka, Japan 2001.

[245] EAG-GTD 2002: Empfehlungen zur Anwendung geosynthetischer Tondichtungsbahnen. Hrsg: Deutsche Gesellschaft für Geotechnik, Ernst & Sohn, Berlin 2002.

[246] ALBERTS, D.: Korrosionsschäden und Nutzungsdauerabschätzung an Stahlspundwänden und -pfählen im Wasserbau, 1. Tagung „Korrosionsschutz in der maritimen Technik" Germanischer Lloyd, Hamburg, Dec. 2001.

[247] ALBERTS, D., SCHUPPENER, B.: Comparison of ultrasonic probes for the measurement of the thickness of sheet-pile walls, Field Measurements in Geotechnics, Sørum (ed.), Balkema, Rotterdam 1991.

[248] HEISS, P., MÖHLMANN, F. UND RÖDER, H.: Korrosionsprobleme im Hafenbau am Übergang Spundwandkopf zum Betonüberbau, HTG-Jahrbuch, Vol. 47, 1992.

[249] BINDER, G., GRAFF, M.: Mikrobiell verursachte Korrosion an Stahlbauten, Materials and Corrosion 46, 1995, pp. 639–648.

[250] GRAFF, M., KLAGES, D., BINDER, G.: Mikrobiell induzierte Korrosion (MIC) in marinem Milieu, Materials and Corrosion 51, 2000, pp. 247–254.

[251] CUDMANI, R. O.: Statische, alternative und dynamische Penetration in nichtbindigen Böden, Veröffentlichung des Instituts für Bodenmechanik und Felsmecanik der Universität Karlsruhe, Karlsruhe 2001.

[252] HOLEYMAN, A.: HYPERVIBIIa, A detailed numerical model for future computer implementation to evaluate the penetration speed of vibratory driven sheet piles, BBRI, 1993.

[253] MIDDENDORP, P.: VDPWAVE, TNO Profound, 2001.

[254] Unfallverhütungsvorschrift „Sprengarbeiten" der Steinbruchs-Berufsgenossenschaft vom April 1985, Fassung vom 1.1.1997,Kommentar zum § 48 (Bohrlöcher), p. 89.

[255] ZDANSKY, VLADAN: Kaimauern in Blockbauweise, Bautechnik Nr. 79, No. 12, Ernst & Sohn, Berlin 2002.

[256] WEISSENBACH, A.: Baugruben, Teil III: Berechnungsverfahren, Ernst & Sohn, Berlin 2001.

[257] PIANC-Report „Effect of earthquakes on port structures", MarCom Report of WG 34, 2001.

[258] PIANC-Report „Inspection, maintenance and repair of maritime structures exposed to material degradation caused by a salt water environment (revised report)", MarCom Report of WG 17, 2004.

I.3 Technical provisions

The standards (EN and DIN, including pre-standards), the Regulations of Deutsche Bahn AG (DS), the Regulations of the German Committee for Steel Constructions (DASt-Ri) and the Steel–Iron Materials Table of the "Verein Deutscher Eisenhüttenleute" (SEW) are valid in their respective latest version. (P = part, S = Supplement.)
Standards referred to in the recommendations are cited without quoting the valid edition. The following list represents the editions current at October 2004.

I.3.1 Standards

DIN Report 101	Actions on bridges
DIN EN 206 P1	Concrete – Part 1: Specification, performance, production and conformity
DIN EN 288 P3	Specification and approval of welding procedures for metallic materials – Part 3: Welding procedure tests for the arc welding of steel
DIN EN 440	Welding consumables: wire electrodes and deposits for gas-shielded metal arc welding of non-alloy and fine grain steels, classification
DIN EN 499	Welding consumables: covered electrodes for manual metal arc welding of nonalloy and fine grain steels, classification
DIN 536 P1/2	Crane rails
DIN EN 729 P3	Quality requirements for welding – Fusion welding of metallic materials – Part 3: Standard quality requirements
DIN EN 756	Welding consumables: solid wires, solid wire-flux and tubular coned electrode-flux combinations for submerged arc welding of nonalloy and fine grain steels – classification
DIN EN 996	Piling equipment – safety requirements
DIN 1045 P1–4	Concrete, reinforced and prestressed concrete
DIN 1048 P1/2/4/5	Testing concrete
E DIN 1052	Design of timber structures
DIN 1054	Subsoil: verification of the safety of earthworks and foundations
DIN 1055 P1–10/100	Action on structures
DIN 1076	Engineering structures in connection with roads – inspection and test
DIN 1080 P1–9	Terms, symbols and units used in civil engineering
DIN 1301 P1–3	Units
DIN EN 1536	Execution of special geotechnical work – bored piles
DIN EN 1537	Execution of special geotechnical works – ground anchors
DIN EN 1538	Execution of special geotechnical work – diaphragm walls
DIN 1681	Cast steels for general engineering purposes; technical delivery conditions
DIN EN ISO 1872 P1/2	Plastics – polyethylene (PE) moulding and extrusion materials
DIN EN 1990	Basis of structural design
DIN V ENV 1991	Basis of design and actions on structures
DIN V ENV 1992	Design of concrete structures

DIN V ENV 1993	Design of steel structures
DIN V ENV 1994	Design of composite steel and concrete structures
DIN V ENV 1995	Design of timber structures
DIN V ENV 1996	Design of masonry structures
DIN V ENV 1997	Geotechnical design
DIN V ENV 1998	Design provisions for earthquake resistance of structures
DIN V ENV 1999	Design of aluminium structures
E DIN 4017	Soil: calculation of design bearing capacity of soil beneath shallow foundations
DIN 4018	Subsoil: calculation of the bearing pressure distribution under spread foundations
DIN V 4019-100	Soil: analysis of settlements – Part 100: analysis in accordance with partial safety factor concept
DIN 4020	Geotechnical investigations for civil engineering purposes
DIN 4021	Subsoil; exploration by excavation and borings as well as by sampling
DIN 4022 P1–3	Subsoil and groundwater; classification and description of soil types and rock
DIN 4030 P1–2	Assessment of water, soils and gases for their aggressiveness to concrete
DIN 4049 P1–3	Hydrology
DIN 4054	Corrections of waterways; terms
E DIN 4084	Soil; calculation of embankment failure and overall stability of retaining structures
E DIN 4085	Soil; calculation of earth pressure
DIN 4093	Ground treatment by grouting; planning, grouting procedure and testing
DIN 4094 P1–5	Subsoil – field investigations
DIN 4102 P1–9/11–19/21–22	Fire behaviour of building materials and components
DIN V 4126-100	Diaphragm walls – Part 100: analysis in accordance with the partial safety factor concept
DIN 4127	Foundation engineering: diaphragm wall clays for supporting liquids; requirements, testing, supply, inspection
DIN 4128	Small diameter injection piles (in-situ concrete piles and composite piles); Construction procedure, design and permissible loading
E DIN 4149	Buildings in German earthquake areas; design loads, analysis and structural design of buildings
DIN 4150 P2/3	Vibrations in buildings
DIN ISO 9613 P2	Acoustics – attenuation of sound during propagation outdoors – Part 2: General method of calculation
E DIN EN 10 025 P1–6	Hot rolled products of structural steels
DIN EN 10 028 P1–7	Flat products made of steels for pressure purposes
DIN EN 10 113 P1–3	Hot-rolled products in weldable fine-grain structural steels
E DIN EN 10 204	Metallic products: types of inspection documents
E DIN EN 10 219 P1/2	Cold-formed welded structural hollow sections of non-alloy and fine-grain steels

DIN EN 10 248 P1/2	Hot-rolled sheet piling of non alloy steels, technical delivery conditions
DIN EN 10 249 P1/2	Cold-formed sheet piling of non alloy steels, technical delivery conditions
DIN EN 12 063	Execution of special geotechnical work, sheet piling construction
DIN EN 12 699	Execution of special geotechnical work – displacement piles
DIN EN 12 715	Execution of special geotechnical work – grouting
DIN EN 12 716	Execution of special geotechnical works – jet grouting
DIN EN ISO 12 944 P1–8	Paints and varnishes – corrosion protection of steel structures by protective paint systems
DIN EN 13 253	Geotextiles and geotextile-related products – Required characteristics for use in external erosion control systems
E DIN EN 14 199	Execution of special geotechnical works – micropiles
DIN 15 018 P1	Cranes; steel structures; verification and analyses
DIN 16 972	Compression moulded plates made of polyethylene high density (PE-UHMW), (PE-HMW), (PE-HD) – technical specifications
DIN 18 122 P1/2	Soil, investigation and testing
DIN 18 125 P1/2	Soil, investigation and testing
DIN 18 126	Soil; investigation and testing – determination of density of non-cohesive soils for maximum and minimum compactness
DIN 18 127	Soil, investigation and testing – Proctor-test
DIN 18 130 P1/2	Soil; determination of water permeability; laboratory tests
DIN 18 134	Soil; testing procedures and testing equipment; plate load test
E DIN 18 135	Soil – investigation and testing – oedometer consolidation test
DIN 18 136	Soil; test procedures and testing equipment; unconfined compression test
DIN 18 137 P1–3	Soil; determination of shear strength
DIN 18 195 P1–4/6/8–10	Water proofing of buildings
DIN 18 196	Earthworks and foundation; soil classification for civil engineering purposes
DIN 18 300	VOB order for placing of contracts for construction work, Part C: General technical contractual conditions for construction work, earthworks
DIN 18 311	VOB, Part C: General technical contract procedures for building works, dredging works
DIN 18 540	Sealing of externer wall joints in building using joint sealants
DIN 18 800 P1–5/7	Steel structures
DIN 18 801	Structural steel in building, design and construction
DIN 19 666	Drain pipes and percolation pipes – general requirements
DIN 19 702	Stability of solid structures in water engineering
DIN 19 703	Locks for waterways for inland navigation – principles for dimensioning and equipment
DIN 19 704 P1–3	Hydraulic steel structures; Part 1: Criteria for design and calculation
DIN 31 051	Fundamentals of maintenance
DIN 45 669 P1–3	Measurement of vibration immission

DIN 50 929 P1–3	Corrosion of metals; probability of corrosion of metallic materials when subject to corrosion from the outside; general
DIN 51 043	Trass; requirements, tests
DIN 52 108	Testing inorganic, non-metallic materials; wear test with grinding wheel according to Böhme, grinding wheel method
DIN 52 170 P1–4	Determination of composition with hardened concrete
DIN 53 505	Testing of rubber; Shore A and D hardness test
DIN 55 928 P8/9	Protection of steel structures from corrosion by organic and metal coatings; preparation and testing of surfaces

I.3.2 Regulations of Deutsche Bahn AG

| RIL 804 | Railway bridges (and other engineering constructions); design, construction and maintenance |
| RIL 836 | Earthwork structures; design, construction and maintenance |

I.3.3 Regulations of the German Committee for Steel Constructions (DASt-Ri)

DASt-Ri 006	Welding over finished coatings in steel construction
DASt-Ri 007	Delivery, processing and use of weather-resistant structural steels
DASt-Ri 012	Check of safety against bulging for plates; only in combination with DIN 18 800-1 (3/81)
DASt-Ri 014	Recommendations for prevention of terracing failure in welded constructions made of structural Steel
DASt-Ri 015	Steel beams with slender webs
DASt-Ri 016	Dimensioning and design of bearing structures made of thin-walled cold-formed components
DASt-Ri 017	Safety against bulging for shells – special cases just in combination with DIN 18 800 P1 to P4 (11/90)
DASt-Ri 103	National Application Document (NAD) for DIN V ENV 1993 part 1-1 (11/93)
DASt-Ri 104	National Application Document (NAD) for DIN V ENV 1994 part 1-1 (2/94)

I.3.4 Steel-Iron Materials Table of the "Verein Deutscher Eisenhüttenleute" (SEW)

| SEW 088 | Fine-grained structural steels suited for welding; regulations for processing, especially for smelt welding |

Annex II   List of conventional symbols

II.1   Symbols

The following lists specify most of the symbols and abbreviations used in the text, equations and figures. They conform as far as possible to DIN 1080. The units conform to DIN 1301. The water level designations correspond to DIN 4049 and DIN 4054.
All symbols are also explained in the relevant text passages.

Symbol	Definition	Unit
A	Anchor force	kN/m, MN/m or kN, MN
A	Energy absorption capacity (dolphin. fender)	kNm
A	Area, cross-sectional area	m^2
C	Equivalent force (soil reaction)	kN/m
C	Cohesion force along a failure plane	kN/m, MN/m
C_D	Drag coefficient taking into account resistance against flow pressure	1
C_e	Eccentricity factor	1
C_m	Mass factor	1
C_M	Inertia coefficient taking into account resistance to acceleration of water particles	1
C_s	Stiffness factor	1
D	Degree of density of substances	1
D	Pile diameter	m
D_{pr}	Degree of compaction according to Proctor	1
E	Modulus of elasticity (Young's modulus)	MN/m^2
E	Kinetic energy	kNm, MNm
E_a	Active earth pressure	MN/m
E_s	Coefficient of compressibility	MN/m^2
E_p	Passive earth pressure	MN/m
F	Force	kN, MN
F	Cross-sectional area	m^2
G	Dead load of body of soil	kN/m, MN/m
G	Water displacement of ship as mass	t
GRT	Gross Register Tonnage	2.83 m^3
GS	Cast steel to DIN 1681	
H	Total height of retaining wall	m
H	Maximum freeboard height of ship	m
H	Wave height	m
H_b	Wave height at moment of breaking	m
H_d	Height of "design wave"	m
H_m	Mean wave height	m
H_{max}	Maximum wave height	m
H_{rms}	root mean square wave height	m

Symbol	Definition	Unit
$H_{1/3}$	Mean value of 33% of the highest waves	m
$H_{1/10}$	Mean value of 10% of the highest waves	m
$H_{1/100}$	Mean value of 1% of the highest waves	m
I	Moment of inertia	m^4
I_c	Consistency index	1
I_D	Density index	1
I_p	Plasticity index	%
K_a	Coefficient of active earth pressure	1
K_{ac}	Coefficient of active earth pressure due to cohesion	1
K_{ag}	Coefficient of active earth pressure due to dead load of soil	1
K_{ah}	Horizontal component of K_a	1
K_o	Coefficient for determining earth pressure at rest	1
K_p	Coefficient of passive earth pressure	1
K_{pc}	Coefficient of passive earth pressure due to cohesion	1
K_{pg}	Coefficient of passive earth pressure due to dead load of soil	1
K_{ph}	Horizontal component of K_p	1
L	Wavelength	m
L_o	Length of deep water wave	m
$L_{ü}$	Overall length of ship	m
L_S	Limit state in the meaning of DIN 1054	
M	Mass	t
M	Moment	kNm
M_E	Moment of fixity	kNm
M_{span}	Moment in span	kNm
N	Normal force	MN
N_{10}	Number of blows per 10 cm penetration in penetro-meter test	1
N	Newton (unit of force)	N
kN	Kilonewton = $10^3 \cdot N$	kN
MN	Meganewton = $10^6 \cdot N$	MN
NN	Mean sea level (MSL)	m
P	Surcharge, wave load, ice load, force	MN/m or MN
P	Probability	1
P_f	Probability of failure	1
$P_{1...n}$	Pile forces in piles 1...n	kN, MN or kN/m, MN/m
P_{stat}	Static force (dolphin, fender)	MN
P_{impact}	Impact force (dolphin, fender)	MN
Q	Shear force	kN/m, MN/m
Q	Soil reaction force	kN/m, MN/m
Q_a	Soil reaction force at active earth pressure failure plane	kN/m, MN/m
Q_p	Soil reaction force at passive earth pressure failure plane	kN/m, MN/m
Q'	Limit tensile load (pile)	kN/m, MN/m
R	Resistance in the meaning of DIN 1054	kN/m^2

Symbol	Definition	Unit
R_d	Design value of resistances	kN/m^2
Re	REYNOLDS number	1
R_k	Characteristic values of resistances	kN/m^2
S	Actions in the meaning of DIN 1054	kN/m^2
S	Degree of saturation	1
S	Flow force	kN/m
S	1st moment of area (section modulus)	m^3
S_d	Design value of actions	kN/m^2
S_k	Characteristic values of actions	kN/m^2
T	Shear flow	MN/m
T	Wave period	s
TEU	Twenty Feet Equivalent Unit	1
U	Coefficient of uniformity	1
U	Force due to pore water pressure	kN/m, MN/m
V	Vertical load	kN/m, MN/m
W	Water surcharge, water pressure	kN/m, MN/m
W	Modulus of resistance	m^3
W_i	Wind load component	kN
$W_{unsteady}$	Force due to water pressure for unsteady flow	kN/m, MN/m
a	Distance between top of wall and anchor position	m
a	Acceleration	m/s^2
a	Half mean tidal range	m
a	Welding seam thickness	mm
b	Width	m
c	Spring constant	kN/m
c	Wave celerity (or velocity)	m/s
c'	Effective cohesion	kN/m^2
c_c	Apparent cohesion, capillary cohesion	kN/m^2
c_{fu}	Corrected shear strength from vane shear test	kN/m^2
c_{fv}	Measured shear strength from vane shear test	kN/m^2
c_f	Shape coefficient	1
c_u	Cohesion of undrained (non-drained) soil	kN/m^2
d	Thickness, layer thickness	m
d	Pile thickness	m, cm
d	Water depth	m
dB	Decibel	dB
d_b	Limit water depth	m
d_f	Water depth at structure	m
d_s	depth of groundwater or harbour water level	m
d_s	Thickness of soil stratum	m
d_w	Water depth at a full wave length from structure	m
d_{wo}	Depth of external water behind a wall	m
d_{wu}	Depth of external water in front of wall	m
dwt or DWT	Deadweight tonnage in metric tons	ts

Symbol	Definition	Unit
d_{50}	Mean particle size	mm
e	Voids ratio	1
e_A	Initial voids ratio	1
e_{ah}	Horizontal ordinate of active earth pressure	kN/m^2
e_{ph}	Horizontal ordinate of passive earth pressure	kN/m^2
f	Deflection	m
$f_{y,k}$	Yield stress	N/mm^2
g	Gravitational acceleration = 9.81	m/s^2
h	Height, rise of water level, hydraulic head	m
h'	Depth of soil in flow on land side	m
h_F	Hydrostatic level difference at base of sheet piling	m
h_r	Hydraulic head at base of wall	m
h_{so}	Depth of soil in flow behind retaining wall	m
h_{su}	Depth of soil in flow in front of retaining wall	m
h_{wo}	Depth of water behind wall	m
h_{wu}	Depth of water in front of wall	m
$h_{wü}$	Hydrostatic difference in level	m
dh	Hydrostatic level difference between two equipotential flow lines	m
Δh	hydrostatic level difference	m
i	Hydraulic gradient	1
k	Coefficient of permeability	m/s
k	Radius of gyration of the ship	m
k	Wave number	l/m
k_s	Modulus of subgrade reaction	kN/m^3
$k_{s,h}$	Modulus of subgrade reaction for horizontal bedding	kN/m^3
$k_{s,v}$	Modulus of subgrade reaction for vertical bedding	kN/m^3
l	Length	m
l_a	Length of anchor pile	m
l_k	Upper, statically ineffective pile length	m
l_r	Minimum anchoring length	m
l_s	Length of the anchor pile toe	m
n	Percentage wave height	m
n	Degree of porosity, pore number	1
n_{pr}	Pore number with optimum water content in Proctor test	1
p	Surcharge	kN/m^2
Δp	Additional surcharge	kN/m^2
p_{fail}	Mean bottom pressure at soil failure	MN/m^2
p_d	Dynamic wave pressure ordinate	kN/m^2
p_D	Pressure due to water particle velocity	kN/m^2
p_M	Force of inertia	kN/m
q	Load per linear metre of waling	kN/m
q	Rate of flow	$m^3/s\ m$
q_c	Toe resistance	MN/m^2
q_u	Unconfined compressive strength of undrained soil	kN/m^2
r	Radius	m

Symbol	Definition	Unit
s	Subsidence, settlement	cm
s	Travel	m
s	Displacement	mm
s_B	Displacement at failure	mm
t	Embedment depth	m
t	Draught	m
t	Time	s, d, a
t	Temperature	°C
t_L	Temperature of the air	°C
t_E	Temperature of the ice	°C
Δt	Additional embedment required for absorption of equivalent force	m
t_0	Theoretical embedment up to the operation line of the equivalent force C	m
u	Horizontal component of orbital velocity of water particles	m/s
u	Pore pressure	kN/m^2
u	Depth of zero pressure point N below the river bed	m
v	Velocity	m/s
v_r	Resulting velocity	m/s
v_e	Empirical parameter for calculating coefficient of compressibility	1
w	Ordinate of water pressure	kN/m^2
w	Water content	1
w_e	Empirical parameter for calculating coefficient of compressibility	1
$w_{\ddot{u}}$	Water pressure difference	kN/m^2
x	Depth of theoretical toe TF of sheet piling below addition zero point	m
α	Angle of slope of bottom	degree
α	Reduction value for moment in span	1
α	Inclination of the wall	degree
α	Angle of wind direction	degree
α_t	Coefficient of thermal expansion	$°C^{-1}$
β	Angle of slope	degree
β, β_s	Yield point	MN/m^2
β_{wN}	Concrete nominal strength	N/mm^2
γ	Partial safety factor, see tables R 0-1 and R 0-2	1
γ	Mass density of soil	kN/m^3
γ'	Submerged mass density of soil	kN/m^3
γ_W	Mass density of water	kN/m^3
γ_r	Mass density of water-saturated soil	kN/m^3
δ_a	Angle of inclination of active earth pressure	degree
δ_C	Angle of friction of equivalent force C	degree
δ_p	Angle of inclination of passive earth pressure	degree
η	Safety coefficient	1

Symbol	Definition	Unit
ϑ	Angle of failure plane	degree
ϑ	Phase angle	degree
ϑ_a	Angle of slip plane, active earth pressure	degree
ϑ_p	Angle of slip plane, passive earth pressure	degree
κ_R	Coefficient of reflection	1
ξ	Breaker coefficient	1
υ	Kinematic viscosity	m^2/s
ρ	Density	t/m^3
ρ_d	Oven-dry density	t/m^3
ρ_{pr}	Oven-dry density with optimum water content (Proctor test)	t/m^3
ρ_w	Density of water	t/m^3
σ	Normal stress	kN/m^2
σ_{at}	Atmospheric pressure	kN/m^2
σ'	Effective normal stress	kN/m^2
σ_v	Equivalent stress	kN/m^2
τ	Shear stress	kN/m^2
τ_f	Ultimate shear strength	kN/m^2
τ_r	Sliding shear strength (residual shear strength)	kN/m^2
τ_u	Shear strength of undrained soil	kN/m^2
φ	Impact factor	1
φ	Angle of internal friction	degree
φ'	Effective angle of internal friction	degree
φ'_f	Effective angle of internal friction at failure	degree
φ'_k	Characteristic value of effective angle of internal friction	degree
φ'_r	Effective angle of internal friction upon sliding	degree
φ_u	Angle of internal friction of undrained soil	degree
ω	Angular velocity	l/s
ω	Wave angular frequency	l/s

II.2   Indices

Symbol	Definition
d	design value
k	characteristic value
wü	due to excess water pressure
G	due to permanent actions
Q	due to varying actions
f	at failure
r	upon sliding

II.3   Abbreviations

Symbol	Definition
abs	absolute
allow	allowable, permissible
approx	approximate
cal	calculation value (calculated)
crit	critical
des	design
eff	effective
exist	existing (actual)
imp	impact
max	maximum
min	minimum
poss	possible (potential)
red	reduced
reqd	required
stat	static

II.4   Symbols for water levels

Symbol	Definition
	Non-tidal water levels
GrW, GW	Groundwater level (water table)
HaW	Normal harbour water level
HHW	Highest water level
HNW	Highest navigation water level
HW	High water level
MHW	Mean high water level
MW	Mean water level
MLW	Mean low water level
MSL	Mean sea level
LHaW	Lowest harbour water level
LW	Low water level
LLW	Lowest water level
	Tidal water levels
HHW	Highest tidal high water level
MHWS	Mean spring tide high water level
MHW	Mean tidal high water level
MW	Mean tidal water level
T½W	Half-tide water level
MLW	Mean tidal low water level
MLWS	Mean spring tide low water level
LAT	Lowest astronomical tide water level
LLW	Lowest tidal low water level
SKN	Marine chart datum (since 2005 LAT is defined as SKN)

Annex III   List of keywords

G

H

I

J

628

631